Edexcel AS/A level

BIOLOGY B

1

Ann Fullick

PEARSON

Published by Pearson Education Limited, 80 Strand, London WC2R 0RL.

www.pearsonschoolsandfecolleges.co.uk

Copies of official specifications for all Edexcel qualifications may be found on the website:
www.edexcel.com

Text © Ann Fullick
Exam-style questions © Pearson Education Limited
Edited by Natalie Bayne and Jo Egre
Designed by Elizabeth Arnoux for Pearson Education Limited
Typeset by Tech-Set Ltd, Gateshead
Original illustrations © Pearson Education Limited 2015
Illustrated by Tech-Set Ltd, Gateshead and Peter Bull Art Studio
Cover design by Elizabeth Arnoux for Pearson Education Limited
Picture research by Caitlin Swain
Cover photo/illustration © Science Photo Library/King's College London

The rights of Ann Fullick and Graham Hartland to be identified as authors of this work have been asserted
by them in accordance with the Copyright, Designs and Patents Act 1988.

First published 2008
Second edition published 2015

19 18 17 16 15
10 9 8 7 6 5 4 3

British Library Cataloguing in Publication Data
A catalogue record for this book is available from the British Library

ISBN 9781447976547

Printed in the UK by CPI

Acknowledgements
Every effort has been made to contact copyright holders of material reproduced in this book. Any
omissions will be rectified in subsequent printings if notice is given to the publishers.

A note from the publisher
In order to ensure that this resource offers high-quality support for the associated Edexcel qualification, it
has been through a review process by the awarding body to confirm that it fully covers the teaching and
learning content of the specification or part of a specification at which it is aimed, and demonstrates an
appropriate balance between the development of subject skills, knowledge and understanding, in addition
to preparation for assessment.

While the publishers have made every attempt to ensure that advice on the qualification and its
assessment is accurate, the official specification and associated assessment guidance materials are the
only authoritative source of information and should always be referred to for definitive guidance.

Edexcel examiners have not contributed to any sections in this resource relevant to examination papers for
which they have responsibility.

No material from an endorsed book will be used verbatim in any assessment set by Edexcel.

Endorsement of a book does not mean that the book is required to achieve this Edexcel qualification, nor
does it mean that it is the only suitable material available to support the qualification, and any resource
lists produced by the awarding body shall include this and other appropriate resources.

Picture credits
The publisher would like to thank the following for their kind permission to reproduce their photographs:

(Key: b-bottom; c-centre; l-left; r-right; t-top)

Alamy Images: Ann and Steve Toon 168cr, Arco Images GmbH 208, BSIP SA 73br, Custom Life Science Images 33, Edwin Remsberg 70br, INTERFOTO 70bl, Jeremy Sutton-Hibbert 102, Mark Conlin 239, Nigel Cattlin 243bl, Patrick J. Endres 150–151, Picture Partners 120cl, Roberto Nistri 186, Scott Camazine 248, Steve Bloom Images 198, Stone Nature Photography 184c, The Natural History Museum 160br, 168cl; **Anthony Short:** 8–9, 10cr, 13, 26br, 110–111, 118, 126–127, 152, 153, 155tl, 155cl, 156t, 159tl, 159bl, 163b, 172–173, 175, 176bl, 179, 184cl, 188, 192–193, 194l, 196b, 204tr, 206cl, 213, 224–225, 232, 233, 238br, 290; **Ardea:** Bill Coster 178br/b, John Mason 178br/t; **Biodiversity Institute of Ontario:** 159r; **Copperhead Institute:** Chuck Smith 119; **Corbis:** Carolina Biological/Visuals Unlimited 116, CDC/PHIL 160cr, Viaframe 36–37; **David Grémillet:** 177tl, 177bl; **DK Images:** Lucy Claxton 156bl; **FLPA Images of Nature:** Fritz Polking 206cr; **Fotolia. com:** Simone Werner-Ney 202, tomatito26 26bl; **Getty Images:** Chris Jackson 205, De Agostini Picture Library 288, Dr. Brad Mogen 292, Glyn Kirk/AFP 204br, Matt Cardy 203, moodboard 133, Peter Tsai 54–55, Photolibrary 76cl, Photolibrary/Ed Reschke 83tl, Ralph Slepecky/Visuals Unlimited, Inc. 95tr, Roland Birke 226, Vetta 29; **Pearson Education Ltd:** 223tl; **Photoshot Holdings Limited:** Oceans-Image 251; **Phototake, Inc:** ISM 83bl; **Professor Legesse Negash:** BBC 207; **Science Photo Library Ltd:** A. Dowsett, Health Protection Agency 106, 70bc, 74, 130/2, 244–245, 258, 264cr, Adrian Bicker 156br, AMI Images 138, Andrew Lambert Photography 20, 31, Asa Thoresen 88bl, Athenais, ISM 264tl, Biology Pics 85, Biophoto Associates 84, 88br, 90bl, 130 (all), 176r, 255br, Chuck Brown 254, CNRI 90tl, 99, D. Phillips 140, 255tr, David M. Phillips 217c, David McCarthy 122, David Scharf 117, Dirk Wiersma 157, Don W. Fawcett 78tl, Dr. P. Marazzi 261, Dr. Yorgos Nikas 143, Dr. Gopal Murti 76bl, 82, 90r, Dr. Jeremy Burgess 86, 87, 237, Dr. John Brackenbury 176cl, Dr. Kari Lounatmaa 75tr, Dr. Keith Wheeler 238bl, Dr. Richard Kessel & Dr. Gene Shih, Visuals Unlimited 243br, Dr. Rosalind King 95br, Dr. Stanley Flegler/Visuals Unlimited, Inc. 217cl, 217bl, Dr. Stanley Flegler, Visuals Unlimited 210–211, Eye of Science 16–17, 101, 182cr, Herve Conge, ISM 124bl, 278, Innerspace Imaging 88bc, J.C. Revy, ISM 217tr, Jackie Lewin, Royal Free Hospital 51, James King-Holmes 146, Juan Gaertner 40, Lee D. Simon 100, Look at Sciences 112, Louise Hughes 56, Louise Murray 155tr, M.I. Walker 121, Martin Oeggerli 92–93, Martin Shields 24, Martyn F. Chillmaid 18, Medimage 23b, Mehau Kulyk 120tl, Michael Abbey 217cr, NIBSC 75bl, Omikron 98, Pascal Goetgheluck 162, PHOTOTAKE Inc. 78tr, Power and Syred 23t, 91, 279, Pr. G. Gimenez-Martin 124tl, Professor P. Motta & D. Palermo 115, Professors P. Motta & T. Naguro 77, Richard J. Green 95cr, Science Source 221, Scott Camazine 285, Sovereign, ISM 265, Steve Gschmeissner 72, 73cr, 231, 253, Ted Kinsman 243bc, Thomas Ames Jr., Visuals Unlimited 243tr, Thomas Deerinck, NCMIR 68–69, Tom Kinsbergen 182tl; **Shutterstock. com:** Alexey Repka 154, Debra James 206tr, Egon Zitter 194c, idreamphoto 12, Image Point Fr 267, Jeff Dalton 155cr, Jim Lopes 284, Joel Blit 163t, Nicky Rhodes 197, Picsfive 272, Svitlana S. 144, Vlad61 194r; **The University of California:** Alex McPherson, Irvine/National Institute of General Medical Sciences 57; **U.S. Department of Agriculture:** Agricultural Research Service 158; **Veer/Corbis:** Backyard Productions 174, enjoylife25 183, gbrouwer 200, goce risteski 128, luchschen 276–277, marilyna 64, Nyker 184cr, prochasson frederic 10bl; **Wellcome Trust Sanger Institute:** 41

Cover images: *Front:* **Science Photo Library Ltd:** King's College London

All other images © Pearson Education Limited

Picture Research by: Caitlin Swain

We are grateful to the following for permission to reproduce copyright material:

Figures
Figure on page 32 from 'Trehalose: an intriguing disaccharide with potential for medical application in ophthalmology', Clinical ophthalmology, 5, 577 (2011), Clinical Ophthalmology by Society for Clinical Ophthalmology (Great Britain) Reproduced with permission of Dove Medical Press Limited in the format Republish in a book via Copyright Clearance Center; Figure on page 203 from the front cover of the DEFRA publication 'What nature can do for you', https://www.gov.uk/government/uploads/system/uploads/attachment_data/file/221097/pb13897-nature-do-for-you.pdf, Published by the Department for Environment, Food and Rural Affairs. © Crown Copyright 2010; Figure on page 240 from http://www.abpischools.org.uk/page/modules/breathingandasthma/asthma7.cfm, ABPI Resources for Schools, Association of the British Pharmaceutical Industry (ABPI) with permission.

Text
Article on page 32 from 'Trehalose: an intriguing disaccharide with potential for medical application in ophthalmology', Clinical ophthalmology, 5, 577 (2011), Clinical Ophthalmology by Society for Clinical Ophthalmology (Great Britain) Reproduced with permission of Dove Medical Press Limited in the format Republish in a book via Copyright Clearance Center; Article on page 106 adapted from 'Deadly Ebola virus "could spread globally" after plane brings it to Nigeria', Daily Mail 28/07/2014 (Nick Fagge), Daily Mail; Article on page 106 adapted from 'Epidemiology and surveillance', http://www.afro.who.int/en/clusters-a-programmes/dpc/epidemic-a-pandemic-alert-and-response/outbreak-news/4236-ebola-virus-disease-west-africa-29-july-2014.html, © Copyright World Health Organization (WHO) – Regional Office for Africa, 2013. All rights reserved. Article on page 106 from http://www.wales.nhs.uk/sitesplus/888/page/74608, Public Health Wales; Extract on page 146 adapted from 'In vitro fertilisation', Heinemann Library (Fullick, A.); Poetry on page 168 from 'Oxford Ragwort' (Short, G.), with permission from Anthony Short; Article on page 188 adapted from 'Quagga rebreeding: a success story', Farmer's Weekly (Harvey, K.), © 2014 Farmer's Weekly Magazine; Article on page 188 from 'A rapid loss of stripes: the evolutionary history of the extinct quagga', September 2005 Volume: 1 Issue: 3 (Jennifer A. Leonard et al), Copyright © 2014, The Royal Society; Article on page 240 from http://www.abpischools.org.uk/page/modules/ breathingandasthma/asthma7.cfm, ABPI Resources for Schools, Association of the British Pharmaceutical Industry (ABPI) with permission; Article on page 290 from Encyclopedia of Life Sciences, John Wiley & Sons, Ltd (Turgor Pressure by Jeremy Pritchard, University of Birmingham 2001) © 2001, John Wiley & Sons, Ltd, Reproduced with permission of Blackwell Publishing.

The Publisher would like to thank Chris Curtis and Wade Nottingham for their contributions to the Maths skills section of this book.

The author would like to acknowledge and thank the teams at Science and Plants for Schools (SAPS), the Wellcome Trust Sanger Institute and the ABPI for their valuable input. The author would also like to thank the following for their support and individual contributions: Dr Jeremy Pritchard; Alice Kelly; Amy Ekins-Coward; Tony Short; William Fullick; Thomas Fullick; James Fullick, Edward Fullick; Chris Short.

Every effort has been made to contact copyright holders of material reproduced in this book. Any omissions will be rectified in subsequent printings if notice is given to the publishers.

Contents

TOPIC 3 Classification

TOPIC 4 Exchange and transport

How to use this book

Welcome to your Edexcel AS/A level Biology B course. In this book you will find a number of features designed to support your learning.

Chapter openers

Each chapter starts by setting the context for that chapter's learning:

- Links to other areas of Biology are shown, including previous knowledge that is built on in the chapter, and future learning that you will cover later in your course.
- The **All the maths you need** checklist helps you to know what maths skills will be required.

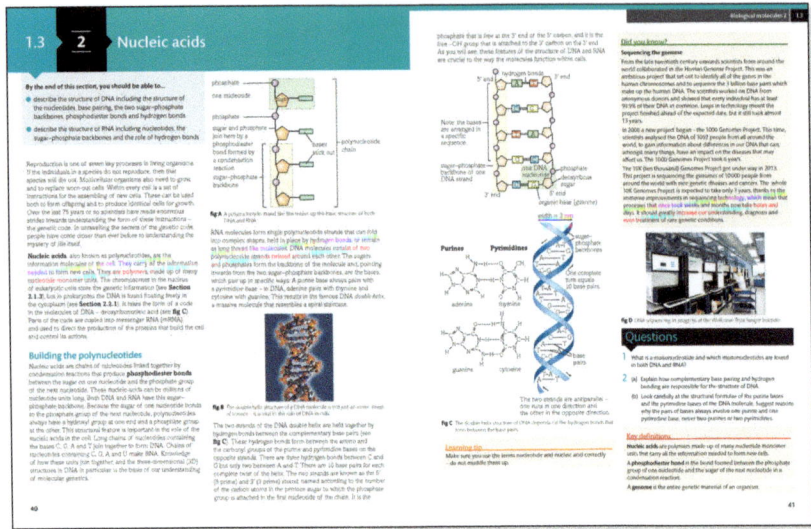

Main content

The main part of each chapter covers all the points from the specification that you need to learn. The text is supported by diagrams and photos that will help you understand the concepts.

Within each section, you will find the following features:

- **Learning objectives** at the beginning of each section, highlighting what you need to know and understand.
- **Key definitions** shown in bold and collated at the end of each section for easy reference.
- **Worked examples** showing you how to work through questions, and how your calculations should be set out.
- **Learning tips** to help you focus your learning and avoid common errors.
- **Did you know?** boxes featuring interesting facts to help you remember the key concepts.
- **Questions** to help you check whether you have understood what you have just read, and whether there is anything that you need to look at again.

Thinking Bigger

The book features a number of **Thinking Bigger** spreads that give you an opportunity to read and work with real-life research and writing about science. The timeline at the bottom of the spreads highlights which of the chapters the material relates to. These spreads will help you to:

- read real-life material that's relevant to your course
- analyse how scientists write
- think critically and consider the issues
- develop your own writing
- understand how different aspects of your learning piece together.

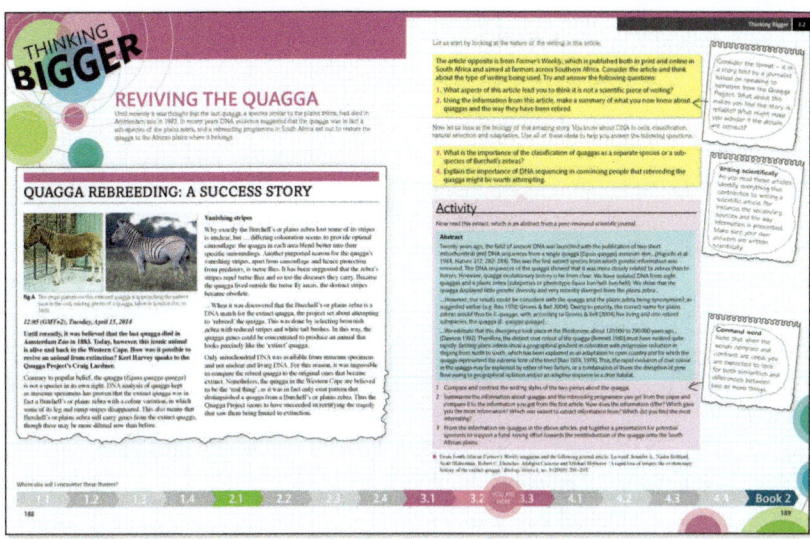

Exam-style questions

At the end of each chapter there are also **exam-style questions** to help you to:

- test how fully you have understood the learning
- practise for your exams.

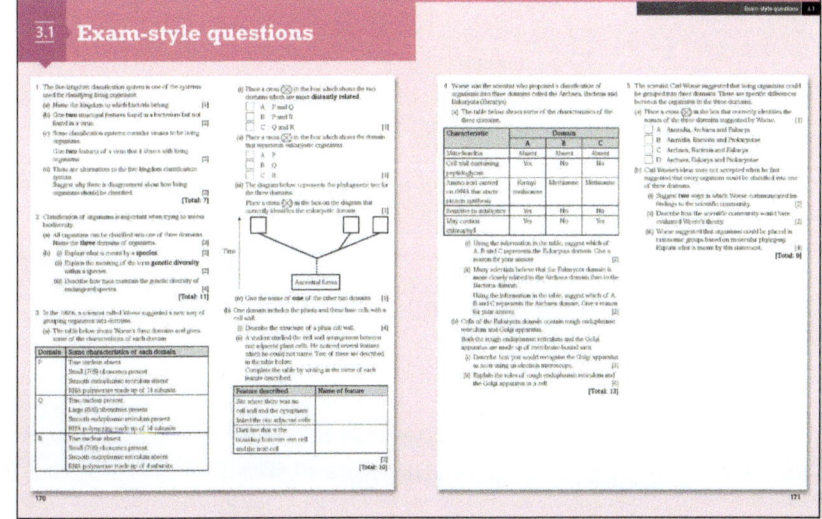

Getting the most from your online ActiveBook

This book comes with 3 years' access to ActiveBook* – an online, digital version of your textbook. Follow the instructions printed on the inside front cover to start using your ActiveBook.

Your ActiveBook is the perfect way to personalise your learning as you progress through your Edexcel AS/A level Biology course. You can:

- access your content online, anytime, anywhere
- use the inbuilt highlighting and annotation tools to personalise the content and make it really relevant to you
- search the content quickly using the index.

Highlight tool

Use this to pick out key terms or topics so you are ready and prepared for revision.

Annotations tool

Use this to add your own notes, for example links to your wider reading, such as websites or other files. Or make a note to remind yourself about work that you need to do.

*For new purchases only. If this access code has already been revealed, it may no longer be valid. If you have bought this textbook secondhand, the code may already have been used by the first owner of the book.

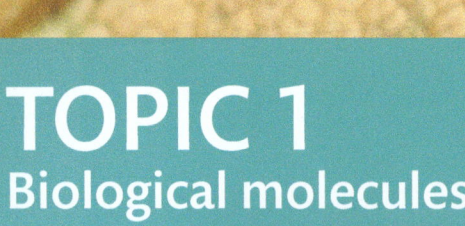

CHAPTER
1.1 Chemistry for life

Introduction

A raft spider *Dolomedes fimbriatus* sits on the surface of the water, hidden by the stems of water plants, waiting for the vibrations in the surface tension that alert her to the presence of her prey. She is large – up to 23 mm across – yet water-repellent hairs enable her to run across the surface to grab her victims. These are usually aquatic invertebrates that also live on or near the water surface. Water is vital for this semiaquatic spider – and for all life on Earth.

Biology is the study of living things. The basic unit of life is the cell, and underpinning all life is – chemistry! The way atoms are bonded together affects the way chemicals work in the cells – and that affects everything, from the way plants make food by photosynthesis to the way your eyes respond to light.

In this chapter you will be looking at some of the key ways in which atoms and molecules interact to make up the chemistry of life. You will be using these basic principles throughout your biology course, because they underpin the structures and functions of all the organisms you will study.

Around two-thirds of the surface of the Earth is covered in water, around two-thirds of your body is water, the oceans, rivers and lakes of the world are teeming with life and all the reactions in your cells take place in solution in water. In this chapter you will be applying your knowledge of the basic chemical principles to help you understand just why water is so vital for life.

All the maths you need

- Recognise and make appropriate use of units in calculations (*e.g. millimetres*)
- Use ratios (*e.g. representing the relationships between atoms in an ion or molecule*)

What have I studied before?

- Life processes depend on molecules whose structure is related to their function
- All living things are made up of cells
- Many processes in living cells, including diffusion and osmosis, depend on water
- Reactions in cells take place in solution in water
- Water is needed for photosynthesis
- Water is transported around plants
- That water makes up around two-thirds of the human body

What will I study later?

- How carbohydrates, lipids and proteins are formed by covalent bonding
- The importance of polarity in the structure and function of phospholipids
- The importance of hydrogen bonding in the tertiary and quaternary structure of proteins and in the structure and function of enzymes
- How water is taken into and moved around plants
- The movement of water into and out of cells, tissues and vessels in animals, plants and fungi
- The importance of water in plant movements
- The role of water in the reactions of cellular respiration and photosynthesis (A level)

What will I study in this chapter?

- How ionic and covalent bonding affect the nature of the compound formed
- The formation of anions and cations in ionic bonding
- The formation of dipoles in some covalent molecules leading to intra- and intermolecular bonds, e.g. hydrogen bonds
- The importance of inorganic ions in organisms, from the nitrate ions needed to form proteins to the calcium ions needed for bone formation and muscle contraction in animals and for cell wall formation in plants
- The chemistry of water and how this affects its properties
- The importance of water to living things

By the end of this section, you should be able to...

● explain the role of inorganic ions in plants

● explain the importance of the dipole nature of water in the formation of hydrogen bonds and the significance of some of the properties of water to the organisms

Ionic and covalent bonding

Biology is the study of living things – but living things are made up of chemicals. If you understand some of the basic principles of chemistry, you will also develop a much better understanding of biological systems. The chemical bonds within and between molecules affect the properties of the compounds they form. This in turn affects their functions within the cell and the organism.

fig A All life depends on some very fundamental chemistry.

The single basic unit of all elements is the atom. When the atoms of two or more elements react they form a compound. An atom is made up of a nucleus containing positive protons and neutral neutrons surrounded by negative electrons. We model these electrons as orbiting around the nucleus in shells. When an atom has a full outer shell of electrons it is stable and does not react. However, most atoms do not have a full outer shell of electrons. In chemical reactions, they are involved in changes that give them a stable outer shell. There are two ways they can achieve this:

- **Ionic bonding:** the atoms involved in the reaction donate or receive electrons. The atom, or part of the molecule, gains one or more electrons and becomes a negative ion (**anion**). The other atom, or part of the molecule, loses one or more electrons and becomes a positive ion (**cation**). Strong forces of attraction called ionic bonds hold the oppositely charged ions together.

fig C Animals such as this cow can use a mineral lick to get the salt they need to function.

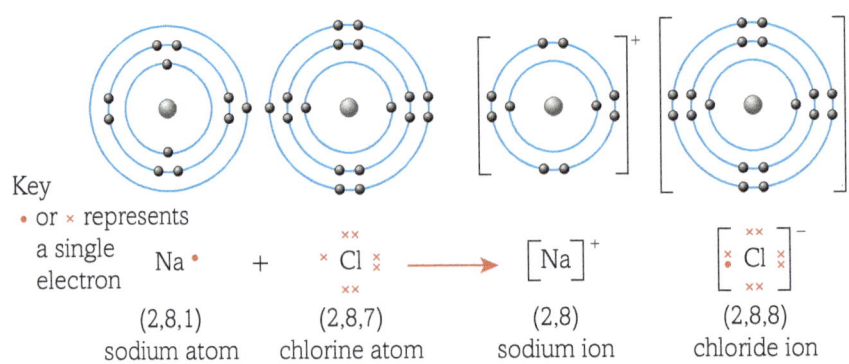

Key
• or × represents a single electron

Na •   +   × Cl ×   →   [Na]⁺   [× Cl ×]⁻

| (2,8,1) | (2,8,7) | (2,8) | (2,8,8) |
| sodium atom | chlorine atom | sodium ion | chloride ion |

fig B The formation of sodium chloride (salt), an inorganic substance that is very important in living organisms, is an example of ionic bonding.

- **Covalent bonding:** the atoms involved in the reaction share electrons. Covalent bonds are very strong and the molecules formed are usually neutral. However, in some covalent compounds, the molecules are slightly polarised. The electrons in the covalent bonds are not quite evenly shared. This means the molecule has a part that is slightly negative and a part that is slightly positive. This separation of charge is called a **dipole**, and the tiny charges are represented as δ^+ and δ^- (see **fig D**). The molecule is described as a **polar molecule**. This polarity is particularly common if one or more hydrogen atoms are involved in the bond.

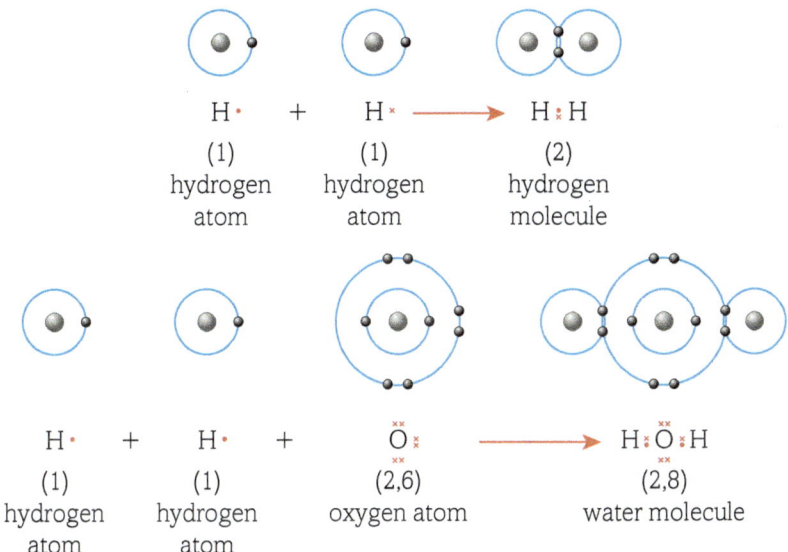

fig D The formation of hydrogen molecules and water molecules are examples of covalent bonding.

The importance of inorganic ions

When ionic substances are dissolved in water, the ions separate. Cells are 60–70% water, and so in living organisms, most ionic substances exist as positive and negative ions. Many of these ions play specialised roles in individual cells and in the functioning of entire organisms. Here are some of the inorganic ions you will meet as you study biology, with an indication of one or more of their roles:

Important anions

- nitrate ions (NO_3^-) – needed in plants for the formation of amino acids and therefore proteins from the products of photosynthesis, and also for the formation of DNA

- phosphate ions (PO_4^{3-}) – needed in all living organisms including plants and animals in the formation of ATP and ADP as well as DNA and RNA

- chloride ions (Cl^-) – needed in nerve impulses, sweating and many secretory systems

- hydrogen carbonate ions (HCO_3^-) – needed for buffering the blood to prevent it from becoming too acidic

Important cations

- sodium ions (Na^+) – needed in nerve impulses, sweating and many secretory systems

- calcium ions (Ca^{2+}) – needed for the formation of calcium pectate for the middle lamella between two cell walls in plants, and for bone formation and muscle contraction in animals

- hydrogen ions (H^+) – needed in cellular respiration and photosynthesis, and in numerous pumps and systems in organisms as well as pH balance

- magnesium ions (Mg^{2+}) – needed for production of chlorophyll in plants

The chemistry of water

Water is the medium in which all the reactions take place in living cells. Without it, substances could not move around the body. Water is one of the reactants in the process of photosynthesis, on which almost all life depends. And water is a major habitat – it supports more life than any other part of the planet. Understanding the properties of water will help you understand many key systems in living organisms.

fig E Water is vital for life on Earth in many different ways.

The importance of water to biological systems is due to the basic chemistry of its molecules. The simple chemical formula of water is H_2O. This tells us that two atoms of hydrogen are joined to one atom of oxygen to make up each water molecule (see **fig F**). However, because the electrons are held closer to the oxygen atom than to the hydrogen atoms, water is a polar molecule**.**

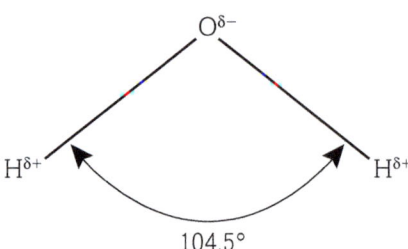

fig F A model of a water molecule.

One of the most important results of this charge separation is that water molecules form **hydrogen bonds**. The slightly negative oxygen atom of one water molecule will attract the slightly positive hydrogen atoms of other water molecules in a weak electrostatic attraction called a hydrogen bond. This means that the molecules of water 'stick together' more than you might otherwise expect, because although each individual hydrogen bond is weak, there are a great many of them (as shown in **fig G**). Water has relatively high melting and boiling points compared with other substances that have molecules of a similar size – it takes more energy to overcome the attractive forces of all the hydrogen bonds. Hydrogen bonds are an important concept in biochemistry – for example they play an important part in protein structure (see **Section 1.2.4**) and in the structure and functioning of DNA (see **Section 1.3.2**).

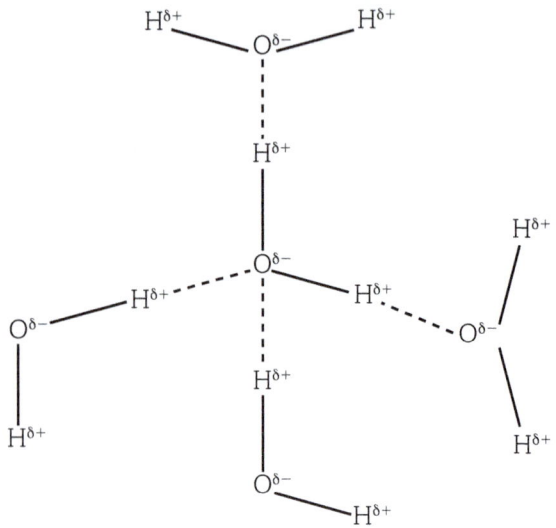

fig G Hydrogen bonding in water molecules.

The importance of water

The properties of water make it very important in biological systems for several reasons:

- Water is a polar solvent. Because water is a polar molecule many ionic substances like sodium chloride will dissolve in it. Many covalently bonded substances are also polar and they too will dissolve in water, although they often do not dissolve in other covalently bonded solvents such as ethanol. Water also carries other substances such as starch that form colloids rather than solutions. As a result most of the chemical reactions within cells occur in water (in aqueous solution).

- Water is an excellent transport medium because so many different substances will dissolve in it.

- As water cools to 4 °C, it reaches its maximum density. As it cools further, the molecules become more widely spaced. As a result, ice is less dense than water and floats, forming an insulating layer and helping to prevent the water underneath it from freezing. It also melts quickly because it is at the top, exposed to the sun. It is very unusual for the solid form of a chemical to be less dense than the liquid, but as a result of this unusual property, organisms can live in water even in countries where it gets cold enough to freeze in winter.

- Water is slow to absorb and release heat – it has a high specific heat capacity. The hydrogen bonds between the molecules mean it takes a lot of energy to separate them. This means the temperature of large bodies of water such as lakes and seas does not change much throughout the year, making them good habitats for living organisms.

- Water is a liquid and so it cannot be compressed. This incompressibility means it is an important factor in many hydraulic mechanisms in living organisms.

- Water molecules are cohesive – the forces between the molecules mean they stick together. This is very important for the movement of water from the roots to the leaves of plants.

- Water molecules are adhesive – they are attracted to other different molecules. This is also important in plant transport systems and in surface tension.

- Water has a very high surface tension because the attraction between the water molecules, including hydrogen bonds, is greater than the attraction between the water molecules and the air. As a result the water molecules hold together forming a thin skin of surface tension. Surface tension is of great importance in plant transport systems, and also affects life at the surface of ponds, lakes and other water masses.

fig H Without surface tension a raft spider could not move across the water in this way.

Questions

1 How do ionic bonds and covalent bonds differ?

2 What are the differences between ionic substances and polar substances?

3 How are hydrogen bonds formed between water molecules and what effect do they have on the properties of water?

4 The properties of water affect its role in living organisms. Discuss.

Key definitions

An **anion** is a negative ion.

A **cation** is a positive ion.

Ionic bonds are formed when atoms give or receive electrons. They result in charged particles called ions.

Covalent bonds are formed when atoms share electrons. Covalent molecules may be polar if the electrons are not shared equally.

A **dipole** is the separation of charge in a molecule when the electrons in covalent bonds are not evenly shared.

A **polar molecule** is a molecule containing a dipole.

Hydrogen bonds are weak electrostatic intermolecular bonds formed between polar molecules containing at least one hydrogen atom.

Exam-style questions

Biology has a lot of application of scientific knowledge, so it's a good idea to remind yourself of the basics learnt at GCSE.

1   Remind yourself of ionic bonds by answering these questions.

(a) Draw a diagram of a sodium atom, including the protons, neutrons and electrons. [2]

(b) Draw a diagram of a chlorine atom, including the protons, neutrons and electrons. [2]

(c) Now show how sodium and chlorine atoms can be turned into sodium and chloride ions to form the ionic bond. [2]

[Total: 6]

2   (a) Draw one water molecule. [1]

(b) Using the atomic structure of oxygen and hydrogen, explain why the electrons are held closer to the oxygen atom. [2]

(c) Explain how a molecule of sodium chloride can dissolve in water. [3]

[Total: 6]

3   Read through the following account about water, then write on the dotted lines the most appropriate word or words to complete the account.

Water molecules are described as ....

because they have a slight positive charge at one end of the molecule and a slight negative charge at the other end. This

makes water a good .... for salts and

substances such as sugars.

Bonds that form between water molecules are called

.... bonds.

Water is a good coolant because it has a high

......................................, which means that it takes a lot of

heat to change it from a liquid to a gas.

Water also has a high ...., which means that a lot of

energy is needed to cause a small rise in its temperature. [5]

[Total: 5]

4   Fill in this table to show which ion is used for which purpose.

Name of ion	Symbol	Function in plants
Nitrate		
	PO^{2-}	
	Ca^{2+}	
		Needed to produce chlorophyll

[4]

[Total: 4]

5   There are many substances important to living organisms. These can be classified as

A   cations

B   anions

C   polar molecules

D   non-polar molecules

Classify the following molecules using one of the terms above.

(a) water

(b) chloride ion (Cl^-)

(c) sodium ion (Na^+)

(d) hydrogen carbonate ion

(e) methane

(f) phosphate ion [6]

[Total: 6]

6   Acids release hydrogen ions (H^+) into solution. Explain how hydrogen carbonate ions (HCO_3^-) act to prevent the blood becoming too acidic. [2]

[Total: 2]

7   (a) A student wrote a title to her table of results in a water-based ink, and then underlined in ballpoint pen. Her lab partner then accidentally spilled water over the page. The title smudged, but the underlining didn't. Using your knowledge of the properties of water, explain these observations. [2]

(b) Having done some research, the student decided that it would be more sensible to do her tables of results using a pencil. Use your knowledge of solvents to explain why this is a good idea. [2]

[Total: 4]

8 (a) Draw the electron shells of the following atoms:
       (i) carbon
       (ii) oxygen
       (iii) sodium
       (iv) argon   [4]
   (b) Use the information from the electron shells to state how
       many protons each of the above elements has.   [4]
   (c) Use the information of the number of protons to explain
       why CH_4 is a non-polar molecule but H_2O is a polar
       molecule.   [2]
   (d) Use the periodic table to find the relative atomic mass of
       each element. Why is this number always bigger than the
       proton number?   [1]
   (e) Looking again at the electron shells, explain why carbon
       can form four bonds, oxygen can form two, sodium only
       forms one bond, but argon can form no bonds.   [4]
                                                    [Total: 15]

9 Marion wanted to build a pond to breed fish in the north of
   England. Temperatures in this region can fall below 0 °C in
   the winter. She was advised to make sure the pond was at
   least 3 m deep and held 3 500 000 litres of water. Use your
   knowledge of the properties of water to explain why such a
   large pond was necessary.   [4]
                                                    [Total: 4]

10 Some animals, such as worms, use water as a basis for their
   hydrostatic skeleton. Why is it more energy efficient to use a
   liquid like water rather than a gas like nitrogen?   [2]
                                                    [Total: 2]

11 Pond skaters are insects that can travel on the surface of water.
   Using your knowledge of the properties of water, explain how
   these insects can travel like this.   [3]
                                                    [Total: 3]

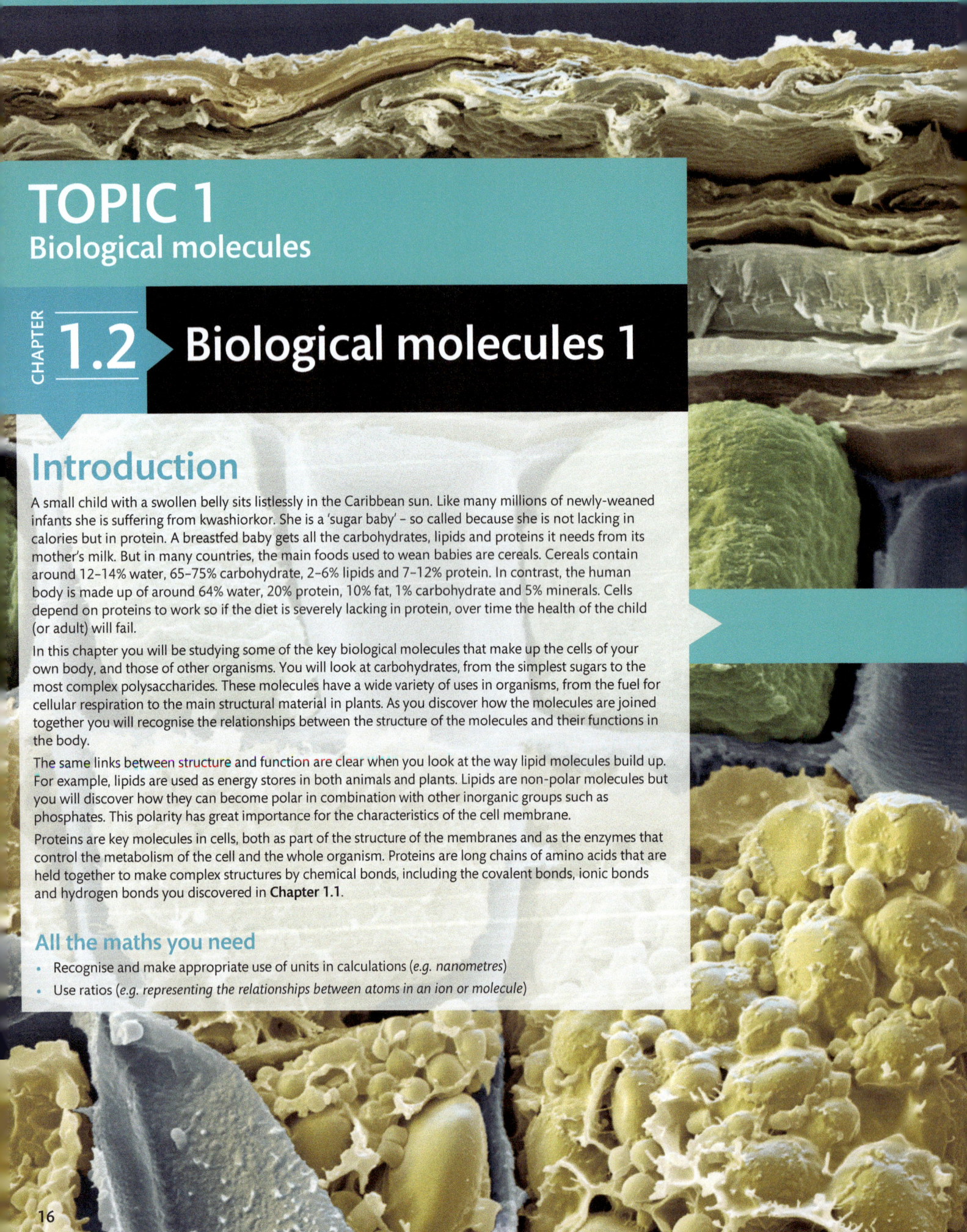

TOPIC 1
Biological molecules

1.2 > Biological molecules 1

Introduction

A small child with a swollen belly sits listlessly in the Caribbean sun. Like many millions of newly-weaned infants she is suffering from kwashiorkor. She is a 'sugar baby' – so called because she is not lacking in calories but in protein. A breastfed baby gets all the carbohydrates, lipids and proteins it needs from its mother's milk. But in many countries, the main foods used to wean babies are cereals. Cereals contain around 12–14% water, 65–75% carbohydrate, 2–6% lipids and 7–12% protein. In contrast, the human body is made up of around 64% water, 20% protein, 10% fat, 1% carbohydrate and 5% minerals. Cells depend on proteins to work so if the diet is severely lacking in protein, over time the health of the child (or adult) will fail.

In this chapter you will be studying some of the key biological molecules that make up the cells of your own body, and those of other organisms. You will look at carbohydrates, from the simplest sugars to the most complex polysaccharides. These molecules have a wide variety of uses in organisms, from the fuel for cellular respiration to the main structural material in plants. As you discover how the molecules are joined together you will recognise the relationships between the structure of the molecules and their functions in the body.

The same links between structure and function are clear when you look at the way lipid molecules build up. For example, lipids are used as energy stores in both animals and plants. Lipids are non-polar molecules but you will discover how they can become polar in combination with other inorganic groups such as phosphates. This polarity has great importance for the characteristics of the cell membrane.

Proteins are key molecules in cells, both as part of the structure of the membranes and as the enzymes that control the metabolism of the cell and the whole organism. Proteins are long chains of amino acids that are held together to make complex structures by chemical bonds, including the covalent bonds, ionic bonds and hydrogen bonds you discovered in **Chapter 1.1**.

All the maths you need

- Recognise and make appropriate use of units in calculations (*e.g. nanometres*)
- Use ratios (*e.g. representing the relationships between atoms in an ion or molecule*)

16

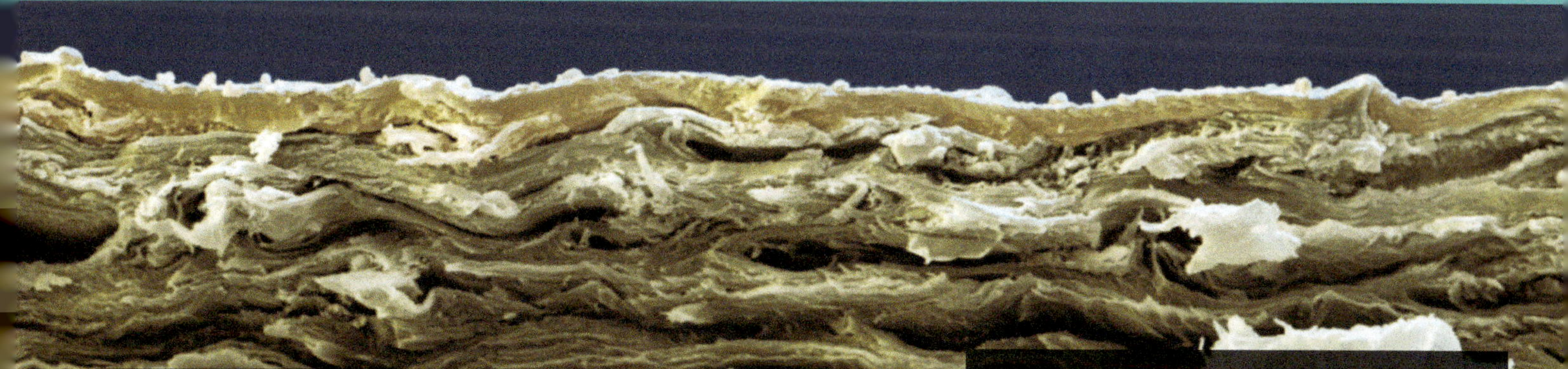

What have I studied before?

- Complex carbohydrates are made up of sugars joined together
- Complex carbohydrates can be broken down to give simple sugars that can be used by cells
- The cell walls of plant cells are made of carbohydrates
- Lipids are made up of fatty acids and glycerol
- Lipids are molecules used to store energy in the bodies of animals and plants
- Proteins are long chains of amino acids
- Enzymes are made of proteins
- Plants make carbohydrates in photosynthesis
- The glucose level in the blood needs to be kept within tight limits

What will I study later?

- How proteins are synthesised on the surface of the ribosomes
- How the tertiary and quaternary structure of proteins is related to their function as enzymes
- How the structure of phospholipids determines many of the characteristics of the cell membrane
- How carbohydrates and proteins act as signalling molecules in and on cell membranes
- How proteins act as carrier systems in cell membranes
- The importance of cellulose in the development of turgor in plants
- How proteins and lipids act as hormones (A level)
- The importance of carbohydrates in cellular respiration (A level)

What will I study in this chapter?

- What makes an organic compound
- The structure of different types of monosaccharides
- The formation of disaccharides by the joining of two monosaccharides in a condensation reaction
- The structure of complex polysaccharides and how their structure is related to their functions as storage molecules and structural compounds
- The importance of lipids as storage molecules in places including the seeds of plants, the hyphae of fungi and the blubber of marine animals such as whales and seals
- The structure of lipids including the formation of ester bonds
- The structure of amino acids, peptides and polypeptides and how they relate to each other
- The formation of peptide bonds between amino acids
- The primary, secondary, tertiary and quaternary structure of proteins and how the structure is related to the function of the protein

Carbohydrates 1 – monosaccharides and disaccharides

By the end of this section, you should be able to...

● describe the difference between monosaccharides and disaccharides

● describe the structure of the hexose glucose (alpha and beta) and the pentose ribose

● explain how monosaccharides join to form disaccharides through condensation reactions forming glycosidic bonds, and how they can be split through hydrolysis reactions

● explain how the structure of glucose relates to its function

What are organic compounds?

Biological molecules are the key to the structure and function of living things. Biological molecules are often organic compounds. Organic compounds all contain carbon atoms. They also contain atoms of hydrogen, oxygen and, less frequently, nitrogen, sulfur and phosphorus. Most of the material in your body that is not water is made up of these organic molecules. An understanding of why organic molecules are special will help you to understand the chemistry of biological molecules including carbohydrates, lipids and proteins.

Each carbon atom can make four bonds and so it can join up with four other atoms. Carbon atoms bond particularly strongly to other carbon atoms to make long chains. The four bonds of a carbon atom usually form a tetrahedral shape and this leads to the formation of branched chains, or rings, or any number of three-dimensional (3D) shapes. In some carbon compounds small molecules (**monomers**) bond with many other similar units to make a very large molecule called a **polymer**. The ability of carbon to combine and make **macromolecules** (large molecules) is the basis of all biological molecules and provides the great variety and complexity found in living things.

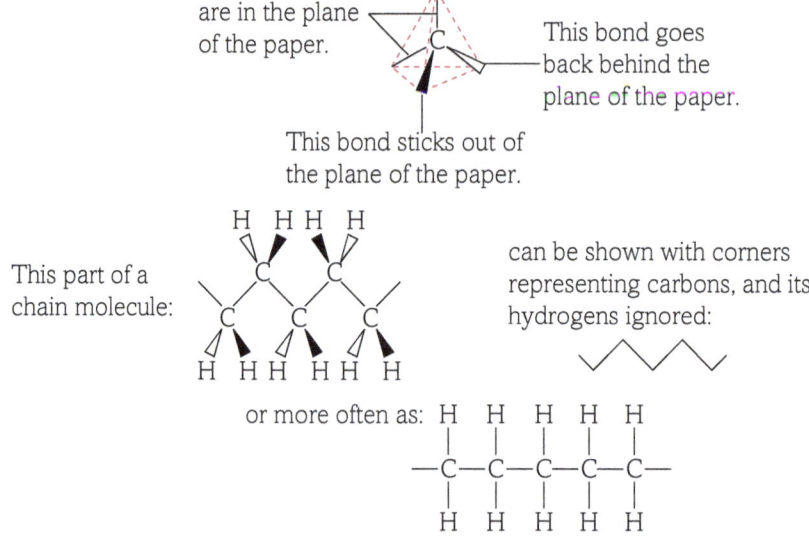

fig A The bonds in a carbon atom have a complicated 3D shape. This is difficult to represent, so in most molecular diagrams we use one of several different ways to draw them 'flat'.

Carbohydrates

Carbohydrates are important in cells as a usable energy source. They are also important for storing energy, and in plants, fungi and bacteria they form an important part of the cell wall. The best known carbohydrates are sugars and **starch**. **Sucrose** is the white crystalline sugar familiar to us all, while **glucose** is the energy supplier in sports and health drinks and starch is in flour and potatoes. But the group of chemicals known as carbohydrates contains many more compounds, as you will discover.

fig B Carbohydrates are important molecules in plants and animals alike – and they also play a major role in the human diet.

The basic structure of all carbohydrates is the same. They are made up of carbon, hydrogen and oxygen. There are three main groups of carbohydrates with varying complexity of molecules: **monosaccharides**, **disaccharides** and **polysaccharides**.

Monosaccharides – the simple sugars

Monosaccharides are simple sugars in which there is one oxygen atom and two hydrogen atoms for each carbon atom present in the molecule. A general formula for this can be written $(CH_2O)_n$. Here n can be any number, but it is usually low:

- **Triose sugars** ($n=3$) have three carbon atoms and the general formula $C_3H_6O_3$. They are important in the mitochondria, where glucose is broken down into triose sugars during respiration.

- **Pentose sugars** ($n=5$) have five carbon atoms and the general formula $C_5H_{10}O_5$. **Ribose** and **deoxyribose** are important in the nucleic acids **deoxyribonucleic acid (DNA)** and **ribonucleic acid (RNA)**, which make up the genetic material (see **Section 1.3.1**).

- **Hexose sugars** ($n=6$) have six carbon atoms and the general formula $C_6H_{12}O_6$. They are the best known monosaccharides, often taste sweet and include glucose, galactose and fructose.

General formulae show you how many atoms there are in the molecule, and what type they are, but they do not tell you what the molecule looks like and why it behaves as it does. To show this you can use displayed formulae. Although these do not follow every wiggle and kink in the carbon chain, they can give you a good idea of how the molecules are arranged in three dimensions. This can reveal all sorts of secrets about why biological systems behave as they do (see **fig D**).

ribose

fig C Pentose sugars such as ribose have 5 carbon atoms.

α-glucose   **fructose**

fig D Hexose sugars have a ring structure. The arrangement of the atoms on the side chains can make a significant difference to the way in which the molecule can be used by the body. We number the carbon atoms so we can identify the different arrangements.

α-glucose and β-glucose

Glucose comes in different forms (**isomers**), including α-glucose and β-glucose. The two isomers result from different arrangements of the atoms on the side chains of the molecule (see **fig E**). The different isomers form different bonds between neighbouring glucose molecules, and this affects the polymers that are made.

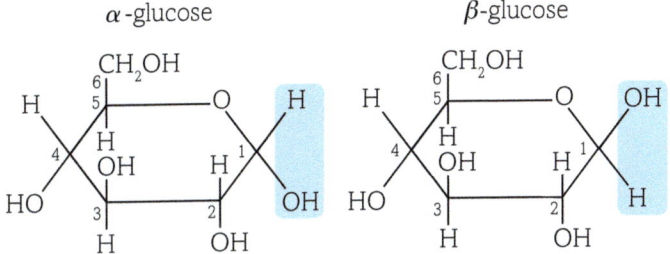

or, even more simply:

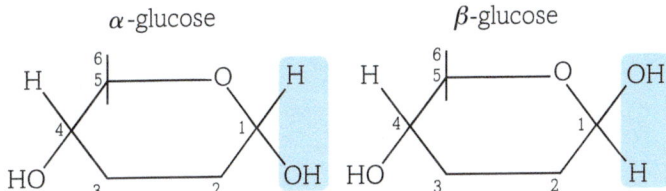

In these diagrams, the positions of carbon atoms are represented by their numbers only.

fig E The difference in structure between α-glucose and β-glucose may seem small, but it has a big impact on the function of the molecule.

Note carefully the different arrangement of atoms around the carbon 1 and carbon 4 atoms in α-glucose and β-glucose. The small difference gives the molecules different properties.

Disaccharides – the double sugars

Disaccharides are made up of two monosaccharides joined together – for example sucrose (ordinary table sugar) is formed by a molecule of α-glucose joining with a molecule of fructose. Two monosaccharides join in a **condensation reaction** to form a disaccharide, and a molecule of water (H_2O) is removed. The link between the two monosaccharides results in a covalent bond known as a **glycosidic bond** (see **fig F**). We use numbers to show which carbon molecules are involved in the bond. If carbon 1 on one monosaccharide joins to carbon 4 on another monosaccharide, we call it a 1,4-glycosidic bond. If the bond is between carbon 1 and carbon 6, it is a 1,6-glycosidic bond.

The joining of monosaccharides gives a different general formula – disaccharides and indeed chains of monosaccharides of any length have the general formula of $(C_6H_{10}O_5)_n$.

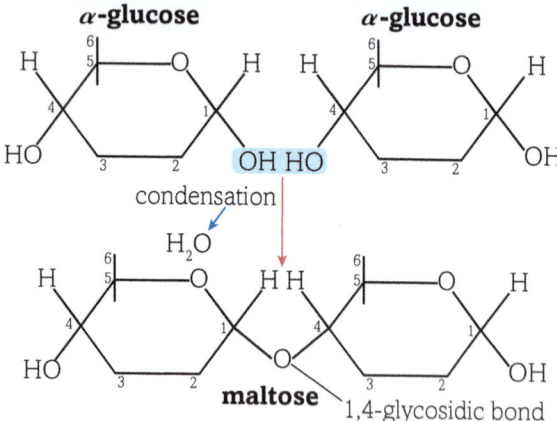

fig F The formation of a glycosidic bond. The condensation reaction between two monosaccharides results in a disaccharide and a molecule of water.

When different monosaccharides join together, different disaccharides result. Many disaccharides taste sweet. **Table A** shows some of the more common ones:

Disaccharide	Source	Monosaccharide
sucrose	stored in plants such as sugar cane	α-glucose + α-fructose
lactose	milk sugar – this is the main carbohydrate found in milk	α-glucose + β-galactose
maltose	malt sugar – found in germinating seed such as barley	α-glucose + α-glucose

table A Three common disaccharides.

Did you know?

Testing for sugars

- Benedict's solution is a chemical test for **reducing sugars**. It is a bright blue solution that contains copper(II) ions. Some sugars react readily with this solution when heated gently and reduce the copper(II) ions to copper(I) ions, forming a precipitate and giving a colour change from blue to orange. They are known as reducing sugars. All of the monosaccharides and some disaccharides are reducing sugars.
- Some sugars do not react with Benedict's solution. They are known as **non-reducing sugars**. You can heat a non-reducing sugar such as sucrose with a few drops of hydrochloric acid, allow it to cool and then neutralise the solution with sodium hydrogen carbonate to hydrolyse the glycosidic bonds. This produces the monosaccharide units of the sugar, which will now give a positive Benedict's test.

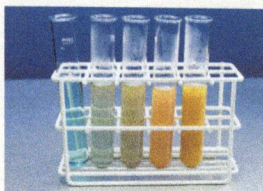

fig G Benedict's test for reducing sugars.

Questions

1 What are the properties of organic compounds that make them so important in living organisms?

2 Describe how a glycosidic bond is formed between two monosaccharides to form a disaccharide.

Key definitions

A **monomer** is a small molecule that is a single unit of a larger molecule called a polymer.

A **polymer** is a long chain molecule made up of many smaller, repeating monomer units joined together by chemical bonds.

A **macromolecule** is a very large molecule often formed by polymerisation.

Starch is a long chain polymer formed of glucose monomers.

Sucrose is a sweet tasting disaccharide formed by the joining of glucose and fructose by a glycosidic bond.

Glucose is a hexose sugar.

A **monosaccharide** is a single sugar monomer.

A **disaccharide** is a sugar made up of two monosaccharide units joined by a glycosidic bond, formed in a condensation reaction.

A **polysaccharide** is a polymer made up of long chains of monosaccharide units joined by glycosidic bonds.

A **triose sugar** is a sugar with three carbon atoms.

A **pentose sugar** is a sugar with five carbon atoms.

Ribose is a pentose sugar that makes up part of the structure of RNA.

Deoxyribose is a pentose sugar that makes up part of the structure of DNA.

Deoxyribonucleic acid (DNA) is a nucleic acid that acts as the genetic material in many organisms.

Ribonucleic acid (RNA) is a nucleic acid which can act as the genetic material in some organisms and is involved in protein synthesis.

A **hexose sugar** is sugar with six carbon atoms.

Isomers are molecules that have the same chemical formula, but different molecular structures.

A **condensation reaction** is a reaction in which a molecule of water is removed from the reacting molecules as a bond is formed between them.

A **glycosidic bond** is a covalent bond formed between two monosaccharides in a condensation reaction, which can be broken down by a hydrolysis reaction to release the monosaccharide units.

Reducing sugars are sugars that react with blue Benedict's solution and reduce the copper(II) ions to copper(I) ions giving an orangey-red precipitate.

Non-reducing sugars are sugars that do not react with Benedict's solution.

By the end of this section, you should be able to...

● explain how monosaccharides join to form polysaccharides through condensation reactions forming glycosidic bonds; and how these can be split through hydrolysis reactions

● explain how the structure of polysaccharides relates to their functions

The most complex carbohydrates are the polysaccharides. They are made of many monosaccharide units joined by condensation reactions that form glycosidic bonds (see **Section 1.2.1**, **fig E**). Molecules with 3–10 sugar units are known as **oligosaccharides**, while molecules containing 11 or more monosaccharides are known as true polysaccharides. Polysaccharides do not have the sweet taste of many mono- and disaccharides, but these complex polymers form some very important biological molecules.

The structure of polysaccharides makes them ideal as storage molecules within a cell:

• They can form very compact molecules, which take up little space.

• They are physically and chemically inactive, so they do not interfere with the other functions of the cell.

• They are not very soluble in water, so have little effect on water potential within a cell and cause no osmotic water movements.

The glycosidic bonds in the polysaccharide can be broken to release monosaccharide units for cellular respiration.

The glycosidic bond between two glucose units is split by a process known as **hydrolysis** (see **fig A**). The hydrolysis reaction is the opposite of the condensation reaction that formed the molecule, so water is added to the bond. Starch and glycogen are gradually broken down into shorter and shorter chains and eventually single sugars are left. Disaccharides break down to form two monosaccharides. Hydrolysis takes place during digestion in the gut, and also in the muscle and liver cells when the carbohydrate stores are broken down to release sugars for use in cellular respiration.

Learning tip

Remember that glycosidic bonds are formed with the removal of a molecule of water in condensation reactions and broken with the addition of a molecule of water in hydrolysis reactions.

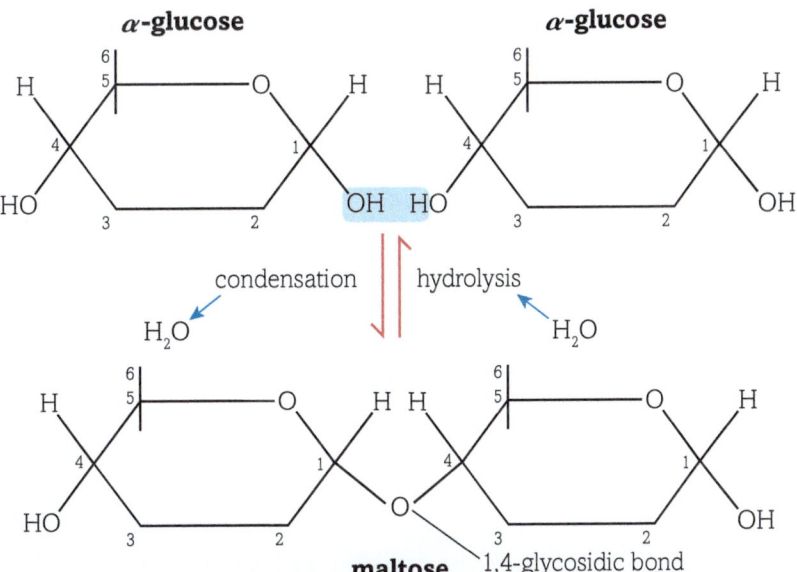

fig A Glycosidic bonds are made by condensation reactions and broken down by hydrolysis.

Carbohydrates as energy stores

Starch

Starch is particularly important as an energy store in plants. The sugars produced by photosynthesis are rapidly converted into starch, which is insoluble and compact but can be broken down rapidly to release glucose when it is needed. Storage organs such as potatoes are particularly rich in starch.

Starch is made up of long chains of α-glucose. But if you look at it more closely you will see that it is actually a mixture of two compounds:

Amylose: an unbranched polymer made up of between 200 and 5000 glucose molecules. As the chain lengthens the molecule spirals, which makes it more compact for storage.

Amylopectin: a branched polymer of glucose molecules. The branching chains have many terminal glucose molecules that can be broken off rapidly when energy is needed.

Amylose and **amylopectin** are both long chains of α-glucose molecules – so why are the molecules so different? It all depends on the carbon atoms involved in the glycosidic bonds.

Amylose is made up purely of 1,4-glycosidic bonds, which is why the molecules are long unbranched chains.

In amylopectin many of the glucose molecules are joined by 1,4-glycosidic bonds, but there are also a few 1,6-glycosidic bonds. This results in the branching chains that change the properties of the molecule.

So starch has a combination of straight chain amylase and branched chain amylopectin molecules. This combination explains why carbohydrate foods like pasta are so good for you when you are doing sport. The amylopectin releases glucose for cellular respiration rapidly when needed. Amylose releases glucose more slowly over a longer period, keeping you going longer.

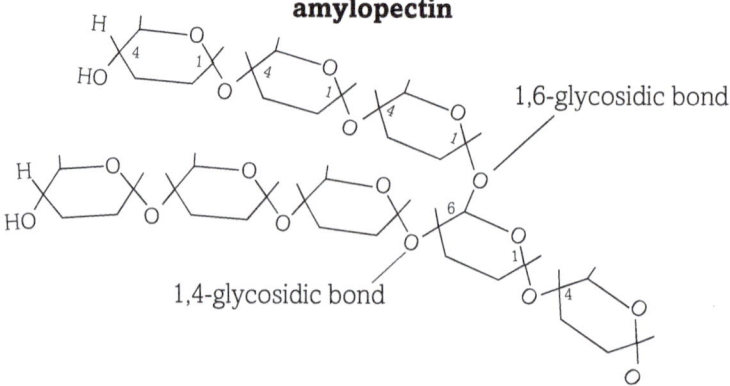

fig B Amylose and amylopectin – a small difference in the position of the glycosidic bonds in the molecule makes a big difference to the properties of the compounds.

Glycogen

Glycogen is sometimes referred to as 'animal starch' because it is the only carbohydrate energy store found in animals (see **fig C**). It is also an important storage carbohydrate in fungi. Chemically, glycogen is very similar to the amylopectin molecules in starch, and is also made up of many α-glucose units. Like starch, it is very compact, but the glycogen molecule has more 1,6-glycosidic bonds, giving it many side branches. As a result, glycogen can be broken down very rapidly. This makes it an ideal source of glucose for active tissues with a constantly high rate of cellular respiration, such as muscle and liver tissue.

Carbohydrates in plants

Polysaccharides are very important in plants. Starch is the main energy storage material in plants. A typical starch grain in a plant cell contains 70–80% amylopectin, with the rest being amylose.

(a) starch grains in a plant cell

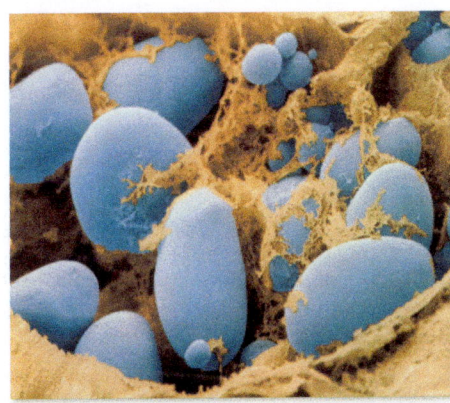

(b) glycogen granules in liver cells

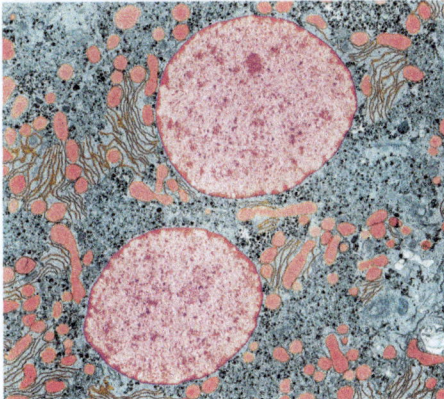

fig C Storage carbohydrates in plant and animal cells.

Cellulose is an important structural material in plants. The cell wall (see **Section 2.1.5**) is an important feature that gives plants their strength and support. It is made up largely of insoluble cellulose. Cellulose has much in common with starch and glycogen. It consists of long chains of glucose joined by glycosidic bonds. However, as you will remember, there are two structural isomers of glucose, α-glucose and β-glucose.

In starch, the monomer units are α-glucose. In cellulose, they are β-glucose and are held together by 1,4-glycosidic bonds where one of the monomer units has to be turned round (inverted) so the bonding can take place. This linking of β-glucose molecules means that the hydroxyl (–OH) groups stick out on both sides of the molecule (see **fig D**). This means hydrogen bonds can

form between the partially positively charged hydrogen atoms of the hydroxyl groups and the partially negatively charged oxygen atoms in other areas of the glucose molecules. This is known as cross-linking and it holds neighbouring chains firmly together.

Many of these hydrogen bonds form, making cellulose a material with considerable strength. Cellulose molecules do not coil or spiral – they remain as very long, straight chains. In contrast, starch molecules, with 1,4- and 1,6-glycosidic bonds between α-glucose monomers, form compact globular molecules that are useful for storage.

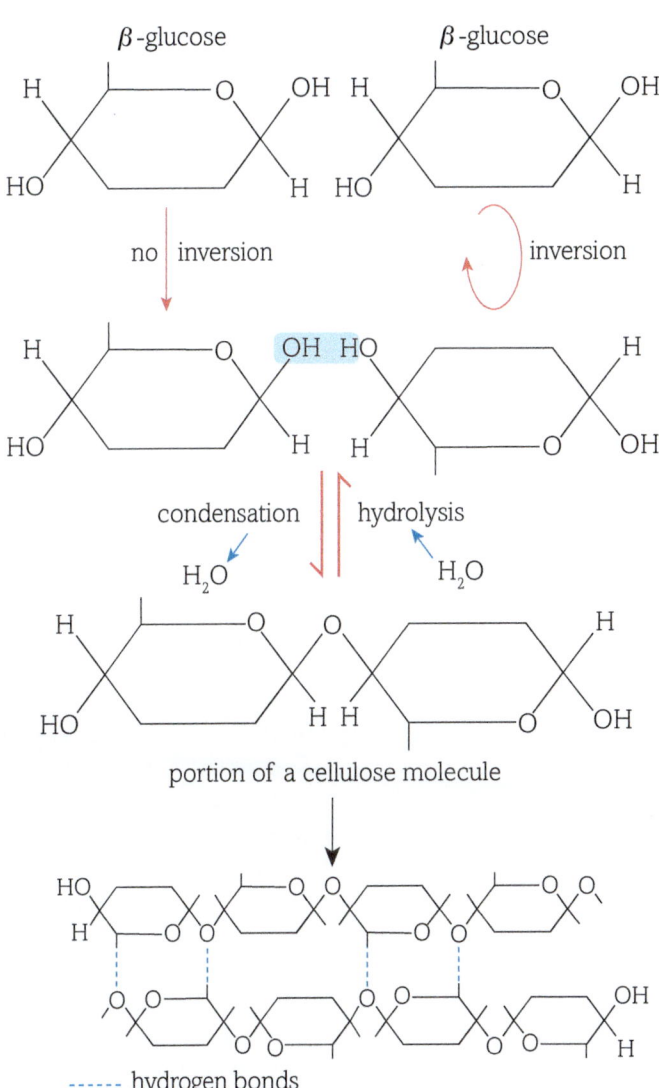

fig D Cellulose molecules consist of β-glucose monomers joined together by 1,4-glycosidic bonds.

This difference in structure between starch and cellulose gives them very different properties and functions. Starch is an important source of energy in the diet for many animals. However, most animals do not possess the enzymes needed to break the 1,4-glycosidic bonds between the molecules of β-glucose and so they cannot digest cellulose. Ruminants such as cows and sheep use the cellulose-digesting enzymes from bacteria, fungi and protozoa living in their gut to digest their food. It is the cellulose in plant food that acts as roughage or fibre in the human diet – an important part of a healthy diet even though you cannot digest it.

fig E The iodine test for starch.

Questions

1 Explain how the structure of carbohydrates is related to their function as storage molecules providing the fuel for cellular respiration in animals and plants.

2 Explain how the chemical structure of cellulose differs from that of starch and how this affects the way they can be used to supply energy in animals.

starch

cellulose

Key definitions

Oligosaccharides are molecules with 3–10 monosaccharide units.

Hydrolysis is a reaction in which bonds are broken by the addition of a molecule of water.

Amylose is a complex carbohydrate containing only glucose monomers joined together by 1,4-glycosidic bonds so the molecules form long unbranched chains.

Amylopectin is a complex carbohydrate made up of glucose monomers joined by both 1,4-glycosidic bonds and 1,6-glycosidic bonds so the molecules branch repeatedly.

Glycogen is made up of many α-glucose units joined by 1,4-glycosidic bonds but also has 1,6-glycosidic bonds, giving it many side branches.

Cellulose is a complex carbohydrate with β-glucose monomers held together by 1,4-glycosidic bonds. It is very important in plant cell walls.

By the end of this section, you should be able to...

● explain the synthesis of a triglyceride, including the formation of ester bonds during condensation reactions between glycerol and three fatty acids

● describe the differences between unsaturated and saturated fatty acids

● explain how the structure of lipids relates to their role in energy storage, waterproofing and insulation

● explain the structure and properties of phospholipids in relation to their function in the cell membranes

The **lipids** are another group of organic chemicals that play a vital role in organisms. They form an integral part of all cell membranes and are also used as an energy store. Many plants and animals convert spare food into oils or fats to use when they are needed. For example, the seeds of plants contain lipids to provide energy for the seedling when it starts to grow, which is why seeds are such an important food source for many animals.

Fats and oils

Fats and oils are important groups of lipids. Chemically they are extremely similar, but fats, such as butter, are solids at room temperature and oils, such as olive oil, are liquids. Like carbohydrates, the chemical elements that make up all lipid molecules are carbon, hydrogen and oxygen. However, lipids contain a considerably lower proportion of oxygen than carbohydrates. Fats and oils are made up of two types of organic chemicals, **fatty acids** and **glycerol** (propane-1,2,3-triol). They are combined using **ester bonds**. Glycerol has the chemical formula $C_3H_8O_3$ (see **fig A**).

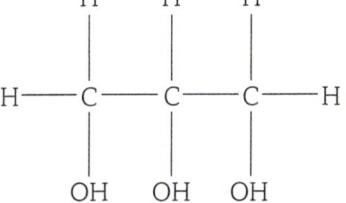

fig A Displayed formula of glycerol (propane-1,2,3-triol).

All fatty acids have a long hydrocarbon chain – a pleated backbone of carbon atoms with hydrogen atoms attached, and a carboxyl group (–COOH) at one end.

Living tissues contain more than 70 different kinds of fatty acids. Fatty acids vary in two ways:

• The length of the carbon chain can differ (although often 15–17 carbon atoms long in organisms).

• The fatty acid may be a **saturated fatty acid** or **unsaturated fatty acid**.

In a saturated fatty acid, each carbon atom is joined to the one next to it by a single covalent bond. A common example is stearic acid (see **fig B**). In an unsaturated fatty acid, the carbon chains have one or more double covalent bonds in them. A **monounsaturated fatty acid** has one double bond and a **polyunsaturated fatty acid** has more than one double bond (see **fig C**). Linoleic acid is an example of a polyunsaturated fatty acid. It is an essential fatty acid in our diet because we cannot make it from other chemicals.

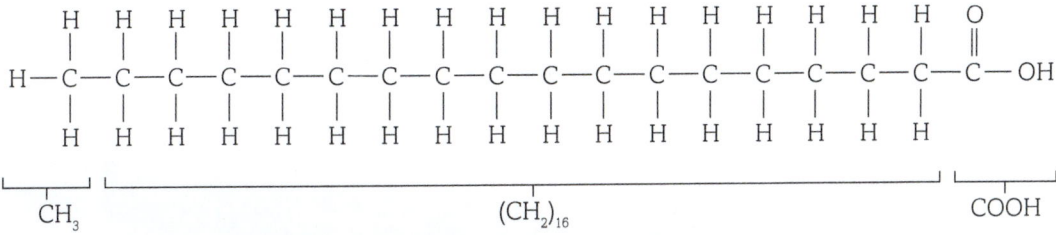

fig B Displayed formula of stearic acid, a saturated fatty acid found in both plant and animal fats.

$$H-C-C-C-C-C-C=C-C-C=C-C-C-C-C-C-C-C-C-OH$$

fig C Displayed formula of linoleic acid, a polyunsaturated fatty acid.

Forming ester bonds

A fat or oil results when glycerol combines with one, two or three fatty acids to form a monoglyceride, a diglyceride or a triglyceride. A bond is formed in a condensation reaction between the carboxyl group (–COOH) of a fatty acid and one of the hydroxyl groups (–OH) of the glycerol. A molecule of water is removed and the resulting bond is known as an ester bond. This type of condensation reaction is called **esterification** (see **fig D**). The nature of the lipid formed depends on which fatty acids are present. So, for example, lipids containing saturated fatty acids are more likely to be solid at room temperature than those containing unsaturated fatty acids.

For simplicity, fatty acids are represented by this general formula where 'R' represents the hydrocarbon chain. The fatty acids below are drawn in reversed form.

$$R-C-OH$$

glycerol   **3 fatty acids**   **triglyceride**

$3H_2O$

hydrolysis

condensation

∿∿ ester bond

Note: there are only 6 atoms of oxygen in a triglyceride molecule.

fig D Formation of ester bonds.

The nature of lipids

Lipids contain many carbon-hydrogen bonds and little oxygen. When lipids are oxidised in respiration, the bonds are broken and carbon dioxide and water are the ultimate products. This reaction can be used to drive the production of a lot of ATP (see **Section 1.3.1**). Lipids, especially triglycerides, store about three times as much energy as the same mass of carbohydrates.

The hydrophobic nature of lipids is a key feature of their role in waterproofing organisms. Oils are important in waterproofing the fur and feathers of mammals and birds, while insects and plants use waxes for waterproofing their outer surfaces (see **fig E**). Lipids are good insulators – a fatty sheath insulates your nerves so the electrical impulses travel faster. They also insulate animals against heat loss – the thick layer of blubber in whales is a good example. Lipids have a very low density, so the body fat of water mammals helps them to float easily. All lipids dissolve in organic solvents, but are insoluble in water, so lipids do not interfere with the many water-based reactions that go on in the cytoplasm of a cell.

(a)   (b)

fig E Oil on the feathers of birds and the waxy layer on the surface of these ginkgo leaves makes them very waterproof.

Phospholipids

Inorganic phosphate ions ($-PO_4^{3-}$) are present in the cytoplasm of every cell. Sometimes one of the hydroxyl groups of glycerol undergoes an esterification reaction with a phosphate group instead of with a fatty acid, and a simple **phospholipid** is formed. Phospholipids are important because the lipid and the phosphate parts of the molecule give it very different properties.

The fatty acid chains of a phospholipid are neutral and insoluble in water. In contrast, the phosphate head carries a small negative charge and is soluble in water. When these phospholipids come into contact with water, the two parts of the molecule behave differently. The polar phosphate part is **hydrophilic** and dissolves readily in water (see **fig F**). The lipid tails are **hydrophobic**, so they do not dissolve in water. If the molecules are tightly packed in water they either form a **monolayer**, with the hydrophilic heads in the water and the hydrophobic lipid tails in the air, or clusters called **micelles**. In a micelle, all the hydrophilic heads point outwards and all the hydrophobic tails are inside (see **fig G**).

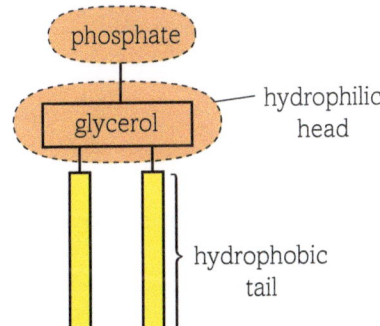

fig F A phospholipid.

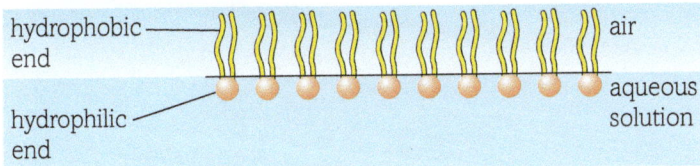

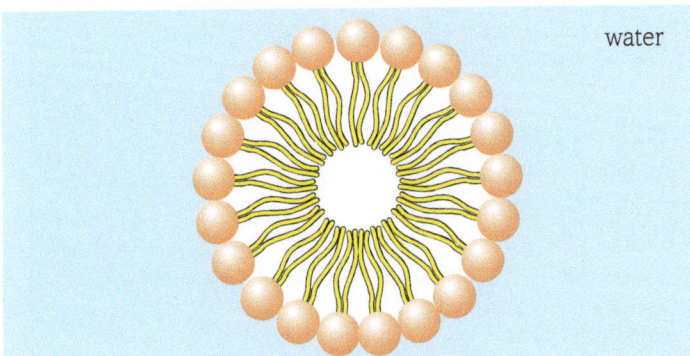

fig G Phospholipids form a monolayer at an air/water junction and a micelle when water surrounds them.

A phospholipid monolayer may form at a surface between air and water, but this is a fairly rare situation in living cells where there are water-based solutions on either side of the membranes. With water on each side, the phospholipid molecules form a **bilayer** with the hydrophilic heads pointing into the water, protecting the hydrophobic tails in the middle (see **fig H**). This structure, the **unit membrane**, is the basis of all membranes.

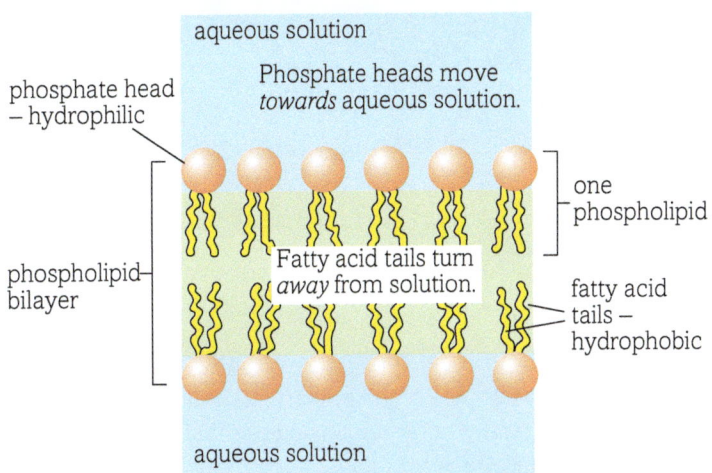

fig H A lipid bilayer is the backbone of all membrane structures in a cell.

Questions

1 Describe the main difference between a saturated and an unsaturated lipid, and the effect of this difference on the properties of the lipid.

2 Explain how triglycerides are formed.

Key definitions

Lipids are a large family of organic molecules that are important in cell membranes and as an energy store in many organisms. They include triglycerides, phospholipids and steroids.

A **fatty acid** is an organic acid with a long hydrocarbon chain.

Glycerol is propane-1,2,3-triol, an important component of triglycerides.

An **ester bond** is a bond formed in a condensation reaction between the carboxyl group (–COOH) of a fatty acid and one of the hydroxyl groups (–OH) of glycerol.

A **saturated fatty acid** is a fatty acid in which each carbon atom is joined to the one next to it in the hydrocarbon chain by a single covalent bond.

An **unsaturated fatty acid** is a fatty acid in which the carbon atoms in the hydrocarbon chain have one or more double covalent bonds in them.

A **monounsaturated fatty acid** is a fatty acid with only one double covalent bond between carbon atoms in the hydrocarbon chain.

A **polyunsaturated fatty acid** is a fatty acid with two or more double covalent bonds between carbon atoms in the hydrocarbon chain.

Esterification is the formation of ester bonds.

A **phospholipid** is a chemical in which glycerol bonds with two fatty acids and an inorganic phosphate group.

Hydrophilic molecules dissolve readily in water.

Hydrophobic molecules will not dissolve in water.

A **monolayer** is a single closely packed layer of atoms or molecules.

A **micelle** is a spherical aggregate of molecules in water with hydrophobic areas in the middle and hydrophilic areas outside.

A **bilayer** is a double layer of closely packed atoms or molecules.

A **unit membrane** is a bilayer structure formed by phospholipids in an aqueous environment, with the hydrophobic tails in the middle and the hydrophilic heads on the outside.

By the end of this section, you should be able to...

● outline the structure of an amino acid

● explain the formation of polypeptides and proteins and the nature of the bonds in proteins

● explain the significance of the primary, secondary, tertiary and quaternary structure of protein in determining the properties of fibrous and globular proteins

● explain how the structure of collagen and haemoglobin is related to their function

About 18% of your body is made up of protein. Proteins form hair, skin and nails, the enzymes needed for metabolism and digestion, and many of the hormones that control various body systems. They enable muscle fibres to contract, form antibodies that protect you from disease, help clot your blood and transport oxygen in the form of **haemoglobin**. Understanding the structure of proteins helps you develop an insight into the detailed biology of cells and organisms. Like carbohydrates and lipids, proteins contain carbon, hydrogen and oxygen. In addition they all contain nitrogen and many proteins also contain sulfur.

Proteins are another group of macromolecules made up of many small monomer units called **amino acids** joined together by condensation reactions. Amino acids combine in long chains to produce proteins. There are about 20 different naturally occurring amino acids that can combine in different ways to form a vast range of different proteins.

Amino acids

All amino acids have the same basic structure, which is represented as a general formula. There is always an amino group (–NH$_2$) and a carboxyl group (–COOH) attached to a carbon atom (see **fig A**). The group known as the R group varies between amino acids. This is where sulfur and selenium are found in the structure of a few amino acids. The structure of the R group affects the way the amino acid bonds with others in the protein, depending largely on whether the R group is polar or not.

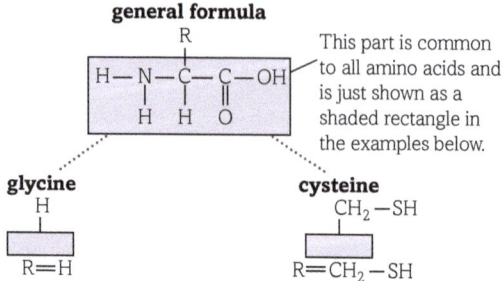

general formula

This part is common to all amino acids and is just shown as a shaded rectangle in the examples below.

fig A Some different amino acids. In the simplest amino acid, glycine, R is a single hydrogen atom. In a larger amino acid such as cysteine, R is much more complex.

Forming proteins from amino acids

Amino acids join together by a reaction between the amino group of one amino acid, and the carboxyl group of another. They join in a condensation reaction and a molecule of water is lost. A **peptide bond** is formed when two amino acids join and a **dipeptide** is the result (see **fig B**). The R group is not involved in this reaction. More and more amino acids join to form **polypeptide** chains, which contain from a hundred to many thousands of amino acids. When the polypeptide folds or coils or associates with other polypeptide chains, it forms a protein.

amino acid 1　　　　　　**amino acid 2** (inverted)

condensation　　hydrolysis

H_2O　　　　　　H_2O

peptide bond

dipeptide

fig B Amino acids are the building blocks of proteins, joined together by peptide bonds.

Bonds in proteins

The peptide bond between amino acids is a strong bond. Other bonds also form between the amino acids in a chain to form the 3D structures of the protein. They depend on the atoms in the R group and include hydrogen bonds, **disulfide bonds** and ionic bonds.

Hydrogen bonds

In amino acids, tiny negative charges are present on the oxygen of the carboxyl groups and tiny positive charges are present on the hydrogen atoms of the amino groups. When these charged groups are close to each other, the opposite charges attract, forming a hydrogen bond. Hydrogen bonds are weak, but they can potentially form between any two amino acids positioned correctly, so there are lots of them holding the protein together very firmly. Hydrogen bonds break easily and reform if pH or temperature conditions change. They are very important in the folding and coiling of the polypeptide chains (see **fig C**).

Disulfide bonds

Disulfide bonds form when two cysteine molecules are close together in the structure of a polypeptide (see **fig C**). An oxidation reaction takes place between the two sulfur containing groups, resulting in a strong covalent bond known as a disulfide bond. These disulfide bonds are much stronger than hydrogen bonds, but they occur much less often. They are important for holding the folded polypeptide chains in place.

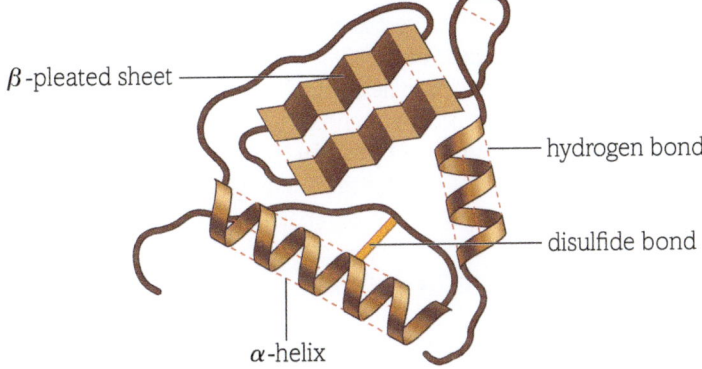

fig C Hydrogen bonds and disulfide bonds maintain the shape of protein molecules and this determines their function.

Ionic bonds

Ionic bonds can form between some of the strongly positive and negative amino acid side chains found buried deep in the protein molecules. These links are known as salt bridges. They are strong bonds, but they are not as common as the other structural bonds.

fig D Straightening your hair changes the arrangement of the hydrogen bonds so the hair curls in a different direction.

Your hair is made of the protein keratin. Some methods of styling the hair actually change the bonds within the protein molecules. Blow drying or straightening your hair breaks the hydrogen bonds and reforms them with the hair curling in a different way temporarily until the next time – if not before – when the hydrogen bonds reform in their original places.

Perming breaks the disulfide bonds between the polypeptide chains and reforms them in a different place. This effect is permanent – your hair will stay styled in that particular way until it is cut.

Protein structure

Proteins can be described by their primary, secondary, tertiary and quaternary structure (see **fig E**).

- The primary structure of a protein is the sequence of amino acids that make up the polypeptide chain held together by peptide bonds.

- The secondary structure of a protein is the arrangement of the polypeptide chain into a regular, repeating three-dimensional (3D) structure, held together by hydrogen bonds. One example is the right-handed helix (α-helix), a spiral coil with the peptide bonds forming the backbone and the R groups sticking out in all directions. Another is the β-pleated sheet, in which the polypeptide chain folds into regular pleats held together by hydrogen bonds between the amino and carboxyl ends of the amino acids. Most **fibrous proteins** have this sort of structure. Sometimes there is no regular secondary structure and the polypeptide forms a random coil.

- The tertiary structure is another level of 3D organisation imposed on top of the secondary structure in many proteins. The amino acid chain, including any α-helices and β-pleated sheets, is folded further into complicated shapes. Hydrogen bonds, disulfide bonds and ionic bonds between amino acids hold these 3D shapes in place (see **page 30**). **Globular proteins** are an example of tertiary structures.

- The quaternary structure of a protein is only seen in proteins consisting of several polypeptide chains. The quaternary structure describes the way these separate polypeptide chains fit together in three dimensions. Examples include some very important enzymes and the blood pigment haemoglobin. The bonds that hold the 3D shapes of proteins together are affected by changes in conditions such as temperature or pH. Even small changes can cause the bonds to break, resulting in the loss of the 3D shape of the protein. We say that the protein is **denatured.** Because the 3D structure of these proteins is important to the way they work, changing conditions inside the body can cause proteins such as enzymes to stop working properly.

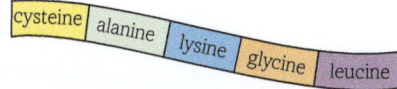

Primary structure – the linear sequence of amino acids in a peptide.

Secondary structure – the repeating pattern in the structure of the peptide chains, such as an α-helix or β-pleated sheets.

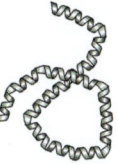

Tertiary structure – the three-dimensional folding of the secondary structure.

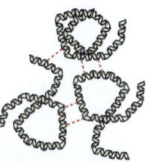

Quaternary structure – the three-dimensional arrangement of more than one tertiary polypeptide.

fig E The 3D structure of proteins.

Fibrous and globular proteins

Fibrous proteins

The complex structures of large protein molecules relate closely to their functions in the body. Fibrous proteins have little or no tertiary structure. They are long, parallel polypeptide chains with occasional cross-linkages that form into fibres. They are insoluble in water and are very tough, which makes them ideally suited to their structural functions within organisms. Fibrous proteins appear in the structure of connective tissue in tendons and the matrix of bones, in the structure of muscles, as the silk of spiders' webs and silkworm cocoons, and as the keratin that makes up hair, nails, horns and feathers.

Collagen is a fibrous protein that gives strength to tendons, ligaments, bones and skin. It is the most common structural protein found in animals – up to 35% of the protein in your body is collagen. Collagen is extremely strong – the fibres have a tensile strength comparable to that of steel. This is due to the unusual structure of the collagen molecule. It is made up of three polypeptide chains, which are each up to 1000 amino acids long. The primary structure of these chains is repeating sequences of glycine with two other amino acids – often proline and hydroxyproline. The three α-chains are arranged in a unique triple helix, held together by a very large number of hydrogen bonds. These collagen molecules, which can be up to several millimetres long, are often found together in fibrils that in turn are held together to form collagen fibres.

Collagen fibres are found combined with the bone tissue, giving it tensile strength rather like the steel rods in reinforced concrete. In the genetic disease osteogenesis imperfecta, the collagen triple helix does not form properly. The bone lacks tensile strength as a result, and it is brittle and breaks very easily.

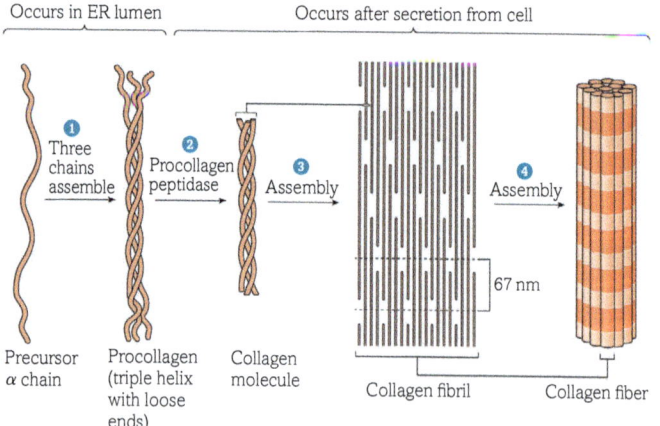

fig F Collagen is a fibrous protein of with an unusual triple helix structure and immense strength

Globular proteins

Globular proteins have complex tertiary and sometimes quaternary structures. They fold into spherical (globular) shapes. The large size of these globular protein molecules affects their behaviour in water.

Because their carboxyl and amino ends give them ionic properties you might expect them to dissolve in water and form a solution. In fact, the molecules are so big that instead they form a colloid. Globular proteins play an important role in holding molecules in position in the cytoplasm. Globular proteins are also important in your immune system – for example, antibodies are globular proteins. Globular proteins form enzymes and some hormones and are involved in maintaining the structure of the cytoplasm (see **Section 1.4.1** for details of proteins as enzymes).

Haemoglobin is one of the best known globular proteins. It is a very large molecule made up of 574 amino acids arranged in four polypeptide chains which are held together by disulfide bonds. Each chain is arranged around an iron-containing haem group. Haemoglobin is a **conjugated protein** as well as a globular protein. It is the iron that enables the haemoglobin to bind and release oxygen molecules, and it is the arrangement of the polypeptide chains that determine how easily the oxygen binds or is released (see **Section 4.3.3**)

Conjugated proteins

The shape of a protein molecule is usually very important in its function. Some protein molecules are joined with or conjugated to another molecule called a **prosthetic group** (see **fig C**). This structural change usually affects the performance and functions of the molecules. You have already looked at haemoglobin, a large protein with iron as the prosthetic group. Chlorophyll, the molecule involved in the capture of light energy in photosynthesis, is another conjugated protein, with magnesium as the prosthetic group.

Glycoproteins are proteins with a carbohydrate prosthetic group. The carbohydrate part of the molecule helps them to hold on to a lot of water and also makes it harder for protein-digesting enzymes (**proteases**) to break them down. Lots of lubricants used by the human body – such as mucus and the synovial fluid in the joints – are glycoproteins whose water-holding properties make them slippery and viscous, which reduces friction. This also helps to explain why the mucus produced in the stomach protects the protein walls from digestion.

Lipoproteins are proteins conjugated with lipids and are very important in the transport of cholesterol in the blood. The lipid part of the molecule enables it to combine with the lipid cholesterol. There are two main forms of lipoproteins in your blood – low-density lipoproteins (LDLs) (around 22 nm in diameter) and high-density lipoproteins (HDLs) (around 8–11 nm in diameter). The HDLs contain more protein than LDLs, which is partly why they are denser – proteins are more compact molecules than lipids.

Learning tip

Remember that amino acids are joined together by peptide bonds to form dipeptides and then polypeptides, but the 3D structures of proteins are the result of hydrogen bonds, disulfide bonds and ionic bonds between amino acids within the polypeptide chains.

Did you know?

Testing for protein

To test for the presence of protein, either add 5% (w/v) potassium or sodium hydroxide solution and 1% (w/v) copper sulfate solution or Biuret reagent, which is the two chemicals ready mixed. When the reagent/s are added to a test solution, a purple colour indicates the presence of protein.

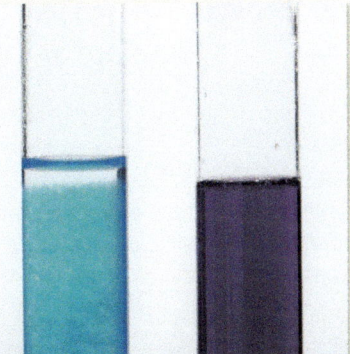

Questions

1 Explain how the order of amino acids in a protein affects the structure of the whole protein.

2 Hydrogen bonds are weaker than disulfide bonds and ionic salt bridges, but they play a much bigger role in maintaining protein structure. Why is this?

3 The body uses many resources to maintain a relatively constant internal environment. With reference to proteins, explain why constant internal conditions are so important.

Key definitions

Amino acids are the building blocks of proteins consisting of an amino group ($-NH_2$) and a carboxyl group ($-COOH$) attached to a carbon atom and an R group that varies between amino acids.

A **peptide bond** is the bond formed by condensation reactions between amino acids.

A **dipeptide** is two amino acids joined by a peptide bond.

A **polypeptide** is a long chain of amino acids joined by peptide bonds.

Fibrous proteins are proteins that have long, parallel polypeptide chains with occasional cross-linkages that form into fibres but with little tertiary structure.

A **disulfide bond** is a strong covalent bond formed as a result of an oxidation reaction between sulfur groups in cysteine or methionine molecules, which are close together in the structure of a polypeptide.

Globular proteins are large proteins with complex tertiary and sometimes quaternary structures, folded into spherical (globular) shapes.

Haemoglobin is a large conjugated protein involved in transporting oxygen in the blood and gives the erythrocytes their red colour.

Collagen is a strong fibrous protein with a triple helix structure.

Denaturation is the loss of the 3D shape of a protein, e.g. as a result of changes in temperature or pH.

A **prosthetic group** is the molecule that is incorporated in a conjugated protein.

A **glycoprotein** is a protein with a carbohydrate prosthetic group.

A **protease** is a protein-digesting enzyme.

A **lipoprotein** is a protein with a lipid prosthetic group.

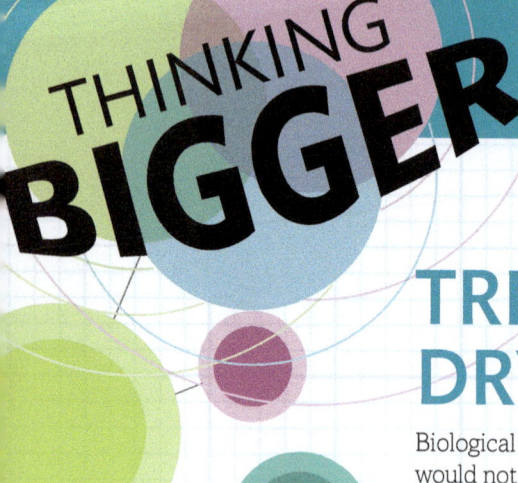

THINKING BIGGER

TREHALOSE – A SUGAR FOR DRY EYES?

Biological molecules have an amazing number of different roles in living organisms, including some you would not expect. In this activity you will discover how current research, which shows that disaccharide trehalose can protect proteins from damage in stressful conditions, is being used to make dry eyes more comfortable – and possibly protect the brain from the damage that can result from ageing.

TREHALOSE: AN INTRIGUING DISACCHARIDE WITH POTENTIAL FOR MEDICAL APPLICATION IN OPHTHALMOLOGY

Jacques Luyckx and Christophe Baudouin

Abstract

Trehalose is a naturally occurring disaccharide comprised of two molecules of glucose. The sugar is widespread in many species of plants and animals, where its function appears to be to protect cells against desiccation, but it is not found in mammals. Trehalose has the ability to protect cellular membranes and labile proteins against damage and denaturation as a result of desiccation and oxidative stress. Trehalose appears to be the most effective sugar for protection against desiccation. Although the exact mechanism by which trehalose protects labile macromolecules and lipid membranes is unknown, credible hypotheses do exist. As well as being used in large quantities in the food industry, trehalose is used in the biopharmaceutical preservation of labile protein drugs and in the cryopreservation of human cells. Trehalose is under investigation for a number of medical applications, including the treatment of Huntington's chorea and Alzheimer's disease. Recent studies have shown that trehalose can also prevent damage to mammalian eyes caused by desiccation and oxidative insult. These unique properties of trehalose have thus prompted its investigation as a component in treatment for dry eye syndrome. This interesting and unique disaccharide appears to have properties which may be exploited in ophthalmology and other disease states.

Trehalose, a naturally occurring alpha-linked disaccharide formed of two molecules of glucose (**fig A**) … is synthesized by many living organisms, including insects, plants, fungi, and micro-organisms as a response to prolonged periods of desiccation. This very useful property, known as anhydrobiosis, confers on an organism the ability to survive almost complete dehydration for prolonged periods and subsequently reanimate.

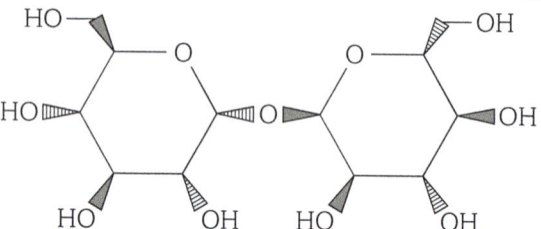

fig A Structure of trehalose. Registry number: 99-20-7; Molar mass: 342.296 g/mol (anhydrous); 378.33 g/mol (dihydrate); molecular structure: α-D-glucopyranosyl α-D-glucopyranoside (α,α-trehalose).

References

1 Iturriaga G, Suárez R, Nova-Franco B. Trehalose metabolism: From osmoprotection to signaling. *Int J Mol Sci*. 2009; 10:3793–3810. […]

8 Elbein AD, Pan YT, Pastuszak I, Carroll D. New insights on trehalose: A multifunctional molecule. *Glycobiology*. 2003; 13:17R–27R.

11 Jain NK, Roy I. Effect of trehalose on protein structure. *Protein Sci*. 2009; 18:24–36. […]

20 Matsuo T. Trehalose protects corneal epithelial cells from death by drying. *Br J Ophthalmol*. 2001; 85:610–612. […]

30 Matsuo T, Tsuchida Y, Morimoto N. Trehalose eye drops in the treatment of dry eye syndrome. *Ophthalmology*. 2002; 109:2024–2029.

31 Matsuo T. Trehalose versus hyaluronan or cellulose in eyedrops for the treatment of dry eye. *Jpn J Ophthalmol*. 2004; 48:321–327.

Where else will I encounter these themes?

Let us start by considering the nature of the writing in this article:

1. This extract comes from a paper published in *Clinical Ophthalmology*, an online journal. Think about the type of writing being used and the audience it is intended for as you try and answer the following questions:

a. What aspects of this writing tell you it is different from a scientific paper rather than a general interest article in a magazine?

b. Choose two words used in the article that you are not familiar with. Find out what they mean and suggest why they have been used by the author of the article.

c. How do you think these ideas about trehalose and the way it may be used to help human health might be presented in a newspaper or on the BBC website? Have a go at writing an article for a public interest website yourself.

d. If trehalose can really help protect people's sight and prevent brain diseases such as Huntington's and Alzheimer's this would make a big difference to people's lives. Notice how cautious the author is. Why are scientific papers so measured in the way they report things?

Think about the level of scientific detail that is suitable for your expected audience. How will you ensure your article is eye-catching and interesting?

Now you are going to think about the science in the article. You will be surprised how much you know already, but if you choose to do so, you can return to these questions later in your course.

2. What do you know about the chemical nature of trehalose from the article?

3. Desiccation (drying out) is a major problem for living organisms. Suggest reasons why drying out is so hard to survive.

4. Scientists think that trehalose protects both lipid membranes and certain proteins from damage, both from drying out and oxidation. Explain why it is so important biologically to protect cell membranes and protein structures.

Think about the chemistry of biological molecules you have learned already and use it to help you understand how trehalose works.

Activity

fig B *Selaginella lepidophylla* is a resurrection plant – it can withstand almost complete dehydration and recover within about 24 hours, thanks to high levels of trehalose in the plant cells.

You can refer to the full version of this paper, to the references listed at the end, to online encyclopedias, to other scientific papers and to books. In each case, judge the reliability of your source before you use it.

Which aspect of trehalose would you like to know more about? The way it prevents desiccation in many groups of organisms? The way it can protect human eyes from damage? The evidence that it could help reduce brain diseases in people?

Choose the area that interests you most and use as many resources as you can to produce a 3-minute presentation about that aspect of trehalose biology. Find interesting images and list all the references to help your colleagues decide if they can rely on the information you present.

- From the following journal article:
  'Trehalose: an intriguing disaccharide with potential for medical application in ophthalmology.' *Clinical ophthalmology* (Auckland, NZ) 5 (2011): 577.

Exam-style questions

1 Place a cross (☒) in the most appropriate box that describes the structure or role of these biological molecules.

(a) Dissacharides can be split by

(i) ☐ hydrolysis of glycosidic bonds

(ii) ☐ condensation of glycosidic bonds

(iii) ☐ hydrolysis of ester bonds

(iv) ☐ condensation of ester bonds   [1]

(b) Amylose is an example of a

(i) ☐ monosaccharide

(ii) ☐ disaccharide

(iii) ☐ polysaccharide

(iv) ☐ trisaccharide   [1]

(c) The role of starch is to

(i) ☐ be a source of energy to plants

(ii) ☐ store energy in all living organisms

(iii) ☐ store energy in plants

(iv) ☐ store energy in animals   [1]

(d) Proteins are polymers of amino acids joined by peptide bonds formed between the

(i) ☐ R groups

(ii) ☐ R group and the amino group

(iii) ☐ R group and the carboxyl group

(iv) ☐ carboxyl group and the amino group   [1]

(e) The three-dimensional structure of a protein is held together by

(i) ☐ peptide, hydrogen and ionic bonds

(ii) ☐ hydrogen, ester and ionic bonds

(iii) ☐ disulfide bridges and ester bonds

(iv) ☐ disulfide bridges, hydrogen and ionic bonds   [1]

(f) DNA consists of mononucleotides joined togther by bonds between

(i) ☐ two pentose sugars

(ii) ☐ one ribose sugar and one phosphate group

(iii) ☐ one deoxyribose sugar and one phosphate group

(iv) ☐ two phosphate groups   [1]

(g) Water is described as a polar molecule because it has a

(i) ☐ positively charged hydrogen end and a negatively charged oxygen end

(ii) ☐ positively charged hydrogen end and a positively charged oxygen end

(iii) ☐ negatively charged hydrogen end and a negatively charged oxygen ends

(iv) ☐ negatively charged hydrogen end and a positively charged oxygen end   [1]

[Total: 7]

2 Fill in this table to show the components and bonding within each carbohydrate.

	Lactose	Maltose	Amylose
Component monosaccharides			
Bonds between monosaccharides			

[3]

[Total: 3]

3 A disaccharide can be hydrolysed to its two monosaccharides. Explain the term **hydrolysis**.   [2]

[Total: 2]

4 Read through the following account on lipids, then write on the dotted lines the most appropriate words to complete the account.

Lipids are insoluble in water because they are

....

A triglyceride is one type of lipid. A triglyceride consists of

one .... molecule with three

.... molecules joined to it by

.... bonds. Triglycerides have

important roles in living organisms, including waterproofing

and ....   [5]

[Total: 5]

5 (a) Draw a diagram to show the structure of a phospholipid. Use the symbols shown for each component in your diagram.

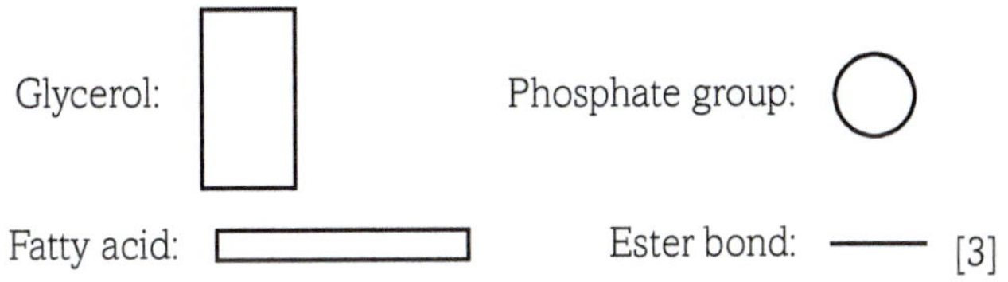

Glycerol: ☐   Phosphate group: ◯

Fatty acid: ▭   Ester bond: —   [3]

(b) The presence of a phosphate makes part of the molecule hydrophilic. Explain what is meant by the term **hydrophilic**. [1]

(c) Describe the role of phospholipids in the cell surface (plasma) membrane. [2]

[Total: 6]

6 (a) Draw a triglyceride. You may use any component more than once.

☐ glycerol   ∿∿∿ fatty acid   — ester bond

(b) There are four statements about triglycerides given below. If the statement is correct, put a tick (✓) in the box to the right of that statement. If the statement is incorrect, put a cross (✗) in the box to the right of the statement.

Statement	Tick (✓) or cross (✗)
Triglycerides are building blocks of polysaccharides	
Triglycerides can contain a small amount of nitrogen	
Triglycerides can be modified into phospholipids	
Triglycerides release water during hydrolysis	

[4]

(c) Fatty acids can be either saturated or unsaturated. Explain what is meant by the term **saturated** fatty acid. [1]

[Total: 5]

7 Describe the structure of an amino acid. [2]

[Total: 2]

8 (a) Insulin and collagen are both proteins that have a primary structure made up of amino acids joined together by peptide bonds.

   (i) Explain what is meant by the term **primary structure** of a protein. [1]

   (ii) Name the type of reaction that occurs when a peptide bond is broken causing a dipeptide to split into two amino acids. [1]

(b) Insulin and collagen both contain the amino acids glycine and serine. The diagram below shows a dipeptide formed from these two amino acids. Complete the diagram to show the structure of serine when the peptide bond breaks.

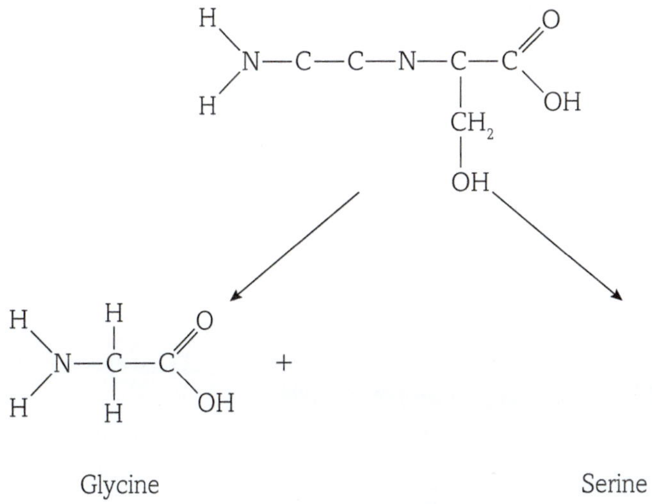

Glycine   Serine

[1]

(c) In the table below give three structural differences between the molecules of collagen and insulin.

	Collagen	Insulin
1		
2		
3		

[3]

[Total: 6]

TOPIC 1
Biological molecules

1.3 > Biological molecules 2

Introduction

In an air-conditioned room near Cambridge, ranks of machines at the Wellcome Trust Sanger Institute sequence the genetic material of thousands of anonymous people, of disease-causing bacteria and of cancers that mutate and grow in spite of chemotherapy. Without noise, without drama, the secrets of life itself are revealed in bars of light as the ever-developing technology identifies the sequences of bases that make up the DNA code. The first complete sequence of the human genome took years of work by scientists in many countries. Now it takes days to complete a human genome and less than 24 hours to sequence the genetic material of a bacterium. The expertise of scientists is still needed to interpret and use the information, which is being produced 24 hours a day. The potential benefits from this rapidly developing area of science, where biology, medicine and computers come together, are almost limitless.

In this chapter you will be studying the nucleotides and some of the molecules in which they play a key role, including DNA.

Adenosine triphosphate (ATP) is the molecule that acts as the universal energy supply in cells of every type. You will look at the structure of the molecule and how this is related to its role in cells. You will be referring to ATP in almost every aspect of your biology studies.

Nucleic acids or polynucleotides are the information molecules of the cell. You will discover the structure of deoxyribonucleic acid (DNA) and crack the code by which it carries the information needed to build an entire new organism.

You will learn about the different types of ribonucleic acid (RNA) and how they work together with the DNA to translate the genetic code into the phenotype of the cell or organism through protein synthesis. You also will build up a model of mutation and see how changes in the genetic code itself can result in changes, which may benefit or damage an organism.

All the maths you need

- Recognise and make use of appropriate units in calculations (*e.g. nanometres*)
- Use ratios (*e.g. representing the relationships between atoms in an ion or molecule*)

What have I studied before?

- DNA is a polymer made up of two strands that form a double helix
- DNA is the genetic material of the cell
- The genome is the entire genetic material of the cell
- Our increasing understanding of the human genome is potentially very important in medicine
- The impact of genome sequencing on how we classify organisms
- Mutations cause variants, most of which have no effect on the phenotype, some of which influence the phenotype and some of which determine the phenotype of the organism. These may be positive or negative in their effect
- Cellular respiration is an exothermic reaction that produces ATP

What will I study later?

- The importance of the enzymes formed during protein synthesis
- The site of the genetic material in the nucleus of the cell
- What happens to the genetic material during mitotic and meiotic cell division
- What happens to the DNA in chromosome mutations and how they can affect the phenotype of the resulting individual
- The impact of DNA sequencing on classification (A level)
- The importance of mutation and genetic variation in evolution by natural selection (A level)
- Biodiversity within the gene pool of a population (A level)
- The role of ATP in cellular respiration during glycolysis and the Krebs cycle (A level)
- The synthesis of ATP by chemosmosis (A level)
- The role of ATP in the light-dependent and the light-independent stages of photosynthesis (A level)
- Gene technology (A level)

What will I study in this chapter?

- The structure of nucleotides
- The structure of ATP related to its function as the universal energy supplier to cells
- The structure of the DNA molecule including the double helix
- The story of how the double helix structure of DNA was discovered
- How the DNA code works and the experimental evidence used to prove it
- Protein synthesis and the roles of the different types of RNA in the process
- Different types of gene mutations and the effect they have on the amino acid sequences of the proteins that are formed

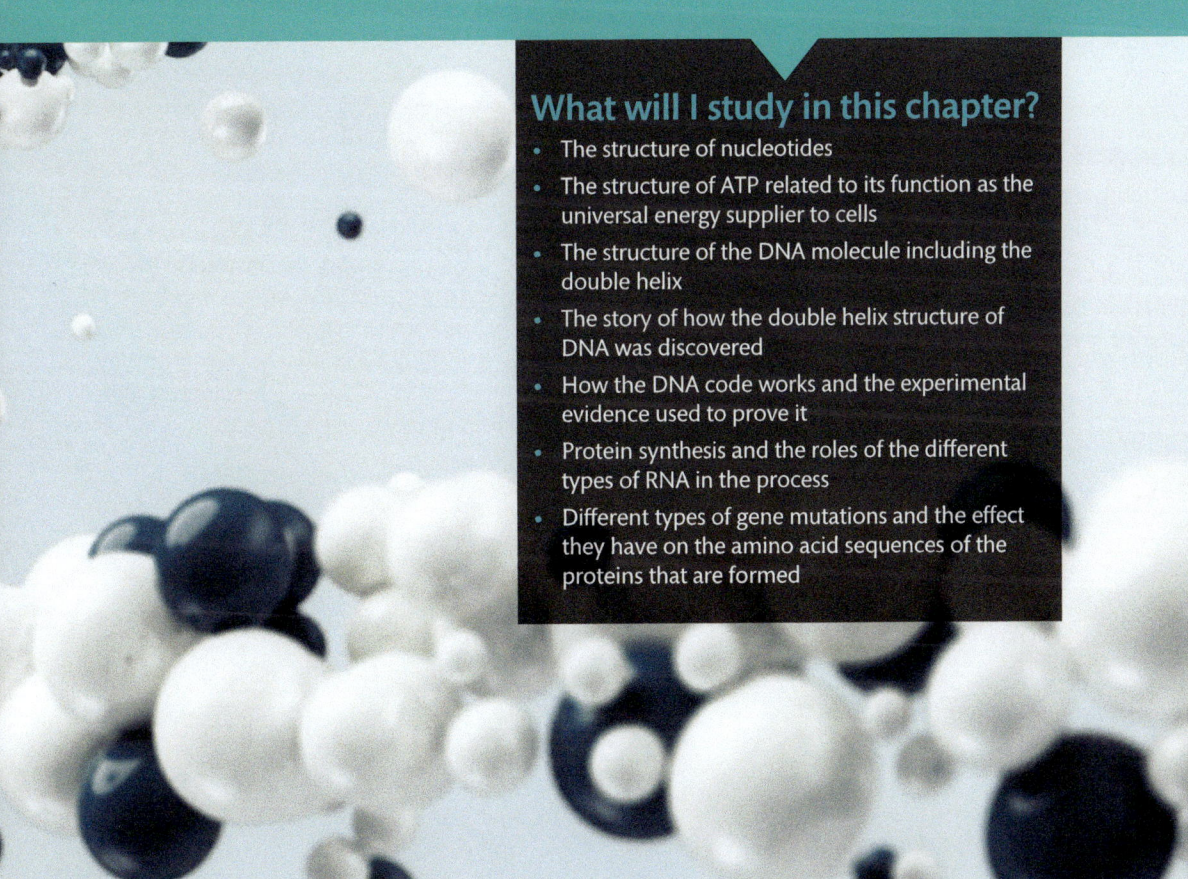

By the end of this section, you should be able to...

● outline the structure of nucleotides, both purines and pyrimidines

● relate the structure and properties of ATP to its function in the cell

Nucleotides

Nucleotides are key molecules in biology. They provide the energy currency of cells in the form of **adenosine triphosphate**, usually referred to as **ATP**. They also provide the building blocks for the mechanism of inheritance in the form of DNA – deoxyribonucleic acid – and RNA – ribonucleic acid.

Each nucleotide has three parts – a 5-carbon pentose sugar, a nitrogen-containing base and a phosphate group. The pentose sugar in RNA is ribose, and in DNA is deoxyribose. Deoxyribose, as its name suggests, contains one fewer oxygen atom than ribose (see **fig A**).

The most common types of nucleotides have either a **purine base**, which has two nitrogen-containing rings, or a **pyrimidine base**, which has only one. Both purines and pyrimidines are weak bases. The most common purines are **adenine** (A) and **guanine** (G) and the most common pyrimidines are **cytosine** (C), **thymine** (T) and **uracil** (U).

A phosphate group ($-PO_4^{3-}$) is the third component of a nucleotide. Inorganic phosphate ions are present in the cytoplasm of every cell (see **Section 1.1.1** for more about inorganic ions). It is as a result of this phosphate group that the nucleotides are acidic molecules and carry a negative charge.

The sugar, the base and the phosphate group are joined together by condensation reactions, with the elimination of two water molecules, to form a nucleotide (see **fig A**).

ATP

Cells are chemical factories, with many reactions continually taking place within the cytoplasm and organelles (see **Sections 2.1.3** and **2.1.5** on cell structures and **Book 2 Chapter 5.1** on cellular respiration). In these reactions, chemical bonds are constantly being broken and energy is always needed to break the bonds. Each cell needs a constantly available and immediately accessible supply of energy to support a multitude of different reactions.

One molecule seems to be the universal energy supplier in cells – adenosine triphosphate (ATP). ATP is found in all living organisms in exactly the same form. Anything which interferes with the production or breakdown of ATP is fatal to the cell and ultimately destroys the whole multicellular organism.

ATP is a nucleotide with three phosphate groups attached (see **fig B**). It is the potential energy in the phosphate bonds, that is made available to cells for use in breaking bonds in chemical reactions.

Fig B shows the structure of ATP. When energy is needed in the cell, the third phosphate bond in the molecule is broken in a hydrolysis reaction. This is catalysed by the enzyme **ATPase**. The products of the reaction are **adenosine diphosphate (ADP)**, another nucleotide, and a free inorganic phosphate group (P_i). One phosphate bond is broken as the ATP is split – this uses energy. Two further bonds are made to produce the ADP and the stable phosphate group and this releases the energy that is needed to drive other reactions. About 34 kJ of energy are released per mole of ATP hydrolysed. Some of this energy is always lost to the system as heat, but the rest is used for any energy-requiring biological activity in the cell such as building up new molecules, active transport (**see Section 4.1.4**), nerve impulses or muscle contraction.

Note: DNA and RNA contain both phosphate and nitrogen (in the base).

ribose   deoxyribose

phosphate

pentose sugar — organic nitrogenous base

fig A The structure of a nucleotide. The properties of the nucleotide molecule are crucial to the roles of ATP, DNA and RNA.

(a)

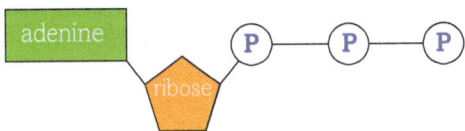

adenine | ribose | 3 × phosphates

(b)

adenine — ribose — P — P — P

fig B The structure of ATP.

The breakdown of ATP into ADP is a reversible reaction. ATP can be synthesised from ADP and a phosphate group in a reaction that requires an input of energy (30.5 kJ per mole of ATP produced). ATPase catalyses this reaction. The direction of the reaction depends on conditions in the cell. The energy needed to drive the synthesis of ATP usually comes from catabolic (breakdown) reactions or from **reduction/oxidation (redox) reactions**. As a result, the ATP molecule provides an immediate supply of energy, ready for use when needed.

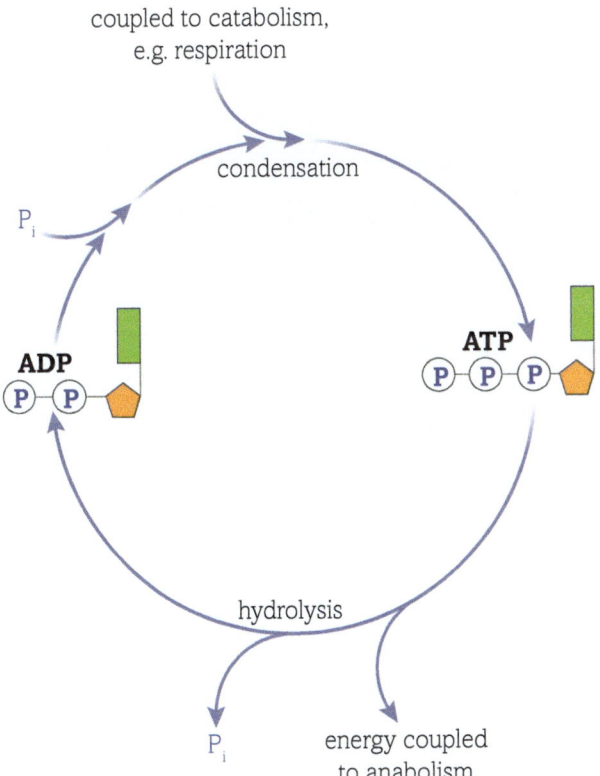

fig C The energy released in catabolic reactions drives the production of ATP. When needed, this energy can then be used to drive anabolic reactions in the cell.

Questions

1 Describe the structure of a nucleotide.

2 ATP is regarded as the universal energy supply molecule. Why is this and how does its structure relate to its role in cells?

Key definitions

Nucleotides are molecules with three parts – a 5-carbon pentose sugar, a nitrogen-containing base and a phosphate group – joined by condensation reactions.

Adenosine triphosphate (ATP) is a nucleotide that acts as the universal energy supply molecule in cells. It is made up of the base adenine, the pentose sugar ribose and three phosphate groups.

A **purine base** is a base found in nucleotides that has two nitrogen-containing rings.

A **pyrimidine base** is a base found in nucleotides that has one nitrogen-containing ring.

Adenine is a purine base found in DNA and RNA.

Guanine is a purine base found in DNA and RNA.

Cytosine is a pyrimidine base found in DNA and RNA.

Thymine is a pyrimidine base found in DNA.

Uracil is a pyrimidine base found in RNA.

ATPase is an enzyme that catalyses the formation and the breakdown of ATP, depending on conditions.

Adenosine diphosphate (ADP) is a nucleotide formed when ATP loses a phosphate group and provides energy to drive reactions in the cell.

Reduction/oxidation (redox) reactions are reactions in which one reactant loses electrons (is oxidised) and another gains electrons (is reduced).

By the end of this section, you should be able to...

- describe the structure of DNA including the structure of the nucleotides, base pairing, the two sugar–phosphate backbones, phosphodiester bonds and hydrogen bonds

- describe the structure of RNA including nucleotides, the sugar–phosphate backbones and the role of hydrogen bonds

Reproduction is one of seven key processes in living organisms. If the individuals in a species do not reproduce, then that species will die out. Multicellular organisms also need to grow, and to replace worn-out cells. Within every cell is a set of instructions for the assembling of new cells. These can be used both to form offspring and to produce identical cells for growth. Over the last 75 years or so scientists have made enormous strides towards understanding the form of these instructions – the genetic code. In unravelling the secrets of the genetic code, people have come closer than ever before to understanding the mystery of life itself.

Nucleic acids, also known as polynucleotides, are the information molecules of the cell. They carry all the information needed to form new cells. They are polymers, made up of many nucleotide monomer units. The chromosomes in the nucleus of eukaryotic cells store the genetic information (see **Section 2.1.3**), but in prokaryotes the DNA is found floating freely in the cytoplasm (see **Section 2.2.1**). It takes the form of a code in the molecules of DNA – deoxyribonucleic acid (see **fig C**). Parts of the code are copied into messenger RNA (mRNA) and used to direct the production of the proteins that build the cell and control its actions.

Building the polynucleotides

Nucleic acids are chains of nucleotides linked together by condensation reactions that produce **phosphodiester bonds** between the sugar on one nucleotide and the phosphate group of the next nucleotide. These nucleic acids can be millions of nucleotide units long. Both DNA and RNA have this sugar–phosphate backbone. Because the sugar of one nucleotide bonds to the phosphate group of the next nucleotide, polynucleotides always have a hydroxyl group at one end and a phosphate group at the other. This structural feature is important in the role of the nucleic acids in the cell. Long chains of nucleotides containing the bases C, G, A and T join together to form DNA. Chains of nucleotides containing C, G, A and U make RNA. Knowledge of how these units join together, and the three-dimensional (3D) structures in DNA in particular, is the basis of our understanding of molecular genetics.

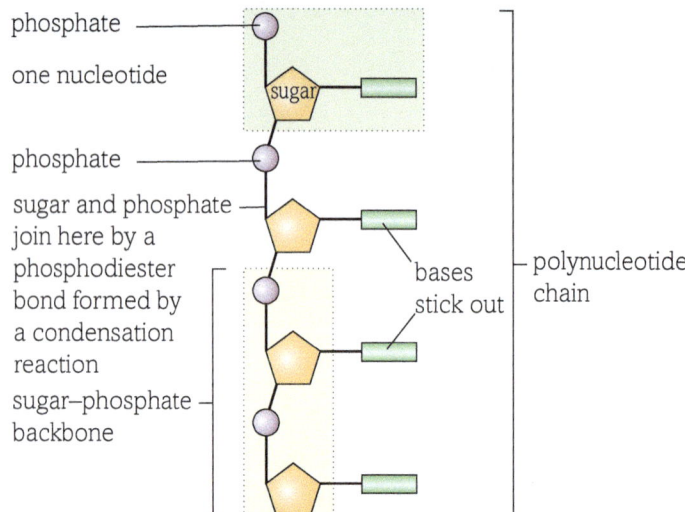

fig A A polynucleotide strand like this makes up the basic structure of both DNA and RNA.

RNA molecules form single polynucleotide strands that can fold into complex shapes, held in place by hydrogen bonds, or remain as long thread-like molecules. DNA molecules consist of two polynucleotide strands twisted around each other. The sugars and phosphates form the backbone of the molecule and, pointing inwards from the two sugar–phosphate backbones, are the bases, which pair up in specific ways. A purine base always pairs with a pyrimidine base – in DNA, adenine pairs with thymine and cytosine with guanine. This results in the famous DNA *double helix*, a massive molecule that resembles a spiral staircase.

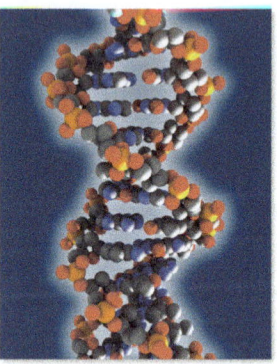

fig B The double helix structure of a DNA molecule is not just an iconic image of science – it is vital to the role of DNA in cells.

The two strands of the DNA double helix are held together by hydrogen bonds between the complementary base pairs (see **fig C**). These hydrogen bonds form between the amino and the carbonyl groups of the purine and pyrimidine bases on the opposite strands. There are three hydrogen bonds between C and G but only two between A and T. There are 10 base pairs for each complete twist of the helix. The two strands are known as the 5′ (5 prime) and 3′ (3 prime) strand, named according to the number of the carbon atoms in the pentose sugar to which the phosphate group is attached in the first nucleotide of the chain. It is the phosphate that is free at the 5' end of the 5' carbon, and it is the free –OH group that is attached to the 3' carbon on the 3' end. As you will see, these features of the structure of DNA and RNA are crucial to the way the molecules function within cells.

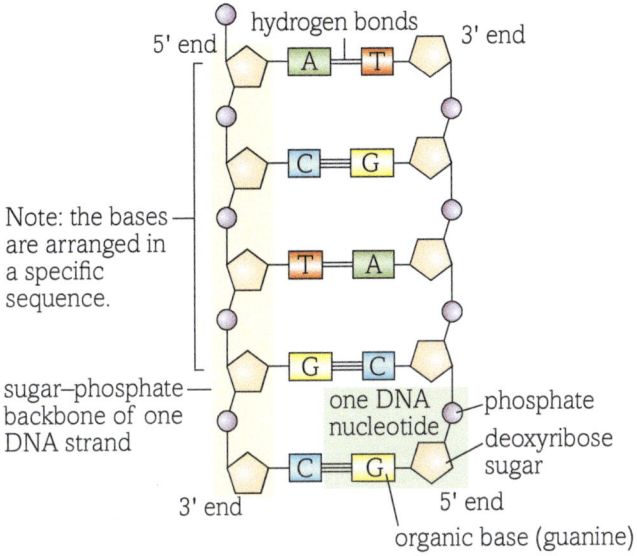

Note: the bases are arranged in a specific sequence.

sugar–phosphate backbone of one DNA strand

one DNA nucleotide

phosphate

deoxyribose sugar

organic base (guanine)

width = 2 nm

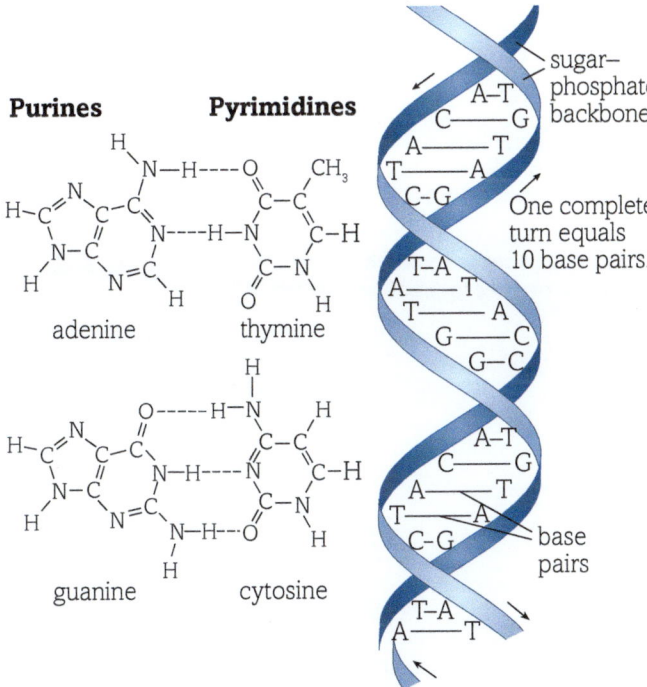

Purines

adenine

guanine

Pyrimidines

thymine

cytosine

sugar–phosphate backbones

One complete turn equals 10 base pairs.

base pairs

The two strands are antiparallel – one runs in one direction and the other in the opposite direction.

fig C The double helix structure of DNA depends on the hydrogen bonds that form between the base pairs.

Did you know?

Sequencing the genome

From the late twentieth century onwards scientists from around the world collaborated in the Human Genome Project. This was an ambitious project that set out to identify all of the genes in the human chromosomes and to sequence the 3 billion base pairs which make up the human DNA. The scientists worked on DNA from anonymous donors and showed that every individual has at least 99.9% of their DNA in common. Leaps in technology meant the project finished ahead of the expected date, but it still took almost 13 years.

In 2008 a new project began – the 1000 Genomes Project. This time, scientists analysed the DNA of 1092 people from all around the world, to gain information about differences in our DNA that can, amongst many things, have an impact on the diseases that may affect us. The 1000 Genomes Project took 6 years.

The 10K (ten thousand) Genomes Project got under way in 2013. This project is sequencing the genomes of 10 000 people from around the world with rare genetic diseases and cancers. The whole 10K Genomes Project is expected to take only 3 years, thanks to the immense improvements in sequencing technology, which mean that processes that once took weeks and months now take hours and days. It should greatly increase our understanding, diagnosis and even treatment of rare genetic conditions.

fig D DNA sequencing in progress at the Wellcome Trust Sanger Institute.

Questions

1 What is a mononucleotide and which mononucleotides are found in both DNA and RNA?

2 (a) Explain how complementary base pairing and hydrogen bonding are responsible for the structure of DNA.

  (b) Look carefully at the structural formulae of the purine bases and the pyrimidine bases of the DNA molecule. Suggest reasons why the pairs of bases always involve one purine and one pyrimidine base, never two purines or two pyrimidines.

Key definitions

Nucleic acids are polymers made up of many nucleotide monomer units that carry all the information needed to form new cells.

A **phosphodiester bond** is the bond formed between the phosphate group of one nucleotide and the sugar of the next nucleotide in a condensation reaction.

A **genome** is the entire genetic material of an organism.

By the end of this section, you should be able to...

- explain how DNA replicates semiconservatively including the role of DNA helicase, polymerase and ligase

- relate the structure of the DNA molecule to the way in which it replicates

One of the most important features of the DNA molecule is that it can **replicate**, or copy itself, exactly. This is the characteristic above all others, that means it can pass on genetic information from one cell or generation to another.

Uncovering the mechanism of replication

After Watson and Crick had produced their double helix model for the structure of DNA, it took scientists some years to work out exactly how the molecule replicates itself.

There were two main ideas about how replication happens: conservative and semiconservative replication. In the conservative replication model, the original double helix remained intact and in some way instructed the formation of a new, identical double helix, made up entirely of new material.

The semiconservative replication model assumed that the DNA 'unzipped' and new nucleotides aligned along each strand. Each new double helix contained one strand of the original DNA and one strand made up of new material. This was the Watson and Crick hypothesis – they felt the double helix would unzip along the hydrogen bonds in their structural model, allowing semiconservative replication to take place. It took a classic piece of practical investigation to settle the argument.

Experimental evidence

As the result of a very elegant set of experiments carried out by Matthew Meselson (1930–) and Franklin Stahl (1929–) at the California Institute of Technology in the late 1950s, semiconservative replication became the accepted model of DNA replication.

- They grew several generations of the gut bacteria *Escherichia coli* (*E. coli*) in a medium where their only source of nitrogen was the radioactive isotope ^{15}N from $^{15}NH_4Cl$. Atoms of ^{15}N are denser than those of the isotope usually found, ^{14}N. The bacteria grown on this medium took up the radioactive isotope to make the cell chemicals, including proteins and DNA. After several generations, the entire bacterial DNA was labelled with ^{15}N ('heavy' nitrogen).

- They moved the bacteria to a medium containing normal $^{14}NH_4Cl$ as their only nitrogen source, and measured the density of their DNA as they reproduced.

- Meselson and Stahl predicted that if DNA reproduced by conservative replication, some of the DNA would have the density expected if it contained nothing but ^{15}N (the original strands), and some of it would have the density expected if it contained nothing but ^{14}N (the new strands). However, if DNA reproduced by semiconservative replication, then all of the DNA would have the same density, half-way between that of ^{15}N- and ^{14}N-containing DNA.

- They found that all DNA had the same density, half-way between that of ^{15}N- and ^{14}N-containing DNA – and so DNA must replicate semiconservatively.

Conservative replication, where the double helix remains intact and new strands form on the outside, would give:

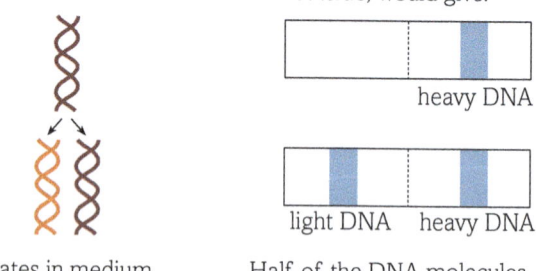

Replicates in medium containing only light nitrogen.

heavy DNA

light DNA   heavy DNA

Half of the DNA molecules have 2 light strands and half have 2 heavy strands.

Semiconservative replication, where the double helix unzips and each strand replicates to produce a second, new strand, would give:

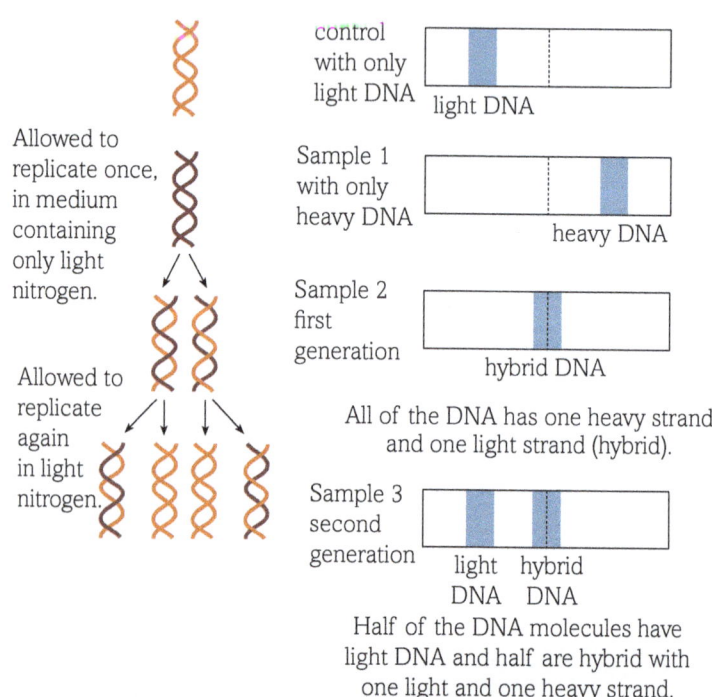

Allowed to replicate once, in medium containing only light nitrogen.

Allowed to replicate again in light nitrogen.

control with only light DNA

light DNA

Sample 1 with only heavy DNA

heavy DNA

Sample 2 first generation

hybrid DNA

All of the DNA has one heavy strand and one light strand (hybrid).

Sample 3 second generation

light   hybrid
DNA   DNA

Half of the DNA molecules have light DNA and half are hybrid with one light and one heavy strand.

fig A The results of these experiments by Meselson and Stahl put an end to the theory of conservative replication of DNA.

How DNA makes copies of itself

A careful look at the process of the semiconservative replication of DNA shows clearly the importance of the structure and properties of the DNA molecule to its role as the genetic material of the cell.

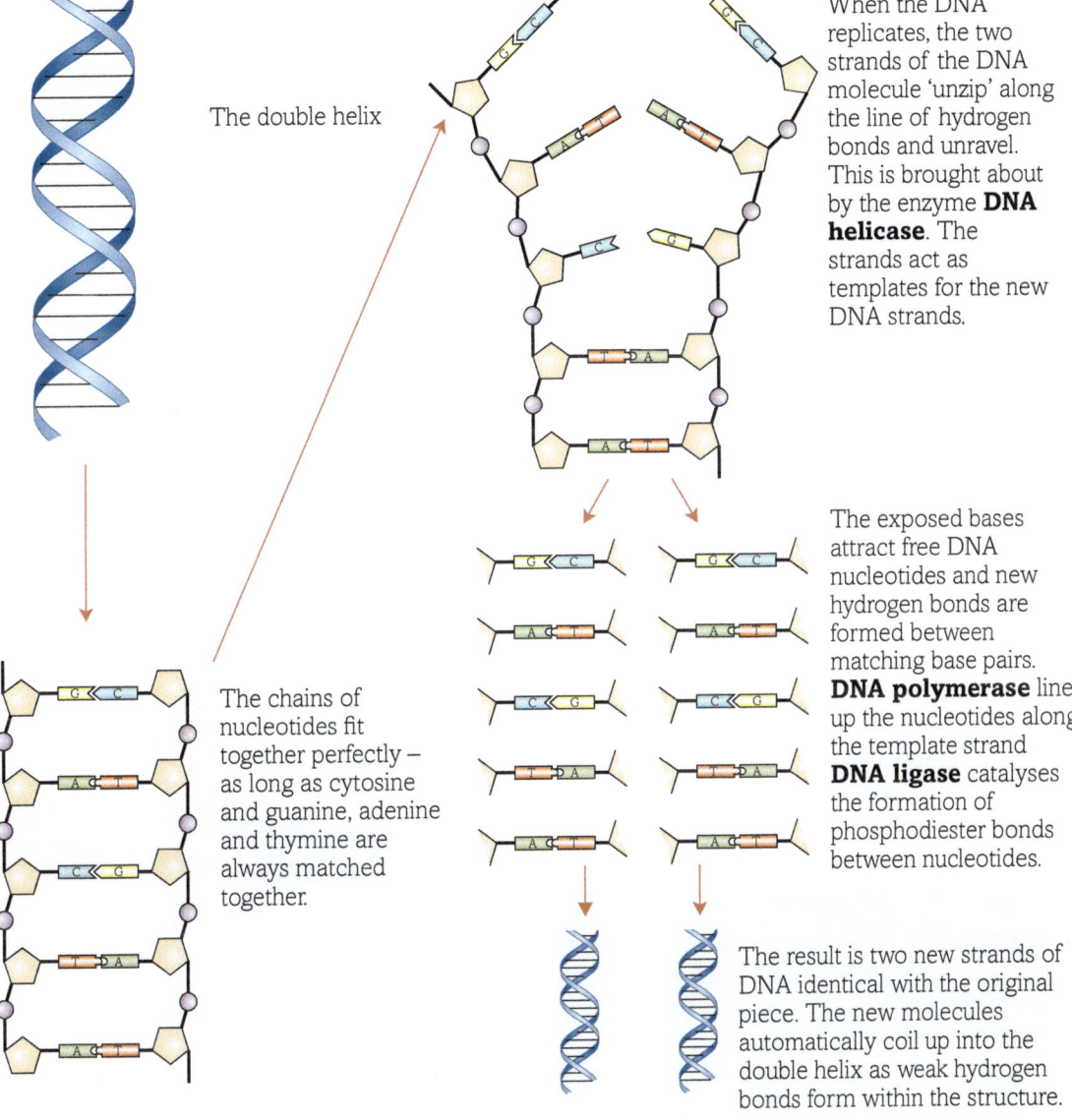

The double helix

When the DNA replicates, the two strands of the DNA molecule 'unzip' along the line of hydrogen bonds and unravel. This is brought about by the enzyme **DNA helicase**. The strands act as templates for the new DNA strands.

The chains of nucleotides fit together perfectly – as long as cytosine and guanine, adenine and thymine are always matched together.

The exposed bases attract free DNA nucleotides and new hydrogen bonds are formed between matching base pairs. **DNA polymerase** lines up the nucleotides along the template strand **DNA ligase** catalyses the formation of phosphodiester bonds between nucleotides.

The result is two new strands of DNA identical with the original piece. The new molecules automatically coil up into the double helix as weak hydrogen bonds form within the structure.

fig B The semiconservative replication of DNA.

Questions

1 Make a flow diagram to describe the replication of DNA.

2 How did the work of Meselson and Stahl destroy support for the model of the conservative replication of DNA?

Key definitions

When a DNA molecule **replicates**, it copies itself exactly.

DNA helicase is an enzyme involved in DNA replication that unzips the two strands of the DND molecules.

DNA polymerase is an enzyme involved in DNA replication that lines up the new nucleotides along to DNA template strands.

DNA ligase is an enzyme involved in DNA replication that catalyses the formation of phosphodiester bonds between nucleotides.

The genetic code

By the end of this section, you should be able to...

- define a gene as a sequence of bases on a DNA molecule coding for a sequence of amino acids in a polypeptide chain
- explain the nature of the genetic code including triplets, codons, the degenerate and non-overlapping nature of the code and that not all of the genome codes for proteins

We know that DNA has a double helix structure and can replicate itself exactly. But how does DNA act as the genetic material and carry the information needed to make new cells and whole new organisms? The key is the link between DNA and proteins. DNA controls protein synthesis and so the DNA instructions control not only how the cell is built, but also how it works.

Proteins are made up of amino acids. Using the DNA code, the 20 naturally occurring amino acids are joined together in countless combinations to make an almost infinite variety of proteins. This process of **translation** happens on the surface of the **ribosomes** (see **Section 1.3.6** on protein synthesis).

What is the genetic code?

In the DNA double helix, the components that vary are the bases. So scientists guessed that it was the arrangement of the bases that carries the genetic code – but how? There are only four bases, so if one base coded for one amino acid there could be only four amino acids. Even two bases do not give enough amino acids – the possible arrangements of four bases into groups of two is $4 \times 4 = 16$. However, a **triplet code** of three bases gives $4 \times 4 \times 4 = 64$ possible combinations – more than enough for the 20 amino acids that are coded for.

Cracking the code

The genetic code is based on **genes**. We can define a gene as a sequence of bases on a DNA molecule coding for a sequence of amino acids in a polypeptide chain, that affect a characteristic in the phenotype of the organism. By the early 1960s it had been proved that a triplet code of bases was the cornerstone of the genetic code. Each sequence of three bases along a strand of DNA codes for something very specific. Most code for a particular amino acid, but some triplets signal the beginning or the end of one particular amino acid sequence.

A sequence of three bases on the DNA or RNA is known as a **codon**. The codons of the DNA are difficult to work out because the molecule is so large, so most of the work was done on the codons of the smaller molecule mRNA. This mRNA is formed as a **complementary strand** to the DNA, so it is like a reverse image of the original base sequence. Once we know the RNA sequence, we can work out the DNA sequence because of the way bases always pair: T/U with A, and G with C. Sequencing tasks like this have become much easier in the twenty-first century as technology has advanced.

The result of all this work is a sort of dictionary of the genetic code (see **table A**). Much of the original work, done in the 1960s, used the gut bacteria *E. coli*, but all the studies done since suggest that the genetic code is identical throughout the living world.

Large parts of the DNA do not code for proteins. Scientists think the non-coding DNA sequences are very important – 98% of the human DNA is non-coding. They know they are involved in regulating the protein-coding sequences – effectively turning genes on or off. Many organisms have similar non-coding sequences, which suggests they are useful, but in many cases we still do not know exactly what they do.

In the 2% of the human DNA that codes for proteins, some codons code for a particular amino acid, while others code for the beginning or the ending of a particular amino acid sequence. We now know that the genetic code is not only a triplet code, it is non-overlapping and degenerate as well.

Did you know?

DNA code-breakers

The first breakthrough in decoding the genetic code came in 1961 when M.W. Nirenberg (1927–2010) and J.H. Matthaei (1929–) in the United States prepared artificial mRNA where all the bases were uracil. They added their polyU – chains reading UUUUUUUUUUUU … – to all the other ingredients needed for protein synthesis (ribosomes, tRNAs, amino acids, etc). When they analysed the polypeptides made, they found chains of a single type of amino acid, phenylalanine. UUU appeared to be the mRNA codon for phenylalanine. So the DNA codon would be AAA. The scientists soon showed that CCC codes for proline and AAA for lysine. Evidence for the triplet code – three non-overlapping bases with some degeneracy – built up swiftly from this early work. It was also shown that the minimum length of artificial mRNA that would bind to a ribosome was three bases long – a single codon. This would then bind with the corresponding tRNA. From this point on it was a case of careful and precise work to identify all of the codons and their corresponding amino acids.

second letter of the codon

first letter	A	G	T	C	third letter
A	AAA / AAG phenylalanine AAT / AAC leucine	AGA / AGG / AGT / AGC serine	ATA / ATG tyrosine ATT / ATC stop codon	ACA / ACG cysteine ACT stop codon ACC tryptophan	A G T C
G	GAA / GAG / GAT / GAC leucine	GGA / GGG / GGT / GEC proline	GTA / GTG histidine GTT / GTC glutamine	GCA / GCG / GCT / GCC arginine	A G T C
T	TAA / TAG / TAT isoleucine TAC methionine; start codon	TGA / TGG / TGT / TGC threonine	TTA / TTG asparagine TTT / TTC lysine	TCA / TCG serine TCT / TCC arginine	A G T C
C	CAA / CAG / CAT / CAC valine	CGA / CGG / CGT / CGC alanine	CTA / CTG aspartic acid CTT / CTC glutamic acid	CCA / CCG / CCT / CCC glycine	A G T C

table A The triplet code that underpins all work on genetics. The first table shows the DNA code.

second letter of the codon

first letter	U	C	A	G	third letter
U	UUU / UUC phenylalanine UUA / UUG leucine	UCU / UCC / UCA / UCG serine	UAU / UAC tyrosine UAA / UAG stop codon	UGU / UGC cysteine UGA stop codon UGG tryptophan	U C A G
C	CUU / CUC / CUA / CUG leucine	CCU / CCC / CCA / CCG proline	CAU / CAC histidine CAA / CAG glutamine	CGU / CGC / CGA / CGG arginine	U C A G
A	AUU / AUC isoleucine AUA AUG methionine; start codon	ACU / ACC / ACA / ACG threonine	AAU / AAC asparagine AAA / AAG lysine	AGU / AGC serine AGA / AGG arginine	U C A G
G	GUU / GUC valine GUA / GUG	GCU / GCC / GCA / GCG alanine	GAU / GAC aspartic acid GAA / GAG glutamic acid	GGU / GGC glycine GGA / GGG	U C A G

table B The triplet code that underpins all work on genetics. This second table shows the RNA code for the same amino acids.

A non-overlapping code...

Once scientists had worked out that the genetic code was based on triplets of DNA bases, they wanted to find out how the code was read. Do the triplets of bases follow each other along the DNA strand like beads on a necklace, or do they overlap? For example, the mRNA sequence UUUAGC could code for two amino acids, phenylalanine (UUU) and serine (AGC). On the other hand, if the code overlaps, it could code for four: phenylalanine (UUU), leucine (UUA), a nonsense or stop codon (UAG) and serine (AGC).

An overlapping code would be very economical – relatively short lengths of DNA could carry the instructions for many different proteins. However, it would also be very limiting, because the amino acids that could be coded for side by side would be limited. In the example given, only leucine out of the 20 available amino acids could ever follow phenylalanine, because only leucine has an mRNA codon starting with UU–.

Scientists rely on experimental observations to help decide whether the genetic code is overlapping or not. If a codon consists of three nucleotides and is completely overlapping, and a single nucleotide is altered by a point mutation, then three amino acids will be affected by that single change. If the code is only partly overlapping, then a single point mutation would result in two affected amino acids. But if the codons do not overlap at all, then a change in a single nucleotide mutation would affect only one amino acid, which is what has been observed, for example in sickle cell disease. All the evidence available suggests that the code is not overlapping and this is generally accepted among scientists.

...and a degenerate code

When you look at the genetic code, it appears that the code is degenerate, also known as redundant. In other words, it contains more information than it needs. If you look carefully at **table B**, you will see that often only the first two of the three nucleotides in a codon seem to matter in determining which amino acid results. This may seem a rather useless feature at first, but if each amino acid was produced by only one codon, then any error or mutation could cause havoc. With a degenerate code, if the final base in the triplet is changed, this mutation could still produce the same amino acid and have no effect on the organism. Only methionine and tryptophan are represented by only one codon.

Mutations can happen any time the DNA is copied – the degenerate code at least partly protects living organisms from their effects.

Questions

1 Explain what is meant by the genetic code.

2 What is non-coding DNA?

3 What are the benefits to an organism of having:
   (a) a non-overlapping code?
   (b) a degenerate code?

Key definitions

Translation is the process by which proteins are produced, via RNA, using the genetic code found in the DNA. It takes place on the ribosomes.

Ribosomes are the site of protein synthesis in the cell.

A **triplet code** is the code of three bases, and is the basis of the genetic information in the DNA.

A **gene** is a sequence of bases on a DNA molecule. It contains coding for a sequence of amino acids in a polypeptide chain that affect a characteristic in the phenotype of the organism.

A **codon** is a sequence of three bases in DNA or mRNA.

A **complementary strand** is the strand of RNA formed that complements the DNA acting as the coding strand.

By the end of this section, you should be able to...

● describe the structure of mRNA including nucleotides, the sugar phosphate backbone and the role of hydrogen bonds

● describe the structure of tRNA including nucleotides, the role of hydrogen bonds and the anticodon

● explain the processes of transcription in the nucleus and translation at the ribosome, including the role of sense and antisense DNA, mRNA, tRNA and the ribosomes

In eukaryotes, the DNA that codes for the individual proteins is in the nucleus of the cell. The ribosomes where proteins are synthesised are in the cytoplasm. DNA from the nucleus has never been detected in the cytoplasm, so the message cannot be carried directly. RNAs (ribonucleic acids) carry the information from the nuclear DNA to the active synthetic enzymes on the ribosomes.

Different types of RNA

RNA is closely related to DNA (see **Section 1.3.2**). However, it contains a different sugar (ribose) and a different base (uracil instead of thymine). It consists of a single helix and does not form enormous and complex molecules like DNA. The sequence of bases along a strand of RNA relates to the sequence of bases on a small part of the DNA in the nucleus. RNA enables DNA to act as the genetic material. It carries out three main functions in the process of protein synthesis:

• It carries the instructions for a polypeptide from the DNA in the nucleus to the ribosomes where proteins are made.

• It picks up specific amino acids from the protoplasm and carries them to the surface of the ribosomes.

• It makes up the bulk of the ribosomes themselves.

To perform these three very different functions, there are three different types of RNA.

Messenger RNA

Messenger RNA (mRNA) is formed in the nucleus. Whereas a double helix of DNA carries information about a vast array of proteins, a piece of mRNA usually has instructions for one polypeptide. The messenger RNA forms on the 3′ coding or **antisense strand** of the DNA. The mRNA formed is a **sense** strand that codes for a polypeptide. The 5′ non-coding strand of DNA is known as the sense strand of DNA. Any mRNA formed on this strand would be nonsense and would not code for a protein so it is not used.

Parts of the DNA molecule unravel and are transcribed onto strands of mRNA by an enzyme called **DNA-directed RNA polymerase**. This enzyme is often known as **RNA polymerase**, but the full name tells you it polymerises nucleotide units to form RNA in a

sequence determined by the DNA. The complementary bases in the nucleotides of the DNA and RNA line up alongside each other. RNA nucleotides from the nucleoplasm line up alongside the exposed DNA. Initially hydrogen bonds hold the complementary RNA bases in place. Then DNA-directed RNA polymerase catalyses the formation of phosphodiester bonds between the sugars and phosphate groups of the bases, to form a strand of mRNA. Hydrogen bonds maintain the helical structure of the RNA molecule. Just as in the DNA, the bases of the mRNA form a triplet code and each triplet of bases is a codon. The relatively small mRNA molecules pass easily through the pores in the nuclear membrane, carrying the instructions from the genes in the nucleus to the cytoplasm. They then move to the surface of the ribosomes, where protein synthesis takes place (see **fig A**).

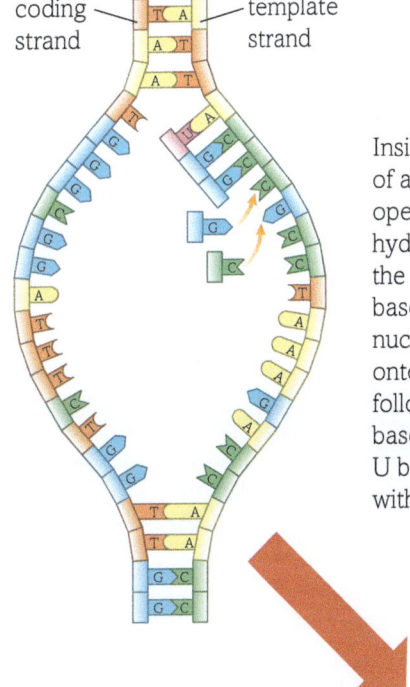

coding strand — template strand

Inside the nucleus a portion of a DNA molecule opens up by breaking hydrogen bonds to reveal the sequence of nucleotide bases. Free RNA nucleotides hydrogen-bond onto the exposed bases, following complementary base pairing rules so U bonds with A, A with T, C with G and G with C.

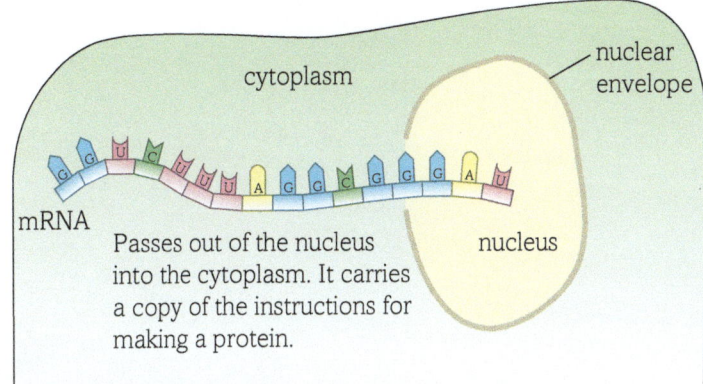

cytoplasm

nuclear envelope

mRNA

nucleus

Passes out of the nucleus into the cytoplasm. It carries a copy of the instructions for making a protein.

fig A The transcription of the DNA message. Any mistakes in this process can have fatal consequences for the cell or even the whole organism if the wrong protein is made.

Transfer RNA

Transfer RNA (tRNA) is found in the cytoplasm. It has a complex shape, often described as a clover leaf, that enables it to carry out its function (see **fig B**). This shape is the result of hydrogen bonding between different bases. One part of the tRNA molecule has a sequence of three bases that matches the genetic code of the DNA and corresponds to one particular amino acid. This sequence of three bases is called the **anticodon**. Each tRNA molecule also has a binding site with which it picks up one particular amino acid from the vast numbers always free in the cytoplasm.

The tRNA molecules, each carrying a specific amino acid, line up alongside the mRNA on the surface of the ribosome. The anticodons of the tRNA line up with the codons of the mRNA on the surface of the ribosome, held in place by hydrogen bonds between the corresponding bases. Because the anticodon has a sequence of bases that align with the corresponding bases in the mRNA on the ribosomal surface, the correct sequence of amino acids is assembled. Once the amino acids are lined up together, peptide bonds form between them, building up a long chain of amino acids.

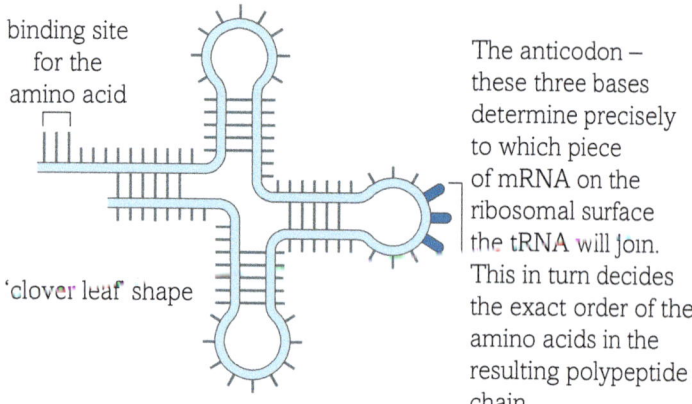

binding site for the amino acid

'clover leaf' shape

The anticodon – these three bases determine precisely to which piece of mRNA on the ribosomal surface the tRNA will join. This in turn decides the exact order of the amino acids in the resulting polypeptide chain.

fig B There are 61 types of tRNA molecules available to carry all the necessary amino acids to the surface of the ribosomes ready for synthesis into protein molecules.

Ribosomal RNA

Ribosomal RNA (rRNA) makes up about 50% of the structure of a ribosome and is the most common form of RNA found in cells. It is made in the nucleus, under the control of the nucleoli, and then moves out into the cytoplasm where it binds with proteins to form ribosomes. The ribosomes consist of a large and a small subunit. They surround and bind to the parts of the mRNA that are being actively translated, and then move along to the next codon. Their job is to hold together the mRNA and tRNA and act as enzymes controlling the process of protein synthesis.

Protein synthesis

In the process of protein synthesis the genetic code of the DNA of the nucleus is transcribed onto messenger RNA. This mRNA moves out of the nucleus into the cytoplasm and becomes attached to a ribosome. Molecules of transfer RNA carry individual amino acids to the surface of the ribosome. The tRNA anticodon lines up alongside a complementary codon in the mRNA, held in place by hydrogen bonds while enzymes link the amino acids together. The tRNA then breaks away and returns to the cytoplasm to pick up another amino acid. The ribosome moves along the molecule of mRNA until it reaches the end, leaving a completed polypeptide chain. The message may be read again and again.

Protein synthesis, like many other events in living things, is a continual process. However, it makes it simpler to understand if we look at the two main aspects of it separately. The events in the nucleus involve the transcription of the DNA message (see **fig A**). In the cytoplasm that message is translated into polypeptide molecules and hence into proteins (see **fig C**).

Mass production

The cytoplasm of cells contains many **polysomes**. These are groups of ribosomes joined by a thread of mRNA, and they appear to be a form of mass production of particular proteins. Instead of one ribosome moving steadily along a strand of mRNA and producing its polypeptide and then repeating the process, ribosomes attach in a steady stream to the mRNA and move along one after the other producing lots of identical polypeptides.

This is how the genetic code carried on the DNA is translated into living material by the synthesis of proteins.

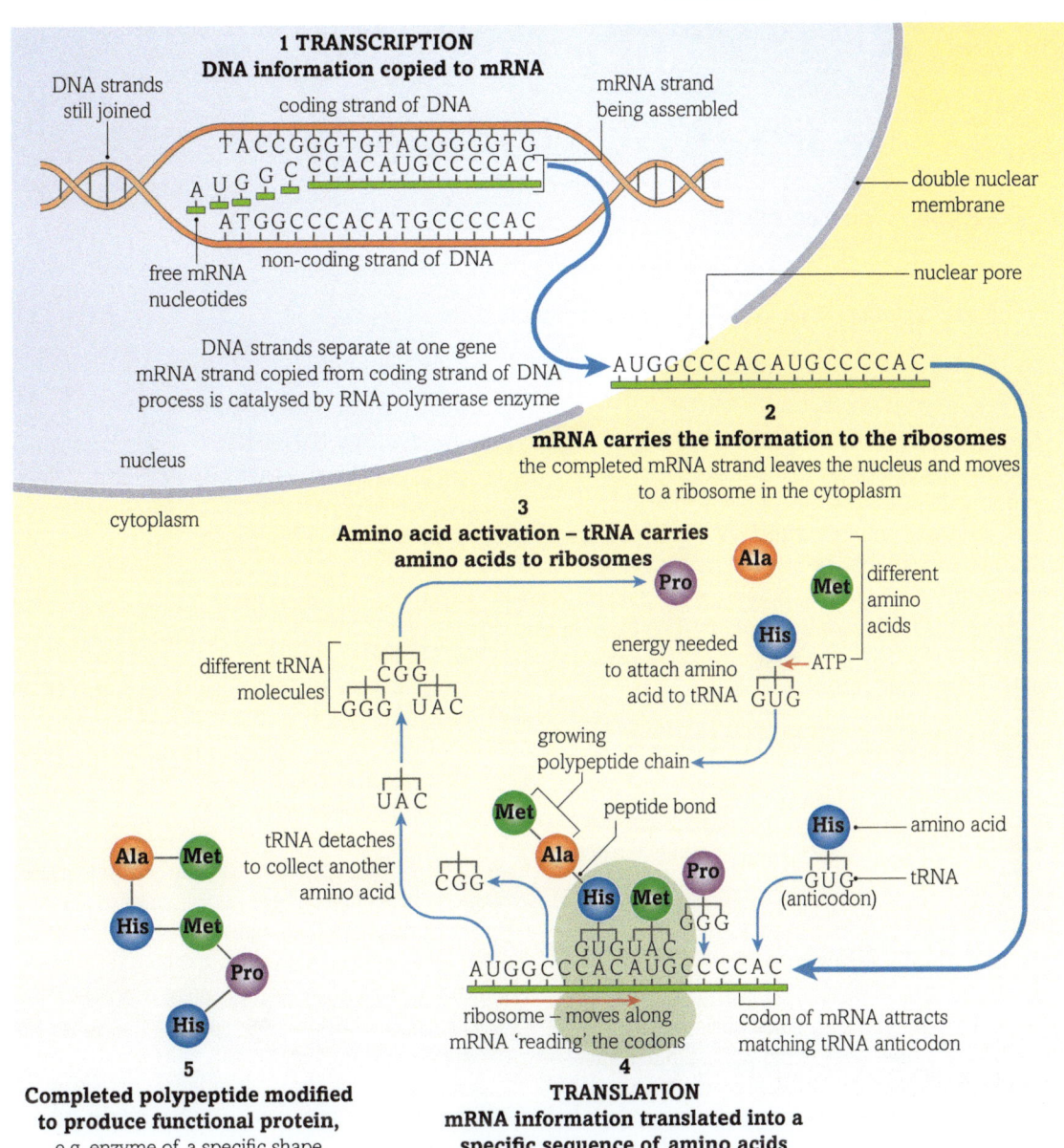

fig C A simplified diagram to show how the information held in the DNA sequence in the nucleus is translated into a sequence of amino acids in a polypeptide chain. In reality, the mRNA strand and the amino acid chain may be thousands of units long.

Questions

1 DNA and RNA are the information molecules of the cell. Explain the differences in the basic structures of these two molecules.

2 In many organisms the DNA is in the nucleus of the cells and the proteins for which it codes are in the cytoplasm. Explain carefully the roles of the following in translating the genetic code into an active enzyme in the cytoplasm of a cell:

(a) DNA

(b) messenger RNA

(c) transfer RNA

(d) ribosomal RNA

Key definitions

Messenger RNA (mRNA) is the RNA formed in the nucleus that carries the genetic code out into the cytoplasm.

The **antisense strand** is the DNA strand that codes for proteins.

DNA-directed RNA polymerase (RNA polymerase) is the enzyme that polymerises nucleotide units to form RNA in a sequence determined by the antisense strand of DNA.

Transfer RNA (tRNA) molecules are small units of RNA that pick up particular amino acids from the cytoplasm and transport them to the surface of the ribosome to align with the mRNA.

The **anticodon** is a sequence of three bases on tRNA that correspond to the bases in the mRNA codon.

Ribosomal RNA (rRNA) is RNA that makes up about 50% of the structure of the ribosome.

Polysomes are groups of ribosomes, joined by a thread of mRNA, that can produce large quantities of a particular protein.

By the end of this section, you should be able to...

● explain the term gene mutation and describe base deletions, insertions and substitutions

● explain the effect of point mutations on amino acid sequences as illustrated by sickle cell disease in humans

The genetic code carried on the DNA is translated into living cellular material through protein synthesis. If a single codon is changed or misread during the process, then the amino acid for which it codes may be different. As a result the whole polypeptide chain and indeed the final protein may be altered. A change like this is known as a **mutation**. A mutation is a permanent change in the DNA of an organism. A mutation can happen when the **gametes** (sex cells) form, although they also occur during the division of somatic (body) cells.

A tiny alteration at this molecular level may have no noticeable effect at all – but it may have devastating effects on the whole organism. Many human genetic diseases are the result of random mutations in the genetic material of the gametes, including thalassaemia, in which the blood proteins are not manufactured correctly, or cystic fibrosis, in which a membrane protein does not function properly.

Different types of mutations

Gene mutations involve changes in the bases making up the codons. The chance of a mutation taking place during DNA replication is around 2.5×10^{-8} per base, although estimates vary widely as it is very difficult to measure. Fortunately the body also has its own DNA repair systems. Specific enzymes cut out or repair any parts of the DNA strands that become broken or damaged. In spite of this, some mutations remain and are copied from the DNA when new proteins are made.

Some mutations occur when just one or a small number of nucleotides are miscopied during transcription. These are **point** or **gene mutations**. If you think of the amino acids produced from each codon as the equivalent of the letters of the alphabet, the result of a point mutation is like changing a letter in one word. It may well still make an acceptable word, but the meaning will probably be different. These gene mutations include **substitutions**, where one base substitutes for another, **deletions** where a base is completely lost in the sequence or **insertions**, when an extra base is added, which may be a repetition of one of the bases already there or a different base entirely.

Chromosomal mutations involve changes in the positions of genes within the chromosomes. This is like rearranging the words within a sentence – if you are lucky they still make sense, but it will not mean the same as the original sentence. Finally there are **whole-chromosome mutations**, where an entire chromosome is either lost during meiosis, which is cell division to form the sex cells, or duplicated in one cell by errors in the process.

This is like the loss or repetition of a whole sentence. For example, Down's syndrome is caused by a whole-chromosome mutation at chromosome 21 – affected individuals have three copies of this chromosome instead of the usual two.

How gene mutations can affect the phenotype

Mutations can be a source of variation within an organism. If the different arrangements of nucleotides code for the same amino acid (see **Section 1.3.5**) a point mutation will have no effect. Very occasionally, a mutation occurs that results in the production of a new and superior protein. This may help the organism gain a reproductive advantage so that it leaves more offspring than other individuals of that species particularly if environmental conditions change. Most mutations are neutral, meaning that they neither improve nor worsen the chances of survival. Some mutations cause great damage, disrupting the biochemistry of the entire organism. If a base mutation change is in a protein that plays an important role in a cell – for example, the active site of an enzyme – the effect can be catastrophic.

Sickle cell disease – when the code goes wrong

Sickle cell disease is a genetic disease that affects the protein chains making up the haemoglobin in the red blood cells. It is the result of a point mutation. A change of one base in one codon changes a single amino acid in a chain of 147 amino acids – but that change alters the nature of the protein. As a result, the haemoglobin molecules stick together to form rigid rods that give the red blood cells a sickle shape. They do not carry oxygen very efficiently and block the smallest blood vessels. This single tiny change in one nucleotide is enough to cause people affected severe pain and even death.

Sequence for healthy haemoglobin								
ATG	GTG	CAC	CTG	ACT	CCT	GAG	GAG	TCT
Start	Val	His	Leu	Thr	Pro	Glu	Glu	Ser
Sequence for sickle cell haemoglobin								
ATG	GTG	CAC	CTG	ACT	CCT	GTG	GAG	TCT
Start	Val	His	Leu	Thr	Pro	Val	Glu	Ser

table A The change in the single codon that causes sickle cell disease (the first nine codons only shown).

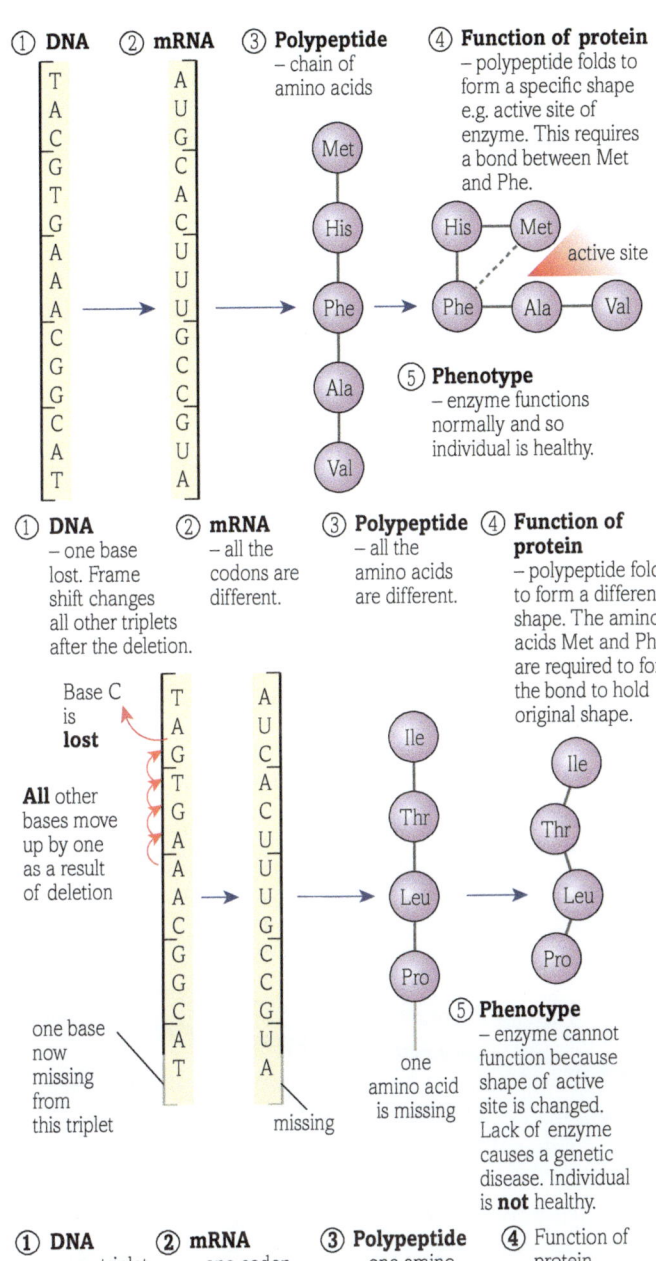

① **DNA** – one base lost. Frame shift changes all other triplets after the deletion.

② **mRNA** – all the codons are different.

③ **Polypeptide** – all the amino acids are different.

④ **Function of protein** – polypeptide folds to form a different shape. The amino acids Met and Phe are required to form the bond to hold original shape.

Base C is **lost**

All other bases move up by one as a result of deletion

one base now missing from this triplet

missing

one amino acid is missing

⑤ **Phenotype** – enzyme cannot function because shape of active site is changed. Lack of enzyme causes a genetic disease. Individual is **not** healthy.

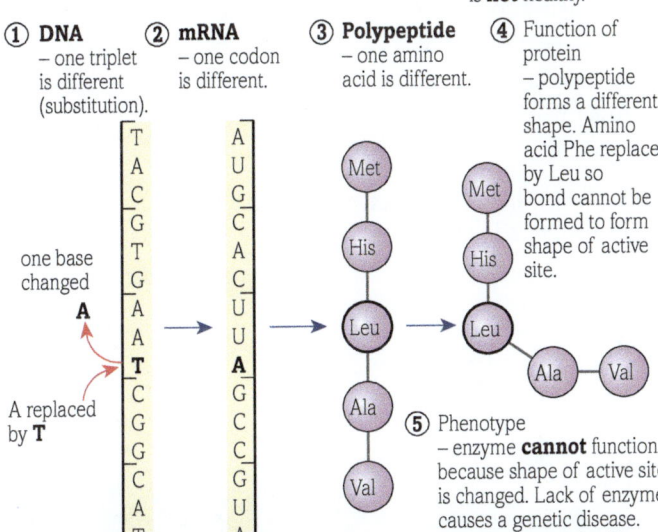

① **DNA** – one triplet is different (substitution).

② **mRNA** – one codon is different.

③ **Polypeptide** – one amino acid is different.

④ Function of protein – polypeptide forms a different shape. Amino acid Phe replaced by Leu so bond cannot be formed to form shape of active site.

one base changed **A**

A replaced by **T**

⑤ Phenotype – enzyme **cannot** function because shape of active site is changed. Lack of enzyme causes a genetic disease. Individual is **not** healthy.

Note: a change in only one base can change the function of the protein.

fig A Two examples of how a change in a single nucleotide can affect the structure and function of the protein formed and hence the phenotype of the individual.

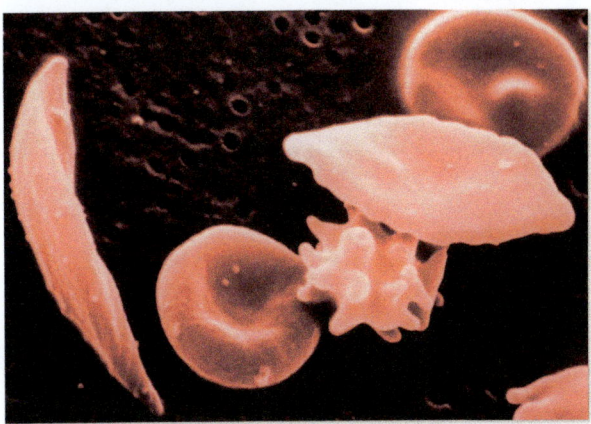

fig B The rigid shape of the red blood cells in sickle cell disease as a result of changes in the haemoglobin molecules prevents them from functioning properly in the body.

Mutations can happen to any cell at any time, though they occur most commonly during the copying of DNA for cell division. Mutations in the body cells can cause problems such as cancer. The most damaging mutations occur in the gametes because they will be passed on to future offspring. These are the mutations that give rise to genetic diseases. Exposure to **mutagens**, such as X-rays, ionising radiation and certain chemicals, increases the rate at which mutations occur. For this reason it is better to keep exposure to these mutagens to a minimum.

Questions

1 Some base mutations will have as big an impact on the way the body works as any chromosomal or whole-chromosome mutation. Others have no effect at all on the organism. Explain.

2 Explain how a change in a single base in the sickle cell mutation has such a dramatic effect on affected individuals.

Key definitions

A **mutation** is a permanent change in the DNA of an organism.

Gametes are haploid sex cells produced by meiosis that fuse to form a new diploid cell (zygote) in sexual reproduction.

A **point mutation (gene mutation)** is a change in one or a small number of nucleotides affecting a single gene.

A **substitution** is a type of point mutation in which one base in a gene is substituted for another.

A **deletion** is a type of point mutation in which a base is completely lost.

An **insertion** is a type of point mutation in which an extra base is added into a gene, which may be a repeat or a different base.

Chromosomal mutations are changes in the position of entire genes within a chromosome.

A **whole-chromosome mutation** is the loss or duplication of a whole chromosome.

Sickle cell disease (sickle cell anaemia) is a human genetic disease affecting the protein chains making up the haemoglobin in the red blood cells.

A **mutagen** is anything that increases the rate of mutation.

1 (a) The diagram below shows the structure of a mononucleotide from a DNA molecule.

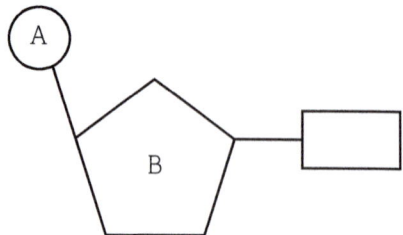

Name the parts of the mononucleotide labelled A and B. [1]

(b) The table below shows the percentage of different bases present in the DNA from a cow.

Percentage of each base present			
Adenine	Guanine	Thymine	Cytosine
		29	

(i) Complete the table to show the percentage of adenine, guanine and cytosine in the DNA of the cow. [1]
(ii) Explain how you worked out the percentage of guanine present in the DNA of a cow. [3]

[Total 5]

2 The diagram below shows part of a DNA molecule.

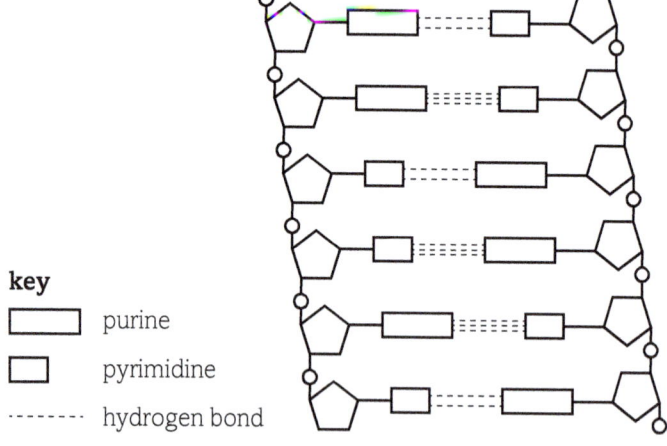

key

☐ purine

☐ pyrimidine

------- hydrogen bond

(a) Draw a ring around **one** mononucleotide. [1]
(b) Name the two **purine** bases found in DNA. [1]
(c) (i) State where transcription takes place in eukaryotic cells. [1]
   (ii) During transcription, part of a DNA molecule unwinds and the DNA strands separate.
   Describe the events that follow to produce a messenger RNA (mRNA) molecule. [3]

(d) Oligonucleotides are short chains of nucleotides. Some of these are man-made and have been used as drugs to treat a wide variety of diseases. They work by binding to mRNA or DNA and inhibiting protein synthesis. The drugs are described as antisense drugs when they bond to mRNA and triplex drugs when they bind to DNA.
   (i) State which stage of protein synthesis will be inhibited by each of the following
   • Antisense drugs
   • Triplex drugs [1]
   (ii) The table below shows the sequence of bases in part of a molecule of mRNA.
   Complete the table to show the sequence of bases in the antisense drug that will bind to this part of the mRNA molecule.

Base sequence on DNA	A	G	U	C	A	U
Base sequence in antisense drug						

[1]

[Total 8]

3 (a) Work by Nachman and Crowell (*Genetics*, September 1, 2000 vol. 156 no. 297–304) using human DNA has estimated an average mutation rate of 2.5×10^{-8} per base. Assuming there are 7×10^9 base pairs in a human diploid cell, calculate the average number of mutations formed each time the cell divides. [2]

(b) If a cell divides 50 times, calculate the total number of mutations accumulated by an average cell. [1]

(c) Explain why most cells still produce proteins that work properly despite this number of mutations. [3]

[Total 6]

4 The diagram below shows some cell structures involved in protein synthesis in eukaryotic cells.

Nucleus   Rough endoplasmic reticulum

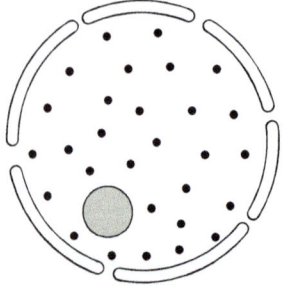
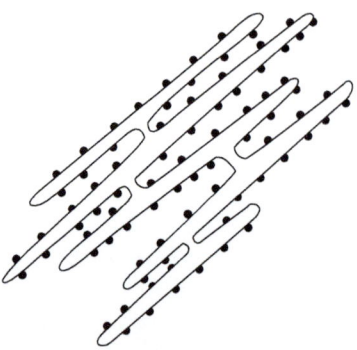

(a) Describe the events that occur inside the nucleus to produce a molecule of messenger RNA (mRNA). [4]

(b) Describe the role of the ribosomes in protein synthesis. [3]

(c) The table below gives some of the base triplets on DNA that code for some amino acids and stop signals.

Base triplet on DNA	Amino acid/stop signal
CCC	Glycine
AAA or AAG	Phenylalanine
AGA or AGC	Serine
GCG	Arginine
TTT	Lysine
ATT or ATC or ACT	Stop signal

The diagram below shows the final base triplets of a gene, labelled T1 to T5, and the complementary messenger RNA (mRNA).

The sequence of amino acids at the end of the protein produced is also shown.

	T1	T2	T3	T4	T5
Last part of the DNA strand:	CCC	GCG	AGC	TTT	
Complementary mRNA:					
Amino acid sequence:	Glycine				

(i) Write in the codons found on the mRNA complementary to the base triplets T1, T2, T3, T4 on the diagram above. [2]

(ii) Using the information in the timetable, complete the amino acid sequence shown in the diagram above. The first one has been done for you. [2]

(iii) Use the information in the table to suggest a base triplet for T5 on the DNA strand. [1]

[Total 12]

6 In DNA, the type of bond that joins deoxyribose sugar to a phosphate group is

(a) a phosphodiester bond

(b) a hydrogen bond

(c) a peptide bond

(d) a glycosidic bond [1]

[Total 1]

7 The sequence **CCGAAACGACTC** on a DNA strand when transcribed would form which mRNA sequence?

(a) CCGUUUCGUCAC

(b) GGCUUUGCUGAG

(c) CCGAAACGACUC

(d) GGCAAAGCAGTG [1]

[Total 1]

8 The strand of DNA that is transcribed is called the

(a) sense strand

(b) antisense strand

(c) same sense strand

(d) missense strand [1]

[Total 1]

9 Meselson and Stahl's experiment showed that DNA replicated semiconservatively. Bacteria grown on a medium containing ^{15}N over many generations would have DNA containing this one isotope of nitrogen. They were then transferred to grow on a medium that contained ^{14}N.

(a) Describe how these enzymes are used in DNA replication
  (i) DNA helicase [2]
  (ii) Polymerase [3]
  (iii) Ligase [2]

(b) Calculate the proportion of bacterial DNA that contains exclusively ^{14}N after the third round of replication. [2]

(c) The strands of DNA were separated according to their density. Explain why it was necessary that ^{15}N is a stable isotope of nitrogen. [1]

(d) If DNA replicated conservatively then the results after the first round of replication would have differed from the observed result with semiconservative replication.

Describe how the DNA strands formed by one round of semiconservative replication differ from those that might have formed from one round of conservative replication. [2]

[Total 12]

TOPIC 1
Biological molecules

Enzymes

Introduction

Around the world, people have observed albino forms in almost every species of vertebrate, including human beings. True albino animals are very striking – they have pure white hair, fur or feathers, usually with pale skin and red or pale blue eyes. Albinism is genetic – it is inherited in the genes passed on from parents to their offspring. But why do albino animals lack colour? It is all down to the lack of a single enzyme – tyrosinase. This enzyme is a key factor in the production of the pigment melanin (and other pigments) from the amino acid tyrosine. Differing amounts of melanin, combined with other pigments, result in a wide range of hair, skin and feather colours. If the enzyme is lacking, the animal lacks pigment and is albino.

In this chapter you will be looking at enzymes – their structure, how they work and what happens when they are inhibited. You will look at the varying roles of enzymes in the body and how they are named. You will go on to discover how enzymes work and how their mechanism of action is related to the shape of the active site produced within the tertiary structure of the protein itself. By looking at the evidence you will see how our models of enzyme action have changed from the relatively simplistic lock-and-key theory to the more complex induced-fit model.

Measuring the rate of enzyme controlled reactions is key to understanding the factors that affect them. You will be considering the practical difficulties of doing this and how they can be overcome. Looking at the factors that affect the rate of an enzyme controlled reaction helps to build up our model of how enzymes act as catalysts in biological systems.

Understanding how enzyme action can be inhibited by other molecules is another way of developing an understanding of how enzymes work. You will consider competitive, non-competitive and irreversible inhibition, as well as considering the situation when the end products of a long chain of reactions inhibit an enzyme earlier in the process.

All the maths you need

- Recognise and make use of appropriate units in calculations (*e.g. the units for the rate of reaction of an enzyme*)
- Use of percentages (*e.g. calculating percentage yields in different enzyme controlled reactions*)
- Use of appropriate number of significant figures (*e.g. understand that results for enzyme rate experiments can be reported only to the limits of the least accurate measurement*)
- Find arithmetic means (*e.g. the mean of a range of data when investigations are repeated*)
- Plot a range of data in an appropriate format (*e.g. enzyme activity over time represented on a graph*)
- Solve algebraic equations (*e.g. calculate the rate of enzyme reactions*)
- Plot two variables from experimental or other data (*e.g. select an appropriate format for presenting data from experimental investigations into enzyme controlled reactions*)
- Understand that $y = mx + c$ represents a linear relationship and predict or sketch the shape of a graph with a linear relationship (*e.g. the effect of substrate concentration on the rate of an enzyme controlled reaction with excess enzyme*)
- Calculate the rate of change from a graph showing a linear relationship (*e.g. the rate of an enzyme controlled reaction*)
- Draw and use the slope of a tangent to a curve as a measure of rate of change and use this method to measure the gradient at a point on a curve (*e.g. amount of product formed plotted against time when the concentration of enzyme is fixed*)

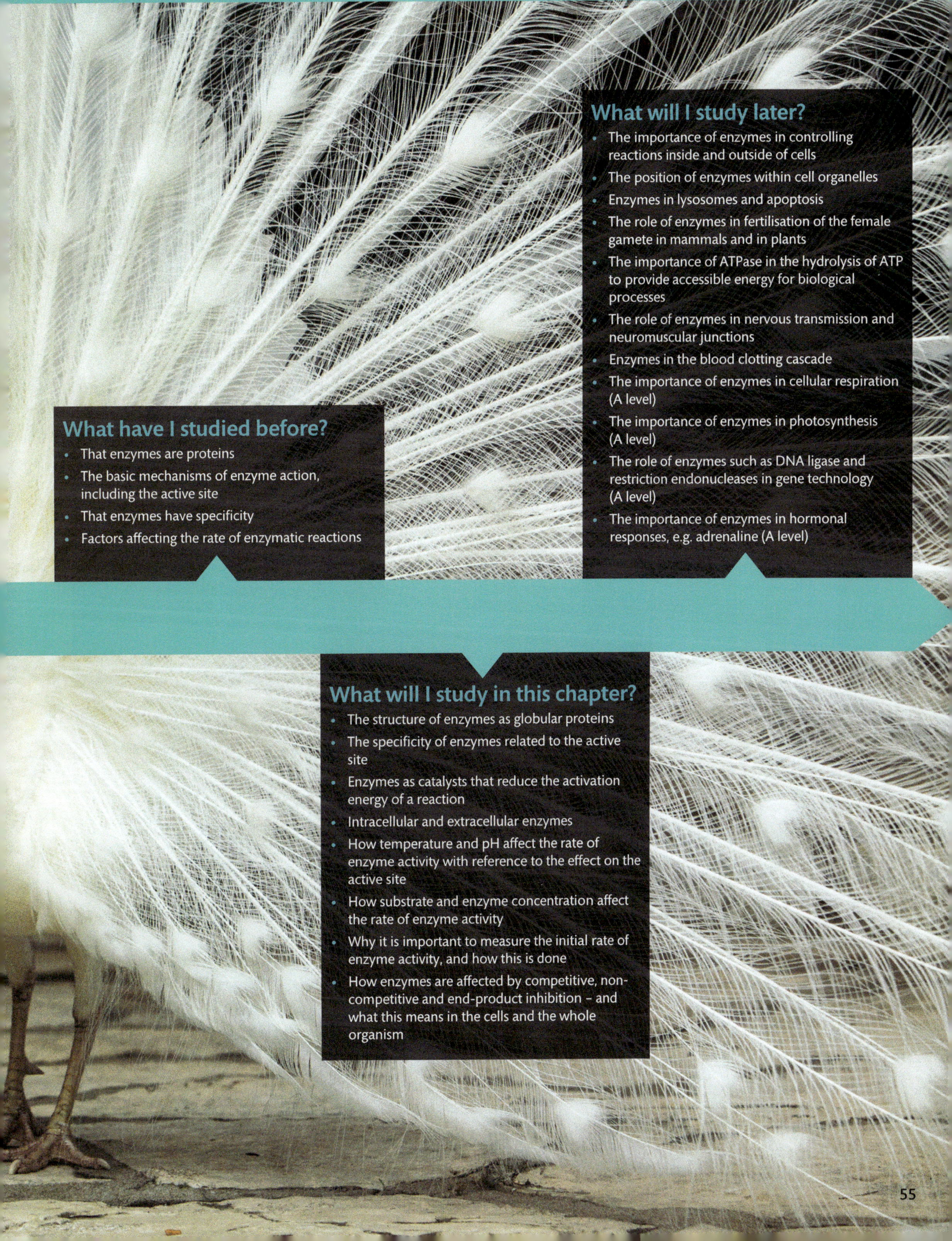

What will I study later?

- The importance of enzymes in controlling reactions inside and outside of cells
- The position of enzymes within cell organelles
- Enzymes in lysosomes and apoptosis
- The role of enzymes in fertilisation of the female gamete in mammals and in plants
- The importance of ATPase in the hydrolysis of ATP to provide accessible energy for biological processes
- The role of enzymes in nervous transmission and neuromuscular junctions
- Enzymes in the blood clotting cascade
- The importance of enzymes in cellular respiration (A level)
- The importance of enzymes in photosynthesis (A level)
- The role of enzymes such as DNA ligase and restriction endonucleases in gene technology (A level)
- The importance of enzymes in hormonal responses, e.g. adrenaline (A level)

What have I studied before?

- That enzymes are proteins
- The basic mechanisms of enzyme action, including the active site
- That enzymes have specificity
- Factors affecting the rate of enzymatic reactions

What will I study in this chapter?

- The structure of enzymes as globular proteins
- The specificity of enzymes related to the active site
- Enzymes as catalysts that reduce the activation energy of a reaction
- Intracellular and extracellular enzymes
- How temperature and pH affect the rate of enzyme activity with reference to the effect on the active site
- How substrate and enzyme concentration affect the rate of enzyme activity
- Why it is important to measure the initial rate of enzyme activity, and how this is done
- How enzymes are affected by competitive, non-competitive and end-product inhibition – and what this means in the cells and the whole organism

By the end of this section, you should be able to...

● describe the structure of enzymes as globular proteins

● explain the concept of specificity

● recognise that enzymes catalyse a wide range of intracellular reactions as well as extracellular ones

What is an enzyme?

A **catalyst** is a substance that changes the rate of a reaction without changing the substances produced. The catalyst is unaffected at the end of the reaction and can be used again. **Enzymes** are biological catalysts, which control the rate of the reactions that take place in individual cells and in whole organisms. Under the conditions of temperature and pH found in living cells, most of the reactions that provide cells with energy and produce new biological material would take place very slowly – too slowly for life to exist. Enzymes make life possible by speeding up the chemical reactions in cells without changing the conditions in the cytoplasm.

Enzymes are globular proteins (see **Section 1.2.4**), produced during protein synthesis as the mRNA transcribed from the DNA molecule is translated (see **Section 1.3.6**). They have a very specific shape as a result of their primary, secondary, tertiary and quaternary structures (see **Section 1.2.4**), and this means each enzyme will only catalyse a specific reaction or group of reactions. We say enzymes show great **specificity**. Changes in temperature and pH affect the efficiency of an enzyme because they affect the intramolecular bonds within the protein that are responsible for the shape of the molecule.

Within any cell many chemical reactions are going on at the same time. Those reactions that build up new chemicals are known as **anabolic reactions** ('ana' means up, as in 'build up'). Those that break substances down are **catabolic reactions** ('cata' means down). The combination of these two processes results in the complex array of biochemistry that we refer to as **metabolism**. Most of the reactions of metabolism occur not as single events but as part of a sequence of reactions known as a **metabolic chain** or **metabolic pathway**. We usually think of enzymes speeding up reactions but sometimes they act to slow them down, or stop them completely.

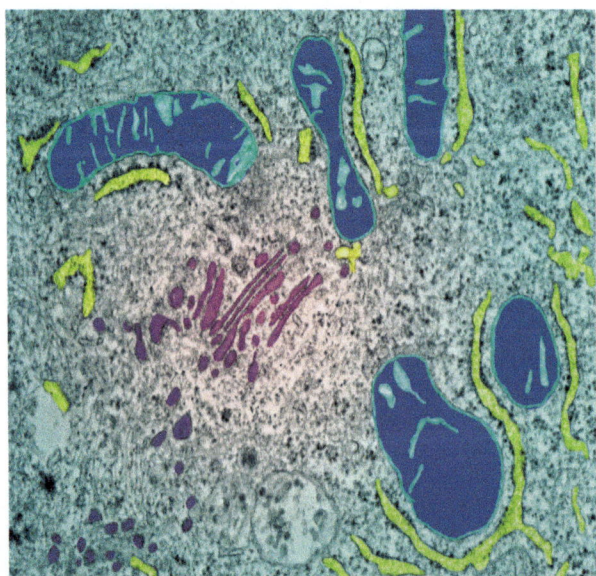

fig A Each cell contains several hundred different enzymes to control the multitude of reactions going on inside.

Naming enzymes

In the study of biology, in medicine, in cellular and genetic research and in industries that use biotechnology, it is important to be able to refer to the action of specific enzymes. To do this we need to understand how enzymes are named.

Many of the enzymes found in animals and plants work inside the cells. These are known as **intracellular enzymes**, for example DNA polymerase and DNA ligase. Cells secrete other enzymes that have an effect beyond the boundaries of the cell membrane. These are **extracellular enzymes**. The digestive enzymes and lysozyme, the enzyme in your tears, are well-known examples of these.

Most enzymes – both intracellular and extracellular – have several names including:

- a relatively short recommended name, which is often the name of the molecule that the enzyme works on (the substrate) with '-ase' on the end, or the substrate with an indication of what it does, e.g. creatine kinase

- a longer systematic name describing the type of reaction being catalysed, e.g. ATP:creatine phosphotransferase

- a classification number, e.g. EC 2.7.3.2.

Some enzymes, such as urease, ribonuclease and lipase, are known by their recommended names. But there are still some enzymes that are known by common but uninformative names – trypsin and pepsin for example. However, the names of most enzymes give you useful information about the role of the enzyme in the cell or the body.

fig B Pure urease does not look very exciting, but the ability to isolate and extract enzymes has revolutionised our understanding of biology and the way we can use enzymes in industry.

Did you know?

The discovery of enzymes

In 1835 people noticed that starch is broken down to sugars more effectively by malt (sprouting barley) than by sulfuric acid.

People also suspected there were 'ferments' in yeast (a single-celled fungus) that turned sugar to alcohol and in 1877 the name enzyme (literally 'in yeast') was introduced. In 1897 Eduard Buchner (1860–1917) extracted the enzyme responsible for fermenting sugar from yeast cells, and showed it worked outside a living cell.

In 1926 James B. Sumner (1887–1955) extracted the first pure, crystalline enzyme from jack beans. It was urease, the enzyme that catalyses the breakdown of urea. Sumner found the crystals were protein and concluded that enzymes must therefore be proteins. Unfortunately no-one believed the young researcher at the time, because many established scientists had been trying and failing to isolate enzymes for years. However, 20 years later Sumner received a Nobel Prize for his ground-breaking work.

Questions

1. From which organisms were the first enzymes isolated?

2. What is the difference between an intracellular enzyme and an extracellular enzyme?

3. Investigate Sumner's work and discover which scientists were particularly against his ideas and why.

Key definitions

A **catalyst** is a substance that speeds up a reaction without changing the substances produced or being changed itself.

Enzymes are proteins that have a very specific shape as a result of their primary, secondary, tertiary and quaternary structures. They act as biological catalysts and each enzyme will only catalyse a specific reaction or group of reactions.

Specificity is the characteristic of enzymes that means that, as a result of the very specific shapes resulting from their tertiary and quaternary structures, each enzyme will only catalyse a specific reaction or group of reactions.

An **anabolic reaction** is the reaction that builds up (synthesises) new molecules in a cell.

A **catabolic reaction** is a reaction which breaks down substances within a cell.

Metabolism is the sum of the anabolic and catabolic processes in a cell.

A **metabolic chain (metabolic pathway)** is a series of linked reactions in the metabolism of a cell.

Intracellular enzymes are enzymes that catalyse reactions within the cell.

Extracellular enzymes are enzymes that catalyse reactions outside of the cell in which they were made.

By the end of this section, you should be able to...

● explain how enzymes act as catalysts by reducing the activation energy of reactions

● explain how the initial rate of enzyme activity can be measured and why this is important

● explain how different factors affect the rate of enzyme activity

For a chemical reaction to take place, the reacting molecules must have enough energy to break the chemical bonds that hold them together. A simple model is that the reaction has to get over an 'energy hill', known as the **activation energy**, before it can get started.

Raising the temperature increases the rate of a chemical reaction by giving more molecules sufficient energy to react. However, living cells could not survive the temperatures needed to make many cellular reactions fast enough – and the energy demands to produce the heat would be enormous. Enzymes solve the problem by lowering the activation energy needed for a reaction to take place (see **fig A**).

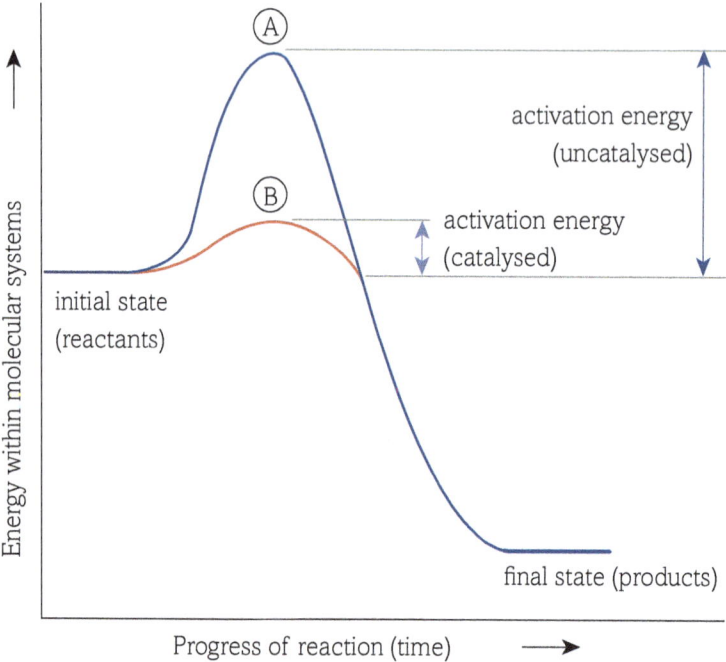

(A) = Energy of transition state in uncatalysed reaction.

(B) = Energy of transition state, i.e. enzyme/substrate complex, during catalysed reaction.

fig A Energy diagram to show the difference between an uncatalysed and a catalysed reaction.

How do enzymes work?

To lower the activation energy and catalyse a reaction, enzymes form a complex with the **substrate** or substrates of the reaction. A simple picture of enzyme action in a catabolic reaction is:

substrate + enzyme ⇌ enzyme/substrate complex ⇌ enzyme + products

Once the products of the reaction are formed they are released and the enzyme is free to form a new complex with more substrate. How does this relate to the structure of the enzyme? The **'lock-and-key hypothesis'** gives us a simple model that helps us understand what happens (see **fig B**). Within the globular protein structure of each enzyme is an area known as the **active site** that has a very specific shape. Only one substrate or type of substrate will fit the shape of the gap, and it is this that gives each enzyme its specificity. Just as a key fits into a lock, so the enzyme and substrate slot together to form a complex.

The formation of the enzyme/substrate complex lowers the activation energy of the reaction. The active site affects the bonds in the substrate, making it easier for them to break, and the reacting substances are brought close together, making it easier for bonds to form between them. Once the reaction is complete the products are no longer the right shape to stay in the active site and the complex breaks up, releasing the products and freeing the enzyme for further catalytic action.

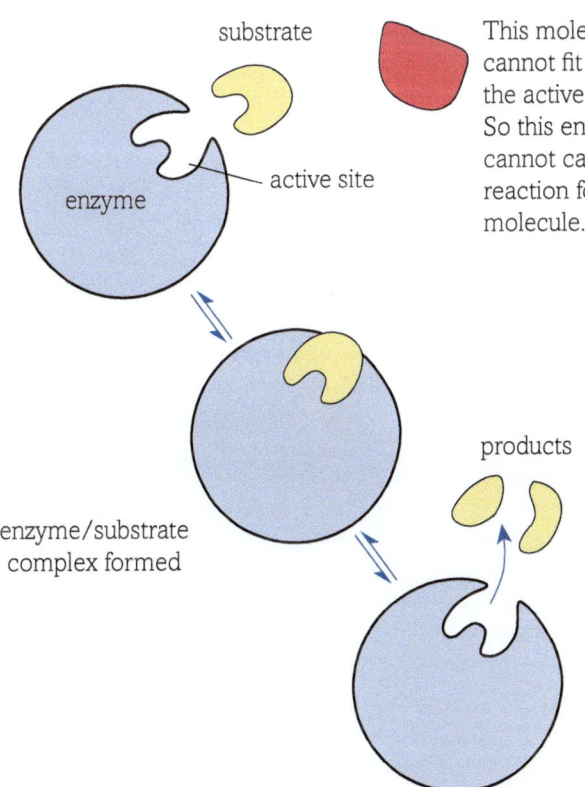

fig B The lock-and-key hypothesis underpins our understanding of how enzymes work.

The lock-and-key hypothesis fits most of our evidence about enzyme characteristics. However, it is now thought to be an over-simplification. Evidence from X-ray crystallography, chemical analysis of active sites and other techniques suggests that the active site of an enzyme is not simply a rigid shape. In the **induced-fit hypothesis**, generally accepted as the best current model of enzyme action, the active site still has a distinctive shape and arrangement, but it is a flexible one. Once the substrate enters the active site, the shape of the site is modified around it to form the active complex. Once the products have left the complex the enzyme reverts to its inactive, relaxed form until another substrate molecule binds (see **fig C**).

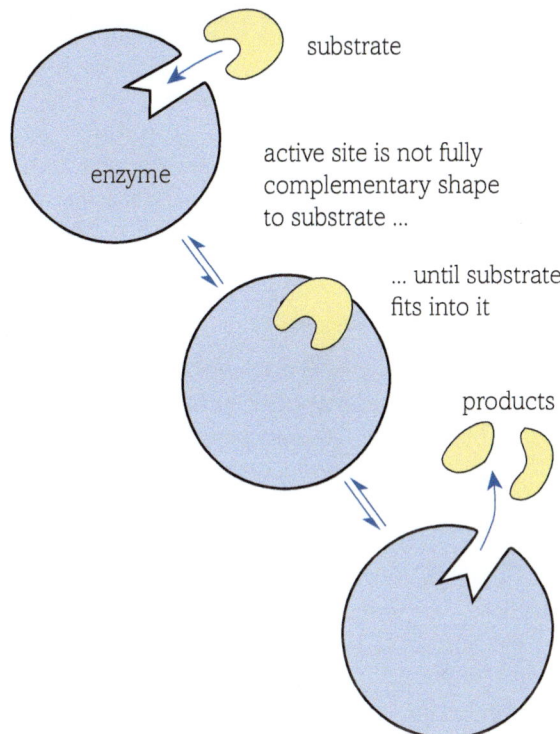

fig C The induced-fit theory of enzyme action proposes that the catalytic groups of the active site are not brought into their most active positions until a substrate is bound to the site, inducing a change in shape.

Measuring reaction rate

When scientists are investigating enzymes and how they act as catalysts, they frequently measure the reaction rate. For example, one practical way of demonstrating the effect of an enzyme on a reaction is to measure the rate of the reaction with and without the enzyme. Using this method it has been shown that when urea breakdown is catalysed by urease extracted from the jack bean, the rate of the reaction increases by a factor of 10^{14}. Enzymes are such efficient catalysts that they generally increase reaction rates by factors from 10^8 to 10^{26}. This is why only tiny amounts of most enzymes are needed.

Much of the evidence for the structure of enzymes and the way this relates to their functions comes from practical investigations into the effect of different factors on the rate of enzyme-catalysed reactions. To investigate the way a factor affects the rate of reaction, biologists measure the **initial rate of reaction** each time the independent variable is changed. Every other factor must be kept the same so that any changes are the result of changing the one variable.

It is important to provide a large excess of substrate in enzyme experiments, unless the effect of substrate concentration is under investigation. The enzyme–substrate complexes develop quickly and the reaction rapidly takes place at a steady rate. As soon as this point is reached, the reaction rate is recorded accurately by measuring the amount of product over a relatively short period of time. Measuring the initial rate only, with an excess of substrate, means other factors such as build up of products, lack of substrate and changes in pH will not have time to influence the rate.

What do we know about enzymes?

Our current model of enzymes is that they are globular proteins (see **Section 1.2.4**), which contain an active site that is vital to the functioning of the enzyme. The active site is a small depression on the surface of the molecule that has a specific shape because of the way the whole large molecule is folded. Anything affecting the shape of the protein molecule affects its ability to do its job, which indicates that the three-dimensional (3D) nature of the molecule is important to the way it works. A change in shape changes the shape of the active site as well – and so the enzyme can no longer function.

Enzymes change only the rate of a reaction. They do not change or contribute to the end products that form, or affect the equilibrium of the reaction. They act purely as catalysts and not as modifying influences in any other way.

Evidence for the relationship between the structure and functions of enzymes

Observing the factors that affect the rate of enzyme activity gives an insight into the relationship between the structure of an enzyme and the way it functions.

* Enzymes speed up reactions to such an extent that only minute amounts of them are needed to catalyse the reaction of many substrate molecules into products. This is described by the **molecular activity** or **turnover number** of an enzyme, which measures the number of substrate molecules transformed per minute by a single enzyme molecule. The number of molecules of hydrogen peroxide catalysed by the enzyme catalase extracted from liver cells is 6×10^6 in 1 minute. Most enzymes would catalyse thousands of molecules per minute rather than millions. If every enzyme molecule is involved in a reaction, it will not go any faster unless there is an increase in the enzyme concentration. In other words, enzyme controlled reactions are affected by the concentration of the enzyme.

* Enzymes are very specific to the reaction that they catalyse. Inorganic catalysts such as platinum frequently catalyse many different reactions, often only at extremes of temperature and pressure. In comparison, some enzymes are so specific that they will catalyse only one particular reaction. Others are specific to a particular group of molecules that are all of similar shape, or to a type of reaction that always involves the same groups. This suggests that there is a physical site within the enzyme with a particular shape into which a specific substrate will fit.

* The number of substrate molecules present (the concentration of the substrate) affects the rate of an enzyme-catalysed reaction. Take a simple reaction where substrate A is converted to product Z. If the concentration of A increases, the rate of the enzyme-catalysed reaction A → Z increases – but only for so long. Then the enzyme becomes saturated – all of the active sites are occupied by substrate molecules – and a further increase in substrate concentration will not increase the rate of the reaction further (see **fig D**). At this point only an increase in enzyme concentration will increase the rate of the reaction.

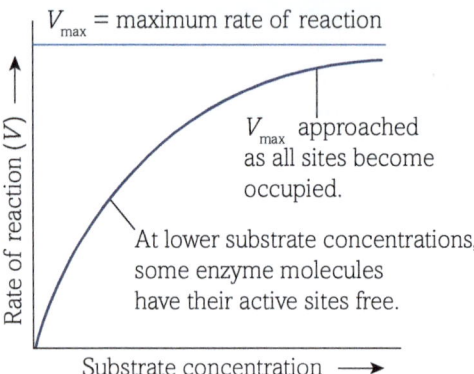

fig D The effect of substrate concentration on an enzyme-catalysed reaction, showing how the enzyme becomes saturated with substrate molecules.

* Temperature affects the rate of an enzyme-catalysed reaction in a characteristic way. Temperature affects all reactions because the number of successful collisions leading to a reaction increases at higher temperatures. The effect of temperature on the rate of any reaction can be expressed as the **temperature coefficient, Q_{10}**. This is expressed as:

$$Q_{10} = \frac{\text{rate of reaction at } (x + 10)\,°C}{\text{rate of reaction at } x\,°C}$$

Between about 0 °C and 40 °C, Q_{10} for any reaction is 2 – the rate of the reaction doubles for every 10 °C rise in temperature. However, outside this range, Q_{10} for enzyme-catalysed reactions in human beings decreases markedly, whilst Q_{10} for other reactions changes only slowly. The rate of enzyme-catalysed reactions in human beings falls as the temperature rises, and at about 60 °C the reaction stops completely in most cases. At temperatures over 40 °C most proteins, including most enzymes, start to lose their tertiary and quaternary structures – they denature. When enzymes denature, the shape of the active site changes and so they lose their ability to catalyse reactions. There are some exceptions to this rule. For example, the enzymes of thermophilic bacteria, which live in hot springs at temperatures of up to 85 °C, are able to work at very high temperatures. They are made of temperature-resistant proteins that contain a very high density of hydrogen bonds and disulfide bonds, which hold them together even at high temperatures (see **Section 1.2.4**). However, the optimum temperature of the enzymes of many organisms, including cold water fish and many plants, is much lower than 40 °C.

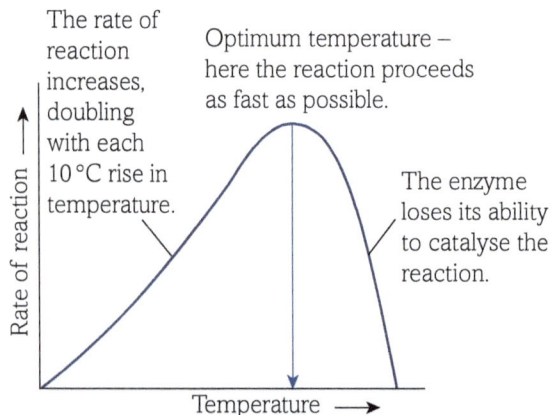

fig E The effect of temperature on the rate of a typical enzyme-catalysed reaction. All other factors must be kept constant.

- pH also has a major effect on enzyme activity by affecting the shape of protein molecules. Different enzymes work in different ranges of pH, because changes in pH affect the formation of the hydrogen bonds and disulfide bonds that hold the 3D structure of the protein together. The optimum pH for an enzyme is not always the same as the pH of its normal surroundings. This seems to be one way in which cells control the effects of their intracellular enzymes, increasing or decreasing their activity by minute changes in the pH.

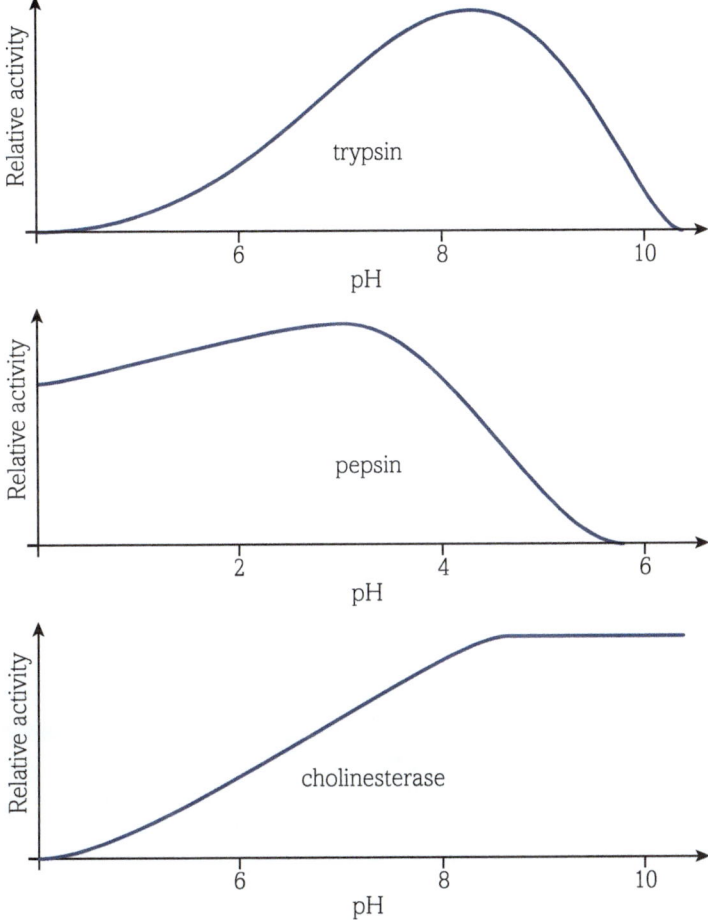

fig F Different enzymes work best at different pH levels. All other factors must be kept constant.

Did you know?

RuBisCo – a key but inefficient enzyme for life

Ribulose bisphosphate carboxylase/oxygenase (known as RuBisCo) is vitally important in photosynthesis. It is the enzyme that catalyses the fixing of carbon dioxide from the air into the biochemistry of sugar formation. Without this enzyme, life as we know it would not exist. But RuBisCo is a remarkably inefficient enzyme.

- Most enzymes catalyse about 1000 reactions per second, RuBisCo only catalyses about 3. Plant cells overcome this by making very large quantities of RuBisCo – about half of the protein in a photosynthetic plant cell is this enzyme.
- The active site of most enzymes is very specific. RuBisCo binds to carbon dioxide molecules in photosynthesis but it can also bind to oxygen molecules in a process called photorespiration. Its affinity for carbon dioxide is about 80 times greater than for oxygen – but there is much more oxygen available so about 25% of RuBisCo binds to oxygen.

Scientists believe RuBisCo evolved in an atmosphere containing very little oxygen and much more carbon dioxide than it does today so oxygen-binding was not a disadvantage at the time and so it was not selected against in evolution.

Questions

1 (a) Summarise the characteristics of enzymes.

  (b) Explain how each characteristic of enzymes provides evidence for the induced-fit hypothesis.

2 Plan a practical investigation into the effect of temperature on enzyme activity.

Key definitions

Activation energy is the energy needed for a reaction to get started.

A **substrate** is the molecule or molecules on which an enzyme acts.

The **lock-and-key hypothesis** is the model that explains enzyme action by an active site in the protein structure that has a very specific shape. The enzyme and substrate slot together to form a complex as a key fits in a lock.

An **active site** is the area of an enzyme that has a specific shape into which the substrate(s) of a reaction fit.

The **induced-fit hypothesis** is a modified version of the lock-and-key hypothesis for enzyme action where the active site is considered to have a more flexible shape. Once the substrate enters the active site, the shape of that site is modified around it to form the active complex. Once the products have left the complex, the enzyme reverts to its inactive, relaxed form.

The **initial rate of reaction** is the measure taken to compare the rates of enzyme controlled reactions under different conditions.

Molecular activity (turnover number) is the number of substrate molecules transformed per minute by a single enzyme molecule.

The **temperature coefficient (Q_{10})** is the measure of the effect of temperature on the rate of a reaction.

By the end of this section, you should be able to...

- describe how enzymes can be affected by competitive, non-competitive and end product inhibition

We can learn more about enzymes and how they work by looking at evidence from substances that stop the enzymes from working. These are called **enzyme inhibitors**. When we look at how inhibitor molecules interfere with the catalytic powers of an enzyme, we can get more evidence about the way they carry out their functions. There are two main types of inhibition, **reversible inhibition** and **irreversible inhibition**.

Reversible inhibition of enzymes

When an inhibitor affects an enzyme in a way that does not permanently damage it, this is reversible inhibition. When a reversible inhibitor is removed, the enzyme can function normally again. Reversible inhibition is a common feature of metabolic pathways, and it provides a key way of controlling reactions, as you will see. There are two major forms of reversible inhibition – **competitive inhibition** and **non-competitive inhibition**.

Competitive inhibition

In competitive, reversible inhibition, the inhibitor molecule is similar in shape to the substrate molecule. It competes with the substrate for binding at the active sites of the enzymes, forming an enzyme/inhibitor complex. If the amount of inhibitor is fixed, the percentage of inhibition can be reduced by increasing the substrate concentration. The two molecule types are competing for the same active site. The more substrate molecules there are, the less likely it is that inhibitor molecules will bind to the active site.

Non-competitive inhibition

In non-competitive reversible inhibition, the inhibitor may form a complex with either the enzyme itself or with the enzyme/substrate complex. This shows that the inhibitor is not competing for the active site. It joins to the enzyme molecule elsewhere. This is confirmed by the fact that only the concentration of inhibitor affects the level of inhibition. The concentration of the substrate makes no difference to how much inhibition occurs. The best model for how this inhibition works is that the presence of the inhibitor on the enzyme or enzyme/substrate complex deforms or changes the shape of the active site so that it can no longer catalyse the reaction. **Fig A** shows the differences between competitive and non-competitive inhibition.

competitive inhibition

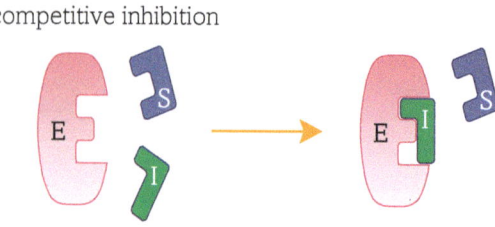

non-competitive inhibition

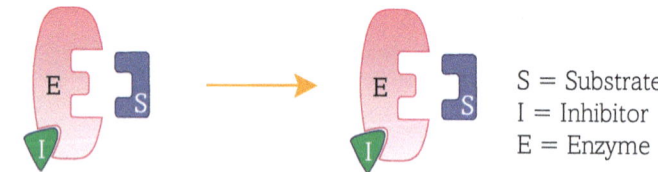

S = Substrate
I = Inhibitor
E = Enzyme

fig A Competitive inhibitors bind at the active site, non-competitive inhibitors do not.

Irreversible inhibition of enzymes

In irreversible inhibition the inhibitor combines with the enzyme by permanent covalent bonding to one of the groups vital for catalysis to occur. It changes the shape and structure of the molecule in such a way that it cannot be reversed – the enzyme is inactivated permanently. Irreversible inhibition tends to occur more slowly than the other forms of inhibition, but its effects are much more devastating and are never used within the cells to control metabolism.

Arsenic, cyanide and mercury are poisonous because they exert irreversible inhibition on enzyme systems. Some of the nerve gases used in chemical warfare also work in this way. They combine with and completely inactivate enzymes such as acetyl cholinesterase that break down chemicals used to transfer impulses from the nervous system to the muscles of the body. The normal function of acetylcholinesterase is to destroy the neurotransmitter called acetylcholine at the junctions between neurones and muscle cells. It does this as soon as an impulse has been passed from a nerve to a muscle. When the enzyme is inhibited the impulse continues. The muscles go into prolonged spasms causing death because breathing and swallowing become impossible.

End-product inhibition and the regulation of the cell

As you know, hundreds of chemical reactions are going on within a cell at any one time, their rate controlled by the action of enzymes. A similar number of reactions occurring in a very small space in a laboratory would, without doubt, end in total chaos if not a large explosion. So how do cells manage their reactions in such a controlled way? There are many factors involved. Membrane compartments keep reactions apart. Variations in pH can change the rate of enzyme-catalysed reactions, and the amount of substrate available is another mechanism at work. But one of the most important methods of control is that exerted by the regulatory enzymes.

Regulatory enzymes often have a site, seperate from the active site, to which another molecule can bind and bring about non-competitive inhibition. They are widely found in complex metabolic pathways such as photosynthesis and respiration.

In **end-product inhibition** the regulatory enzyme is found near the beginning of the pathway. It is inhibited by one of the end products of the chain. There are some very important examples of end-product inhibition in the pathways of cellular respiration in all organisms. Phosphofructokinase (PFK) is an enzyme involved in the production of ATP in the process of glycolysis in cellular respiration (see **Section 5.1.2**). PFK controls the rate of respiration by end-product inhibition. It is inhibited by ATP, which binds non-competitively and changes the shape of the active site. If the ATP concentration goes up, PFK is inhibited and cellular respiration slows down. As ATP levels fall, ATP molecules detach from PFK and the enzyme becomes active again. Rates of celluar respiration – and so ATP production – increase.

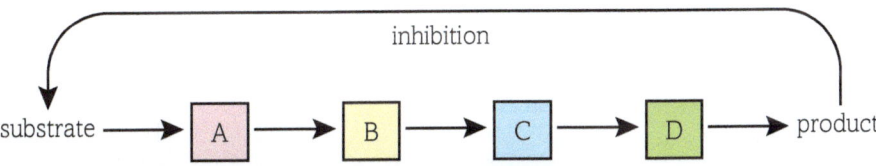

fig B Feedback control gives a simple and effective way of controlling the rate of several reactions at once.

Questions

1 What is the difference between reversible and irreversible enzyme inhibition?

2 What are the main differences between reversible competitive and non-competitive inhibition in enzymes and how does this affect the control of reactions within a cell?

Key definitions

Enzyme inhibitors are substances that slow down enzymes or stop them from working.

Reversible inhibition is inhibition of the action of an enzyme by an inhibitor that does not permanently affect the functioning of the enzyme and can be removed from the enzyme. It is often used to control reaction rates within a cell.

Irreversible inhibition is inhibition of the action of an enzyme that is permanent and cannot be undone. It is never used within cells to control the rate of reactions.

Competitive inhibition is inhibition in which the inhibitor molecule is similar in shape to the substrate molecule and competes with it for the active site of the enzyme (affected by both inhibitor and substrate concentrations).

Non-competitive inhibition is inhibition in which the inhibitor does not compete for the active site but forms a complex with the enzyme or enzyme/substrate complex and changes the shape of the active site so it can no longer catalyse the reaction (affected only by concentration of inhibitor).

Allosteric enzymes are enzymes that have a site separate to the active site where another molecule can bind to have either an activating or inhibitory effect.

End product inhibition is a control system in many metabolic pathways in which an enzyme at the beginning of the pathway is inhibited by one of the end products of the reaction.

THINKING BIGGER

RAW ENZYMES – REALLY?

The enzymes made by the cells of your body are vitally important. Inside your cells, they control all the reactions of life. Outside your cells they are particularly important in the digestion of your food. The internet is a great source of information but not all of it is reliable. Read the following texts about food and enzymes, based on a number of different websites promoting 'good health'…

Site 1

Each person is born with a limited enzyme-producing capacity. Your life expectancy depends on how well you preserve this enzyme potential. You need to take in enzymes from the food you eat. If you don't take in enough enzymes, it imposes a great strain on your digestive system because it has to produce all the enzymes you need. This in turn reduces the numbers of enzymes available for the metabolic reactions taking place in your cells – and this is the root cause of most chronic health problems. The solution is simple: eat at least 75% of your food raw to make use of the enzymes in the food, eat less, chew your food well and don't chew gum!

Site 2

When food is cooked, enzymes are destroyed by the heat. Enzymes help us digest our food. Enzymes are proteins, and they work because they have a very specific 3D structure in space. Once they are heated much above 118 degrees, this structure can be changed so they no longer work. Cooked foods contribute to chronic illness, because their enzyme content is damaged and so we have to make our own enzymes to process the food. This uses up valuable metabolic enzymes. It takes a lot more energy to digest cooked food than raw food – the evidence being that raw food passes through the digestive tract about 50% faster than cooked food. Eating enzyme-dead (cooked!) foods overworks and eventually exhausts your pancreas and other organs. Many people progressively lose the ability to digest their food after years of eating cooked and processed food.

Site 3

Enzymes are an essential part of a healthy diet. As an expert explains, 'Science cannot duplicate enzymes. Only raw food has functional living enzymes. The chain reaction generated by enzymes helps to send fats to where they are needed in our body, instead of being stored'.

fig A The cells of raw fruit and vegetables are full of enzymes – but how much use are they to you?

Where else will I encounter these themes?

Let us start by considering the nature of the writing in these articles:

1. The information given above comes from a number of different web sites that promote 'healthy living'. Think about the way they are using scientific information as you try and answer the following questions:

   a. Who do you think these web resources are aimed at?

   b. Do you think that the people producing these resources are writing objectively? Explain your answer.

   c. What tactics are used to try to persuade people that eating raw food provides you with useful enzymes and that cooking food is bad for you?

> **Command word**
> If the word explain is used in a question your answer should clearly describe the thing you are trying to explain. You'll need to use your reasoning and maybe examples to support your point.

Now let us have a look at the biology. Your knowledge of biochemistry is now at a level that allows you to read this article with a scientific mind!

2. Make a table to identify the biologically correct information and the biologically suspect information in the articles.

3. Do you think the people writing these web resources are real biologists or doctors? Explain your opinion.

4. Write a blog post describing the dangers of articles like these and putting right all of the biological misconceptions you found in question 2.

Activity

Enzymes are vital for life. A healthy diet provides your body with the materials it needs to make enzymes – but you do not directly use the enzymes in the food that you eat.

Prepare a three minute talk for a debate titled 'Raw food – the only healthy way to support your enzymes.'

Choose whether you want to support this idea or oppose it.

Focus on the biology of enzymes and of the compounds that make up your food. Whichever side you choose your argument must be backed up by good scientific evidence.

> Consider what you have learned about enzymes and their roles in the cells and in the digestive systems of organisms, including people. You can also do research to find out more, but make sure that your sources are reliable!

● Based on a number of different websites promoting 'good health'.

1 Amylase is an enzyme that breaks down starch to maltose. A student carried out an investigation to determine the effect of copper ions on the activity of this enzyme. She added different concentrations of copper ions and timed how long it took the amylase to break down starch.

The results of this investigation are shown in the graph below.

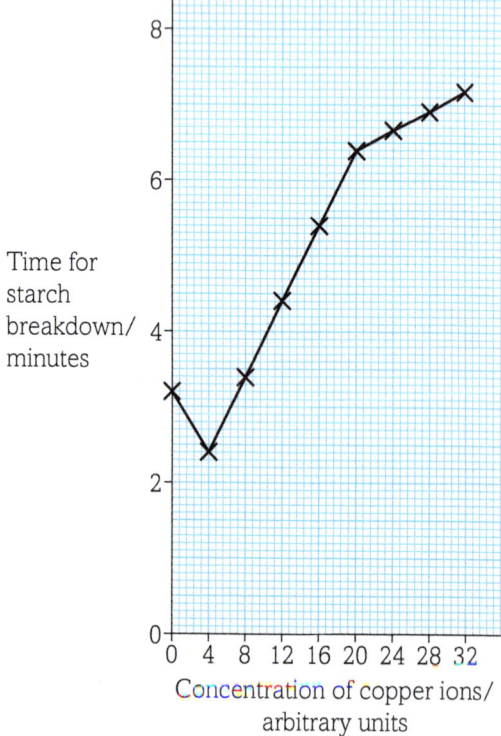

Time for starch breakdown/ minutes

Concentration of copper ions/ arbitrary units

(a) Describe a test that could be used to show that starch has been broken down. [3]

(b) Describe the effect that an increase in the concentration of copper ions has on the **activity** of amylase. [3]

(c) The student suggested that the copper ions were acting as an active site-directed inhibitor at concentrations above 4 arbitrary units. Explain what is meant by the term **active site-directed inhibition**. [3]

(d) The student then investigated the initial rate of reaction using amylase and different concentrations of starch. She did this first with copper ions present and then with no copper ions present. The results are shown in the graph below.

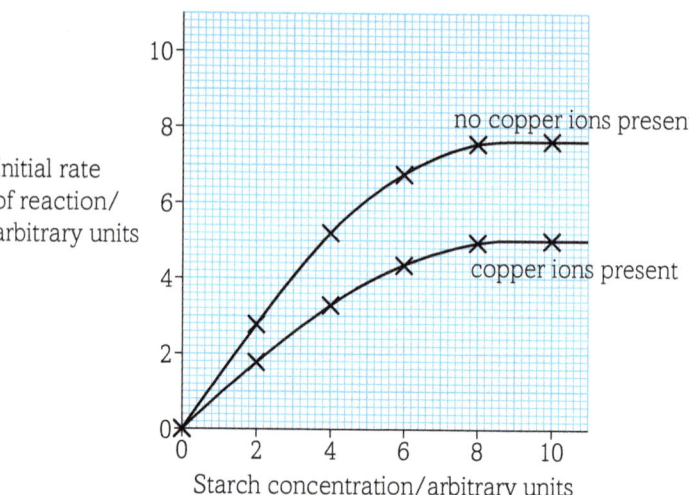

Initial rate of reaction/ arbitrary units

Starch concentration/arbitrary units

(i) Suggest why the **initial** rate of reaction was measured in this investigation. [2]

(ii) State why the results do **not** support the hypothesis that copper ions are an active site-directed inhibitor of amylase. [1]

[Total: 12]

2 (a) The graph below shows the change in energy that takes place during a chemical reaction.

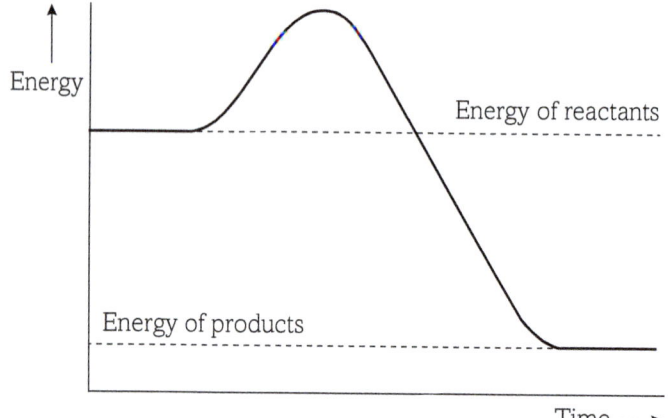

Energy

Energy of reactants

Energy of products

Time ⟶

(i) With reference to enzyme activity, explain the meaning of each of the following terms.
   • Activation energy
   • Catalyst [4]

(ii) On the graph above, draw the energy changes that would take place if the same chemical reaction was catalysed by an enzyme. [2]

(b) An experiment was carried out to determine the effect of temperature on the activity of a protein-digesting enzyme (a protease). Solutions of the protease were incubated with a protein called gelatine at three temperatures: 20 °C, 30 °C, and 40 °C. The concentration of amino acids were measured over a 48-hour period. The results are shown in the graph below.

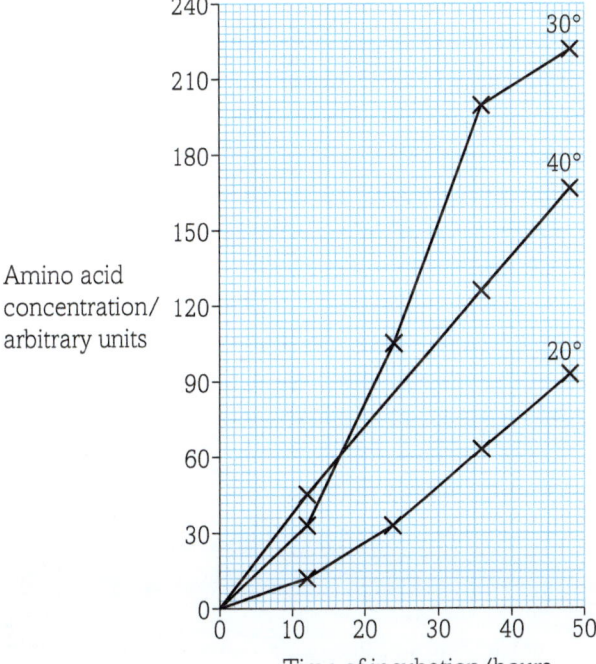

(i) Name the type of reaction catalysed by this protease. [1]

(ii) Name the bond broken by the protease. [1]

(iii) Calculate the mean rate of production of amino acids at 40 °C during the first 36 hours of incubation. Show your working and give your answer in arbitrary units hr⁻¹. [2]

(iv) The optimum temperature for this reaction is 30 °C. Explain the shape of the curve at this temperature. [2]

[Total: 12]

3 Trypsin is a protease enzyme that catalyses the breakdown of proteins. An investigation was carried out to study the effect of pH on the activity of trypsin. The source of protein in this investigation was milk powder mixed in distilled water. This gives a white, cloudy suspension.

A 2% solution of trypsin was prepared and placed in a waterbath at 45 °C. A 10% suspension of milk powder was prepared separately and 3 cm^3 samples of this suspension were mixed with 3 cm^3 of a pH buffer solution and placed in a waterbath at 45 °C. Buffer solutions between pH 5 and pH 9 were used.

When all the mixtures had reached a temperature of 45 °C, 0.5 cm^3 of the trypsin solution was added to the suspension and the time taken for the suspension to clear was recorded.

The results of this experiment are shown in the table below.

pH of suspension	Time taken for suspension to clear/min
5	10.82
6	6.68
7	1.38
8	0.71
9	1.42
10	7.80

(a) (i) Explain how an enzyme, such as trypsin, digests (breaks down) the protein in the milk powder. [3]

(ii) Suggest why the cloudy suspension goes clear when mixed with the enzyme. [1]

(b) (i) Use the information in the table to describe the effect of pH on the activity of trypsin. [2]

(ii) Explain how pH affects the activity of enzymes. [3]

(c) Suggest why the experiment was carried out in a water bath at 45 °C. [2]

[Total: 11]

TOPIC 2
Cells and viruses

Eukaryotic cells

Introduction

In October 1951, Henrietta Lacks, a young, 31-year-old American black woman, died of an aggressive cancer of the cervix. Before she died, doctors took a sample of her cells, which became the first human cells to grow successfully in culture. Known as HeLa cells (for Henrietta Lacks), her cells are still alive and well. Ever since they were first cultured, HeLa cells have played an important part in cell biology and medical research. The cells have been infected with viruses, and played a big role in the development of the first polio vaccine. HeLa cells reproduce themselves quickly and reliably and they have travelled all over the world. They have been used to investigate radiation damage and they have even gone up into space.

In this chapter you will be discovering more about eukaryotic cells like Henrietta's and the sub-cellular membrane-bound organelles they contain. The animals, plants and fungi in the world around you are all eukaryotic. Without microscopes we could not even see most cells, and so you will learn something of how both light and electron microscopes work.

Membranes are key to understanding the compartmentalisation of the functions of a cell. The structure of a cell membrane is closely linked with its function and you will be looking at the link between the two.

By looking at the many different types of organelles within a cell you will discover both their structure, their function and how they work together. Key organelles in all cells are the mitochondria – so important they have their own DNA that enables them to divide independently. This is where the reactions of cellular respiration take place.

Plant cells have most if not all of the features found in animal cells – and a few of their own. You will be considering the structure and function of these features, which highlight at a cellular level some of the major differences between plants and animals.

All the maths you need

- Carry out calculations using numbers in standard and ordinary form (*e.g. use of magnification*)
- Use expressions in standard form (*e.g. when applied to areas such as the size of organelles*)
- Use scales for measuring (*e.g. graticule to measure size of cells*)
- Make order of magnitude calculations (*e.g. use and manipulate the magnification formula: magnification = size of image/size of real object*)
- Use and manipulate equations, including changing the subject of an equation (*e.g. magnification*)
- Calculate the circumference, surface areas and volumes of regular shapes (*e.g. calculate the surface area or volume of a cell*)

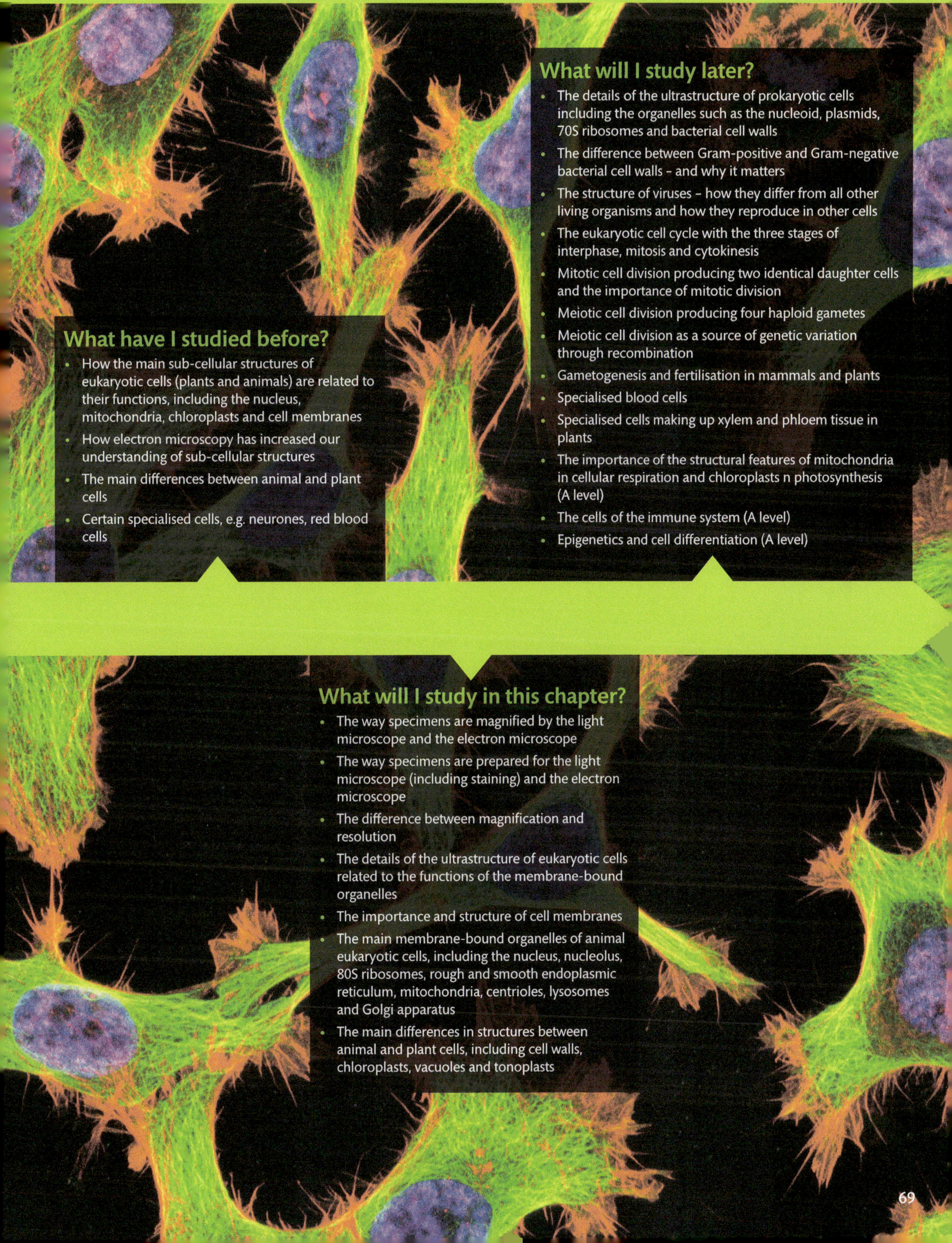

What will I study later?

- The details of the ultrastructure of prokaryotic cells including the organelles such as the nucleoid, plasmids, 70S ribosomes and bacterial cell walls
- The difference between Gram-positive and Gram-negative bacterial cell walls – and why it matters
- The structure of viruses – how they differ from all other living organisms and how they reproduce in other cells
- The eukaryotic cell cycle with the three stages of interphase, mitosis and cytokinesis
- Mitotic cell division producing two identical daughter cells and the importance of mitotic division
- Meiotic cell division producing four haploid gametes
- Meiotic cell division as a source of genetic variation through recombination
- Gametogenesis and fertilisation in mammals and plants
- Specialised blood cells
- Specialised cells making up xylem and phloem tissue in plants
- The importance of the structural features of mitochondria in cellular respiration and chloroplasts n photosynthesis (A level)
- The cells of the immune system (A level)
- Epigenetics and cell differentiation (A level)

What have I studied before?

- How the main sub-cellular structures of eukaryotic cells (plants and animals) are related to their functions, including the nucleus, mitochondria, chloroplasts and cell membranes
- How electron microscopy has increased our understanding of sub-cellular structures
- The main differences between animal and plant cells
- Certain specialised cells, e.g. neurones, red blood cells

What will I study in this chapter?

- The way specimens are magnified by the light microscope and the electron microscope
- The way specimens are prepared for the light microscope (including staining) and the electron microscope
- The difference between magnification and resolution
- The details of the ultrastructure of eukaryotic cells related to the functions of the membrane-bound organelles
- The importance and structure of cell membranes
- The main membrane-bound organelles of animal eukaryotic cells, including the nucleus, nucleolus, 80S ribosomes, rough and smooth endoplasmic reticulum, mitochondria, centrioles, lysosomes and Golgi apparatus
- The main differences in structures between animal and plant cells, including cell walls, chloroplasts, vacuoles and tonoplasts

By the end of this section, you should be able to...

● explain cell theory as a unifying concept that states that cells are a fundamental unit of structure, function and organisation in all living organisms

● describe how magnification and resolution can be achieved using light and electron microscopy

● explain the importance of staining specimens in microscopy

Cells are discussed in the media on an almost daily basis in relation to topics such as cancer, stem cells and DNA testing. However, in spite of the fact that we have known about cells for over 300 years, most people have only a vague idea about what they are and how they function.

Discovering cells

In 1665 Robert Hooke (1635–1703), an English architect and natural philosopher, designed and put together one of the first working optical microscopes. Amongst the many objects he examined were thin sections of cork, made up of tiny, regular compartments that he called cells, as they reminded him of the monks' cells in a monastery. In 1676 Anton van Leeuwenhoek (1632–1723), a Dutch draper who ground lenses in his spare time to check the weave of his fabrics, used his lenses to observe a wide variety of living unicellular organisms in drops of water, which he called 'animalcules'. At the same time the English plant scientist Nehemiah Grew (1641–1712) was one of the first scientists to publish accurate drawings of 'tissues'. By the 1840s we understood that cells are the basic units of life, an idea that was first expressed by Matthias Schleiden (1804–1881) and Theodor Schwann (1810–1882) in their cell theory of 1839. Cell theory is now accepted as a unifying concept in biology. It states that cells are a fundamental unit of structure, function and organisation in all living organisms. Improvements in the quality of lenses, new staining techniques and the introduction of new technologies such as electron and confocal microscopes, have allowed us to see cells in increasing detail and so develop our understanding of both their structure and function.

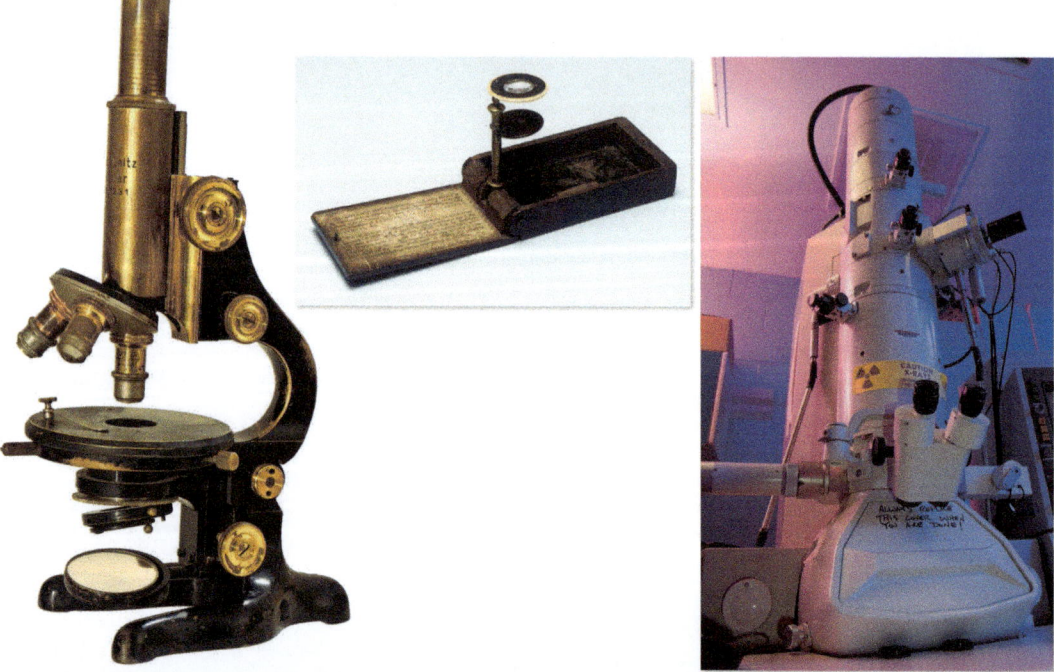

fig A As microscopes have developed, more and more has been revealed about cells, the key to understanding biology.

Microscopes

We can see some cells easily with the naked eye, for example the ovum in an unfertilised bird's egg is a single cell. But we need some kind of magnification to enable us to see most cells.

The **light microscope** or **optical microscope** has been the main tool for observing cells over the years and it is still widely used. A good light microscope can magnify to 1500 times and still give a clear image. At this magnification an average person would appear to be 2.5 km tall.

Since the mid-twentieth century the **electron microscope** has given scientists an even greater insight into the inner workings of cells. An electron microscope can give a magnification of up to 500 000 times, making an average person appear over 830 km tall!

Did you know?

Magnification and **resolution** are the two features of any microscope that determine how clear the image is:

- Magnification is a measure of how much bigger the image you see is than the real object, e.g. ×40, ×1000 or ×500 000.
- Resolution or **resolving power** is a measure of how close together two objects must be before we can see them as one. For example, the resolution of the naked eye is around 0.1 mm. Two objects closer together than 0.1 mm cannot be seen as separate objects. The resolution of a light microscope is around 0.2 μm (200 nm), and the resolution of an electron microscope is around 0.1–1 nm.

The light microscope

A specimen or thin slice of biological material is placed on the stage of a light microscope (see **fig B**) and illuminated from underneath, either by sunlight reflected with a mirror or by a built-in light source. The objective lens produces a magnified and inverted image, which the eyepiece lens focuses at the eye. The total magnification of the specimen is calculated:

magnification of objective lens	×	magnification of eyepiece lens	=	total magnification
e.g. ×10	×	×10	=	×100

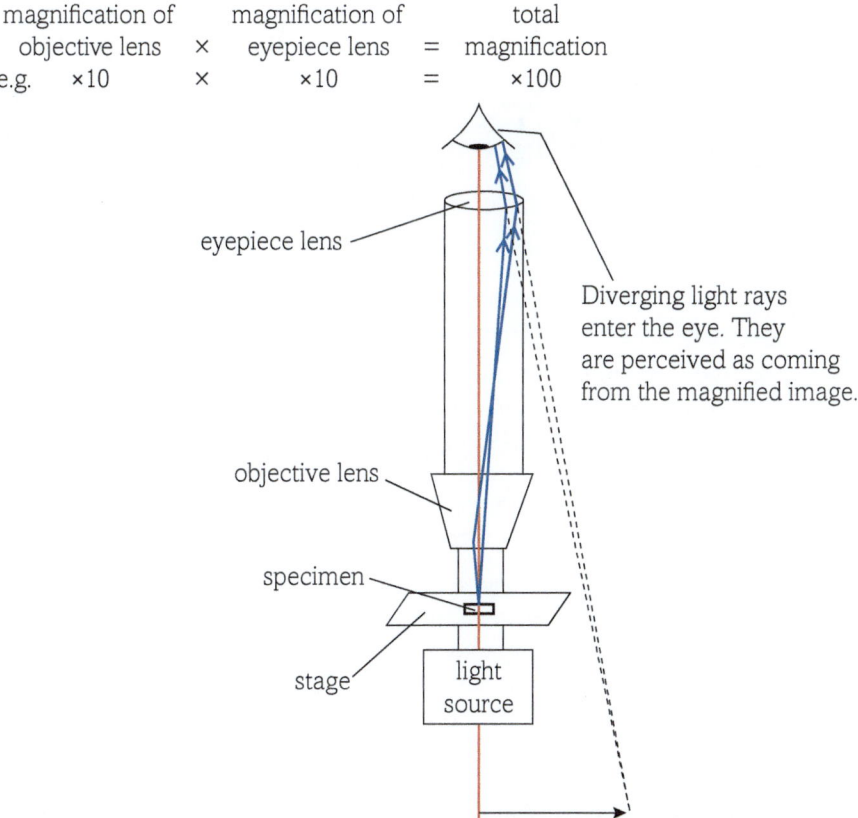

eyepiece lens

Diverging light rays enter the eye. They are perceived as coming from the magnified image.

objective lens

specimen

stage

light source

apparent size of specimen after magnification

fig B Light passes through the specimen and on through the lenses to give an image that is magnified and upside down.

Using the equation:

image size = actual size × magnification

You can work out the size of a specimen by measuring it under the microscope, as long as you always record the magnification you are using.

For example, the diameter of a cell measured under the light microscope at magnification ×400 is 0.1 cm (10 mm):

$$\frac{\text{image size}}{\text{magnification}} = \text{actual size}$$

$$\frac{10}{400} = 0.025 \text{ mm } (25 \text{ μm})$$

You can look at living organisms, tissues and cells under the light microscope. However, most of the specimens will be dead, stained, specially preserved and sectioned (very thinly sliced) before they are mounted on a slide. The staining is used to make it easier to identify particular types of cell, or particular parts of the cells, under the microscope. Some of the stains you may come across include:

- haematoxylin – stains the nuclei of plant and animal cells purple, blue or brown
- methylene blue – stains the nuclei of animal cells blue
- acetocarmine – stains the chromosomes in dividing nuclei in both plant and animal cells
- iodine – stains starch-containing material in plant cells blue-black.

There are big advantages to using light microscopes, but there are some disadvantages too:

Advantages of the light microscope	Disadvantages of the light microscope
• Can see living plants and animals, or parts of them, directly. This is useful in itself and allows you to compare prepared slides with living tissue. • Relatively cheap so are available in schools and universities, hospitals, industrial labs and research labs. • Relatively light and portable so we can use them almost anywhere, e.g. identifying malaria in the field.	• Preservation and staining tissue can produce artefacts in the tissues being observed, so what we see may be the result of preparation rather than a true representation of the living tissue. • Limited powers of resolution and magnification.

Developments including the confocal microscope mean the information we can get from light microscopes continues to increase.

The electron microscope

The electron microscope uses a beam of electrons to form an image. The electrons are scattered by the specimen in much the same way as light is scattered in the light microscope. In an electron microscope, the electrons effectively behave like light waves with a very tiny wavelength. Electromagnetic or electrostatic lenses focus the electron beam to form an image. Resolving power increases as the wavelength gets smaller, so the electron microscope can resolve detail down to less than 0.00001 μm, about 10 000 times better than the light microscope.

For the electron microscope to work, the specimens have to be in a vacuum, so they are always dead. The preparation of a specimen for the electron microscope is a very complex process that may involve chemical preservation, freeze drying, freeze fracturing, removing the water (dehydration), embedding, sectioning and mounting on a metal grid. Specimens for electron microscopy are often stained using heavy metal ions such as lead and uranium. This is not to identify particular tissues, but to improve the scattering of the electrons and make greater contrast in the image, making it clearer and easier to interpret. The image is displayed on a monitor or computer screen.

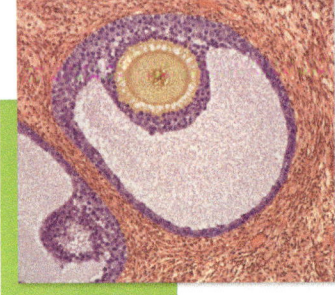

fig C A good light micrograph of tissue with staining that shows up the different types of cells can provide us with a lot of information, as demonstrated by this section through ovarian tissue.

Be very clear about the difference between magnification and resolution.

Make sure you are able to calculate the size, magnification or image size of any specimen.

There are two main types of electron micrographs. **Transmission electron micrographs (TEMs)** are two-dimensional (2D) images like those from a light microscope. **Scanning electron micrographs (SEMs)** have a lower magnification, but are three-dimensional (3D) and can be very striking. Sometimes electron micrographs are given false colours to make it easier to identify the different cells, but these are not stains. They are added after the image has been taken.

There are big advantages to using electron microscopes, but there are some disadvantages too:

Advantages of the electron microscope	Disadvantages of the electron microscope
• Huge powers of magnification and resolution. Many details of cell structure have been seen for the first time since they were developed.	• All specimens are examined in a vacuum – air would scatter the electrons and make the image of the tissue fuzzy – so it is impossible to look at living material.
	• Specimens undergo severe treatment that is likely to result in artefacts. Preparing specimens for the electron microscope is very skilled work.
	• Extremely expensive.
	• Large, have to be kept at a constant temperature and pressure and need to maintain an internal vacuum. Relatively few scientists outside research laboratories have easy access to such equipment.

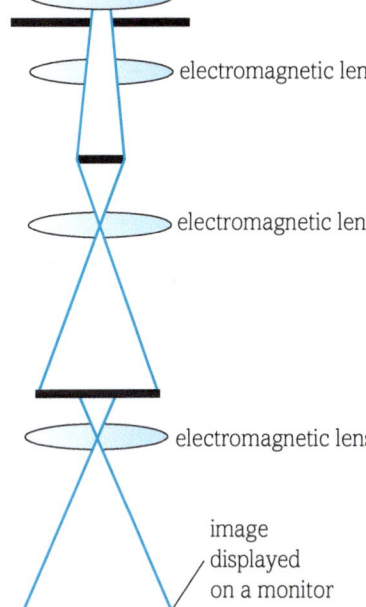

electron source

electromagnetic lens

electromagnetic lens

electromagnetic lens

image displayed on a monitor or computer

fig D A beam of electrons passes through the specimen and on through electromagnetic or electrostatic lenses to give a greatly magnified image.

Questions

1 Why is high magnification alone not enough to give us biological details of cells?

2 Both light and electron micrographs can be brightly coloured. Explain the differences and similarities between the way colour is used in light and electron microscopy.

3 A student measured the diameter of three cells of the same type under the microscope. Measurement 1 was taken with a magnification of ×40, and measurements 2 and 3 with a magnification of ×100. Work out the mean diameter of the cells.

Measurement 1 = 5 mm   Measurement 2 = 12 mm   Measurement 3 = 1 mm

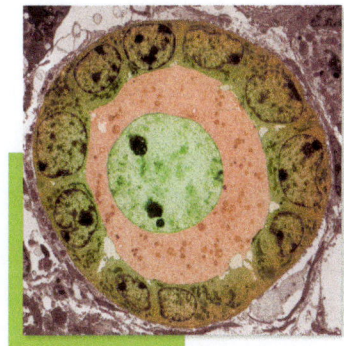

fig E A transmission electron micrograph of a cell gives you much more detailed information than the light micrograph in **fig C**.

A **light microscope (optical microscope)** is a tool that uses a beam of light and optical lenses to magnify specimens up to 1500 times life size.

An **electron microscope** is a tool that uses a beam of electrons and magnetic lenses to magnify specimens up to 50 000 times life size.

Magnification is a measure of how much bigger the image you see is than the real object.

Resolution (resolving power) is a measure of how close together two objects must be before they are seen as one.

Transmission electron micrographs (TEMs) are micrographs produced by the electron microscope that give 2D images like those from a light microscope, but magnified up to 500 000 times.

Scanning electron micrographs (SEMs) are micrographs produced by the electron microscope that have a lower magnification than TEMs, but produce a 3D image.

fig F Scanning electron micrographs open up a whole new, 3D world of biology on a small scale.

By the end of this section, you should be able to...

- define eukaryotic and prokaryotic cells

- recognise that cell membranes are common to eukaryotes and prokaryotes

- explain how the structure and properties of phospholipids relate to their function in cell membranes

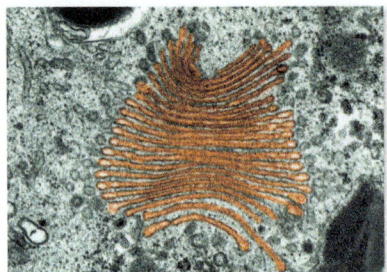

fig A One of the functions of this stack of membranes in a plant cell (the Golgi apparatus) is to package secretions into vesicles. It is just one of the membrane-rich organelles you can find in a eukaryotic cell.

Most of the familiar organisms in the world around you have the same sort of cells. Animals, plants, protists, algae and fungi have cells with the genetic material contained in a membrane-bound nucleus. The cells also contain a number of other membrane-bound **organelles** such as mitochondria and chloroplasts. These organisms are called the **eukaryotes**. But there is another ancient group of organisms, including the bacteria and cyanobacteria, known as the **prokaryotes**. They have cells of a very different type that lack much of the structure and organisation of the eukaryotic cells, but they do have a **cell surface membrane**. You will look at both eukaryotic and prokaryotic cells in this topic, but you will begin by studying the structure of the membranes that are common to all.

Membranes in cells

There are many membranes within cells, such as those that surround organelles like the nucleus and mitochondria. But the most obvious membrane is the cell surface membrane, also known as the outer cell membrane, which forms the boundary of all cells – controlling what passes into and out of the cell and allowing the fluids either side of them to have different compositions. Membranes within cells make it possible to have the right conditions for a particular reaction in one part of a cell and different conditions to suit other reactions elsewhere in the same cell.

Membranes perform many other functions too. Many chemical processes take place on membrane surfaces. For example, the reactions of respiration in eukaryotic cells take place on the mitochondrial membrane. Enzymes and any other factors are held closely together so that the reaction processes can proceed smoothly. The cell surface membrane must also be flexible to allow the cell to change shape very slightly as its water content changes, or quite dramatically, for example when a white blood cell engulfs a bacterium. Chemical secretions made by the cell are packaged into membrane bags known as **vesicles**, so some membranes must be capable of breaking and fusing together readily.

The structure of membranes

Our current model of the structure of membranes has been worked out over many years. The model developed as microscopy improved, from light to electron and then scanning electron microscopes. In time there may well be further refinements to the model presented here, but the overall picture seems unlikely to change dramatically. The membrane is made up mainly of two types of molecules – lipids and proteins – arranged in a very specific way.

The phospholipid bilayer

The lipids in the membrane are of a particular type called **polar lipids**. These are lipid molecules with one end joined to a polar group. Many of the polar lipids in the membrane are phospholipids, with a phosphate group forming the polar part of the molecule (see **Section 1.2.3**). With water or aqueous solutions on each side, phospholipid molecules form a bilayer with their hydrophilic heads pointing into the water while the hydrophobic tails stay protected in the middle. This structure is known as a unit membrane.

However, a simple lipid bilayer alone would not explain either the microscopic appearance of membranes or the way in which they behave. A simple lipid bilayer allows fat-soluble organic molecules to pass through it, but many vital chemicals needed in cells are ionic. Whilst these dissolve in water they cannot dissolve in or pass through lipids, even polar lipids. They can enter cells because the membrane consists not only of lipids, but also of proteins and other molecules.

The membrane proteins

The best model of a membrane we have today sees the basic bilayer of phospholipid as a fluid system, with many proteins and other molecules floating within it like icebergs whilst others are fixed in place (see **fig B**). The proportion of phospholipids containing unsaturated fatty acids (see **Section 1.2.3**) in the bilayer seems to affect how freely the moving proteins float about in the membrane. The more unsaturated fatty acids, the more fluid the membrane. Many of the proteins have a hydrophobic part, which is buried in the lipid bilayer, and a hydrophilic part, which can be involved in a variety of activities. Some proteins penetrate all the way through the lipid, while others only go part of the way through the bilayer.

One of the main functions of the membrane proteins is to help substances move across the membrane. The proteins can form pores or channels – some permanent, some temporary – that allow specific molecules to move through the pores. Some of these channels can be open or shut, depending on conditions in the cell. These are known as **gated channels**. Some of the protein pores are active carrier systems using energy to move molecules, as you will see later. Others are simply gaps in the lipid bilayer that allow ionic substances to move through the membrane in both directions.

Proteins may act as specific receptor molecules – for example, making cells sensitive to a particular hormone. They may be enzymes, particularly on any internal cell membranes, to control reactions linked to that membrane. Some membrane proteins are glycoproteins, proteins with a carbohydrate part added to the molecule. These are very important on the surface of cells as part of the way cells recognise each other.

This model of the floating proteins in a lipid sea is known as the **fluid mosaic model** and was first proposed by S. Jonathan Singer (1924–) and Garth Nicholson (1943–) in 1972.

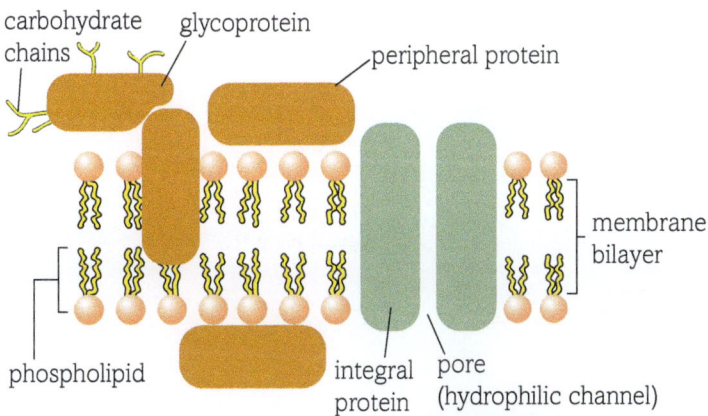

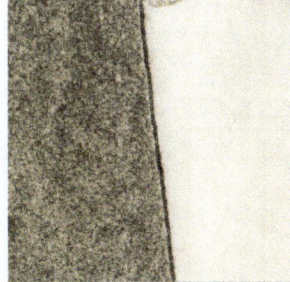

fig B Whether acting as the boundary of a cell or as part of its internal make-up, the complex structure of the membrane is closely linked to its wide variety of functions.

Did you know?

Evidence for the fluid mosaic model

Techniques such as X-ray diffraction and advanced electron microscopy have added to our knowledge of the structure of cell membranes, giving us more details of the layers, the pores and the carrier molecules.

However, even under the electron microscope, cell membranes are very small. Microscopes have helped scientists develop our current membrane model, but other techniques are also important. We can

identify proteins that appear to have a specific function of transporting particular ions into or out of the cell through the membrane. Cystic fibrosis is a genetic disease that affects transport across the membranes of the glands, the digestive system, the respiratory system and the reproductive system. If an individual inherits a faulty allele (variant) from each parent, the protein needed to transport chloride ions across the membranes (cystic fibrosis transmembrane regulatory channel protein) does not form properly. This affects the movement of water out of the cells and leads to the formation of sticky mucus, which can lead to serious chest infections, digestive problems and infertility. Identifying protein channels in the membrane that have a very specific function that can be measured helps to confirm our model of cell membrane structure and function.

fig C The membrane pores through which mRNA leaves the nucleus are clearly visible in this freeze-etched electron micrograph of the nuclear membrane of a cell.

Questions

1 Summarise the main functions of membranes in cells.

2 Which kinds of molecule make up the structure of a membrane and how do their properties affect the properties of the membrane itself?

3 Discuss why the flexible structure of the cell membrane is well adapted for its functions in the cell.

Key definitions

Organelles are sub-cellular bodies found in the cytoplasm of cells.

Eukaryotes are a group of organisms with cells that have the genetic material contained in a membrane-bound nucleus and also contain a number of membrane-bound organelles such as mitochondria and chloroplasts.

Prokaryotes are a group of organisms including bacteria and blue-green algae (cyanobacteria) that have few organelles and do not have the genetic material contained in a membrane-bound nucleus.

The **cell surface membrane** is the membrane that forms the outer boundary of the cytoplasm of a cell and controls the movement of substances into and out of the cell.

Vesicles are membrane 'bags' that hold secretions made in cells.

Polar lipids are lipids with one end attached to a polar group, e.g. a phosphate group that makes one end of the molecule hydrophilic and one end hydrophobic.

Gated channels are protein channels through the lipid bilayer of a membrane that are opened or closed, depending on conditions in the cell.

The **fluid mosaic model** is the current model of the structure of the cell membrane including floating proteins forming pores, channels and carrier systems in a lipid bilayer.

Eukaryotic cells 1 – common cellular structures

By the end of this section, you should be able to...

- describe eukaryotic cells

- describe the ultrastructure of eukaryotic cells and the functions of organelles including the nucleus, nucleolus, mitochondria, centrioles and vacuoles

Most microscope images, apart from those of living material or from a scanning electron microscope, make cells appear flat and two-dimensional (2D). But cells are actually spheres, cylinders or asymmetrical three-dimensional (3D) shapes – so try to use your imagination when you look at cells and visualise them in three dimensions.

The characteristics of eukaryotic cells

In eukaryotic organisms such as animals, plants and fungi there is a very wide range of different types of cell, each with a different function. But there are certain cell features that turn up again and again, and we can put these together as a 'typical' plant or animal cell. Remember that this typical cell does not really exist, but acts as a useful guide to what to look for in any eukaryotic cell.

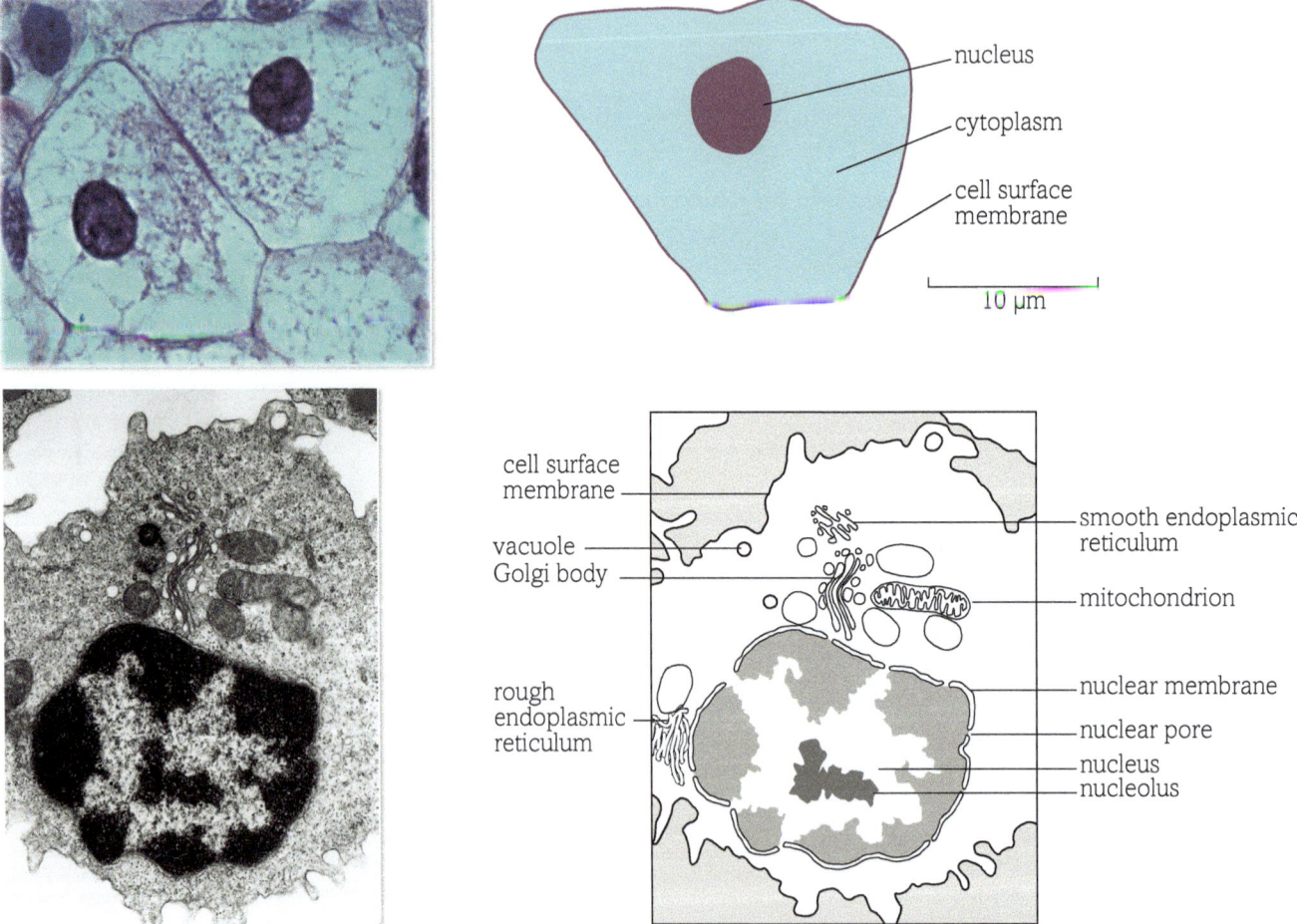

fig A These images show clearly how the introduction of the electron microscope increased our detailed knowledge and understanding of structures within cells.

The typical animal cell

A typical animal cell contains many things that are common to all eukaryotic cells, including plants and fungi. A membrane known as the cell surface membrane surrounds the cell. Inside this membrane is a jelly-like liquid called the **cytoplasm**, containing a **nucleus** – the two together are known as the **protoplasm**. The cytoplasm contains most of what is needed to carry out the day-to-day tasks of living, whilst the nucleus is vital to the long-term survival of the cell, because it contains the information needed to produce all the chemicals that make up the cell. This basic pattern gives rise to an enormous number of variations suited for the different functions that arise within the animal kingdom. The various parts of the cell have complex and detailed structures, which we can see more clearly when an electron microscope is used. The structures that can only be observed in detail using the electron microscope are known as the **ultrastructure** of the cell. The structure of each part of the cell relates closely to the job it has to do.

Membranes

Membranes in a cell are important both as an outer boundary to the cell and in the multitude of internal (**intracellular**) membranes. In **Section 2.1.2** you looked at the importance of cell membranes for controlling the movement of substances, but membranes inside the cell also have other functions. They localise enzymes in reaction pathways, for example respiration in mitochondria and photosynthesis in chloroplasts, and they compartmentalise chemicals, for example hydrolytic enzymes in lysosomes. You will learn more about membrane functions throughout this section as you consider the various structures that make up the cell.

The protoplasm

When the light microscope was the only tool biologists had to observe cells, they thought that the cytoplasm was a relatively structureless, clear jelly. But the electron microscope revealed the cytoplasm to be full of all manner of structures, known as organelles, some of which are described below.

The nucleus

The nucleus is usually the largest organelle in the cell (1–20 µm) and it can be seen with the light microscope. Electron micrographs show that the nucleus, which is usually spherical in shape, is surrounded by a double nuclear membrane containing holes or pores. Chemicals can pass in and out of the nucleus through these pores so that the nucleus can control events in the cytoplasm. Inside the nuclear membrane, or envelope as it is sometimes called, are two main substances, nucleic acids and proteins. The nucleic acids are deoxyribonucleic acid (DNA) and ribonucleic acid (RNA) (see **Chapter 1.3**).

When the cell is not actively dividing, the DNA is bonded to the protein to form **chromatin**, which looks like tiny granules. Also in the nucleus there is at least one **nucleolus** – an extra-dense area of almost pure DNA and protein. The nucleolus is involved in the production of ribosomes. Recent research also suggests that the nucleolus plays a part in the control of cell growth and division.

Mitochondria

The name **mitochondrion** simply means 'thread granule' and describes the tiny rod-like structures that are 1 µm wide by up to 10 µm long, seen in the cytoplasm of almost all cells under the light microscope. In recent years, by using the electron microscope we have been able to understand not only their complex structure, but also their vital functions.

The mitochondria are the 'powerhouses' of the cell. Here, in a series of complicated biochemical reactions, simple molecules are oxidised in the process of cellular respiration, producing ATP (see **Chapter 1.3**) that can be used to drive the other functions of the cell and indeed the organism. The number of mitochondria present can give you useful information about the functions of a cell. Cells that require very little energy, for example white fat storage cells, have very few mitochondria. Any cell with an energy demanding function, for example muscle cells or cells that carry out a lot of active transport such as liver cells, will contain large numbers of mitochondria.

An outer and inner membrane surrounds the mitochondria. They also contain their own genetic material, so that when a cell divides, the mitochondria replicate themselves under the control of the nucleus. This mitochondrial DNA is part of the whole genome of the organism.

Mitochondria have an internal arrangement adapted for their function (see **fig B**). The inner membrane is folded to form **cristae**, which give a very big surface area, surrounded by a fluid matrix. This structure is closely integrated with the events in cellular respiration that take place in the mitochondrion (see **Book 2 Sections 5.1.3** and **5.1.4**). Backed by evidence that shows that mitochondria have their own DNA, scientists think that mitochondria and chloroplasts originated as symbiotic **eubacteria** living inside early cells. Over millions of years of evolution they have become an integral part of the cell (see **Section 3.1.4**). This is the **endosymbiotic theory** of the evolution of eukaryotic cells.

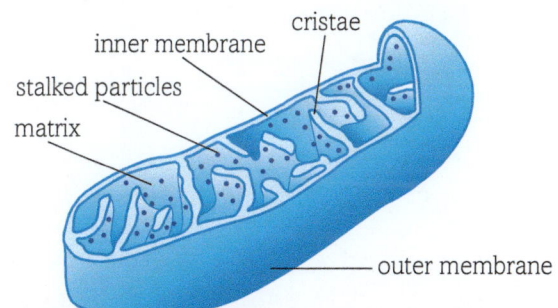

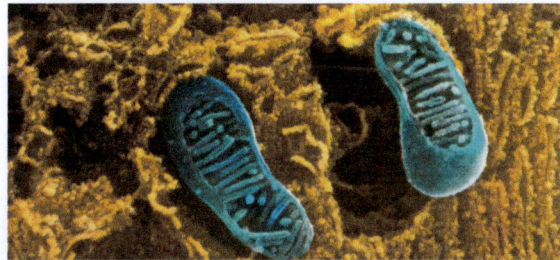

fig B The 3D structure of the mitochondria (blue) is closely related to their functions in cellular respiration.

The centrioles

In each cell there is usually a pair of **centrioles** near the nucleus (see **fig C**). Each centriole is made up of a bundle of nine tubules and is about 0.5 μm long by 0.2 μm wide. The centrioles are involved in cell division. When a cell divides, the centrioles pull apart to produce a **spindle** of microtubules that are involved in the movement of the chromosomes, as you will see later in this chapter.

(a)

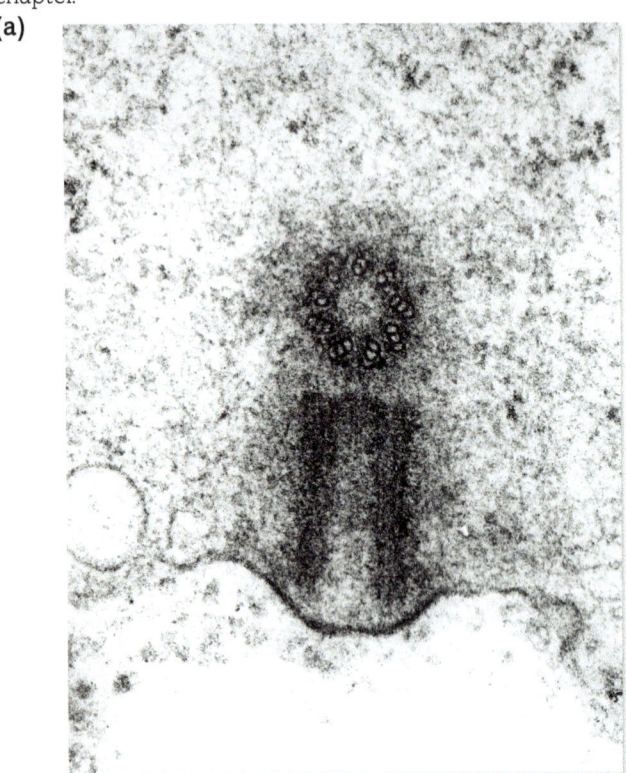

(b)

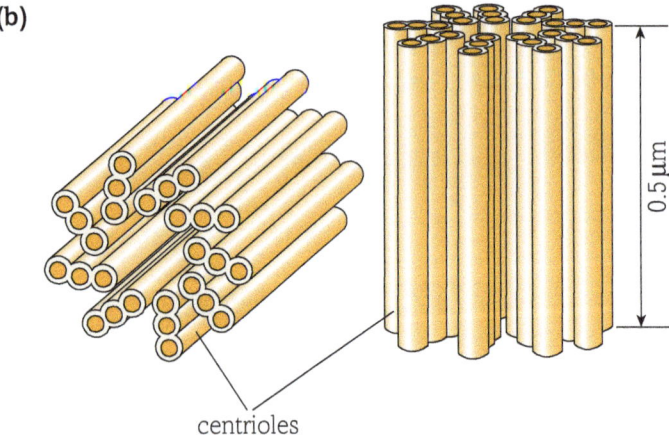

centrioles

fig C (a) Transition electron micrograph of centrioles; and (b) Diagram of centrioles.

The cytoskeleton

A cellular skeleton may seem a contradiction in terms, yet work in recent years has shown that a **cytoskeleton** is a feature of all eukaryotic cells. It is a dynamic, 3D web-like structure that fills the cytoplasm (see **fig D**). It is made up of **microfilaments**, which are protein fibres, and **microtubules**, tiny protein tubes about 20 nm in diameter. Microtubules are found both singly and in bundles throughout the cytoplasm. These microtubules consist mainly of the globular protein tubulin. The cytoskeleton performs several functions. It gives the cytoplasm structure and keeps the organelles in place. Many of the proteins in the microfilaments are related to actin and myosin, the contractile proteins in muscle, and the cytoskeleton is closely linked with cell movements and transport within cells. Recent research has shown that a cytoskeleton is also a feature of some prokaryotic cells.

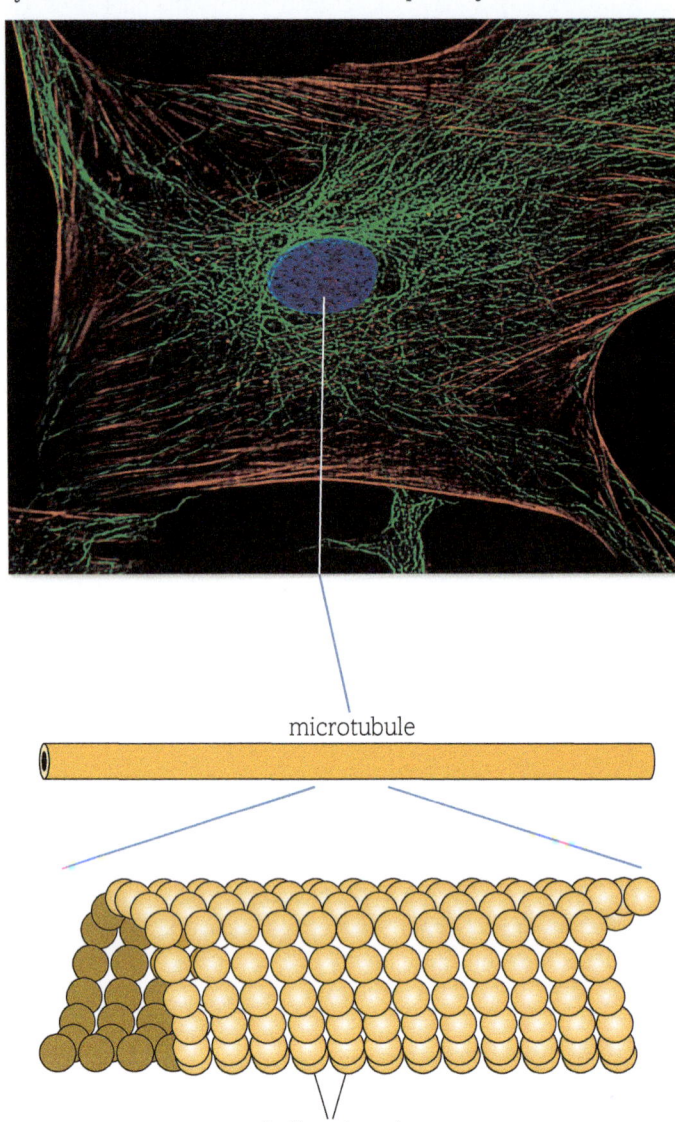

microtubule

tubulin sub-units

fig D The cytoskeleton forms a tangled web of structural and contractile fibres that hold the organelles in place and enable cell movement to occur.

Vacuoles

Vacuoles are not a permanent feature in animal cells. These membrane-lined enclosures are formed and lost as needed. Many simple animals make food vacuoles around the prey they engulf. White blood cells in higher animals form similar vacuoles around engulfed pathogens. **Contractile vacuoles** are an important feature in simple animals that live in fresh water because they allow the water content of the cytoplasm to be controlled. But in spite of these examples, vacuoles are not a major feature of animal cells and permanent vacuoles are never seen.

Learning tips

Remember that different types of electron microscopy provide very different types of information:

*   The scanning EM can show intact organelles, allowing detailed measurements of the outer dimensions to be taken or it can take 3D images along fracture lines.
*   The transmission EM provides clear images of the internal structures of the organelles.

Together the information is useful to produce a detailed image of the ultrastructure of a cell.

Questions

1   What is the role of the cytoskeleton in the cytoplasm and why has its importance only recently been recognised?

2   Explain the importance of organelles in eukaryotic cells.

3   Look at the different images that result from transmission and scanning electron microscopes in **Section 2.1.3** and describe how they differ. Suggest the advantages of each type of image and give examples where each would be more appropriate to use.

Key definitions

Cytoplasm is a jelly-like liquid that makes up the bulk of the cell and contains the organelles.

The **nucleus** is an organelle containing the nucleic acids DNA (the genetic material) and RNA, as well as protein, surrounded by a double nuclear membrane with pores.

Protoplasm is the cytoplasm and nucleus combined.

The **ultrastructure** is the detailed organisation of the cell, only visible using the electron microscope.

Intracellular means inside the cell.

Chromatin is the granular combination of DNA bonded to protein found in the nucleus when the cell is not actively dividing.

A **nucleolus** is an extra dense area of almost pure DNA and protein found in the nucleus involved in the production of ribosomes and control of growth and division.

Mitochondria are rod-like structures with inner and outer membranes that are the site of aerobic respiration.

Cristae are the infoldings of the inner membrane of the mitochondria which provide a large surface area for the reactions of aerobic respiration.

Eubacteria are true bacteria (prokaryotic organisms).

The **endosymbiotic theory** is a theory that suggests that mitochondria and chloroplasts originated as independent prokaryotic organisms that began living symbiotically inside other cells as endosymbionts.

Centrioles are bundles of tubules found near the nucleus and involved in cell division by the production of a spindle of microtubules that move the chromosomes to the ends of the cell.

A **spindle** is a set of overlapping protein microtubules running the length of the cell, formed as the centrioles pull apart in mitosis and meiosis.

The **cytoskeleton** is a dynamic, 3D web-like structure made up of microfilaments and microtubules that fills the cytoplasm and gives it structure, keeping the organelles in place and enabling cell movements and transport within the cell.

Microfilaments are protein fibres that make up part of the structure of the cytoskeleton.

Microtubules are tiny protein tubes about 20 nm in diameter that make up part of the structure of the cytoskeleton.

A **vacuole** is a fluid-filled cavity within the cytoplasm of a cell surrounded by a membrane enclosing food, water or air.

Contractile vacuoles are vacuoles that can fill and empty to help control the concentration of the cytoplasm of simple freshwater animals.

Eukaryotic cells 2 – protein transport

By the end of this section, you should be able to...

● describe the ultrastructure of eukaryotic cells and the functions of organelles including the rough and smooth endoplasmic reticulum, 80S ribosomes, Golgi apparatus and lysosomes

The cytoplasm of the cell contains the **endoplasmic reticulum (ER)**, a three-dimensional (3D) network of cavities bounded by membranes. The electron microscope reveals that some of the cavities are sac-like and some are tubular, and that the ER spreads extensively through the cytoplasm. The ER network links with the membrane around the nucleus, and makes up a large part of the transport system within a cell as well as being the site of synthesis of many important chemicals. It has been calculated that $1\,cm^3$ of liver tissue contains about $11\,m^2$ of endoplasmic reticulum. Electron microscopes also helped scientists to work out the functions of the endoplasmic reticulum, by showing up the different forms – the rough and the smooth endoplasmic reticulum.

Another useful technique is to provide cells with radioactively labelled chemicals that are building blocks for specific modules, for example labelled amino acids for the synthesis of proteins, and then find out where they appear in the cell. The labelled products can be tracked using microscopy. Another method of locating them is to break the cells open and then spin the contents in a centrifuge. The different parts of the cell can be separated out and the regions containing the radioactively labelled substances identified.

80S and 70S ribosomes

In **Section 1.3.6** you met ribosomes, the organelles on which protein synthesis takes place in the cytoplasm of the cell. Ribosomes are made from ribosomal RNA and protein, and consist of a large subunit and a small subunit. The main type of ribosomes in eukaryotic cells are **80S ribosomes**. The 'S' stands for Svedberg, a unit used to measure how quickly particles settle in a centrifuge. The rate of sedimentation depends on the size and shape of the particle. When 80S ribosomes are broken into their two units, they are made up of a 40S small subunit and a 60S large subunit. The ratio of RNA : protein in 80S ribosomes is 1 : 1.

However, eukaryotic cells also contain another type of ribosome. Scientists have discovered **70S ribosomes** in the mitochondria, and in the chloroplasts of plant cells. These ribosomes are usually found in prokaryotic cells (bacteria and cyanobacteria). They are made up of a small 30S subunit and a larger 50S subunit and the ratio of RNA : protein in 70S ribosomes is 2 : 1.

These 70S ribosomes are reproduced in the mitochondria and chloroplasts independently when a cell divides. This is seen as good evidence for the endosymbiotic theory that mitochondria and chloroplasts evolved from bacteria caught inside eukaryotic cells very early on in the process of evolution.

Rough and smooth endoplasmic reticulum

Electron micrographs show that much of the outside of the endoplasmic reticulum membrane is covered with granules, which are 80S ribosomes, so this is known as **rough endoplasmic reticulum (RER)** (see **fig A**). The function of the ribosomes is to make proteins and the RER isolates and transports these proteins once they have been made. Some proteins, such as digestive enzymes and hormones, are not used inside the cell that makes them, so they have to be secreted without interfering with the cell's activities. This is an example of **exocytosis**.

Many other proteins are needed within the cell. The RER has a large surface area for the synthesis of all these proteins, and it stores and transports them both within the cell and from the inside to the outside. Cells that secrete materials, such as those producing the digestive enzymes in the lining of the gut, have a large amount of RER.

Not all endoplasmic reticulum is covered in ribosomes (see **fig A**). **Smooth endoplasmic reticulum (SER)** is also involved in synthesis and transport, but in this case of steroids and lipids. For example, lots of SER is found in the testes, which make the steroid hormone testosterone, and in the liver, which metabolises cholesterol amongst other lipids. The amount and type of endoplasmic reticulum in a cell give an idea of the type of job the cell does.

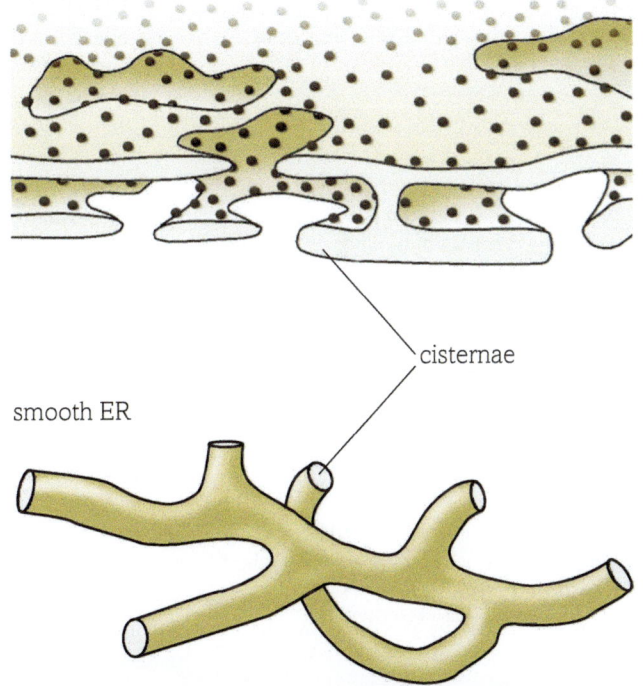

fig A Rough and smooth endoplasmic reticulum. Smooth ER is more tubular than rough ER and also lacks ribosomes on the surface.

The Golgi apparatus

Under the light microscope the **Golgi apparatus** looks like a rather dense area of cytoplasm. An electron microscope reveals that it is made up of stacks of parallel, flattened membrane pockets called cisternae, formed by vesicles from the endoplasmic reticulum fusing together (see **Section 2.1.2, fig A**).

The Golgi apparatus has a close link with, but is not joined to, the RER. It has taken scientists a long time to discover exactly what the Golgi apparatus does. Materials have been radioactively labelled and tracked through the cell to try and find out exactly what goes on inside it. Proteins are brought to the Golgi apparatus in vesicles that have pinched off from the RER where they were made. The vesicles fuse with the membrane sacs of the Golgi apparatus and the protein enters the Golgi stacks. As the proteins travel through the Golgi apparatus they are modified in various ways.

Carbohydrate is added to some proteins to form glycoproteins such as mucus. The Golgi apparatus also seems to be involved in producing materials for plant and fungal cell walls and insect cuticles. Some proteins in the Golgi apparatus are digestive enzymes. These may be enclosed in vesicles to form an organelle known as a **lysosome**. Alternatively, enzymes may be transported

through the Golgi apparatus and then in vesicles to the cell surface membrane where the vesicles fuse with the membrane to release extracellular digestive enzymes. The Golgi apparatus was first reported over 100 years ago, in April 1898. The flattened stack of membranes was observed by the Italian scientist Camillo Golgi (1843–1926) through a light microscope. For more than 50 years scientists argued over its function. Some thought it was an artefact from the process of fixing and staining during tissue preparation. The arrival of the electron microscope in the 1950s allowed the detailed structure of the Golgi apparatus to be seen.

The electron microscope has been central in showing details of the internal structure of the Golgi apparatus. In addition, a number of techniques have been developed that have allowed more detailed understanding. The most important of these has been the process of labelling specific enzymes so they can be seen using the electron microscope. The inner areas of the Golgi apparatus, nearer to the RER, have been shown to be very rich in enzymes that modify proteins in various ways. This is where most enzymes or membrane proteins are converted into the finished product. In contrast, in the outer regions of the Golgi apparatus you find lots of finished protein products, but not many of the enzymes that make them. The movement of cell membrane proteins through the Golgi apparatus is very complex. Areas of the protein that need to be on the outside of the cell membrane, such as receptor binding sites, are orientated by the Golgi apparatus so that when they arrive at the membrane they are inserted facing in the right direction.

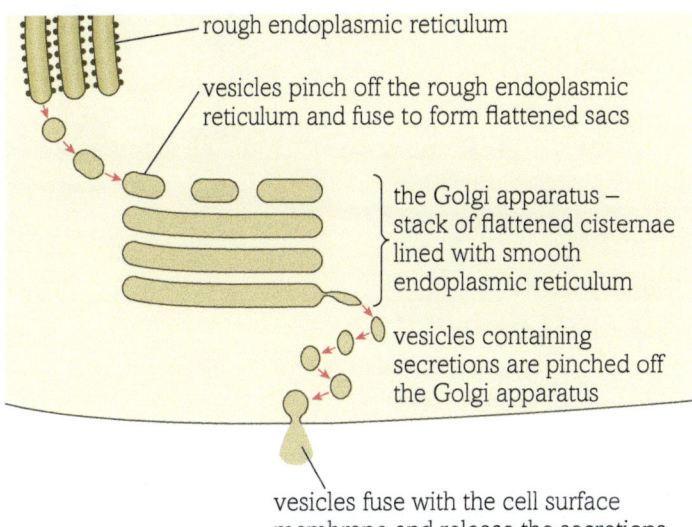

fig B The Golgi apparatus takes proteins from the RER, assembles and packages them and then transports them to where they are needed. This may be the surface of the cell or different regions inside it.

Lysosomes

Food taken into the cell of single-celled protists such as *Amoeba* must be broken down into simple chemicals that can then be used. Organelles in the cells of your body that are worn out need to be destroyed. These jobs are the function of the lysosomes. The word lysis, from which they get their name, means 'breaking down'. Lysosomes appear as dark, spherical bodies in the cytoplasm of most cells and they contain a powerful mix of digestive enzymes. They frequently fuse with each other and with a membrane-bound vacuole containing either food or an obsolete organelle. Their enzymes then break down the contents into molecules that can be reused. A lysosome may fuse with the outer cell membrane to release its enzymes outside the cell as extracellular enzymes, for example to destroy bacteria or in digestion.

Lysosomes can also self-destruct. If an entire cell is wearing out, needs to be removed during development, has a mutation or is under stress, its lysosomes may rupture, releasing their enzymes to destroy the entire contents of the cell. This programmed, controlled cell death is known as **apoptosis**.

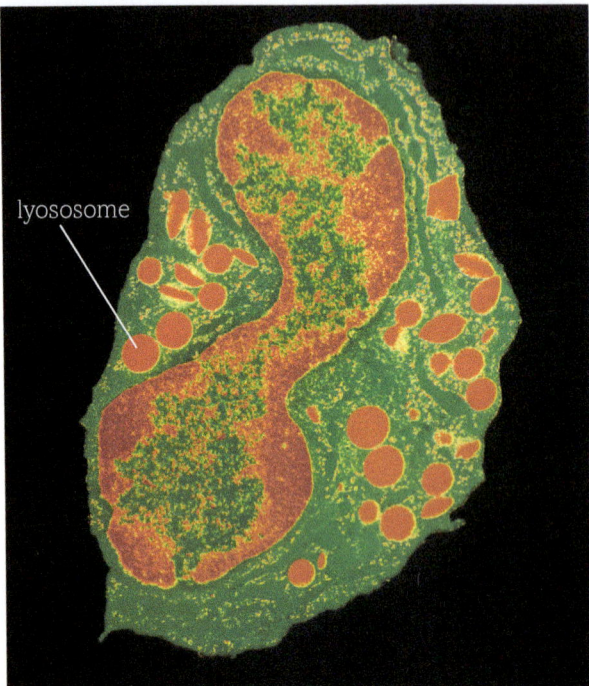

lyososome

fig C Good microscopic evidence of lysosomes helped scientists work out their functions in the cell.

Did you know?

Apoptosis and disease

Apoptosis or 'cell suicide' is vital to the maintenance of a healthy body. Lysosomes rupture and their enzymes are released to kill cells that are old and coming to the end of their healthy life, or cells that need to be removed during development, for example the webbing that initially forms between the fingers and toes of a fetus in the uterus. Lysosomes may also destroy cells in which the DNA replication system is not functioning properly. But if apoptosis stops working properly – if too many cells are destroyed, or not enough lysosomes rupture so that cell death no longer takes place – this can have serious consequences for your health. For example, cancer is often thought of as a disease of uncontrolled cell growth. But scientists are increasingly convinced that uncontrolled growth is not the whole story. Cancer cells also fail to die by apoptosis. As a result they propagate the genetic mutations that allow them to reproduce uncontrollably. Excessive apoptosis also causes problems. It leads to the damage seen in the heart after a heart attack, and is linked to the death of T killer cells in HIV/AIDS. This is covered in more detail in **Book 2**. The excessive rupturing of lysosomes may also be involved in autoimmune diseases such as rheumatoid arthritis, when cartilage tissue in joints self-destructs, and possibly in other conditions such as osteoporosis and retinitis pigmentosa.

Questions

1 What type of questions would scientists ask when they set out to investigate the functions of the endoplasmic reticulum, and how might they have set about finding the answers?

2 Describe the role of the RER and the Golgi body in the production of both intracellular and extracellular enzymes, and explain the importance of packaging products within a cell.

3 Why is it important that apoptosis does not occur more or less than it should? Investigate examples of diseases that are caused at least in part by apoptosis.

Key definitions

The **endoplasmic reticulum** is a 3D network of membrane-bound cavities in the cytoplasm that links to the nuclear membrane and makes up a large part of the cellular transport system as well as playing an important role in the synthesis of many different chemicals.

80S ribosomes are the main type of ribosome found in eukaryotic cells, consisting of ribosomal RNA and protein, made up of a 60S and 40S subunit. They are the site of protein synthesis.

70S ribosomes are the ribosome found in the mitochondria and chloroplasts of eukaryotic cells and in prokaryotic organisms.

Rough endoplasmic reticulum (RER) is endoplasmic reticulum that is covered in 80S ribosomes and which is involved in the production and transport of proteins.

Exocytosis is the energy-requiring process by which a vesicle fuses with the cell surface membrane so the contents are released to the outside of the cell.

Smooth endoplasmic reticulum (SER) is a smooth tubular structure similar to RER, but without the ribosomes, which is involved in the synthesis and transport of steroids and lipids in the cell.

Golgi apparatus consists of stacks of membranes that modify proteins made elsewhere in the cell and package them into vesicles for transport, and also produce materials for plant cell walls and insect cuticles.

A **lysosome** is an organelle full of digestive enzymes used to break down worn out cells or organelles, or digest food in simple organisms.

Apoptosis is cell suicide – the breakdown of worn out, damaged or diseased cells by the lysosomes.

By the end of this section, you should be able to...

● describe the ultrastructure of the cell wall in eukaryotic cells and relate its structure to its functions

Plants, like animals, are eukaryotes. A typical plant cell has many features in common with a typical animal cell (see **Sections 2.1.3** and **2.1.4**). They have many membranes and contain cytoplasm and a nucleus. Rough and smooth endoplasmic reticulum spread throughout the cytoplasm, along with active Golgi apparatus. Mitochondria produce ATP, which is as vital to the working of the plant cell as it is to the animal cell. However, there are several quite fundamental differences between plant and animal cells. They contain several kinds of organelle that are not found in animal cells, including permanent vacuoles and chloroplasts.

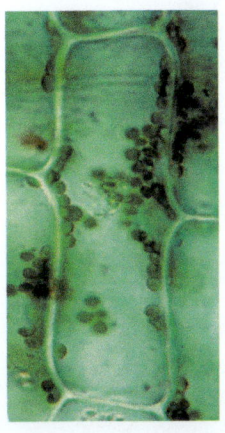
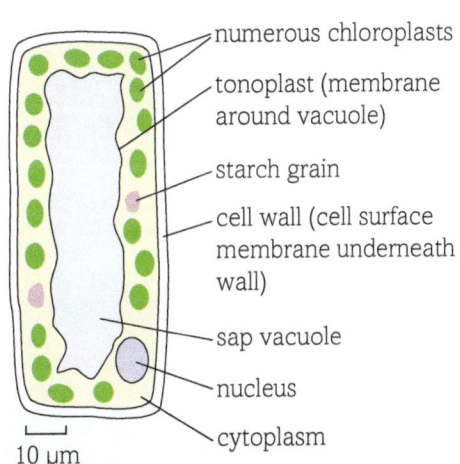

numerous chloroplasts

tonoplast (membrane around vacuole)

starch grain

cell wall (cell surface membrane underneath wall)

sap vacuole

nucleus

cytoplasm

10 μm

(a) a light micrograph and drawing of a plant cell ×250

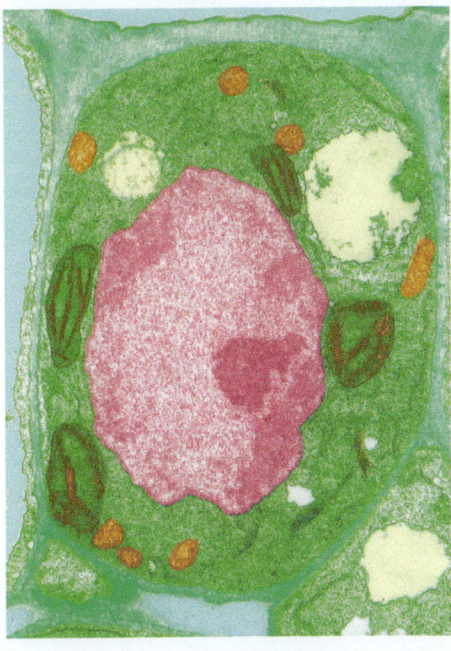
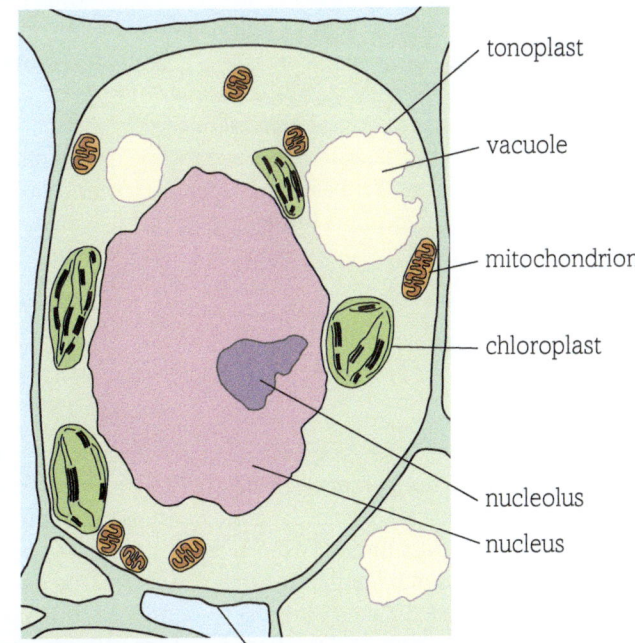

tonoplast

vacuole

mitochondrion

chloroplast

nucleolus

nucleus

cell wall (with cell surface membrane beneath)

(b) an electron micrograph and drawing of a plant cell ×5000

fig A The light microscope gives us the major features of a plant cell; the electron microscope reveals many more details.

The plant cell wall

Animal cells can be almost any shape. Plant cells tend to be more regular and uniform in their appearance. This is largely because each cell is bounded by a **cell wall**. You can visualise a plant cell as a jelly-filled balloon inside a shoe box. The cell wall (shoe box) is an important feature that gives plants their strength and support. It is made up largely of insoluble cellulose (see **Section 1.2.2**). The plant cell wall is usually freely permeable to everything that is dissolved in water – it does not act as a barrier to substances getting into the cell. However, the cell wall can become impregnated with **suberin** in cork tissues, or with **lignin** to produce wood. These compounds affect the permeability of the cell wall so that water and dissolved substances cannot pass through it.

fig B These cellulose microfibrils are made up of thousands of cellulose chains held together by hydrogen bonds. Their orientation and packing changes from primary to secondary cell walls, affecting both flexibility and strength.

The plant cell wall consists of several layers. The **middle lamella** is the first layer to form when a plant cell divides into two new cells. It is made largely of **pectin**, a polysaccharide that acts like glue and holds the cell walls of neighbouring plant cells together. Pectin has lots of negatively charged carboxyl (–COOH) groups and these combine with positive calcium ions to form calcium pectate. This binds to the cellulose that forms on either side. The cellulose microfibrils and the matrix build up on either side of the middle lamella. To begin with, these walls are very flexible, with the cellulose microfibrils all orientated in a similar direction. They are known as **primary cell walls**. As the plant ages, secondary thickening may take place. A **secondary cell wall** builds up, with the cellulose microfibrils laid densely at different angles to each other. This makes the composite material much more rigid. Hemicelluloses harden it further. In some plants, particularly woody perennials, lignin is then added to the cell walls to produce wood, which makes the structure even more rigid. Within the structure of a plant there are many long cells with cellulose cell walls that have been heavily lignified. These are known as **plant fibres** and people use them in many different ways including clothing, building material, ropes and paper.

Plasmodesmata

In spite of being encased in cellulose cell walls, plant cells seem to be in close communication with each other.

Intercellular exchanges seem to take place through special cytoplasmic bridges between the cells known as **plasmodesmata** (see **fig C**). The plasmodesmata appear to be produced as the cells divide – the two cells do not separate completely, and threads of cytoplasm remain between them. These threads pass through gaps in the newly formed cell walls and signalling substances can pass from one cell to another through the cytoplasm. The interconnected cytoplasm of the cells is known as the **symplast**. Scientists are still working hard to discover exactly how plant cells communicate through plasmodesmata. One clear piece of evidence showing that these intercellular junctions are vital in the life of plants comes from work with plant grafts. If we graft a rose onto a hardy root stock, the graft tissue only starts healthy cell division and growth once plasmodesmata bridges are established between the host tissue and the graft tissue.

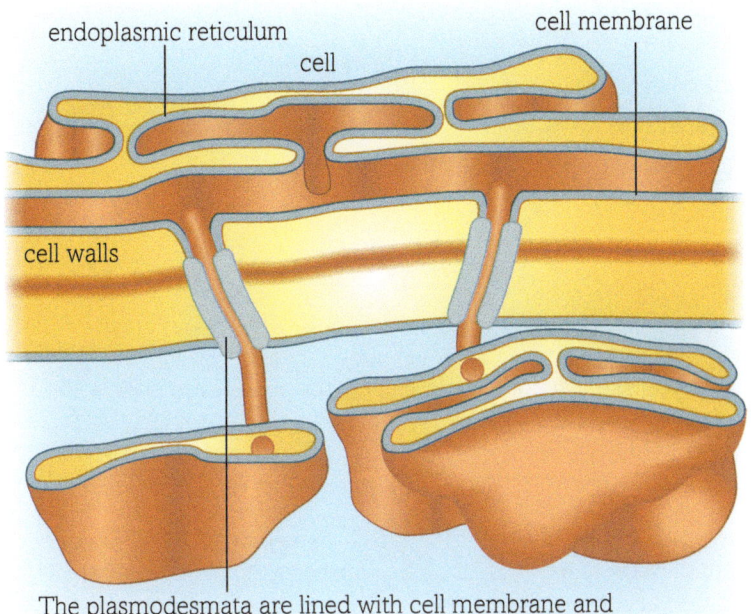

endoplasmic reticulum   cell membrane

cell

cell walls

The plasmodesmata are lined with cell membrane and molecules pass freely from cell to cell through these canals.

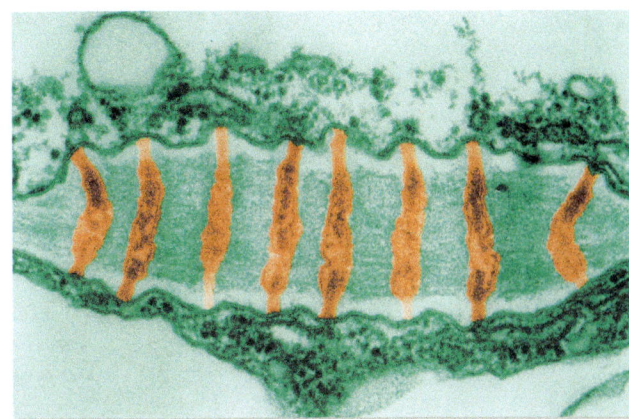

fig C Plasmodesmata provide a route for communication between plant cells, but scientists are still trying to find out exactly how it works.

Questions

1 What role do cell walls play in the structure of a plant, and how is their structure related to their function?

2 How does the plant cell wall change as the cell grows and develops, and how does this affect the cell?

3 Explain why plasmodesmata are an important feature of plant cell structure.

Key definitions

A **cell wall** is a freely permeable wall around plant cells, made mainly of cellulose.

Suberin is a chemical that impregnates cellulose cell walls in cork tissues and makes them impermeable.

Lignin is a chemical that impregnates cellulose cell walls in wood and makes it impermeable.

The **middle lamella** is the first layer of the plant cell wall to be formed when a plant cell divides, made mainly of calcium pectate (pectin) that binds the layers of cellulose together.

Pectin is a polysaccharide that holds cell walls of neighbouring plant cells together and is part of the structure of the primary cell wall.

The **primary cell wall** is the first very flexible plant cell wall to form, with all the cellulose microfibrils orientated in a similar direction.

The **secondary cell wall** is the older plant cell wall in which the cellulose microfibrils have built up at different angles to each other making the cell wall more rigid.

Plant fibres are long cells with cellulose cell walls that have been heavily lignified so they are rigid and very strong.

Plasmodesmata are cytoplasmic bridges between plant cells that allow communication between the cells.

The **symplast** is all of the material (cytoplasm, vacuole, etc.) contained within the surface membrane of a plant cell.

By the end of this section, you should be able to...

● describe the ultrastructure of the chloroplast, vacuole and tonoplast in eukaryotic cells, and relate these structures to their functions

Plant cells contain several kinds of organelle that are not found in animal cells. These include permanent vacuoles and chloroplasts.

Permanent vacuole

A vacuole is any fluid-filled space inside the cytoplasm surrounded by a membrane. Vacuoles occur quite frequently in animal cells, but they are only temporary, being formed and destroyed when needed. In non-woody plant cells the vacuole is a permanent structure with an important role. The vacuole can occupy up to 80% of the volume of a plant cell. It is surrounded by a specialised membrane called the **tonoplast**. The tonoplast has many different protein channels and carrier systems in it. It controls the movements of substances into and out of the vacuole and so controls the water potential of the cell. The vacuole is filled with **cell sap**, a solution of various substances in water. This solution causes water to move into the cell by **osmosis** (see **Section 4.1.3**), and as a result the cytoplasm is kept pressed against the cell wall. This in turn keeps the cells turgid (swollen) and the whole plant upright. The pressures that can be developed in this way are very large indeed. The pressure in a leaf cell can be up to 1500 kPa – in contrast, the pressure in a human artery when the heart is pumping blood out into the body is only 16 kPa.

As well as fulfilling the important role of maintaining the plant cell shape, the many different types of vacuoles in plants carry out a range of different functions. Vacuoles are used for the storage of a number of different substances. Many vacuoles store pigments; for example the betacyanin pigment of beetroot is normally stored in the vacuoles of the cells and does not leak out into the cytoplasm unless the root is cut. If the tissue is heated, the characteristics of the membrane around the vacuole will change and so pigment will leak out more rapidly. Vacuoles can store proteins in the cells of seeds and fruits, and in some plant cells they contain lytic enzymes and have a function rather like lysosomes in animal cells. Vacuoles often store waste products and other chemicals. For example, digitalis, a chemical found in foxgloves that can act both as a heart drug and a deadly poison, is stored in the vacuoles of the cells.

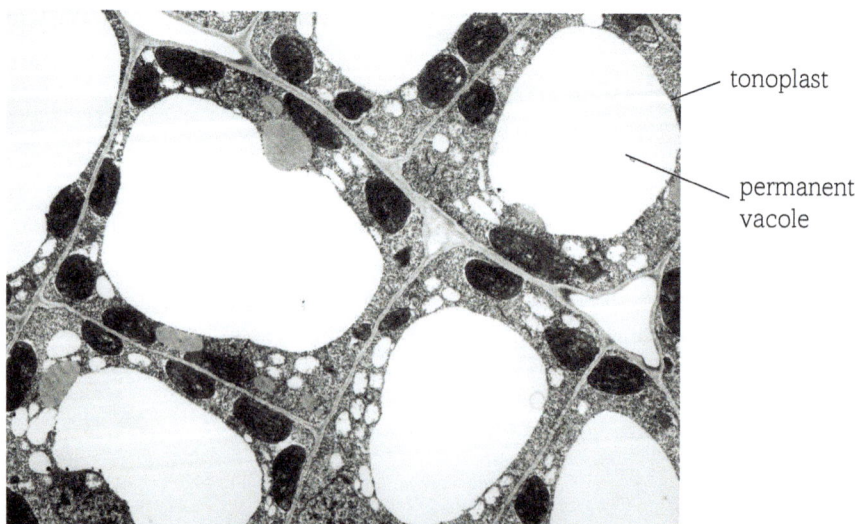

tonoplast

permanent vacole

fig A The tonoplast and the permanent vacuoles are key structures in the support systems of plants, but they have many other functions as well.

Chloroplasts

Of all the differences between plant and animal cells, the presence of **chloroplasts** in plant cells is probably the most important because they enable plants to make their own food. Not all plant cells contain chloroplasts – only those cells from the green parts of the plant. However, almost all plant cells contain the genetic information to make chloroplasts and so in some circumstances different areas of a plant will become green and start to photosynthesise. The exceptions are parasitic plants such as broomrape. Cells in flowers, seeds and roots contain no chloroplasts and neither do the internal cells of stems or the transport tissues. In fact the majority of plant cells do not have chloroplasts, but these organelles are very special and unique to plants.

(a)

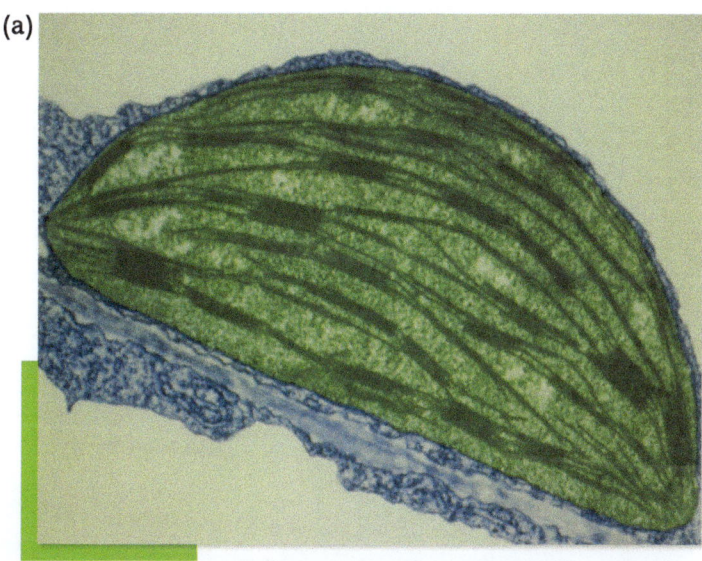

(b)

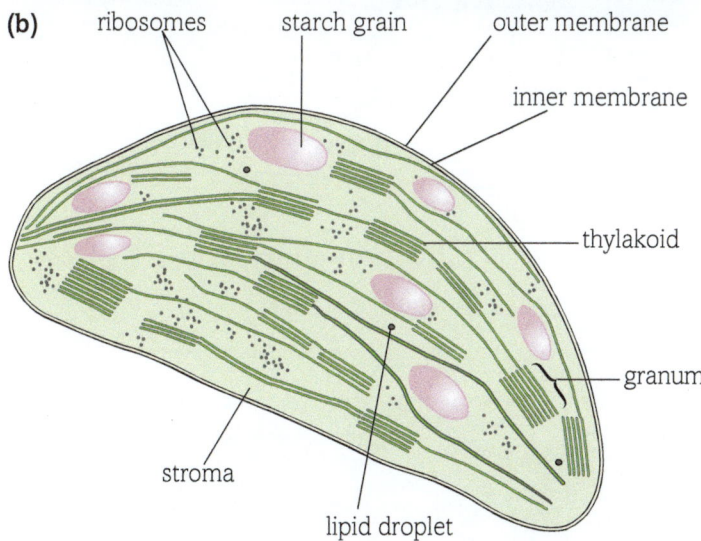

ribosomes   starch grain   outer membrane

inner membrane

thylakoid

granum

stroma

lipid droplet

fig B (a) Micrograph of a chloroplast; and (b) Labelled diagram to show structures in a chloroplast.

There are some clear similarities between chloroplasts and mitochondria. Like mitochondria, chloroplasts:

- are large organelles: they have a biconvex shape with a diameter of 4–10 μm and are 2–3 μm thick
- contain their own DNA

- are surrounded by an outer membrane
- have an enormously folded inner membrane that gives a greatly increased surface area on which enzyme-controlled reactions take place
- are thought to have been free-living prokaryotic organisms that were engulfed by and became part of other cells at least 2000 million years ago.

However, there are also some clear differences. Chloroplasts:

- are the site of photosynthesis
- contain **chlorophyll**, the green pigment that is largely responsible for trapping the energy from light, making it available for the plant to use
- are formed from a type of relatively unspecialised plant 'stem cell' known as a leucoplast.

Amyloplasts

Amyloplasts are another specialised plant organelle and, like chloroplasts, they develop from leucoplasts. They are colourless and store amylopectin, a polysaccharide combined with amylose to form starch (see **Section 1.2.2**). This can then be converted to glucose and used to provide energy when the cell needs it. Amyloplasts are found in large numbers in areas of a plant that store starch, for example potato tubers.

Questions

1 Amyloplasts and chloroplasts come from the same type of unspecialised cell. How do the two structures differ?

2 Compare and contrast the structure of a typical plant cell with the structure of a typical animal cell.

3 Explain why chloroplasts are found only in particular parts of a plant. Suggest what happens to make part of a plant, e.g. a potato tuber, turn green when exposed to light?

Key definitions

The **tonoplast** is the specialised membrane that surrounds the permanent vacuole in plant cells and controls movements of substances into and out of the cell sap.

Cell sap is the aqueous solution that fills the permanent vacuole.

Osmosis is a specialised form of diffusion that involves the movement of solvent molecules down a concentration gradient through a partially permeable membrane.

A **chloroplast** is an organelle adapted to carry out photosynthesis, containing the green pigment chlorophyll.

Chlorophyll is the green pigment that is largely responsible for trapping the energy from light, making it available for the plant to use in photosynthesis.

Amyloplasts are plant organelles that store amylopectin, a polysaccharide used to form starch.

By the end of this section, you should be able to...

● Explain how, in complex organisms, cells are organised into tissues, organs and organ systems

Multicellular organisms are made up of specialised cells but these cells do not operate on their own. The specialised cells are organised into groups of cells known as **tissues**. These tissues consist of one or more types of cells all carrying out a particular function in the body. However, tissues do not operate in isolation. Many tissues are further organised into **organs**.

Tissues

Tissues are groups of similar cells that all develop from the same kind of cell. Although there are many different types of specialised cells, there are only four main tissue types in the human body – epithelial tissue, connective tissue, muscle tissue and nervous tissue. Modified versions of these tissue types containing different specialised cells carry out all the functions of the body. **Fig A** shows some different **epithelial tissues**, which are tissues that form the lining of surfaces both inside and outside of the body. Although some epithelial tissues consist of more than one kind of cell, they all originate from the basement membrane. Cells in epithelial tissues usually sit tightly together and form a smooth surface that protects the cells and tissues below.

Squamous epithelium is commonly found lining the surfaces of blood vessels, and forms the walls of capillaries and the lining of the alveoli. Cuboidal and columnar cells line many other tubes in the body. Ciliated epithelia often contain goblet cells that produce mucus. These epithelia form the surfaces of tubes in the gas exchange system and the oviducts. The regular waving of the cilia from side to side moves materials along inside the tubes. Compound epithelia are found where the surface is continually scratched and abraded, such as the skin. The thickness of the tissue protects what lies beneath as new cells continue to grow from the basement membrane.

There are many other tissues in the body, including muscle tissue, nervous tissue, the collagen tissue and elastin tissue found in artery walls and the glandular tissue that secretes substances from inside the cells. Connective tissue is the main supporting tissue in the body, and includes bone tissue and cartilage tissue as well as packing tissue that supports and protects some of the organs. Some of these tissues are shown in **fig B**.

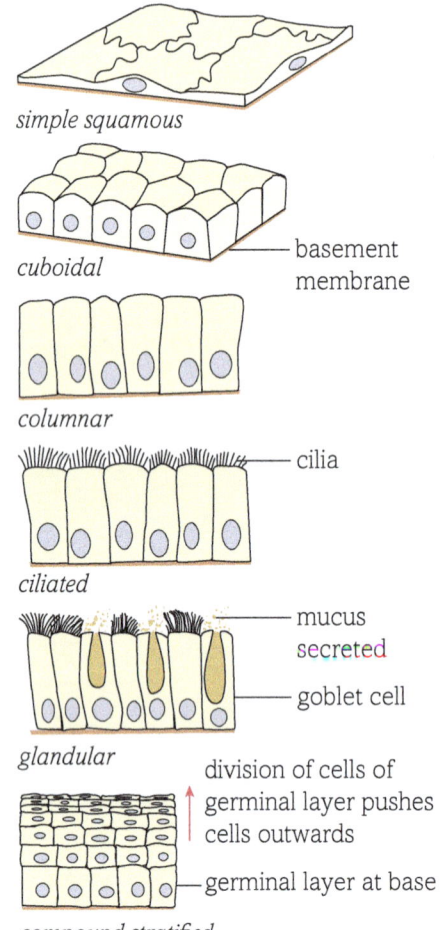

fig A There are many kinds of epithelial tissues inside the human body.

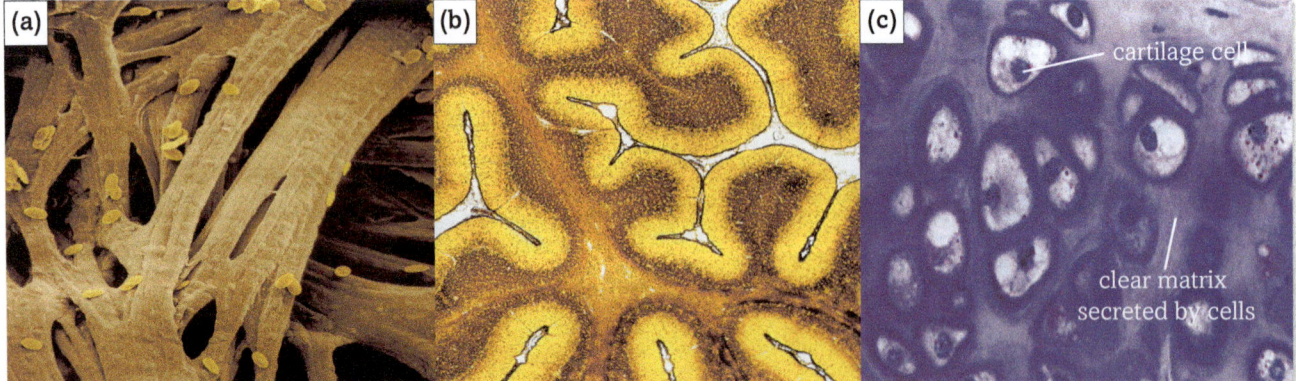

fig B (a) cardiac muscle tissue; (b) brain tissue; (c) cartilage tissue.

Organs

An organ is a structure made up of several different tissues grouped into a structure so that they can work effectively together to carry out a particular function. There are many organs in the human body, some of which are shown in **fig C**. Plants also have cells grouped into tissues and organs. For example the leaf is an organ that is composed of vascular tissue, epithelial tissue and mesophyll tissues as shown in **fig D**.

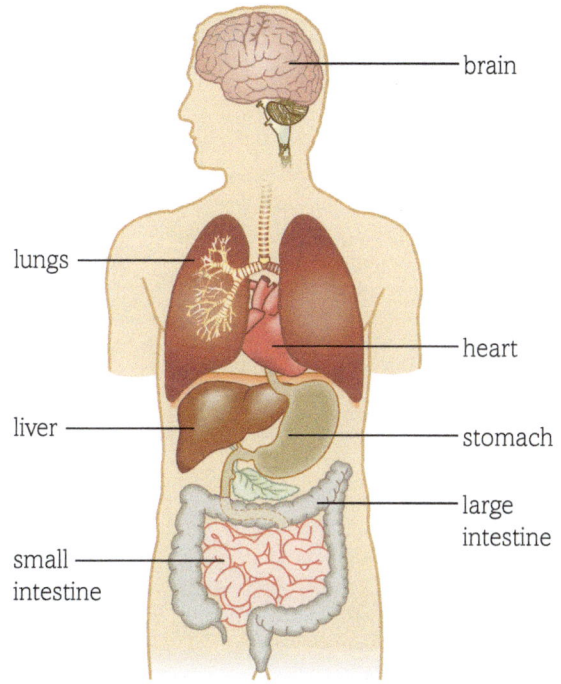

fig C Some of the organs and organ systems of the human body.

Systems

In animals, in many cases a number of organs work together as an **organ system** to carry out large-scale functions in the body. For example the digestive system includes the organs of the stomach, pancreas, small and large intestines, and the nervous system includes the brain, spinal cord and all peripheral nerves.

Most of the cells in tissues, organs and systems have differentiated during development so that they are capable of carrying out their specific function. You will find out more about how this process happens in **Book 2 Chapter 7.2**.

Questions

1 Explain how the structure of the following tissues is related to their function.
   (a) squamous epithelium lining an alveolus
   (b) ciliated epithelium lining a bronchus
   (c) muscle tissue in the biceps muscle.

2 (a) Choose one of the systems in the human body and describe briefly the cells, tissues and organs found within that system.
   (b) Explain why this grouping enables the system to carry out its function effectively.

Key definitions

A **tissue** is a group of specialised cells carrying out a particular function in the body.

An **organ** is a structure made up of several different types of tissues grouped together to carry out a particular function in the body.

Epithelial tissues are tissues that form the lining of surfaces inside and outside the body.

An **organ system** is a group of organs working together to carry out particular functions in the body.

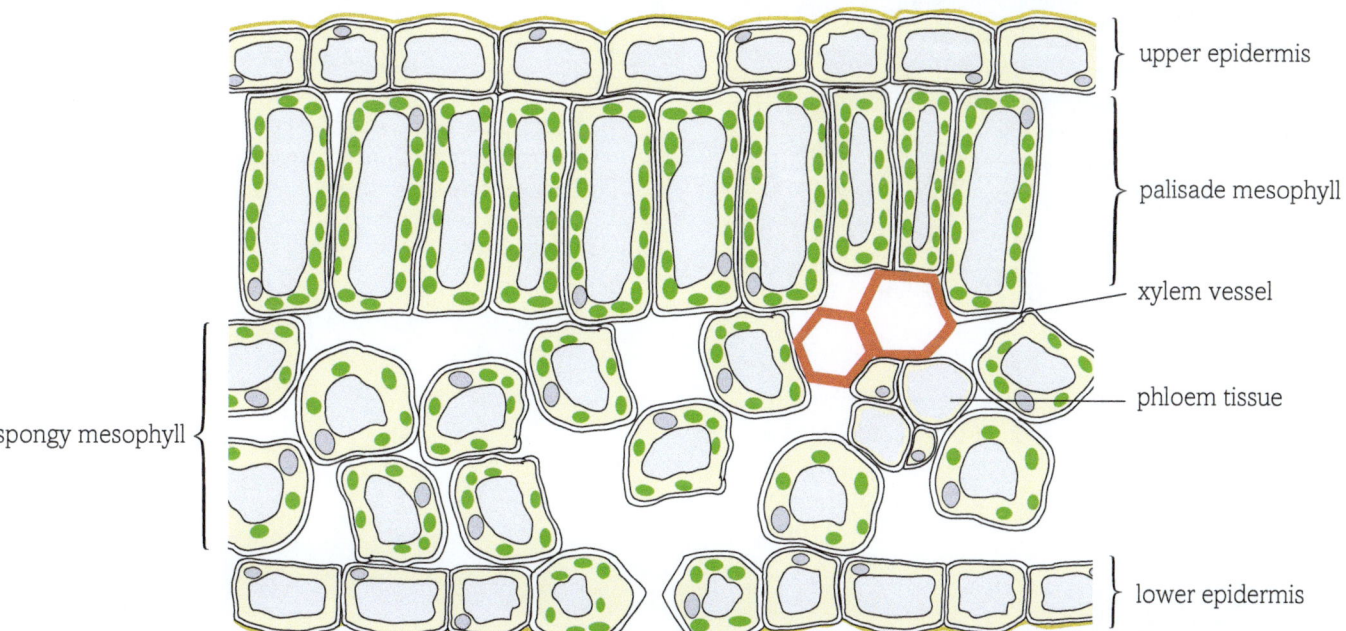

fig D Some of the tissues found in a leaf – the photosynthetic organ of a plant.

Exam-style questions

1 The photograph below shows a mitochondrion as seen using an electron microscope.

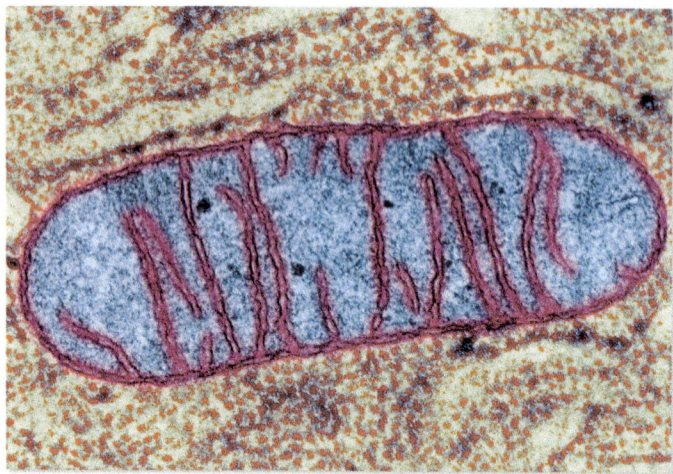

(a) Describe the role of mitochondria in a cell. [2]

(b) Make an accurate drawing of this mitochondrion enlarged ×2. On your drawing label the **matrix** and a **crista**. [4]

[Total: 6]

2 The photograph below shows a chloroplast as seen using an electron microscope.

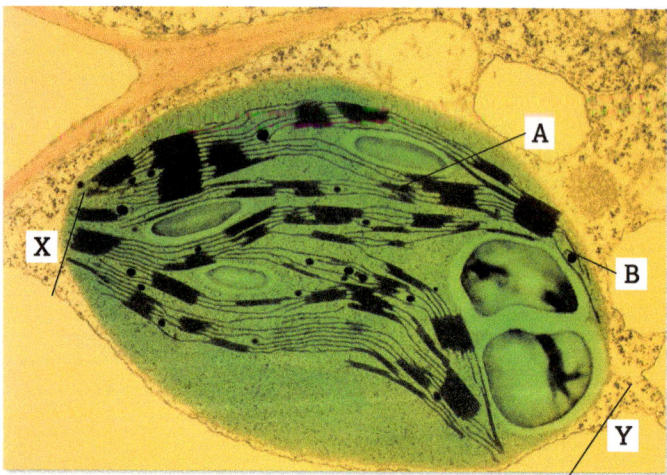

(a) Name the parts labelled **A** and **B**. [2]

(b) The actual length of the chloroplast between **X** and **Y** is 5 µm. Calculate the magnification of this chloroplast. Show your working. [3]

(c) Name **one** type of cell that contains chloroplasts. [1]

[Total: 6]

3 (a) Draw and label a diagram to show the structure of a chloroplast, as seen using an electron microscope. [4]

(b) The photograph below shows a group of mitochondria in a liver cell, as seen using an electron microscope. The magnification is ×50 000.

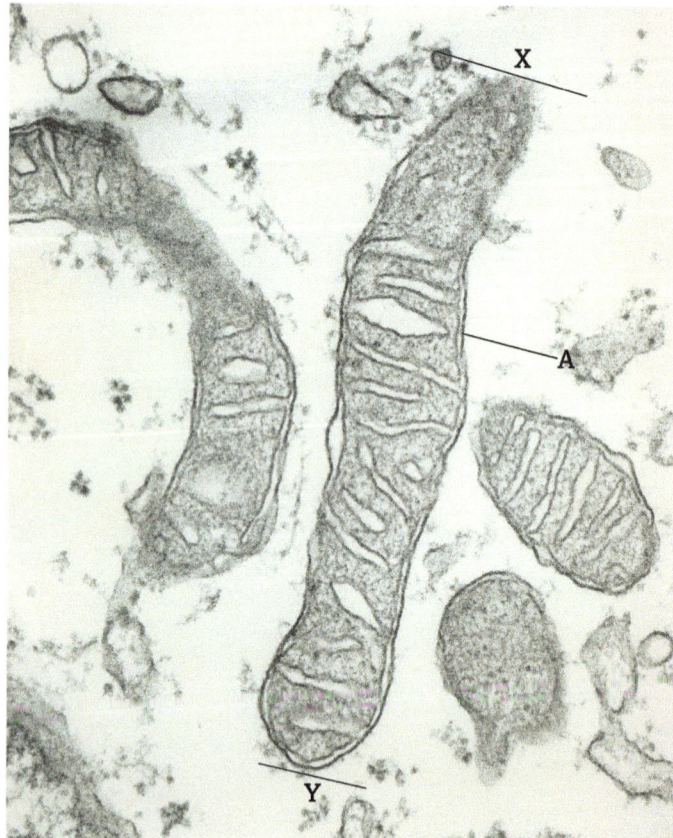

(i) Measure the length of the mitochondrion labelled **A** between **X** and **Y**. Calculate the actual length of this mitochondrion in µm. Show your working. [3]

(ii) Suggest **one** other structure that might be visible in the cytoplasm of this liver cell if the magnification used was higher. [1]

(iii) Suggest one reason why the double membrane is not clearly visible all around the mitochondrion labelled **A**. [1]

[Total: 9]

4 The table below refers to some cell structures. Complete the table by inserting the correct word, words, or diagram in the appropriate boxes. Leave the shaded grey boxes empty.

Name of cell structure	Description of cell structure	Diagram of cell structure
[1]	1 Darkly-stained region in the nucleus. 2 Where ribosomal RNA is made.	
Centrioles		[2]
Lysosome	1 2 [2]	
[1]	1 Hollow cylinders made of protein. 2 Form spindle fibres.	

[Total: 6]

5 The photograph below shows some onion cells as seen using the high power of a light microscope.

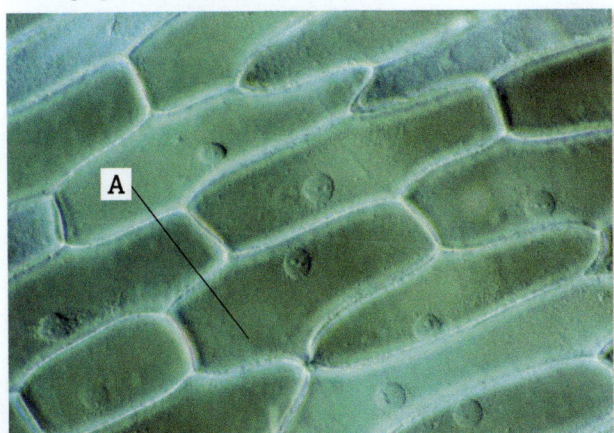

(a) Make an accurate drawing, enlarged ×2, of the cell labelled **A**.
Do **not** label your drawing. [3]

(b) All the onion cells have a cell surface (plasma) membrane. The diagram below shows the structure of this membrane.

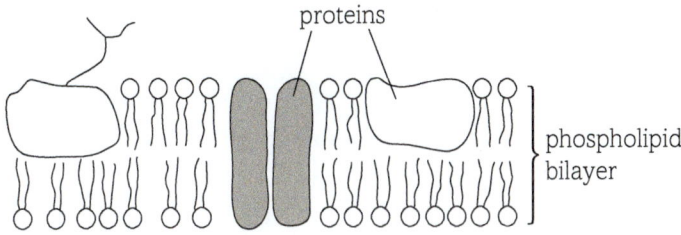

Explain how the properties of phospholipids result in the formation of a bilayer. [3]

[Total: 6]

6 The endosymbiotic theory received fresh impetus from a 1967 paper by Lynne Margulis, who offered evidence from microbiology.
(a) What is a symbiotic relationship? [1]
(b) What do chloroplasts and mitochondria have in common with prokaryotic organisms? [1]
(c) Both chloroplasts and mitochondria have double membranes. When a white blood cell engulfs bacteria it forms a vacuole around them. Suggest how this adds support to the endosymbiotic theory. [1]
(d) Further work by Lynn Margulis in 1981 argued that eukaryotic flagella came from bacteria called spirochaetes. Explain why this has not received much support. [1]
[Total: 4]

7 Scientists using light microscopes were unable to distinguish small organelles like ribosomes.
(a) Name **three** other organelles that were too small for these scientists to see. [3]
(b) Name **two** organelles that would have been seen using the light microscope. [2]
(c) Light microscopes can be used to watch substances move in real time. Suggest why this is impossible in an electron microscope. [2]
[Total: 7]

8 Outline the stages of the production of the primary and secondary cell walls in a perennial plant. [6]
[Total: 6]

CHAPTER

2.2 > Prokaryotic cells

Introduction

The human body contains around 10 times more bacterial cells than it does human cells – on a numbers basis, we are more prokaryotic than eukaryotic! Bacteria are found on our skin, in our nose and most of all in our digestive systems. But bacteria are so small that even the trillions found in the human body have an estimated mass of only between 0.9 and 2.7 kg, which is just 1–3% of the total body mass. In the Human Microbiome Project, scientists have sequenced the genomes of bacteria taken from healthy humans. They have identified around 10 000 species of bacteria in the human ecosystem alone. Many of these bacteria have a direct and positive effect on our health. For example, bacteria help us digest food and absorb nutrients in the digestive system.

The ultrastructure of prokaryotic cells differs from that of eukaryotic cells in a number of fundamental ways, and in this chapter you will be looking at the similarities and differences between these groups of organisms. Prokaryotes do not have membrane-bound organelles so the genetic material is free in the cytoplasm. They may have extra bits of genetic material called plasmids and the ribosomes are different in their chemical make-up. Bacterial cell walls are unique – and they vary considerably. You will discover how stains can be used to identify some of these differences, which are important in pathogenic bacteria.

You will also look at viruses. These are the ultimate parasites, taking over the genetic material of other organisms to replicate and make more viruses. Most naturally occurring viruses cause disease, so it is important to understand as much as we can about their structures and life cycles to help defend ourselves against them.

All the maths you need

- Make use of appropriate units (*e.g. relative sizes of eukaryotic cells, prokaryotic cells and viruses*)
- Carry out calculations using numbers in standard and ordinary form (*e.g. use of magnification*)
- Make order of magnitude calculations (*e.g. use and manipulate the magnification formula:*
  $$\text{magnification} = \frac{\text{size of image}}{\text{size of real object}})$$
- Use and manipulate equations, including changing the subject of an equation (*e.g. magnification*)

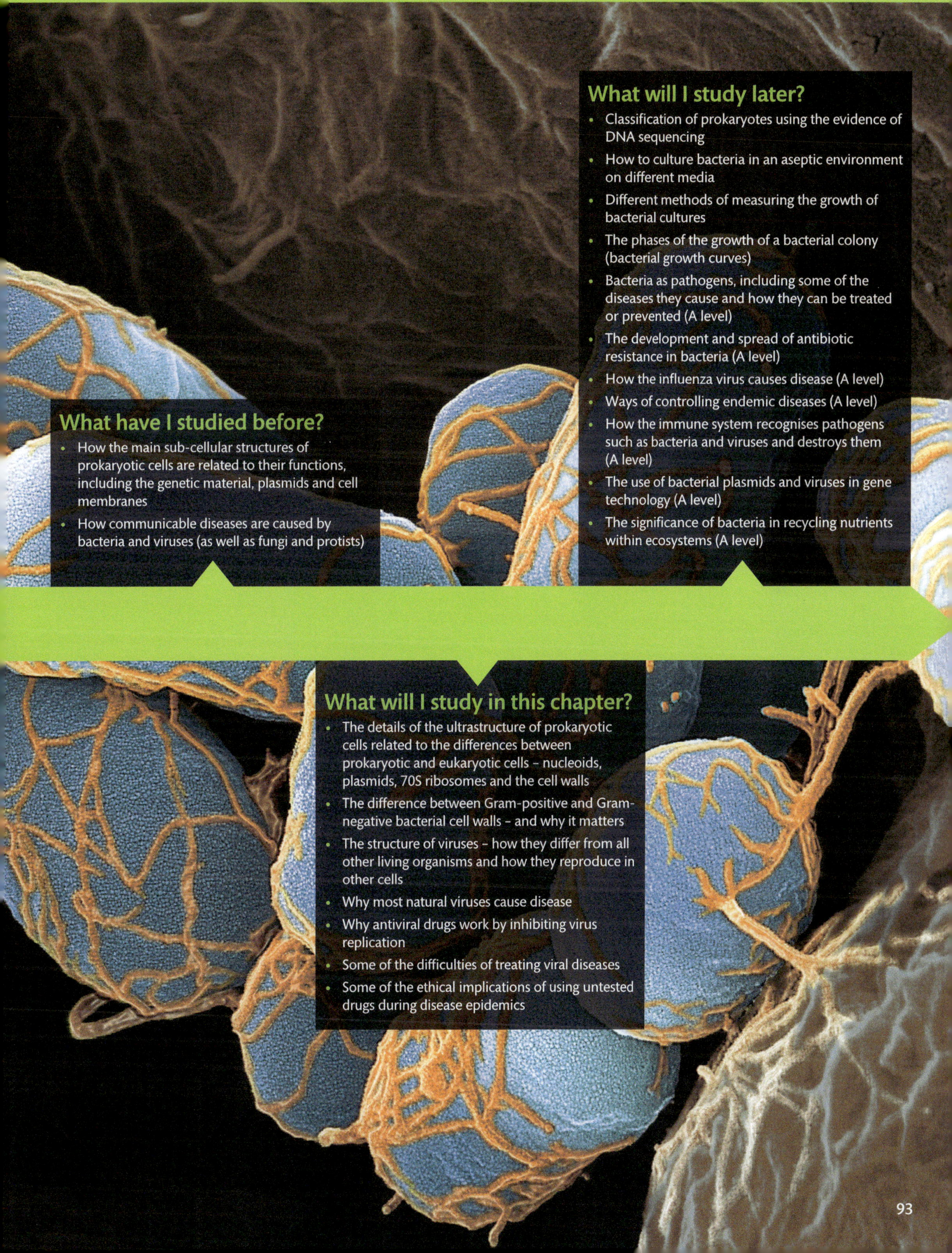

What will I study later?

- Classification of prokaryotes using the evidence of DNA sequencing
- How to culture bacteria in an aseptic environment on different media
- Different methods of measuring the growth of bacterial cultures
- The phases of the growth of a bacterial colony (bacterial growth curves)
- Bacteria as pathogens, including some of the diseases they cause and how they can be treated or prevented (A level)
- The development and spread of antibiotic resistance in bacteria (A level)
- How the influenza virus causes disease (A level)
- Ways of controlling endemic diseases (A level)
- How the immune system recognises pathogens such as bacteria and viruses and destroys them (A level)
- The use of bacterial plasmids and viruses in gene technology (A level)
- The significance of bacteria in recycling nutrients within ecosystems (A level)

What have I studied before?

- How the main sub-cellular structures of prokaryotic cells are related to their functions, including the genetic material, plasmids and cell membranes
- How communicable diseases are caused by bacteria and viruses (as well as fungi and protists)

What will I study in this chapter?

- The details of the ultrastructure of prokaryotic cells related to the differences between prokaryotic and eukaryotic cells – nucleoids, plasmids, 70S ribosomes and the cell walls
- The difference between Gram-positive and Gram-negative bacterial cell walls – and why it matters
- The structure of viruses – how they differ from all other living organisms and how they reproduce in other cells
- Why most natural viruses cause disease
- Why antiviral drugs work by inhibiting virus replication
- Some of the difficulties of treating viral diseases
- Some of the ethical implications of using untested drugs during disease epidemics

By the end of this section, you should be able to...

- describe the ultrastructure of prokaryotic cells and their organelles including nucleoid, plasmids, 70S ribosomes and the cell wall

- distinguish between Gram-positive and Gram-negative bacterial cell walls and explain why each type responds differently to some antibiotics

Bacteria, cyanobacteria and archaebacteria are prokaryotic organisms. Bacteria alone are probably the most common form of life on Earth. Some bacteria are pathogens and cause disease, but the great majority do no harm and many are beneficial to living organisms, for example as gut bacteria and in the cycling of nutrients in the natural world. (see **Book 2 Chapter 10.2**). In this section, you will mainly consider the structure and functions of bacterial cells.

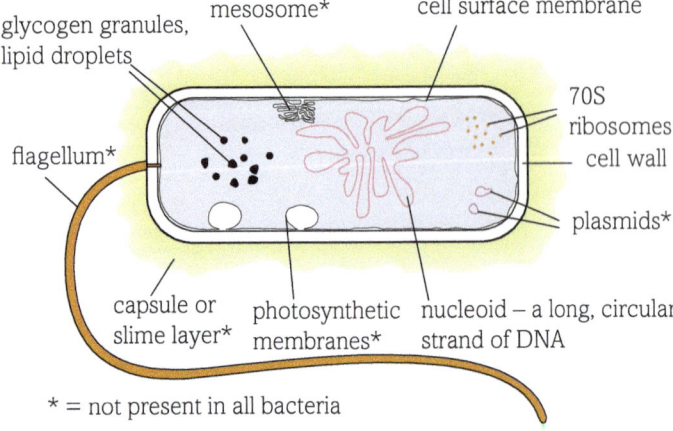

glycogen granules, lipid droplets
mesosome*
cell surface membrane
70S ribosomes
cell wall
flagellum*
plasmids*
capsule or slime layer*
photosynthetic membranes*
nucleoid – a long, circular strand of DNA

* = not present in all bacteria

fig A Structure of a typical bacterium.

The structure of bacteria

All bacterial cells have certain features in common, although these vary greatly between species.

Bacterial cell walls

All bacterial cells have a cell wall. The contents of bacterial cells are usually **hypertonic** to the medium around them, so water tends to move into the cells by osmosis. The cell wall prevents the cell swelling and bursting. It also maintains the shape of the bacterium, and gives support and protection to the contents of the cell. All bacterial cell walls consist of a layer of **peptidoglycan** that is made up of many parallel polysaccharide chains with short peptide cross-linkages forming an enormous molecule with a net-like structure. Some bacteria have a capsule (or slime layer if it is very thin and diffuse) around their cell walls. This may be formed from starch, gelatin, protein or glycolipid, and protects the bacterium from phagocytosis by white blood cells. It also covers the cell markers on the cell membrane that identify the cell. So a capsule can make it easier for a bacterium to be pathogenic (to cause disease) because it is not so easily identified by the immune

system. This is the case for the bacteria that cause pneumonia, meningitis, tuberculosis (TB) and septicaemia. However, many capsulated bacteria do not cause disease. It seems likely that capsules evolved to help the bacteria survive very dry conditions.

Pili and flagellae

Some bacteria have from one to several hundred thread-like protein projections from their surface. These are called the **pili** (or **fimbriae**) and they are found on some well-known bacteria such as *Escherichia coli* (*E. coli*) and *Salmonella* spp. They seem to be used for attachment to a host cell and for sexual reproduction. However, they also make bacteria more vulnerable to virus infections, as a **bacteriophage** can use pili as an entry point to the cell.

Some bacteria can move themselves using **flagella**. These are little bigger than one of the microtubules contained in a eukaryotic flagellum, and are made of a many-stranded helix of the protein flagellin. The flagellum moves the bacterium by rapid rotations – about 100 revolutions per second.

Cell surface membrane

The cell surface membrane in prokaryotes is similar in both structure and function to the membranes of eukaryotic cells. However, bacteria have no mitochondria so the cell membrane is also the site of some of the respiratory enzymes. In some bacterial cells such as *Bacillus subtilis*, a common soil bacterium, the membrane shows infoldings known as **mesosomes**. There is still some debate about their function. Some scientists think they may be an artefact from the process of preparing the cell for an electron micrograph, others believe they are associated with enzyme activity, particularly during the separation of DNA and the formation of new cross walls during replication. It appears that other infoldings of the bacterial cell surface membrane may be used for photosynthesis by some bacterial species.

Plasmid

Some bacterial cells also contain one or more much smaller circles of DNA known as **plasmids**. A plasmid codes for a particular aspect of the bacterial phenotype in addition to the genetic information in the nucleoid, for example the production of a particular toxin or resistance to a particular antibiotic. Plasmids can reproduce themselves independently of the nucleoid. They can be transferred from one bacterium to another in a form of sexual reproduction using the pili.

Nucleoid

The genetic material of prokaryotic cells consists of a single length of DNA, often circular, which is not contained in a membrane-bound nucleus. However, the DNA is folded and coiled to fit into the bacterium. The area in the bacterial cell where this DNA tangle is found is known as the **nucleoid**. In an *E.coli* bacterium it takes about half of the cytoplasm of the bacterium.

70S ribosomes

The bacteria, cyanobacteria and archaebacteria have no membrane-bound organelles, but they do have ribosomes where they carry out protein synthesis. The ribosomes in bacterial cells are 70S, smaller than the 80S ribosomes in eukaryotes. They have two subunits. The smaller is 30S and the larger is 50S (see **Section 2.1.4**). They are involved in the synthesis of proteins in a similar way to eukaryotic ribosomes.

Gram staining and bacterial cell walls

Whilst all bacterial cell walls contain peptidoglycan, there are in fact two main types of bacterial cell wall. These can be distinguished by **Gram staining**, a staining technique developed by Christian Gram (1853–1938) in 1884 and still in use today. It is valuable because different types of bacteria are vulnerable to different types of antibiotics and one of the factors that affects their vulnerability is the type of cell wall.

Before staining, bacteria are often colourless. The cell walls of **Gram-positive bacteria** (e.g. methicillin-resistant *Staphylococcus aureus*, MRSA) have a thick layer of peptidoglycan containing chemicals such as **teichoic acid** within their net-like structure. The crystal violet/iodine complex in the Gram stain is trapped in the thick peptidoglycan layer and resists decolouring when the bacteria are dehydrated using alcohol. As a result it does not pick up the red safranin counterstain, leaving the positive purple/blue colour.

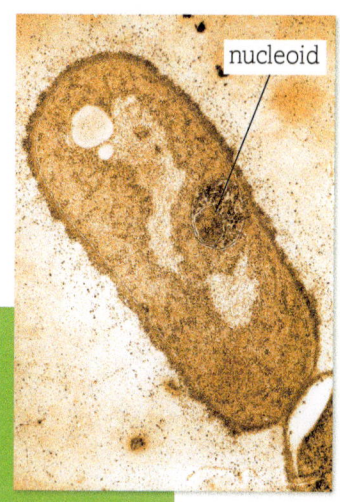

fig B The nucleiod area of a bacterium

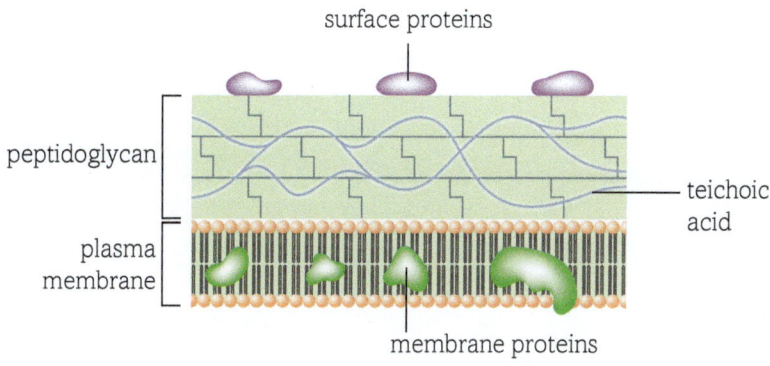

Gram-positive bacterial cell walls

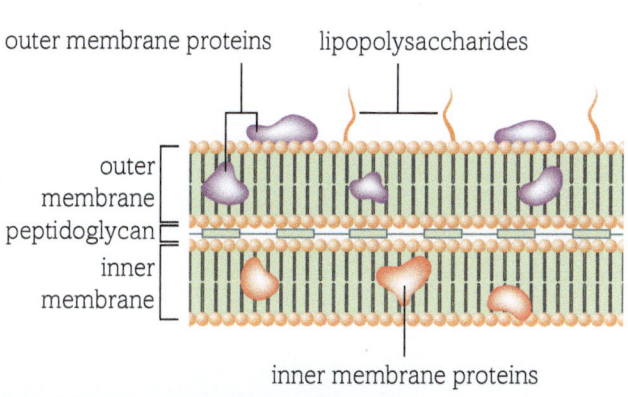

Gram-negative bacterial cell walls

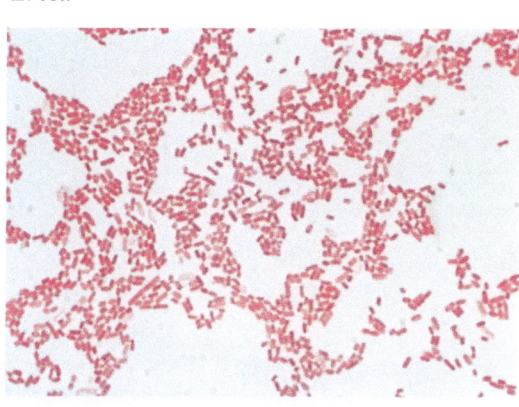

fig C The difference in the cell wall structure of the bacteria results in the different reactions with the Gram stain.

The cell walls of **Gram-negative bacteria** have a thin layer of peptidoglycan with no teichoic acid between the two layers of membranes. The outer membrane is made up of lipopolysaccharides. After the crystal violet/iodine complex is applied, the bacteria are dehydrated in ethanol. The lipopolysaccharide layer dissolves in the ethanol leaving the thin peptidoglycan layer exposed. The crystal violet/iodine complex is washed out and the peptidoglycan takes up the red safranin counterstain, so the cells appear red.

Antibiotics and bacterial cell walls

Antibiotics are drugs that are used against bacterial pathogens. There are a number of different types of antibiotics, each working in different ways. They may work by affecting the bacterial cell walls, the cell membranes, the genetic material, the enzymes or the ribosomes. Antibiotics usually target features of bacterial cells that differ to those of eukaryotic cells, including the bacterial cell walls and the 70S ribosomes.

Different types of bacteria are sensitive to different types of antibiotics. Doctors need to know if a pathogenic bacterium is Gram-positive or Gram-negative as this will affect the choice of antibiotic used to treat the disease.

To pinpoint the actions of an antibiotic, first think about the difference between human cells and bacterial cells, and then about the differences between Gram-positive and Gram-negative bacteria.

Some antibiotics, such as beta-lactam antibiotics (penicillins and cephalosporins), inhibit the formation of the peptidoglycan layer of the bacterial cell wall. As a result they are very effective against Gram-positive bacteria, as they have a thick peptidoglycan layer on the surface of the cell, but less effective against Gram-negative bacteria, as their peptidoglycan layer is hidden and less vital to the wall structure. They don't affect human cells as they don't have a peptidogycan cell wall at all.

Glycopeptide antibiotics, such as vancomycin, are large polar molecules that cannot penetrate the outer membrane layer of Gram-negative bacteria. However, they are very effective against Gram-positive bacteria, even ones that have developed resistance to many other antibiotics.

Polypeptide antibiotics, such as polymixins, are rarely used as they can have serious side effects. They are very effective against Gram-negative bacteria because they interact with the phospholipids of the outer membrane. They do not affect Gram-positive bacteria.

Most other antibiotics affect both Gram-positive and Gram-negative bacteria because they target common processes such as protein synthesis by the ribosomes. They only target prokaryote ribosomes, not eukaryotic ones.

If you are studying A level Biology, you will learn more about antibiotics in Book 2.

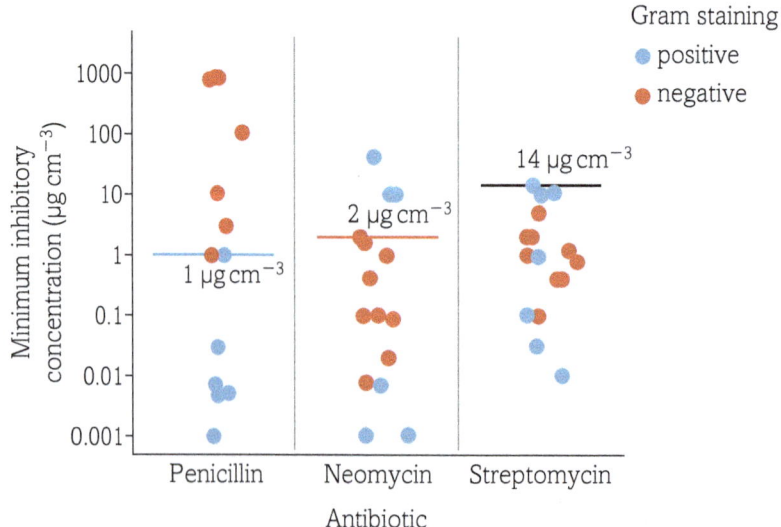

fig D This graph shows that penicillin is effective against Gram-positive bacteria as all the blue circles are on or below 1μg cm⁻³, so all types of Gram-positive bacteria are killed at a relatively low dose. Neomycin is best for Gram-negative bacteria as all the red circles are below 2 μg cm⁻³. Streptomycin is the antibiotic to choose if the Gram status of the bacteria is unknown because it kills all types of bacteria at a dose of only 14 μg cm⁻³.

Alternative ways of classifying bacteria

Grouping bacteria simply by the way their cell walls do or do not take up Gram stains is of limited use in classifying the different types. Another way in which bacteria can be identified is by their shape. Some bacteria are spherical (**cocci**) while the **bacilli** are rod-shaped. Yet others are twisted (**spirilla**) or comma-shaped (**vibrios**).

Bacteria are also sometimes grouped by their respiratory requirements. **Obligate aerobes** need oxygen for respiration. **Facultative anaerobes** use oxygen if it is available, but can manage without it. Many human pathogens fall into this group. **Obligate anaerobes** can only respire in the absence of oxygen – in fact oxygen will kill them.

Questions

1 Make a table to compare and contrast prokaryotic and eukaryotic cells.

2 What is the difference in the structure of the walls of Gram-positive and Gram-negative bacteria?

3 (a) How does the structure of the walls of Gram-positive and Gram-negative bacteria affect the effectiveness of some antibiotics?

   (b) Using the data in **fig D** suggest why streptomycin would be the best of these three antibiotics to use if you did not know whether a bacterial pathogen was Gram-positive or Gram-negative.

Key definitions

A **hypertonic solution** is a solution with a higher concentration of solutes and lower concentration of water (solvent) than the surrounding solution.

Peptidoglycan is a large, net-like molecule found in all bacterial cell walls made up of many parallel polysaccharide chains with short peptide cross-linkages.

Pili (fimbriae) are thread-like protein projections found on the surface of some bacteria.

Bacteriophages are viruses that attack bacteria.

Flagella are many-stranded helices of the contractile protein flagellin found on some bacteria. They move the bacteria by rapid rotations.

Mesosomes are infoldings of the cell membrane of bacteria.

A **nucleoid** is the area in a bacterium where we find the single length of coiled DNA.

Plasmids are small, circular pieces of DNA that code for specific aspects of the bacterial phenotype.

Gram staining is a staining technique used to distinguish types of bacteria by their cell wall.

Gram-positive bacteria are bacteria that contain teichoic acid in their cell walls and stain purple/blue with Gram staining.

Teichoic acid is a chemical found in the cell walls of Gram-positive bacteria.

Gram-negative bacteria are bacteria that have no teichoic acid in their cell walls. They stain red with Gram staining.

Cocci are spherical bacteria.

Bacilli are rod-shaped bacteria.

Spirilla are bacteria with a twisted or spiral shape.

Vibrios are comma-shaped bacteria.

Obligate aerobes are organisms that need oxygen for respiration.

Facultative anaerobes are organisms that use oxygen if it is available, but can respire and survive without it.

Obligate anaerobes are organisms that can only respire in the absence of oxygen and are killed by oxygen.

By the end of this section, you should be able to...

- recognise that viruses are not living cells

- explain the classification of viruses based on structure and nucleic acid types as illustrated by λ (lambda) phage (DNA), tobacco mosaic virus and Ebola (RNA) and human immunodeficiency virus (RNA retrovirus)

- describe the lytic cycle of a virus and explain latency

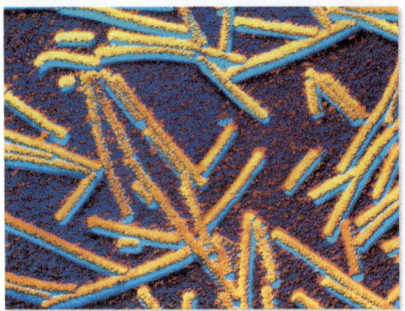

fig A The tiny rod-shaped particles of the tobacco mosaic virus, seen here under the scanning electron microscope, can cause serious damage to a crop.

Viruses are the smallest of all the microorganisms, and range in size from 0.02 μm to 0.3 μm across, about 50 times smaller than the average bacterium. Viruses are not cells. They are arrangements of genetic material and protein that invade other living cells and take over their biochemistry to make more viruses. It is because of this reproduction and the fact that they change and evolve in an adaptive way, that they are classed as living organisms.

Viruses

Most scientists working on viruses class them as obligate intracellular parasites, meaning they can only exist and reproduce as parasites in the cells of other living organisms. Because natural viruses invade and take over living cells to reproduce, they usually all cause damage and disease of some sort. They can withstand drying and long periods of storage whilst maintaining their ability to infect cells. There are very few drugs that have any effect on viruses, and those that do only work in very specific instances; for example, acyclovir can help prevent herpes (cold sores) and genital herpes.

> **Did you know?**
>
> **Discovering viruses**
>
> People suspected the presence of viruses causing disease in the late nineteenth century. They were developed as a model to explain the way certain diseases were passed from one individual to another, but it was not until 1935 that the first virus was identified by Wendell Stanley (1904–71).
>
> The leaves of tobacco plants are prone to an unpleasant blotchy disease that has a devastating effect on the plants, and no-one could find the cause. Stanley pressed the juice from around 1300 kg of diseased tobacco leaves. After extraction and purification, this produced pure, needle-like crystals which, if dissolved in water and painted onto tobacco leaves, caused the symptoms of the disease. The particles were called tobacco mosaic virus (TMV). It was obvious that the crystals were not living in the usual sense of the word, yet they retained the ability to cause disease. Viruses cannot be seen using a light microscope because they are usually smaller than half a wavelength of light. With the development of the electron microscope the TMV particles were shown to be rod-like structures with a protein coat formed around a core of RNA.

The structure of viruses

Viruses usually have geometric shapes and similar basic structures. However, there is considerable variation in the genetic material they possess, the structure of their protein coat and whether or not they have an **envelope**. The protein coat or **capsid** is made up of simple repeating protein units known as **capsomeres**, arranged in different ways. Using repeating units minimises the amount of genetic material needed to code for coat production. It also makes sure that assembling the protein coat in the host cell is as simple as possible. In some viruses the genetic material and protein coat is covered by a lipid envelope, produced from the host cell. The presence of the envelope makes it easier for the viruses to pass from cell to cell, but it does make them vulnerable to substances such as ether, which will dissolve the lipid membrane.

Classifying viruses

Viruses attach to their host cells by means of specific proteins (antigens) known as **virus attachment particles (VAPs)** that target proteins in the host cell surface membrane. Because they respond to particular molecules of the host cell surface, viruses are often quite specific in the tissue they attack.

Viruses are classified by their genome and their mode of replication. Viral genetic material can be DNA or RNA, and the nucleic acid is sometimes double-stranded and sometimes single. The way in which the viral genetic material is used in the host cell to make new viruses depends on which form it is in:

- **DNA viruses**: In these viruses the genetic material is DNA. The viral DNA acts directly as a template for both new viral DNA and for the mRNAs needed to induce synthesis of viral proteins. Examples of DNA viruses include the smallpox virus, adenoviruses, which cause colds, and bacteriophages, viruses which infect bacteria, for example the λ (lambda) phage in **fig B**.

- **RNA viruses**: 70% of viruses have RNA as their genetic material and they are much more likely to mutate than DNA viruses. RNA viruses do not produce DNA as part of their life cycle. The majority of RNA viruses contain a single strand of RNA and are know as ssRNA viruses. Positive ssRNA viruses (also known as positive-sense ssRNA viruses) have

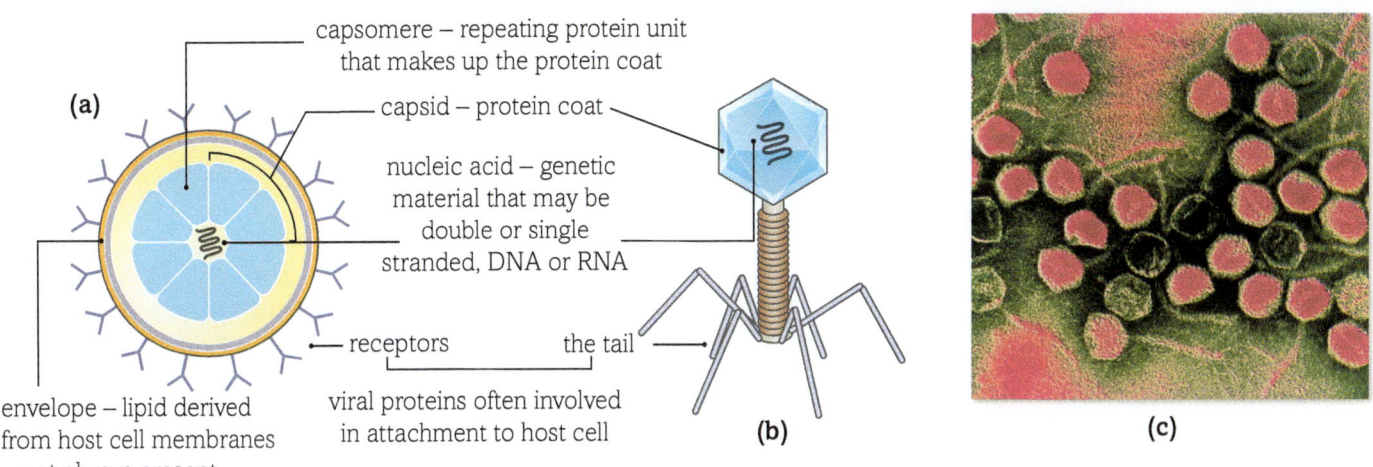

capsomere – repeating protein unit that makes up the protein coat

capsid – protein coat

nucleic acid – genetic material that may be double or single stranded, DNA or RNA

receptors   the tail

envelope – lipid derived from host cell membranes – not always present

viral proteins often involved in attachment to host cell

(b)

(c)

fig B General viral structures: (a) RNA virus; (b) λ (lambda) phage; (c) electron micrograph of λ (lambda) bacteriophages.

RNA that can act directly as mRNA and be translated at the ribosomes. Examples of plant and animal diseases caused by positive ssRNA viruses include tobacco mosaic viruses, SARS, polio and hepatitis C. Negative ssRNA viruses (also known as negative-sense ssRNA viruses) cannot be directly translated. The RNA strand must be transcribed before it is translated at the ribosomes. Examples of diseases caused by negative ssRNA viruses include measles, influenza and Ebola.

- **RNA retroviruses**: Retroviruses are a special type of RNA virus. They have a protein capsid and a lipid envelope. The single strand of viral RNA directs the synthesis of a special enzyme called **reverse transcriptase**. This goes on to make DNA molecules corresponding to the viral genome. This DNA is then incorporated into the host cell DNA and used as a template for new viral proteins and ultimately a new viral RNA genome. HIV (human immunodeficiency virus) is a retrovirus and some forms of leukaemia are also caused by this type of virus.

How viruses reproduce

Natural viruses all cause disease, and they attack every other known type of living organism. There are even viruses that attack bacteria, known as bacteriophages. We are constantly involved in a battle against the viruses that cause disease in ourselves, our animals, our crops and our environment. In order to understand how viruses cause damage to the body, and to be able to try to target drugs effectively, it is important to understand how they reproduce in the human body.

Virus 'life cycles'

Viruses only reproduce within the cells of the body. They attack their host cells in a number of different ways. For example, bacteriophages inject their genome into the host cell, but the bulk of the viral material remains outside the bacterium. The viral DNA forms a circle or plasmid within the bacterium.

The viruses that infect animals get into the cells in several ways. Some types are taken into the cell by endocytosis – either with or without the envelope – and the host cell then digests the capsid, releasing the viral genetic material. Most commonly, the viral envelope fuses with the host cell surface, releasing the rest of the

virus into the cell membrane. Plant viruses usually get into the plant cell using a vector, often using an insect, such as an aphid, to pierce the cell wall.

DNA virus replication

Once a virus is in the host cell there are two different routes of infection:

Latency – the lysogenic pathway

Many DNA viruses are **non-virulent** when they first get into the host cell. They insert their DNA into the host DNA so it is replicated every time the host cell divides. This DNA inserted into the host is called a **provirus**. Messenger RNA is not produced from the viral DNA because one of the viral genes causes the production of a repressor protein that makes it impossible to translate the rest of the viral genetic material. The virus does not affect the host cell or make the host organism ill at this stage in the life cycle. During this period of **lysogeny**, when the virus is part of the reproducing host cells, the virus is said to be **latent**.

The lytic pathway

Sometimes the viral genetic material is replicated independently of the host DNA straight after entering the host. Mature viruses are made and eventually the host cell bursts, releasing large numbers of new virus particles to invade other cells. The virus is said to be **virulent** (disease causing) and the process of replicating and killing cells is known as the lytic pathway. Under certain conditions, such as when the host is damaged, viruses in the lysogenic state are activated. The amount of repressor protein decreases and the viruses enter the lytic pathway and become virulent (see **fig C**).

Some types of virus have both latent and lytic stages in their life cycle, but others move straight to the lytic stage after they have infected a cell.

RNA virus replication

There are a number of different types of RNA viruses and they replicate themselves in different ways.

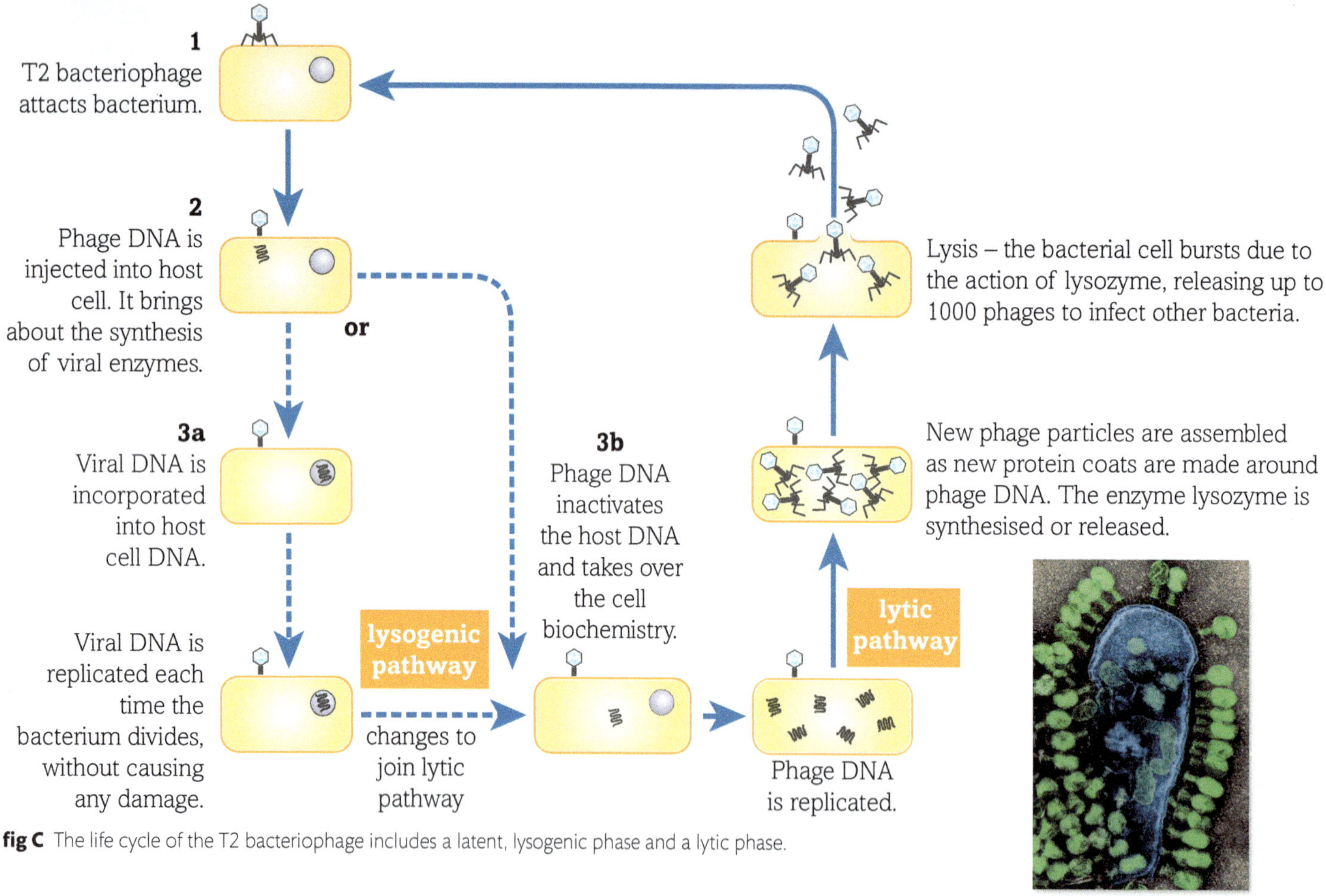

1
T2 bacteriophage attacts bacterium.

2
Phage DNA is injected into host cell. It brings about the synthesis of viral enzymes.

or

3a
Viral DNA is incorporated into host cell DNA.

Viral DNA is replicated each time the bacterium divides, without causing any damage.

lysogenic pathway

changes to join lytic pathway

3b
Phage DNA inactivates the host DNA and takes over the cell biochemistry.

Phage DNA is replicated.

lytic pathway

New phage particles are assembled as new protein coats are made around phage DNA. The enzyme lysozyme is synthesised or released.

Lysis – the bacterial cell bursts due to the action of lysozyme, releasing up to 1000 phages to infect other bacteria.

fig C The life cycle of the T2 bacteriophage includes a latent, lysogenic phase and a lytic phase.

Positive ssRNA viruses

These are viruses that contain a single strand of RNA that is a sense strand. It is used directly as mRNA for translation into proteins at the ribosomes. The proteins made include viral structural proteins and an RNA polymerase, which is used to replicate the viral RNA.

Negative ssRNA viruses

The single strand of RNA in these viruses is an antisense strand. Before it can be used to make viral proteins and more viral RNA it must be transcribed into a sense strand. The virus imports RNA replicase, which uses free bases in the host cell to transcribe the antisense RNA strand and produce a sense strand that can be translated at the ribosomes. Once the RNA strand has been transcribed it acts as mRNA at the ribosomes and codes for viral proteins including RNA replicase. These viral proteins combine with replicated viral RNA to form new viral particles.

RNA retroviruses

Retroviruses, including the HIV virus that causes AIDS and the Rous sarcoma virus that causes cancer in chickens, have a rather different and complex life cycle. They have viral RNA as their genetic material. It cannot be used as mRNA, but is translated into DNA by the specific enzyme reverse transcriptase in the cytoplasm of the cell. This viral DNA passes into the nucleus of the host cell where it is inserted into the host DNA. Host transcriptase enzymes then make viral mRNA and new viral

genome RNA. New viral material is synthesised and the new viral particles leave the cell by exocytosis (see **Section 2.1.4**). The host cell continues to function as a virus-making factory, while the new viruses move on to infect other cells.

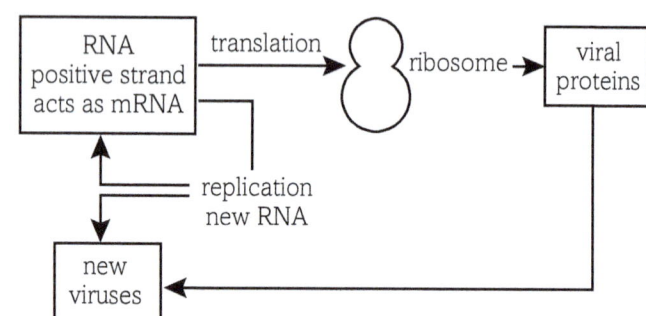

fig D Replication of a positive ssRNA virus.

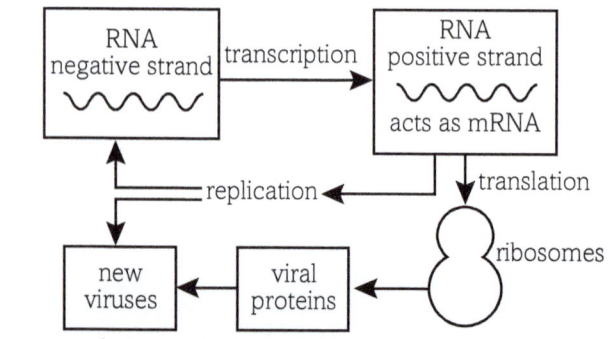

fig E Replication of a negative ssRNA virus.

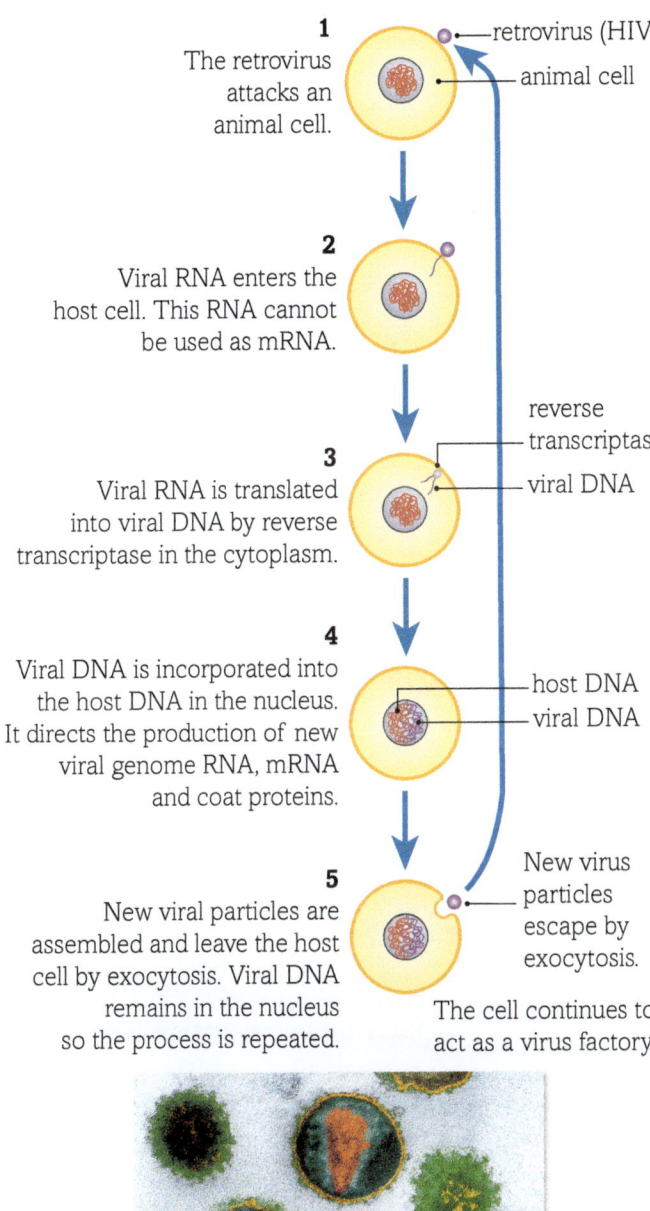

1 The retrovirus attacks an animal cell.

retrovirus (HIV)
animal cell

2 Viral RNA enters the host cell. This RNA cannot be used as mRNA.

3 Viral RNA is translated into viral DNA by reverse transcriptase in the cytoplasm.

reverse transcriptase
viral DNA

4 Viral DNA is incorporated into the host DNA in the nucleus. It directs the production of new viral genome RNA, mRNA and coat proteins.

host DNA
viral DNA

5 New viral particles are assembled and leave the host cell by exocytosis. Viral DNA remains in the nucleus so the process is repeated.

New virus particles escape by exocytosis.

The cell continues to act as a virus factory.

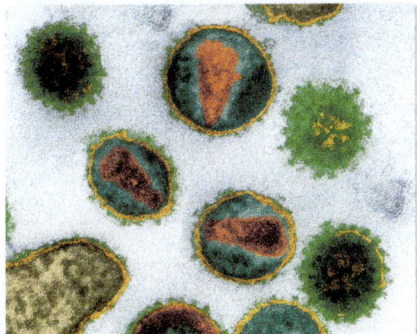

fig F The life cycle of a retrovirus.

Viruses and disease

Viruses cause disease in animals, plants and even in bacteria. They can cause the symptoms of disease by the lysis of the host cells, by causing the host cells to release their own lysosomes (see **Section 2.1.4**) and digest themselves from the inside or by the production of toxins that inhibit cell metabolism.

Viral infections are often specific to particular tissues. For example, adenoviruses, which cause colds, affect the tissues of the respiratory tract, but do not damage the cells of the brain or the intestine. This specificity seems to be due to the presence or absence of cell markers on the surface of host cells. Each type of cell has its own recognition markers and different types of virus can only bind to particular markers. The presence or absence of these markers can even affect whether a group of living organisms is vulnerable to attack by viruses at all. For example, the angiosperms (flowering plants) are vulnerable to viral diseases, but the gymnosperms (conifers and their relatives) are not.

Viruses are well-known for causing diseases like flu, measles, AIDS and Ebola. Research also shows that in some cases they play a role in the development of cancers. Certain animal cancers have been clearly linked to viral infection, and in humans there seems to be a link in certain specific cases. For example, the human papilloma virus responsible for warts on the skin, including genital warts, has been linked with the occurrence of pre-cancerous and cancerous changes in the cells of the cervix, and there is now a vaccine against it.

Questions

1 What adaptations make viruses such successful pathogens?

2 Suggest valid arguments for the case that:
   (a) viruses are living organisms
   (b) viruses are **not** living organisms.

3 What are the main differences between the lytic and lysogenic pathways of infection by DNA viruses?

4 Make a table to compare the different ways in which RNA viruses reproduce.

Key definitions

An **envelope** is a coat around the outside of a virus derived from lipids in the host cell.

The **capsid** is the protein coat of a virus.

Capsomeres are the repeating protein units that make up the capsid of a virus.

Virus attachment particles (VAPs) are specific proteins (antigens) that target proteins in the host cell surface membrane.

DNA viruses are composed of DNA as the genetic material.

RNA viruses are composed of RNA as the genetic material.

Retroviruses are a special type of RNA virus that control the production of DNA corresponding to the viral RNA and insert it into the host cell DNA.

Reverse transcriptase is an enzyme synthesised in the life cycle of a retrovirus that makes DNA molecules corresponding to the viral RNA genome.

Non-virulent is a term used to describe a microorganism that is not disease-causing.

A **provirus** is the DNA that is inserted into the host cell during the lysogenic pathway of reproduction in viruses.

Lysogeny is the period when a virus is part of the reproducing host cell, but does not affect it adversely.

Latent is the state of the non-virulent virus within the host cell.

Virulent is a term used to describe a microorganism that is disease-causing.

By the end of this section, you should be able to...

● Describe how antiviral medicines work by inhibiting viral replication because viruses are not living cells

● Explain how control of viral infections focuses on the prevention of the spread of disease as viral infections are difficult to treat

● Evaluate the ethical implications of using untested drugs during epidemics

As you have seen, the lifecycle of a virus involves the destruction of host cells. As a result of this direct damage, and the response of the host body to infection, viruses usually cause disease in the organisms they infect.

The spread of viral diseases

Viral diseases are spread in many different ways. The key feature is that material carrying viruses from an infected animal or plant comes into contact with vulnerable tissues in another uninfected organism. So, viruses may be spread through infected mucus, droplets of saliva, infected blood or faeces, or simple contact between infected organisms. International travel means that diseases that would once have just caused local outbreaks can now rapidly spread all over the world. Different viral diseases are spread in different ways. For example:

- Foot-and-mouth disease is a serious disease of cloven-hoofed animals such as cattle. It severely weakens adult animals and kills a high percentage of young animals. It is spread through body secretions, such as milk and semen, and transmitted in the breath and the faeces of infected animals. Healthy animals can pick up the virus from contaminated pens, food, water, contact with diseased animals and even from infected meat and animal products if they are eaten.

- **Ebola** is a severe viral illness caused by the Ebola virus. It is often fatal, especially if the symptoms are untreated. It is an animal disease that spreads to humans through the faeces, urine, blood and meat of infected animals. It then spreads easily from person to person by the direct contact of the skin or mucous membranes of a healthy person with blood, faeces and other body secretions of an infected person, or even bedding and surfaces contaminated with fluids from an infected person.

Treating viral diseases

As you have seen, bacterial diseases can be treated with antibiotics. The drugs affect the bacterial cells in one way or another (see **Section 2.2.1**). Viruses, however, are not living cells in the conventional sense. Scientists have not yet developed drugs that can affect the virus particles themselves. Instead, antiviral treatments target virus replication. There are a number of different ways in which they can work. They can:

- target the receptors by which viruses recognise their host cells

- target the enzymes that help to translate or replicate the viral DNA or RNA

- inhibit the protease enzymes that enable new virus particles to bud from host membranes.

So far, scientists have not been able to cure viral diseases, but they have reduced the time a person is sick (see **fig A**) and can delay the development of symptoms after infection (e.g. the cocktail of antiretroviral drugs used to treat HIV/AIDS).

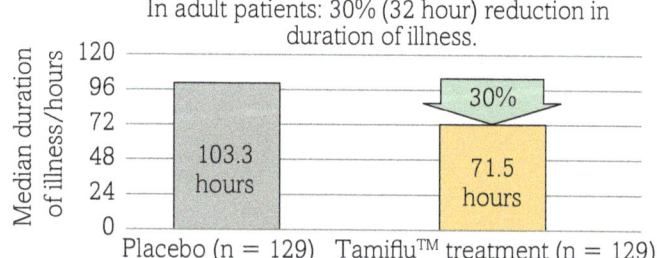

fig A The impact of antiviral medication on the duration of influenza.

Preventing viral disease

Some viral diseases, such as the common cold, are relatively mild and have a very low **mortality rate**. Others, however, are very serious. During 1918–19 an outbreak of influenza killed up to six times more people than the whole First World War. Foot-and-mouth disease has an almost 100% mortality rate in young stock.

In the 2001 UK epidemic of foot-and-mouth disease there was no treatment available and no tests to reliably identify infected animals before they showed symptoms. So all of the cloven-hoofed animals on infected farms were destroyed and burned to try and prevent the spread of the virus to other farms in the area. Over 6 million animals were killed. Veterinary scientists are working on developing sensitive tests to identify infected animals so that control of the disease may be possible in the future without this extensive culling.

fig B Mass culling and burning of possibly infected livestock during the 2001 outbreak of foot-and-mouth disease in the UK was eventually successful in stopping the spread of the disease.

The mortality rate of humans infected with Ebola varies but can be very high with 25–90% of people infected with the disease dying. Mortality depends on the strain of the virus, the health of the infected person and the speed with which they get support and health care. The average mortality rate is around 50%.

Viral diseases like these can be devastating. Because there are no antiviral drugs against most viral infections, disease control focuses on vaccination and reducing the spread of viruses.

Vaccinations

Vaccination plays a major role in the prevention of disease outbreaks. When you are vaccinated against a disease you become immune to it and so will not become infected should you encounter it. Ideally, everyone is vaccinated against serious diseases that may affect them. If an epidemic breaks out and the population is not vaccinated, there is a rush to deliver vaccines to everyone who is not already infected. Usually health care workers, the very young and the elderly are vaccinated first. Unfortunately we have not yet developed fully-tested vaccines against some of the worst viral diseases, such as HIV/AIDS and Ebola. You will learn more about vaccination if you continue to study A level Biology.

Disease control

Understanding the cause of a disease and how it is spread means we can work to control it. Disease control is particularly important when an epidemic occurs. An epidemic is when the levels of people with a particular disease are much higher than expected over a given period of time. When there is a vaccine available, this is the time for mass vaccination of vulnerable people, alongside measures to prevent the spread of disease. In diseases where no vaccine is available, controlling the spread of the disease is key.

Identifying the pathogen early and putting control measures in place can make all the difference to the numbers of people affected (see **fig C**).

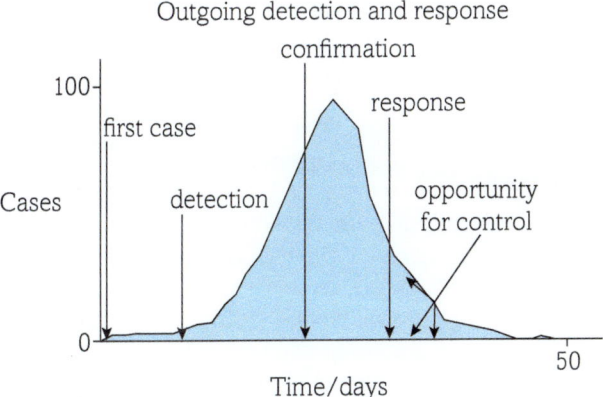

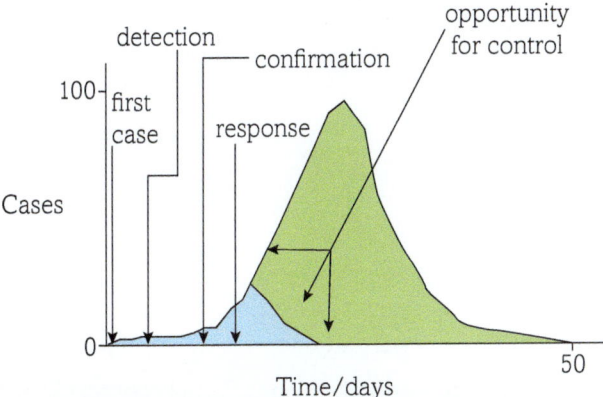

fig C These graphs show the difference that effective detection and response can make to the number of cases that develop.

There are a number of different ways of controlling the spread of a disease. Some are relevant to all diseases, some are only used in more extreme diseases such as Ebola. They include:

- Rapid identification of disease: For example, in West Africa in 2014, it was some time before the Ebola was recognised and effective testing regimes put in place. If the disease is bacterial, it must be identified and an effective antibiotic used.

- Nursing in isolation: This is used for serious infections such as Ebola and *C. difficile* only. It is readily available in countries such as the UK but sealed isolation units are rarely available in developing countries such as those in West Africa. This lack of health infrastructure makes it difficult to isolate people affected by diseases such as Ebola. When ill people are cared for within their families the virus spreads easily. Simple units nursing all infected patients together can help.

- Preventing transmission from one individual to another: Simple measures such as regular hand washing, hand washing before and after every contact with patients by health workers and families, care in handling body fluids and wastes, careful disposal of infected bodily wastes, and frequent disinfecting of surfaces and people are key. Body fluids are very infectious in Ebola cases and good hygiene is vital.

- Sterilising or disposing of equipment and bedding after use: One of the main transmission routes for Ebola at the beginning of the epidemic was through unsterilized needles used in an antenatal clinic.

- The wearing of protective clothing by health workers: When dealing with dangerous and highly infectious viruses such as Ebola, health workers should wear facemasks, gowns, gloves, and goggles to protect the eyes. The slightest contact of infected material with the eyes is enough to lead to infection. The gloves should be washed and disinfected before removal and then the hands washed as well.

- Indentifying contacts: People who have been in contact with infected people need to be monitored so that they can be treated and/or isolated rapidly if they show signs of disease.

Did you know?

Rituals and infection

Many cultures have rituals that are carried out after a death. Family and friends may visit, touch and kiss the body, and they may wash the body and prepare it for burial in the family home. In the case of a disease such as Ebola, the body remains highly infectious after death and so funeral rituals can lead to outbreaks of further infection. It was observed that about a week after the funeral of an Ebola victim, many of the mourners would become sick themselves. Communities accepted that these rituals had to be changed to prevent the spread of Ebola. By sealing bodies in plastic and burying them immediately after death, the spread of the disease was greatly reduced.

In the twenty-first century, in countries such as the UK, we expect to be able to take some medicine and get better if we feel unwell. In an epidemic caused by a virus this isn't always possible due to limited treatment options. If the epidemic is of a potential killer disease, such as flu or Ebola, the pressure to find an effective treatment or vaccine is very high.

The development of new medicines

The development of a new medicine or vaccine takes up to 10 years, involves many different scientists and doctors, and costs millions of pounds. Initial ideas for potential drugs come from a wide range of sources including genome analysis of pathogens, computer modelling, clinical compound banks and medicinal plants. These chemicals have to go through thorough research and testing on cell and tissue cultures, safety analyses and molecular modifications. This is followed by animal testing to ensure the compound works in a whole organism and is safe. This is then followed by three phases of human testing to further ensure safety and that the drug works well. This goes along with complex regulation and licensing procedures until finally, a new drug may reach the doctor's surgery. This process is summarised in **fig D**.

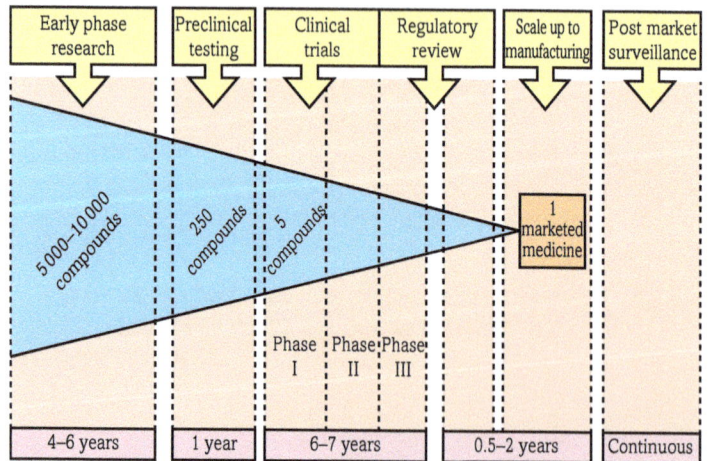

fig D This diagram summarises the main stages in the normal drug development process.

Speeding up the process

When an epidemic develops, some of the final stages of the testing of a new medicine or vaccine may be speeded up to try and save lives and prevent the spread of a deadly disease. Here are two examples.

In 2006 a new strain of H5N1 flu (known as bird flu) caused a global **pandemic.** A vaccine against the new strain was produced very quickly, fast-tracked using existing techniques and technology for producing annual flu vaccines and put through basic clinical trials. They were to be used for health workers if the pandemic hit the UK. The Medical Research Council said they expected the vaccines would give some, if not total, protection whilst a better vaccine was developed. In addition, antiviral medicines such as Tamiflu™ and Relenza™ were stockpiled in spite of concerns that there was incomplete data about their effectiveness. These concerns were raised again when the drugs were used in the 2009 swine flu epidemic and the Cochrane Collaboration, which carries out systematic analysis of the published data on medicines, has since stated that these drugs have not been proven to reduce hospitalisations and serious complication from influenza. You will learn more about influenza in **Book 2**.

The first case of Ebola in this outbreak occurred in late 2013 in West Africa. However, it took until mid-2014 for the world beyond Africa to recognise the size and severity of the outbreak of disease and the speed with which it was spreading. Once the severity of the outbreak was recognised, the World Health Organisation (WHO) and pharmaceutical companies around the world looked for ways to fast-track drugs and vaccines that were already in development and had passed many of the development stages, but which had not completed human trials. In this situation there are two challenges: to make sure the drugs are safe and effective and also to ramp up production to be able to make enough of the medicine or vaccine for it to be useful. Potential treatments included:

- ZMapp™, an experimental drug produced after long-term studies of people who had survived Ebola in previous less-widespread outbreaks. Scientists had genetically modified tobacco plants to produce three antibodies that seem to be associated with surviving the disease. In trials it was effective in treating monkeys, but had not been tried on people. Tiny amounts of the drug were available and used to treat 7 people including African, American, Spanish and British health workers who developed Ebola. Some recovered, but some of the seven died, as you would expect with a disease with around 50% mortality. Scientists are trying to produce more of the substance to run bigger trials on more people.

- Vaccines: Several companies had vaccines in trials that are being fast-tracked for use against Ebola. They are making many doses of the vaccine so that if they are safe to use in humans, many health workers and then people living in epidemic areas can be vaccinated.

Other pharmaceutical companies are supporting the work of companies with drugs and vaccines closest to completion, and are also developing other drugs against the virus itself.

Ethical implications

Historically, doctors and scientists tried out new medicines on themselves, their families or their patients with little or no testing or trials. Today it would be considered completely unethical under normal circumstances to give anyone a medicine or vaccine that had not been through the full process of testing and approval. However, in severe epidemics or pandemics, with thousands of lives at risk, decisions may be made to use drugs that are only part way through the full testing process. Most often this involves drugs that have not completed human trials. Although the media will report these as 'untested', they have in fact already undergone a minimum of five years testing and development, and often will be part way though human trials.

There are a number of factors that have to be evaluated when considering whether a drug should be fast-tracked for use in an epidemic. These include:

- the severity of the disease
- the availability of any other treatments for the disease
- the effectiveness of standard disease control measures in halting the spread of the disease
- transparency about the process and informed consent of those given the treatment
- freedom of choice over participation
- involvement of the affected community – community consent for treatment can be more valuable than individual consent

- collection of clear clinical data from the use of new medicines in this situation so an on-going assessment of the safety and efficacy of the drug or vaccine can be made.

Reasons against using untested drugs include:

- Some people simply feel that it is not ethical under any circumstances to use drugs that have not completed full human trials.

- If an untested drug produces unexpected side effects it can make the situation worse

- Deciding who gets the drug or vaccine can be difficult. For example, in a situation such as the Ebola epidemic, local people might feel they were being used as guinea pigs for Western medicine if they are given the medicine, but might feel resentful if only health workers are treated.

- Informed consent is an issue as it depends on a level of education to understand the drug and how it works and also clarity of thought. People who are dying may grasp at straws but their relatives may then blame the treatment for an inevitable death.

- Issues of trust between individuals or communities and health workers, especially if supplies of a new drug are limited.

Did you know?

No epidemic but more ethical decisions

Fungal infections can kill people if their immune system is not working well, for example, in people suffering from diseases such as leukaemia or HIV/AIDS, or those taking immunosuppressant drugs. In the late 1980s the search was on for a new antifungal medicine. Chris Hitchcock and his team at the pharmaceutical company Pfizer set out to design a new molecule that would be more powerful than the existing fungicides and would also kill fungal pathogens resistant to the antifungal drugs available at the time. A molecule known as voriconazole was discovered in 1990 and the long process of development and trialling began.

In 1997 there was a tragedy at the Maccabiah Games, held every four years in Israel. As the Australian team entered the arena over a foot bridge across the very polluted Yarkon River, it collapsed. Over 100 athletes were injured and four died. Three of these deaths were due to a deadly fungal infection picked up from the river. Sasha Elterman, a talented 15-year-old tennis player, was infected with the fungus that day. It attacked her brain and spine and she was given only a 3% chance of surviving. After several months of treatment with every available antifungal medicine, Sasha was still alive – but only just.

Then her medical team read about voriconazole, but it had only just started clinical trials so was a long way from getting a licence. Sasha's doctors got permission to try it as at this stage she had nothing to lose. Without a different treatment she was going to die.

The improvement in Sasha's condition was almost immediate and after 451 days of treatment with the new anti-fungal drug she was fully recovered.

At the opening ceremony of the 2000 Sydney Olympics, Sasha carried the torch into the arena at the head of the Australian Olympic team. She was alive as a result of the use of a new and incompletely trialled medicine. In 2002 voriconazole finally passed all its clinical trials successfully and was licensed. It is still used effectively to treat life-threatening fungal infections today.

In any epidemic situation, the ethical implications of using a fast-tracked and relatively untested drug have to be evaluated at national and international levels. In the US, the Federal Drug Agency decided that an antiviral called peramivir that had not completed testing could be used intravenously in seriously ill patients in the 2009 H1N1 flu epidemic. The WHO recently decided that ZMapp™, which had had no human trials, could be used in the Ebola epidemic in Western Africa and that at least two vaccines could also be fast-tracked through the process for use. The effectiveness of these interventions is yet to be seen, and only then can a full evaluation be made.

Questions

1 Why is the control of the spread of disease particularly important in viral diseases?

2 What might be a disadvantage of giving people a medicine that reduces the symptoms of a viral disease such as flu?

3 (a) Summarise the main ways in which the spread of an infectious disease can be controlled.

(b) Explain why it was particularly difficult to contain the spread of the Ebola virus in West Africa in 2014.

4 (a) Make a flow chart to show the main stages in the normal drug development process.

(b) Which stages are most likely to be bypassed if a drug is fast-tracked for use in an epidemic?

5 Suggest why the severity of the disease, the availability of other treatments, and the effectiveness of normal disease control measures are such important factors when evaluating whether a new drug should be tried.

6 Write a paragraph supporting the use of a new drug that has not undergone human trials in the 2014 Ebola outbreak, and a similar paragraph against. In each case evaluate the evidence and put forward scientifically sound opinions.

7 Are the ethical considerations for using an unlicensed treatment the same or different when considering the treatment of a single individual like Sasha Elterman or a community such as those in Sierra Leone during an epidemic? Discuss.

Key definitions

Ebola is a highly infectious viral disease that causes fever and internal bleeding and death in about 50% of cases.

The **mortality rate** is a measurement of the number of deaths in a given population or due to a specific cause.

A **pandemic** is an epidemic that takes place in several countries at once.

EBOLA – A DEADLY VIRUS

In 1976 a new and deadly disease appeared in Africa. Ebola virus causes Ebola, a disease that is so damaging to the body tissues that it has a 25–90% death rate. It is both contagious (spread by contact) and infectious (spread by droplets in the air) and symptoms can appear from 2 to 21 days after the initial infection. Reporting on this viral disease varies greatly…

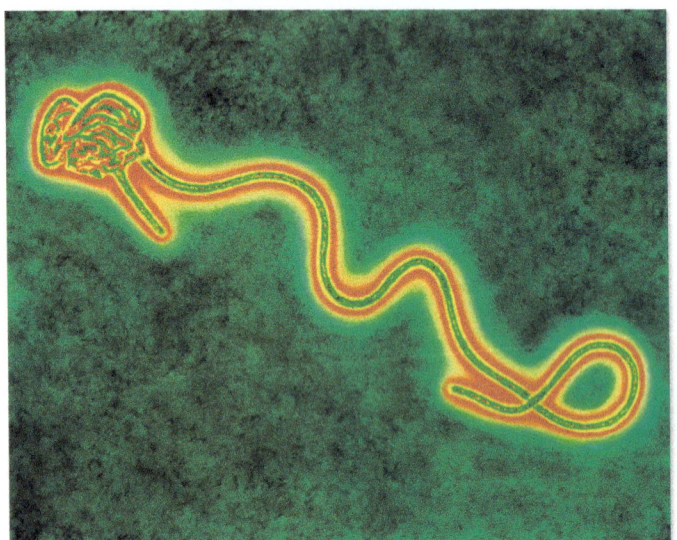

fig A The deadly Ebola virus.

From the *Daily Mail*:

Deadly Ebola virus 'could spread globally' after plane brings it to Nigeria

- Health experts fear other passengers could now be carrying the virus
- It lays dormant in victims for up to three weeks – and 90 per cent die of it

By NICK FAGGE

…The news came as it emerged that an American doctor working for a charity in Liberia had become infected. Dr Kent Brantly, 33, from Texas, had moved to the country for the Samaritan's Purse organisation with his children and wife, Amber, to help contain the disease.

More than 1,000 others have been infected by the virus, which can go unnoticed for three weeks and kills 90 per cent of victims.

Follow us: @MailOnline on Twitter

From the website of the World Health Organization:

Epidemiology and surveillance

WHO continues to monitor the evolution of the Ebola virus disease (EVD) outbreak in Sierra Leone, Liberia and Guinea. The Ebola epidemic trend remains precarious. Between 21 and 23 July 2014, 96 new cases and 7 deaths were reported from Liberia and Sierra Leone. In Guinea, 12 new cases and 5 deaths were reported during the same period. These include suspect, probable and laboratory-confirmed cases. The surge in the number of new EVD cases in Guinea after weeks of low viral activity demonstrates that undetected chains of transmission existed in the community. This phenomenon … calls for stepping up outbreak containment measures, especially effective contact tracing.

From the website of Public Health Wales:

Ebola virus disease: an overview

Ebola virus disease is a serious, usually fatal, disease for which there are no licensed treatments or vaccines. But for people living in countries outside Africa, it continues to be a very low threat.

The current outbreak of the Ebola virus mainly affects three countries in West Africa: Guinea, Liberia and Sierra Leone. …This is the largest known outbreak of Ebola.

So far, there has been just one imported case of Ebola in the UK. Experts studying the virus believe it is highly unlikely the disease will spread within the UK.

What are the symptoms, and what should I do if I think I'm infected?

A person infected with Ebola virus will typically develop a fever, headache, joint and muscle pain, a sore throat, and intense muscle weakness.

These symptoms start suddenly, between two and 21 days after becoming infected, but usually after five to seven days.

If you feel unwell with the above symptoms within 21 days of coming back from Guinea, Liberia or Sierra Leone, you should stay at home and immediately telephone 111 or 999 and explain that you have recently visited West Africa.

These services will provide advice and arrange for you to be seen in a hospital if necessary so the cause of your illness can be determined.

It's really important that medical services are expecting your arrival and calling 111 or 999 will ensure this happens.

Where else will I encounter these themes?

1.1   1.2   1.3   1.4   2.1   2.2 YOU ARE HERE   2.3   2.4

Let us look at the different levels of information given in these pieces of writing, and consider who they are aimed at:

1. The extracts here come from a popular newspaper, the World Health Organization website and the Public Health Wales website.

   a. Which article seems to be the most scientific, and which the least? Explain your answer.

   b. Discuss the different purposes of the three pieces of writing and consider whether you think they are each fit for their purpose.

   c. Each of these extracts shows a bias – they are trying to communicate different things. Comment on what each of the pieces is trying to do in terms of informing the readers.

> Remember that a newspaper has to persuade people to buy copies or pay an online subscription. WHO and Public Health Wales know that anyone visiting their sites has a genuine interest in finding out detailed facts about the topic they are researching!

Now let us examine the biology given in each piece of writing. You already know about viruses and bacteria, so you can answer these questions now. If you are going to continue your biology studies to A level, you may like to revisit these pages after you have learned more about communicable diseases in **Book 2 Topic 6**.

2. Look at the newspaper article and summarise the knowledge about Ebola that you have at the end. How accurate is that information biologically?

3. Summarise what the extract from the Public Health Wales website tells you about Ebola and how you think the virus causes the symptoms of the disease.

4. If you become ill after visiting certain African countries, the Public Health Wales website emphasises the need to inform a doctor or any hospital you attend. Why is this so important?

5. The WHO extract gives little or no information about the Ebola virus itself. What is the focus of this article? Why is this information also important biologically?

Activity

Research is key in preventing the spread of viruses like Ebola.

Find out as much as you can about the Ebola virus, focusing on how it is spread and the way it infects and takes over the cells of the body.

Think carefully about which stage of the viral life cycle you would target to try and prevent the spread of the disease.

Imagine you have to bid for funding to carry out your research. Put together a poster presentation summarising the problem you have identified with Ebola, explaining what you want to research and why you should receive the funding.

> Be clear about the different types of viruses and the ways they reproduce in cells.
>
> Refer to the information in the Public Health Wales extract and visit the original website as well as others and use the information in your textbook. Think very carefully about where and when a virus might be vulnerable to attack during its reproductive cycle.

1 The table below refers to some features of prokaryotic and eukaryotic cells. If the feature is present, place a tick (✔) in the appropriate box and if the feature is absent, place a cross (✗) in the appropriate box.

Feature	Prokaryotic cell	Eukaryotic cell
Nuclear envelope		
Cell surface (plasma) membrane		
Mitochondria		
Golgi apparatus		

[Total: 4]

2 An analysis of the large organic molecules found in a prokaryotic cell was made. The dry mass of the cell is the mass of the cell not including water. The results of the analysis are shown in the table below.

Molecule	Percentage of total dry mass of the cell/%	Number of molecules per cell	Number of different types of molecule
Protein	55.0	2 360 000	1050
Lipid	9.1	22 000 000	4
Glycogen	2.5	4360	1
DNA	3.1	2	1
RNA	20.5	262 480	463

(a) The molecular mass of a substance is the mass of one molecule of that substance. Using information in the table, state which of the molecules has the largest molecular mass.
    Give an explanation for your answer. [2]

(b) Glycogen and protein molecules are both polymers. Explain why there is only one type of glycogen molecule but there are many types of protein molecule. [2]

(c) Explain why many different RNA molecules are found in a cell. [2]

[Total: 6]

3 The table below refers to some of the stages involved in Gram staining and the appearance of Gram-negative and Gram-positive bacteria after each stage. Complete the table by writing the most appropriate word or words in the empty boxes.

Stage of Gram staining	Appearance of Gram-negative bacteria	Appearance of Gram-positive bacteria
Cells heat fixed onto slide	Colourless	Colourless
Slide flooded with crystal violet		
Slide flooded with Gram's iodine		
Slide rinsed with alcohol or acetone		
Slide counterstained with safranin/carbol fuchsin		

[Total: 4]

4 The table below refers to features of λ (lambda) phage, tobacco mosaic virus (TMV) and human immunodeficiency virus (HIV). Complete the table by writing the most appropriate word or words in the boxes.

Feature	λ-phage	TMV	HIV
Type of nucleic acid			
Shape of protein coat			

[Total: 6]

5 The table below refers to some structures of microorganisms. Complete the table by writing the name of the type of microorganism possessing each structure in the empty boxes.

Structure	Type of microorganism
Nucleus	
Capsid	
Flagellum	
Peptidoglycan (murein) cell wall	

[Total: 4]

6 Prokaryotes, mitochondria and chloroplasts have many features in common.

(a) (i) The diagram below shows a mitochondrion. Two of the features labelled are typical of prokaryotes. Place a tick (✓) in each of the **two** boxes that correctly identify these features.

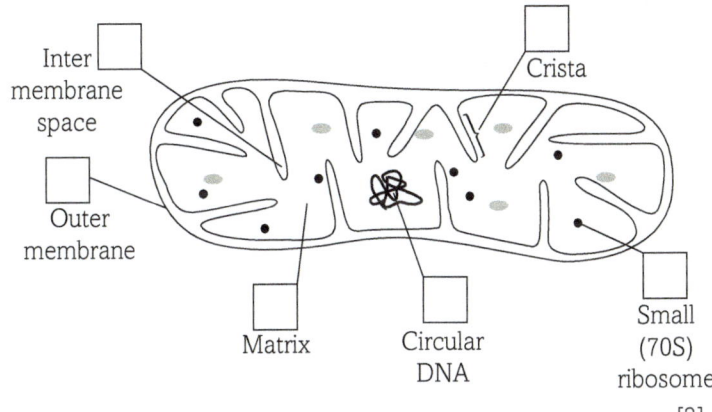

[2]

(ii) The table below shows some features of mitochondria. If the feature is also present in chloroplasts, place a tick (✓) in the box to the right of that feature and if it is absent, place a cross (✗) in the box.

Features present in mitochondria	Feature present (✓) or absent (✗) in chloroplasts
Surrounded by a double membrane	
Crista present	
Circular DNA	
Matrix	
Glycogen granule	
Stalked particles	

[3]

(b) Bacteria can be identified and classified by looking for certain features. Using the information in the passage below, label the five bacteria with the correct letter.
Bacterium P has a single flagellum to enable it to move whilst bacterium Q has several flagella.
Only bacterium R has visible plasmids and bacterium S has an infolding of its cell surface membrane.
Bacterium T has a slime capsule.

Bacterium ....   Bacterium ....

Bacterium ....   Bacterium ....

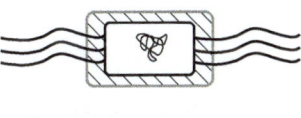

Bacterium ....

[4]

[Total: 9]

CHAPTER

2.3 > Eukaryotic cell division – mitosis

Introduction

Normal growth and division in cells occurs in a cycle that is controlled by literally hundreds of genes acting in different ways to stimulate or repress the process. Cells which mutate are usually removed by apoptosis, programmed cell death brought about at least in part by the action of the lysosomes. But sometimes cells go wrong – as they divide they develop mutations that eventually mean they do not respond to the normal signals controlling the cell cycle. This potentially results in one of the many cancers that can affect almost every tissue and organ in the body. Identifying these changes and controlling them, or preventing them from happening in the first place, is the focus of much research.

In this chapter you will be looking at mitotic cell division in eukaryotic cells. The cell cycle is of great importance and you will discover the main stages of the cycle, and how the speed of the cycle varies at different ages and in different tissues.

You will learn the stages by which cell division – mitosis – takes place. By looking at the way the chromosomes replicate and divide in a graceful 'dance', followed by the rest of the cytoplasm of the cell, you will come to understand how mitosis results in two identical daughter cells. You will carry out practical investigations using plant meristems to see the stages of mitosis under the microscope.

You will also consider the importance of mitosis in living organisms – it is the basis of asexual reproduction in many animals, plants and fungi. Mitosis produces offspring that are identical to the single parent, and it can result in enormous numbers of offspring being produced at one time.

Mitosis is also important for the repair of damaged tissues and for normal growth from infancy to adulthood, and you will look at how this growth can be observed and measured. Growth in animals stops when they reach maturity. In plants, growth continues throughout life, a pattern revealed in the trunks of trees when they are felled.

All the maths you need

- Carry out calculations using numbers in standard and ordinary form (*e.g. use of magnification*)
- Use scales for measuring (*e.g. measuring sizes of cells at different stages of growth*)
- Find arithmetic means (*e.g. measuring sizes of cells at different stages of the cell cycle*)
- Make order of magnitude calculations (*e.g. use and manipulate the magnification formula:*
  $magnification = \dfrac{size\ of\ image}{size\ of\ real\ object}$)
- Use and manipulate equations, including changing the subject of an equation (*e.g. magnification*)

What have I studied before?

- Asexual reproduction in living things
- The basic process of mitosis, including the idea of the cell cycle
- That cancer is the result of changes in cells that lead to uncontrolled growth and division
- The ultrastructure of eukaryotic cells including the nucleus, nuclear membrane, chromosomes, centriole, etc.
- The way in which DNA replicates in the nucleus
- Gene mutations

What will I study later?

- Meiosis – the process of cell division that reduces the number of chromosomes in the nucleus to form the gametes
- The early development of a mammalian embryo to the blastocyst stage, that involves rapid mitosis
- Fertilisation in plants
- The growth of a bacterial colony that will involve understanding binary fission (A level)
- Measuring growth of bacterial colonies (A level)
- The immune response of the body including clonal selection and the rapid mitosis that results in the production of plasma cells and T killer cells as well as T and B memory cells (A level)
- Stem cells in animals and plants (A level)
- The control of cell differentiation (A level)
- Plant responses to environmental stimuli that depend on cell division and growth

What will I study in this chapter?

- The cell cycle as a regulated process made up of interphase, mitosis and cytokinesis, in which cells divide to produce two identical daughter cells
- The replication and division of the genetic material in the main stages of mitosis
- The importance of mitosis in growth, repair of damaged or ageing tissues and asexual reproduction, to produce offspring that are identical to the one parent
- Techniques to make a temporary squash preparation of the cells of a root tip where active mitosis is taking place

By the end of this section, you should be able to...

● describe the cell cycle as a regulated process in which cells divide into two identical daughter cells

● describe the three main stages of the cell cycle as interphase, mitosis and cytokinesis

● explain what happens to the genetic material during the cell cycle

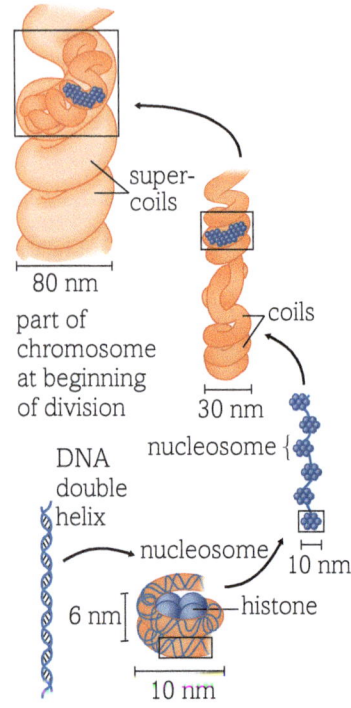

fig A Histones play an important role in the organisation of DNA into orderly chromosomes that can be replicated.

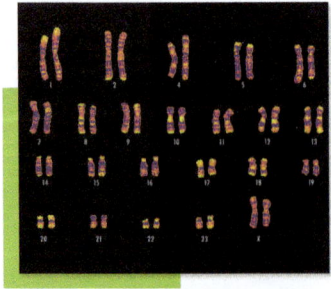

fig B This female human karyotype shows the 22 pairs of autosomes and one pair of sex chromosomes found in every healthy human cell, except the eggs and sperm.

One of the most awe-inspiring processes of life is the way in which organisms reproduce. Like begets like – buttercups produce new buttercups, *Amoeba* produce more *Amoeba* and liver cells generate more liver cells. Most new biological material comes about as a result of the process of nuclear division known as **mitosis**, followed by the rest of the cell dividing. **Asexual reproduction** – the production of genetically identical offspring from a single parent cell or organism – and growth are both the result of mitotic cell division. The production of offspring by **sexual reproduction** is also largely dependent on mitosis to produce new cells after the gametes (sex cells) have fused. In mitosis the chromosomes of a cell are duplicated and the genetic information is then equally shared out between the two daughter cells that result. The formation of the sex cells involves a different process of nuclear division called **meiosis** (see **Section 2.4.1**).

What are chromosomes?

Eukaryotic cell division involves replicating the chromosomes that carry the genetic information. A chromosome is made up of a mass of coiled threads of DNA and proteins. If a chromosome were as long as five consecutive letters on this page, the DNA molecule it contained would stretch the length of a football pitch or more. In a cell that is not actively dividing, the chromosomes are translucent to both light and electrons so we cannot see them easily or identify them as individual structures. When the cell starts to actively divide, the chromosomes condense – they become much shorter and denser. They then take up stains very readily. This is the basis of the name 'chromosome' or 'coloured body' and as a result at this stage of the process we can identify individual chromosomes.

When the DNA molecules condense, they have to be packaged very efficiently. This is achieved with the help of positively charged basic proteins called **histones**. The DNA winds around the histones to form dense clusters known as **nucleosomes** (see **fig A**). These then interact to produce more coiling and then supercoiling to form the dense chromosome structures you can see through the microscope in the nucleus of a dividing cell. In the supercoiled areas the genes are not available to be copied to make proteins.

The cells of every different species possess a characteristic number of chromosomes – in humans this is 46. These chromosomes occur in matching pairs, one of each pair originating from each parent. In mitosis the two cells that result from the division must both receive a full set of chromosomes. So before a cell divides it must duplicate the original set of chromosomes. During mitosis these chromosomes are divided equally between the two new cells so that each has a complete and identical set of genetic information. During the active phases of cell division the chromosomes become very coiled and condensed. In this state they can be photographed to produce a special display or **karyotype** (showing all the chromosomes of the cell).

The cell cycle

Cells divide on a regular basis to bring about growth and asexual reproduction. They divide in a sequence of events known as the **cell cycle**, which involves several different phases (see **fig C**). **Interphase** is a period of non-division when the cells increase in mass and size, carry out normal cellular activities and replicate their DNA ready for division. This is followed by mitosis, a period of active division, and cytokinesis when the new cells separate. The length of the cell cycle is variable. It can be very rapid, taking 24 hours or less, or it can take a few years.

Labels on fig A: super-coils; 80 nm; part of chromosome at beginning of division; coils; 30 nm; DNA double helix; nucleosome; nucleosome; 10 nm; 6 nm; histone; 10 nm

Phases of the cell cycle

- G_1 (gap 1) is the time between the end of the previous round of mitotic cell division and the start of chromosome duplication. The cell assimilates material, grows and develops. This is the time that is most variable. In actively dividing cells, G_1 is very short – a matter of hours or days, but in other cells, it can be months or even years.

- S is the stage when the chromosomes replicate and become double stranded **chromatids** ready for the next cell division.

- G_2 (gap 2) is the time that the organelles and other materials needed for cell division are synthesised – before a cell can divide, it needs two of everything.

- M is mitosis when the cells are actively dividing.

- C is cytokinesis, the final stage of the cell division when the new cells separate.

In multicellular organisms the cell cycle is repeated very frequently in almost all cells during development. However, once the organism is mature, it may slow down or stop completely in some tissues. The cell cycle is controlled by a number of chemical signals made in response to different genes. This control is brought about at a number of checkpoints where the cell cycle moves from one phase to the next. The control chemicals are small proteins called **cyclins**. These build up and attach to enzymes called **cyclin-dependent kinases (CDKs)**. The cyclin/CDK complex that is formed phosphorylates other proteins, changing their shape and bringing about the next stage in the cell cycle. Examples include the phosphorylation of the chromatin in the nucleus, which results in the chromosomes becoming denser, and the phosphorylation of some of the proteins in the nuclear membrane, which leads to the breakdown of the nuclear membrane structure during cell division.

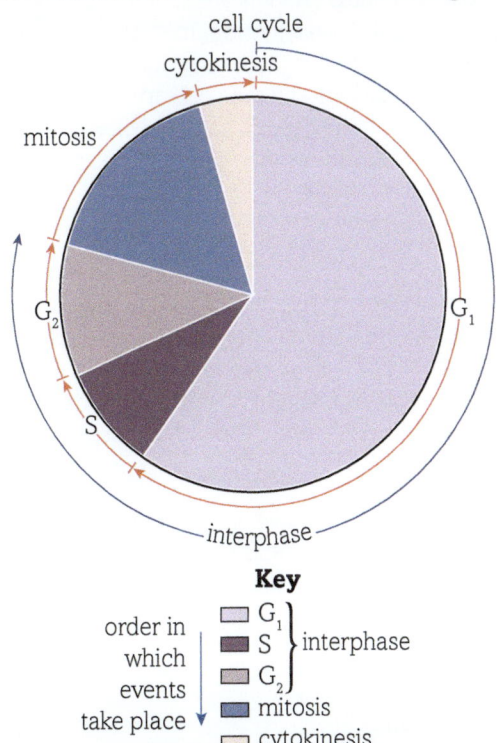

fig C The phases of the cell cycle. In very actively dividing tissue the cycle is repeated as fast as possible, whilst in other tissues the time between successive divisions may be years.

Did you know?

Permanent cells

There are some cells that do not enter the cell cycle once they have formed – they have to last a lifetime. They are known as permanent cells. Examples include nerve cells, the light sensitive cells of the retina, the transparent cells of the lens of the eye and the cardiac muscle – the muscle that makes up your heart.

Questions

1 Why do chromosomes only become visible as a cell goes into mitosis?

2 If a culture of cells is dividing every 48 hours, how long would you expect the different stages of the cycle to take?

Key definitions

Mitosis is the process by which a cell divides to produce two genetically identical daughter cells.

Asexual reproduction is the production of genetically identical offspring from a single parent or organism.

Sexual reproduction is the production of offspring that are genetically different from the parent organism or organisms by the fusing of two sex cells (gametes).

Meiosis is a form of cell division in which the chromosome number of the original cell is halved, leading to the formation of the gametes.

Histones are positively charged proteins involved in the coiling of DNA to form dense chromosomes in cell division.

Nucleosomes are dense clusters of DNA wound around histones.

A **karyotype** is a way of displaying an image of the chromosomes of a cell to show the pairs of autosomes and sex chromosomes.

The **cell cycle** is a regulated process of three stages (interphase, mitosis and cytokinesis) in which cells divide into two genetically identical daughter cells.

Interphase is the period between active cell divisions when cells increase their size and mass, replicate their DNA and carry out normal metabolic activities.

A **chromatid** is one strand of the replicated chromosome pair that is joined to the other chromatid at the centromere.

Cyclins are small proteins that build up during interphase and are involved in the control of the cell cycle by their attachment to cyclin-dependent kinases.

Cyclin-dependent kinases (CDKs) are enzymes involved in the control of the cell cycle by phosphorylating other proteins, activated by attachment to cyclins.

By the end of this section, you should be able to...

● explain what happens to the genetic material during the cell cycle, including the stages of mitosis

A cell is in the interphase stage of the cell cycle for much of its life. This used to be called the resting phase, but nothing could be further from the truth. During interphase the normal metabolic processes of the cell continue and new DNA is produced as the chromosomes replicate. New proteins, cytoplasm and cell organelles are synthesised so that the cell is prepared for the production of two new cells. ATP production is also stepped up at times to provide the extra energy needed as the cells divide. Once all that is needed is present and the parent cell is large enough, interphase ends and mitosis begins.

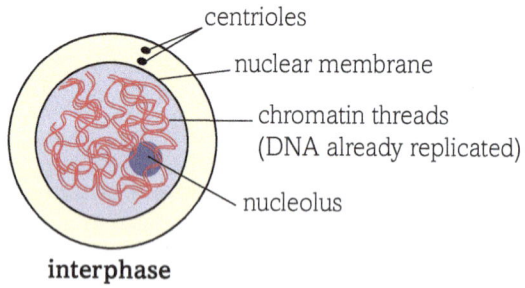

interphase

The stages of mitosis

During the process of cell division the chromosomes duplicated during interphase are divided up with the remaining contents of the cell so two identical daughter cells are formed. Walther Flemming (1843–1905), a German cytologist, was the first to describe what is sometimes called the 'dance of the chromosomes'. It refers to the complex series of movements that occur during cell division as the chromosomes jostle for space in the middle of the nucleus and then pull apart to opposite ends of the cell. The events of mitosis are continuous, but as in the case of so many biological processes it is easier to describe what is happening by breaking events down into phases. These are known as **prophase**, **metaphase**, **anaphase** and **telophase**.

Prophase

Before mitosis begins the genetic material has been replicated to produce exact copies of the original chromosomes. By the beginning of prophase both the originals and the copies are referred to as chromatids. In prophase the chromosomes coil up, can take up stains and become visible. Each chromosome at this point consists of two daughter chromatids that are attached to each other in a region known as the **centromere**. The nucleolus breaks down and the centrioles begin to pull apart to form the spindle.

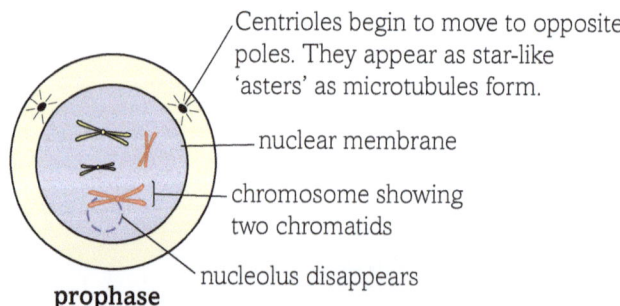

prophase

Metaphase

The nuclear membrane has broken down and the centrioles have moved to opposite poles of the cell, forming a set of microtubules between them that is known as the spindle. The chromatids appear to jostle about for position on the **metaphase plate** or **equator** of the spindle during metaphase. They eventually line up along this plate, with each centromere associated with a microtubule of the spindle.

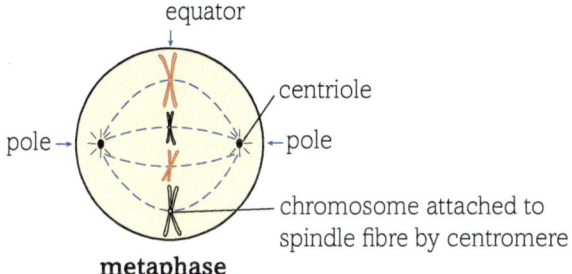

metaphase

Anaphase

The centromeres that have linked the two identical chromatids split, and from then on the chromatids act as completely separate entities. They effectively become new chromosomes. The chromatids from each pair are drawn, centromere first, towards opposite poles of the cell. This separation occurs quickly, taking only a matter of minutes. At the end of anaphase the two sets of chromatids have been separated to opposite ends of the cell. The chromatids cannot move on their own. They rely on the microtubules of the spindle to allow them to move. The spindle was for many years envisaged as a structure running from one end of the cell to the other. It is now known to be made up of overlapping microtubules containing contractile fibres, which are similar to those in animal muscles. Contraction of the overlapping fibres causes the movement of the chromatids. This is an energy-using process, and the energy is supplied by cell respiration.

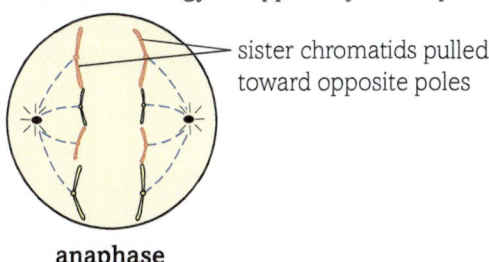

anaphase

Telophase

During telophase the spindle fibres break down and nuclear envelopes form around the two sets of chromosomes. The nucleoli and centrioles are also re-formed. The chromosomes begin to unravel and become less dense and harder to see.

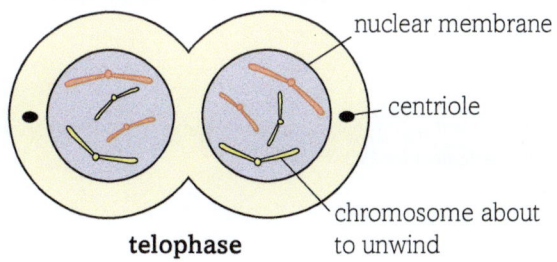

telophase

Cytokinesis

The final phase of the cell cycle is the division of the cytoplasm, sometimes referred to as **cytokinesis**. In animal cells a ring of contractile fibres tightens around the centre of the cell rather like a belt tightening around a sack of flour. These fibres seem to be the same as those found in animal muscles. They continue to contract until the two cells have been separated. In plant cells the division of the cell occurs rather differently, with a cellulose cell wall building up from the inside of the cell outwards. In both cases the end result is the same – two identical daughter cells are formed, which then enter interphase and begin to prepare for the next cycle of division.

(a)

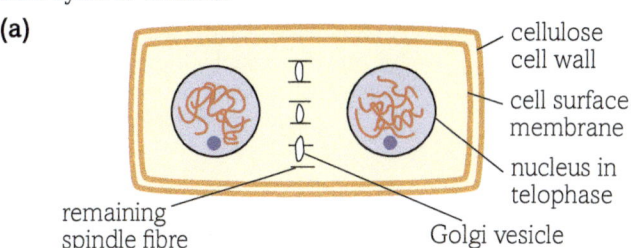

Some spindle fibres remain and guide Golgi vesicles to the equator of the cell.

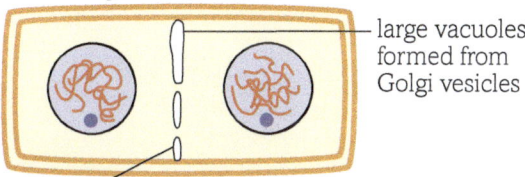

The vesicles enlarge and fuse together, forming a cell plate.

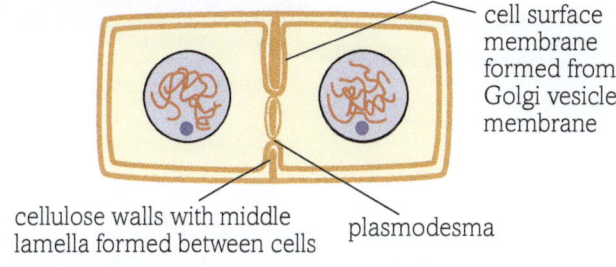

The basic structure of the cell walls forms within each vesicle, and the vesicles fuse to join the cell wall together. Small gaps left between the vesicles form plasmodesmata (see **Section 2.1.1**).

(b)

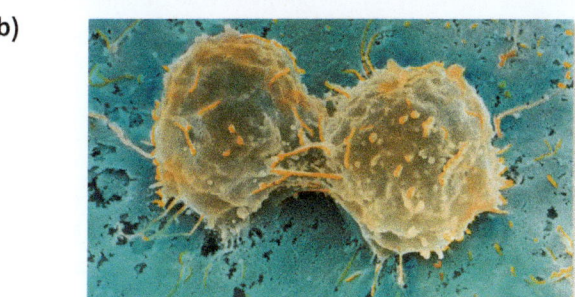

fig A The final stages of the cell cycle in (a) a plant cell; and (b) an animal cell.

Mitosis is the source of all the new cells needed for organisms to grow and to replace worn out cells. It is also the method by which organisms undergo asexual reproduction.

Did you know?

Observing mitosis

The discovery of mitosis depended on the development of the microscope. Walther Flemming published his work on mitosis in 1882. Flemming had also discovered the presence of chromosomes in the cell by using dyes that were taken up by the genetic material. A Belgian scientist, Edouard van Beneden (1846–1910), discovered chromosomes at much the same time. Flemming had not come across Mendel's work on inheritance and so he did not make the connection between what he was seeing and genetic inheritance. In spite of this, Flemming's discoveries are widely regarded as some of the most important work in cell biology.

You can observe mitosis relatively easily in the cells of rapidly dividing tissues such as the meristem at a growing root tip. Using a dye such as acetic orcein, which stains the chromosomes, you can make a temporary tissue squash preparation showing the stages of mitosis. You can also observe mitosis in living tissue, and dramatic recordings of the activity of the chromosomes have been made using time-lapse photography. This has moved our understanding forward considerably. Viewing of the movements of the cell contents during mitosis shows it as a dynamic process in a way which cannot be achieved on the printed page and explains why it is called the 'dance of the chromosomes'.

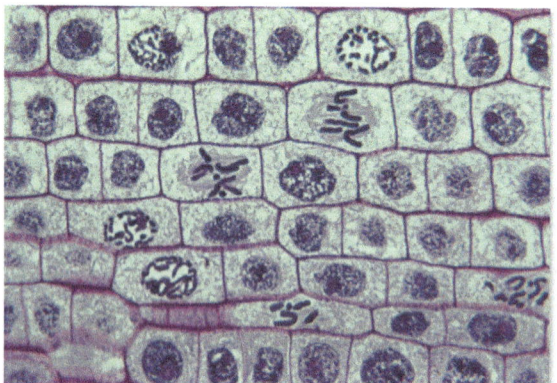

fig B Stained section of a root tip squash showing cells in different stages of the cell cycle, including active mitosis.

Questions

1 Summarise the stages of mitotic cell division in animal cells.

2 Explain why root tips are particularly suitable material to use for preparing slides to show mitosis.

Key definitions

Prophase is the first stage of active cell division where the chromosomes are coiled up and consist of two daughter chromatids joined by the centromere. The nucleolus breaks down.

Metaphase is the second stage of active cell division where a spindle of overlapping protein microtubules forms and the chromatids line up on the metaphase plate.

Anaphase is the third stage of active cell division where the centromeres split so chromatids become new chromosomes. They are moved to the opposite poles of the cell, centromere first, by contractions of the microtubules of the spindle.

Telophase is the fourth stage of active cell division where a nuclear membrane forms around the two sets of chromosomes, the chromosomes unravel and the spindle breaks down.

The **centromere** is the region where a pair of chromatids are joined and which attaches to a single strand of the spindle structure at metaphase.

The **metaphase plate (equator)** is the region of the spindle in the middle of the cell along which the chromatids line up.

Cytokinesis is the final stage of the cell cycle before it enters interphase again – division of the cytoplasm at the end of mitosis to form two independent, genetically identical cells.

3 ▶ Asexual reproduction

By the end of this section, you should be able to...

● explain the importance of mitosis in asexual reproduction

As you have already seen, mitosis is the basis of asexual reproduction. Asexual reproduction involves only one parent individual and it results in genetically identical individuals or **clones**. It has many advantages for an organism. It does not rely on finding a mate and can give rise to large numbers of offspring very rapidly. It also has one big disadvantage – the offspring are almost all genetically identical to the parent organism. This becomes a problem when living conditions change in some way. The introduction of a new disease to an environment, a change in the temperature or human intervention can cause the total destruction of a group of genetically identical organisms, because if one cannot cope, neither can all the others. Many species of plants and fungi undergo both sexual and asexual reproduction as a matter of course. For example, plants may reproduce sexually by flowering, but they also reproduce asexually by methods including bulbs, corms, tubers, rhizomes, runners and suckers.

Strategies for asexual reproduction

There are a variety of strategies for asexual reproduction, all of which are dependent on mitosis and some of which are outlined below.

Binary fission

Fission involves mitosis followed by the splitting of an individual. Two new individuals are usually formed, so it is known as **binary fission**. Bacteria and protists such as *Amoeba* undergo this form of asexual reproduction. Bacteria are capable of enormous increases in numbers under ideal conditions, when they may divide every 20 minutes – one of the shortest known cell cycles. Although fission is limited as a reproductive strategy in the world of multicellular organisms, a similar method is used in cell reproduction for growth and repair in all living things.

Producing spores

Sporulation involves mitosis and the production of asexual spores that are capable of growing into new individuals. These spores can usually survive adverse conditions, and are also easily spread over great distances. This form of asexual reproduction is most common in fungi and plants such as mosses and ferns.

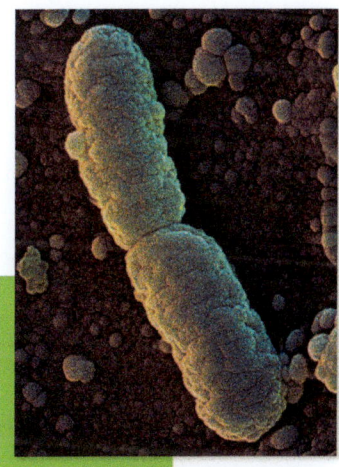

fig A Fission is an effective reproductive strategy for single-celled organisms such as this bacterium.

Regeneration

Regeneration constitutes a very dramatic form of asexual reproduction, occurring when organisms replace parts of the body that have been lost. For example, many lizards shed their tails when attacked and then grow another. This is known as regeneration. Some organisms manage an even more spectacular form of regeneration – they can reproduce themselves asexually from fragments of their original body, a process known as **fragmentation**. For example, certain starfish attack and eat oysters. To protect oyster beds from destruction oyster fishermen have attempted to destroy the starfish, often by chopping them up and throwing them back into the sea. This failed dismally as each fragment can regenerate to form another starfish hungry for oysters.

This type of cloning occurs naturally – some members of groups as diverse as fungi, flatworms, filamentous algae and sponges fragment and then regenerate as a regular method of reproducing. An adaptation of this ability has been developed to allow artificial cloning of plants.

fig B This lizard lost its tail escaping a predator. The tail has not just healed, it is actually regenerating using mitosis.

Producing buds

Budding in a reproductive sense does not mean the production of buds containing flowers or leaves. In reproductive budding there is an outgrowth from the parent organism that produces a smaller but identical individual, produced purely by mitotic cell division. This 'bud' eventually becomes detached from the parent and has an independent existence. Yeast cells, which are single-celled fungi, reproduce by budding. In single-celled organisms like these, the only recognisable difference between budding and binary fission is that in budding the parent cell is larger than the bud. Budding is relatively rare in the animal kingdom. A good example of an animal budding is in *Hydra* (see **fig C**). Asexual budding is only part of the reproductive strategy of *Hydra* – they reproduce sexually as well.

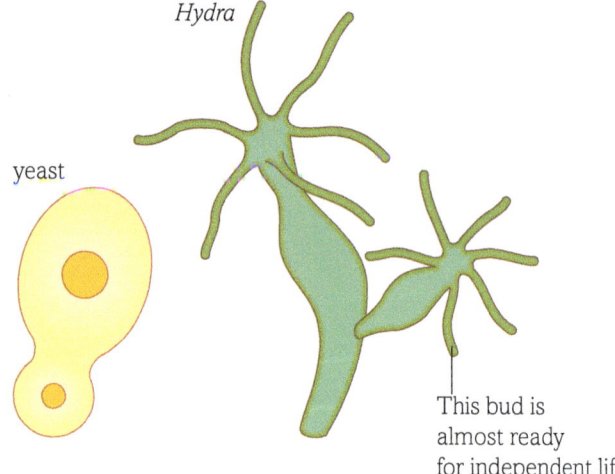

Hydra

yeast

This bud is almost ready for independent life.

fig C Budding in yeast cells and *Hydra*. Even in single-celled yeast the new organism is much smaller than the parent. This asymmetry makes the process different from binary fission.

New plant structures

Vegetative propagation is in some ways a more sophisticated version of reproductive budding and occurs in flowering plants. A plant forms a structure that develops into a fully differentiated new plant, which is identical to the parent, and eventually becomes independent. The new plant may be propagated from the stem, leaf, bud or root of the parent, depending on the type of plant. It involves only mitotic cell division. Vegetative propagation often involves perennating organs. These contain stored food from

photosynthesis and can remain dormant in the soil to survive adverse conditions. They are often not only a means of asexual reproduction, but also a way of surviving from one growing season to the next. Examples include bulbs, corms, runners, suckers, rhizomes, stem tubers and root tubers.

Vegetative propagation is easily exploited by human gardeners to produce new plants. Splitting daffodil bulbs, removing new strawberry plants from their runners and cutting up rhizomes are all easy ways of increasing plant numbers cheaply. As an added advantage the new plants are all clones so they will have exactly the same characteristics as their parents, so they will be the same colour or have the potential to produce fruit that is just as good.

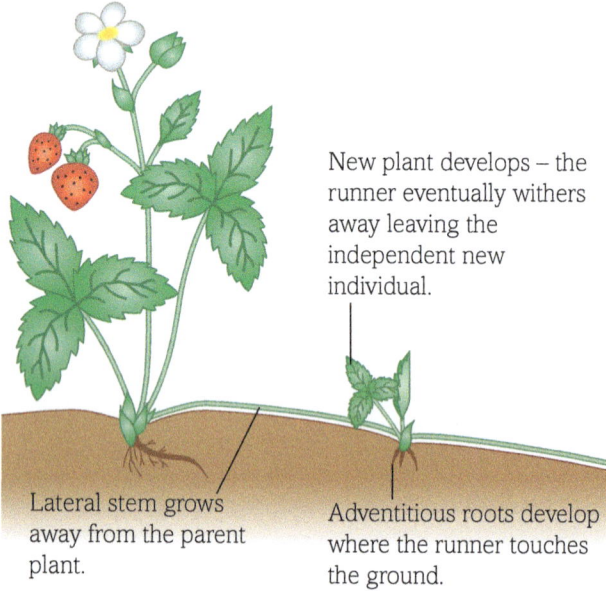

New plant develops – the runner eventually withers away leaving the independent new individual.

Lateral stem grows away from the parent plant.

Adventitious roots develop where the runner touches the ground.

fig D Asexual reproduction in strawberry plants results in identical clones.

Gardeners and farmers take asexual reproduction in plants one step further when they take cuttings. They induce fragmentation artificially. This involves taking a small piece of a plant – often part of the shoot – and planting it to grow on and develop by mitosis into another entire identical plant.

Did you know?

The dangers of cloning crops

Cloning crop plants allows farmers to produce large numbers of productive crops very quickly. Clones are commonplace – potato crops, bananas and grape vines are examples of cloned crops – but because clones are genetically identical, if a disease affects one plant in a crop it will affect them all. The potatoes that were destroyed by fungal blight in the Irish potato famine of the 1840s were all clones. They were lumper potatoes, which had no blight resistance. There are more varieties of modern potatoes, but most potato plants are still clones. Bananas are the clearest example of a global monoculture resulting from artificial asexual reproduction. In the early twentieth century, almost all banana plantations grew a clone or cultivar called Gros Michel. In the 1950s, a fungal disease called Panama disease wiped out almost all the banana plantations in South America and Africa. Now most banana plantations use the Cavendish variety, another banana clone, which so far is proving disease resistant. But it only takes a new disease such as black sigotoka to leave us all without the bananas we love to eat.

Asexual reproduction is common even in complex plants partly because they maintain areas of unspecialised dividing cells throughout their life. In more complex animals, where the cells tend to become specialised, asexual reproduction is much less common.

Did you know?

Komodo dragons and virgin births

Most vertebrates do not reproduce asexually. When they do it is known as **parthenogenesis**. This has been seen in about 80 species of vertebrates, including snakes, fish, a monitor lizard and a turkey. In spite of this, parthenogenesis has always been thought to be very rare in vertebrates. Yet new evidence suggests that parthenogenesis is not as rare as was thought. Komodo dragons are the largest land lizards on Earth. Their natural habitat is Indonesia where there are fewer than 4000 animals remaining. So the Komodo dragons in captivity are part of an important breeding programme. However, two different zoos in the UK, Chester Zoo and London Zoo, have reported that live, apparently healthy young dragons have hatched from eggs laid by females in the complete absence of any male dragons. The females were not related in any way, so it was not a rare family mutation. The young dragons have been DNA tested and their genetic make-up shows they come only from their mother. Both females have also bred sexually. Female Komodo dragons have one W and one Z chromosome, while males have two Zs. Each egg carries either a W or a Z. When parthenogenesis takes place, the single chromosome is duplicated. Any eggs with WW will not develop so there are no parthenogenic female babies, but ZZ eggs can develop into normal male baby lizards.

Richard Gibson is in charge of reptiles at London Zoo and an international expert in Komodo dragons. Richard feels that the arrival of these parthenogenic dragons should lead scientists to rethink their ideas on how common the process of parthenogenesis is, at least in reptiles. The ability to reproduce asexually as well as sexually could have evolved so that animals stranded in an isolated situation can nevertheless breed. If so it would be a very useful adaptation indeed – and therefore it would not be surprising if it was relatively common.

Until recently, all the known vertebrate parthenogenic births had been to animals in captivity. However, in 2012, Warren Booth and his team at the University of Tulsa in the USA used DNA genotyping to demonstrate the first documented cases of parthenogenesis in two closely related species of wild pit viper snakes. This throws open the whole question of the role of parthenogenesis in the lives of reptiles and many other species. There is a lot more research to be done.

fig E This small male snake is the result of parthenogenesis – he was born to the wild mother snake with no male input into his genetic make-up.

Questions

1 Based on the information in this section and further research, make a table to summarise the forms of asexual reproduction most commonly seen in eukaryotic organisms.

2 More living organisms result from asexual reproduction than sexual reproduction. Do you think this statement is accurate? Explain your response.

3 New observations can change long-held scientific ideas. Why have the Komodo dragon hatchlings and the wild-born parthenogenic pit vipers forced scientists to rethink their ideas about asexual reproduction in vertebrates?

Key definitions

Clones are genetically identical offspring produced as a result of natural or artificial asexual reproduction.

Binary fission is the splitting of one individual to form two new individuals as a result of mitosis.

Sporulation is the process involving mitosis in the production of asexual spores that can grow into new individuals.

Regeneration is the use of mitosis to regrow a body part that has been lost.

Fragmentation is the use of mitosis to regenerate a whole organism from a fragment of the original.

Budding is the production by mitosis of an outgrowth from the parent organism that develops into a small independent organism.

Vegetative propagation is the process by which a plant forms a structure by mitosis that develops into a fully differentiated, genetically identical new plant.

Parthenogenesis is the process by which an unfertilised egg cell develops into a new individual.

By the end of this section, you should be able to...

- explain the importance of mitosis in growth and repair

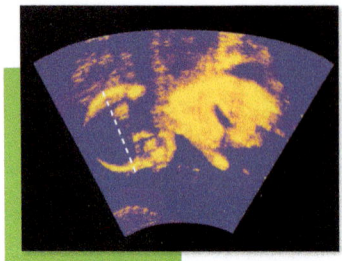

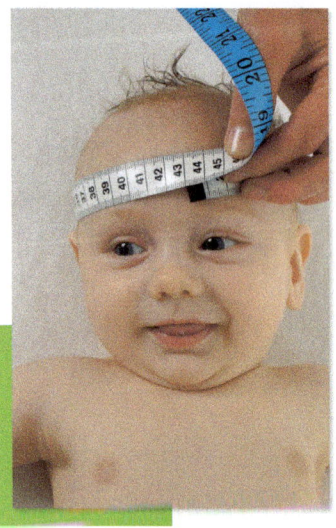

fig A Head circumference is measured at intervals as a child grows, starting before birth, to check that all is well. However, an increase in circumference does not always indicate growth.

Mitosis is not just about asexual reproduction. It plays a vital role in growth as well. Everyone is familiar with the concept of growth, but defining it in a biological sense is not so easy. Growth is a permanent increase in the number of cells, or in the mass or size of an organism. There are three distinct aspects of growth. They are cell division, assimilation and cell expansion. Cell division, or mitosis, is the basis of growth. Once cells have divided they usually get larger before dividing again. The resources needed to produce new cell material come from photosynthesis in plant cells, from feeding in animal cells and from nutrient absorption in fungi. This is what is meant by assimilation, and when these materials are incorporated into cells the result is cell expansion. Cells can expand in other ways, for example by taking in water, but this increase in size may be only temporary. So growth is defined as involving a *permanent* increase in cell number, size or mass – or all three.

How is growth measured?

The measurement of growth is important both scientifically and medically. Growth may be affected by factors such as the availability of food, temperature and light intensity as well as the genetic make-up of the organism. Unfortunately the measurement of growth is not at all easy. Linear dimensions, such as height or head circumference (see **fig A**), can be very deceptive – cake mixture will increase in both height and circumference as it cooks, but it has not grown. Measuring mass also has its problems – the water content of the cells may vary greatly, particularly in fungi and plants, and more complex animals will have varying quantities of faecal material and urine held in their bodies.

Because growth involves an increase in the cell content of an organism, mass is the best and most commonly used measure of growth. However, the water content of organisms can vary greatly so **dry mass** is the most accurate way of measuring growth. The dry mass is the mass of the body of an organism with all the water removed from it. This gives an accurate picture of the amount of biological material present, but has one major drawback. If you remove all the water from an organism you kill it, so that further growth cannot be measured. To get useful results from dry mass measurements you need to grow large samples of genetically identical organisms under similar conditions, then take random samples and dry them to a constant dry mass. This method is very useful for plants, fungi and bacteria, but has obvious limitations for animals. It is not easy or ethical to maintain large colonies of genetically identical vertebrates and then kill and dry them, for example. This means that in most cases scientists use less reliable indicators such as height and wet mass to measure growth when working with animals.

Growth patterns

In spite of the difficulties in measuring growth, we have a good picture of the patterns of growth of many organisms. Growth curves show growth throughout the life of an organism, including when most growth takes place. The growth curve is very similar for most organisms (see **fig B**). In many animals, after an initial relatively slow start there is a rapid period of growth until maturity is reached, when growth slows down and may stop. In most land animals, growth stops completely with maturity because size is limited by the weight of the animal and the ability of its muscles to move it against gravity. In plants, growth often continues throughout life, and the same is true for marine animals, where the mass of the body is supported by the water. This pattern – even when it stops at maturity – is known as continuous growth.

Not all organisms undergo continuous growth. Insects grow in a series of moults. They shed one exoskeleton and then, while the new exoskeleton is soft, they expand the body by taking in air or water and 'grow'. Once the new skeleton has hardened, the air or water can be released and there is room for the tissues of the insect to increase in size and mass. This is known as discontinuous growth. If length is measured the insect appears to grow in a series of steps.

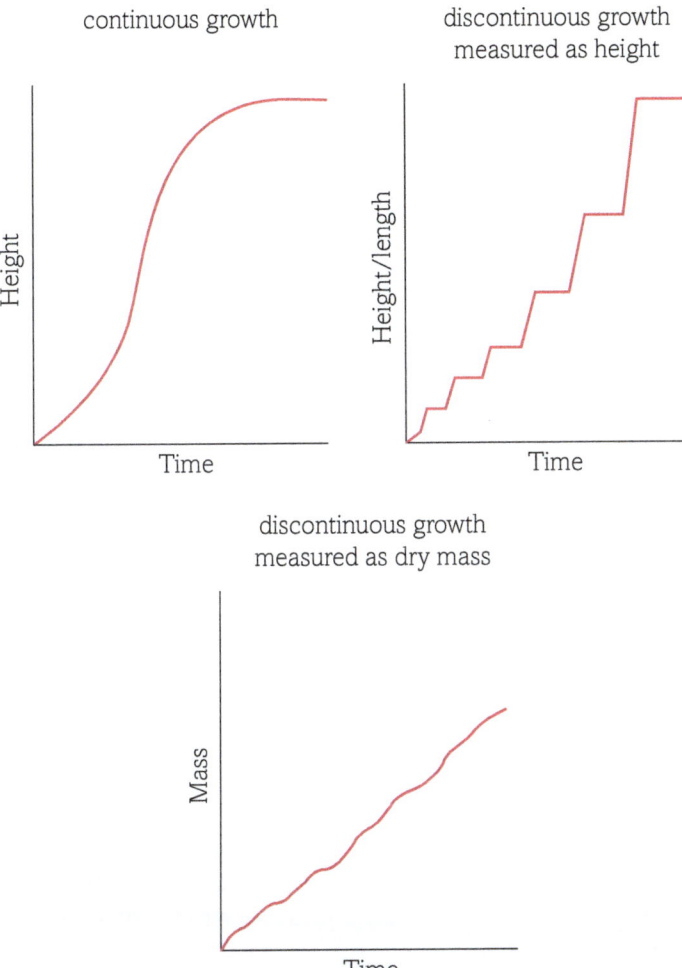

fig B Growth curves are usually measured by linear dimensions. If wet mass is used, the discontinuous growth curve becomes even more pronounced.

The development of the embryo is the time when the largest amount of growth, measured as a percentage of body mass, occurs. Different parts of the organism can grow at very different rates. For example, in the human embryo the nervous system and the head grow much faster than some other areas. Later in life – through puberty, for example – the head stops growing while the long bones and the rest of the body continue. Right at the beginning of life, mitosis takes place at a very rapid rate.

Growth in plants

In plants mitosis continues throughout life. It takes place in regions known as the **meristems**. These are areas just behind the tip of the stem or root where the cells continue to divide actively throughout the life of the plant. After the cells have divided they absorb water into their vacuoles and elongate rapidly before the cellulose cell wall becomes more rigid. These areas of plant growth are particularly sensitive to a variety of stimuli such as light and gravity.

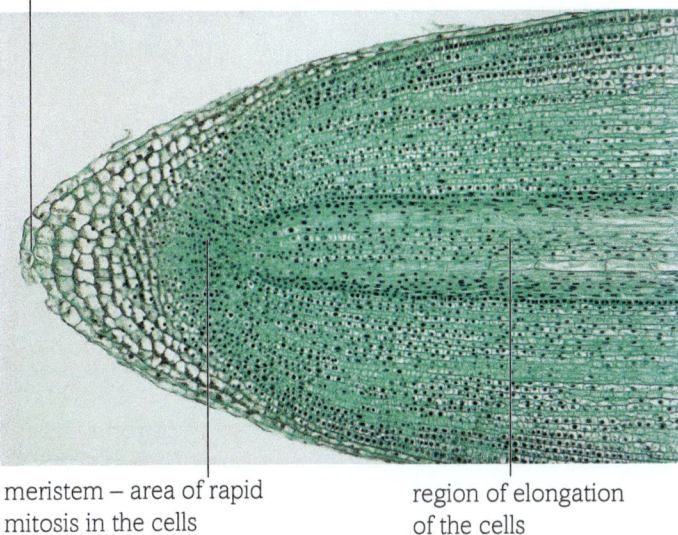

root tip protects dividing cells

meristem – area of rapid mitosis in the cells

region of elongation of the cells

fig C The apical meristems in the root tips and shoots are the sites of growth in a plant.

Mitosis and repair

However, even in organisms where growth slows down or stops completely at maturity, mitosis does not stop. Cells are continually becoming worn out and being replaced by mitotic divisions. In some tissues mitosis occurs at a rapid rate all the time. For example, the entire surface of your skin is replaced every 28 days so mitosis is taking place in skin cells all the time. Each red blood cell only lasts 120 days in the blood so mitosis also occurs rapidly in the red bone marrow of the flat bones such as your ribs to keep up with the demand. This continues until the onset of senescence or old age, when mitosis occurs less frequently and the cells dying begin to outnumber the new cells being formed. When this process reaches a certain point, death of the whole organism will occur.

If the skin is damaged, rapid mitosis is triggered in the cells around the wound. They regenerate the lost skin tissue and heal the wound. Plants can also produce scar tissue to seal damage to their bark. Mitosis is vital for the repair of worn out or damaged tissues.

Questions

1 How does the role of mitosis differ in the processes of growth and repair?

2 Why are the curves for discontinuous growth so different when measured as height and as dry mass?

Key definitions

Dry mass is the mass of the body of an organism with all the water removed from it.

The **meristem** is the region of mitosis and growth in a plant shoot or root.

THINKING BIGGER

CANCER – MITOSIS OUT OF CONTROL

In 2013 an estimated 8.2 million people around the world died from different forms of cancer. The article below suggests answers to some common questions about cancer.

What is cancer?

Cancer is an umbrella term that covers more than 200 different diseases. However, most cancers have some features in common:

- Cancer cells do not respond to the normal mechanisms that control cell growth – they divide rapidly to form a mass of abnormally growing cells called a tumour.
- Malignant tumours split and release small groups of cells into the blood and tissue fluid. They circulate and often lodge in different areas of the body, forming secondary tumours.
- Cancer cells divide more rapidly and usually live longer than normal cells.

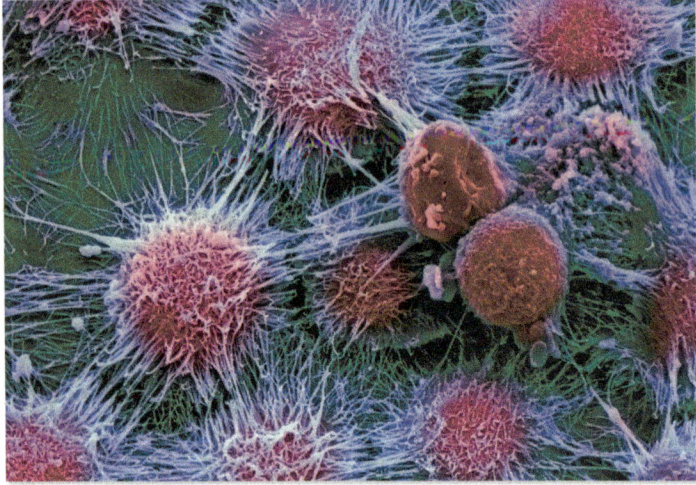

fig A Kidney cancer cells.

Why is control of the cell cycle lost?

About 15% of human cancers are the result of viral infections which cause changes in the cells, affecting the control of the cell cycle and leading to cancer.

Most cancers result from mutations which occur in the DNA of a normal body cell as it reproduces:

- Several different mutations may work together to increase the likelihood of cancer, e.g. by interfering with the accurate replication of the DNA, decreasing the efficiency of the DNA repair mechanisms or increasing the likelihood of the chromosomes breaking during mitosis.
- A single mutation changing a proto-oncogene to an oncogene can cause cancer. Proto-oncogenes code for proteins which stimulate the cell cycle. Oncogenes produce uncontrolled amounts of stimulating proteins.
- Tumour-suppressor genes code for the production of chemicals which suppress the cell cycle, acting as a brake on cell division. If the gene mutates, the brake is removed and the cell just keeps on dividing.

What causes cancer?

Two main factors affect whether an individual will develop a form of cancer:

- Genetics: some people are more likely than others to experience a mutation which results in cancer, and some people are born with a mutation which gives them a very high risk of developing particular cancers.
- Environment: factors such as the tar in cigarette smoke, ionising radiation, the chemicals in alcoholic drinks and asbestos are carcinogenic – they increase the risk of mutations in vulnerable cells which can result in cancer.

If a genetically vulnerable person encounters carcinogenic environmental factors, their risk of developing cancer somewhere in their body increases even more.

Where else will I encounter these themes?

1.1   1.2   1.3   1.4   2.1   2.2   2.3   YOU ARE HERE   2.4

Let us start by considering the nature of the writing in this article.

There are many different ways of getting information across. This article assumes knowledge of cell division in its readers:

1. How well does a question and answer format work for imparting this information?

2. Are bullet points helpful or would continuous text make the explanations more effective?

3. Try a different approach – work with a partner and choose two alternative ways of getting this information across. Each take one method and work it through – then compare notes and decide which you prefer and why.

4. How would you explain cancer to people who are not studying AS/A level Biology and do not know what mitosis and the cell cycle are?

> Alternative ways of getting information across include formal continuous prose, a series of diagrams, flow diagrams, a more relaxed and friendly style of writing … and any others you can think of! For a non-scientific audience you have to think about the words you use – and how pictures might help.

Now let us look at the biology behind these questions and answers. You already know about cell division and mutation, so you can answer these questions now. If you are going to continue your biology studies to A level, you may like to revisit these pages after you have learned more about communicable diseases and the immune system in **Book 2 Topic 6**.

5. At one level there are as many types of cancer as there are types of human cell – any type of cell can become cancerous, although some are more prone to cancer than others. At another level, every cancer is different because each one of us is different and it is our cells that mutate. Describe how the process of cell division can lead to mutations and therefore cancer.

> **Command word**
> If you are asked to describe something, you need to give a clear and concise account of it. Pull out all the important pieces of information as a starting point.

Activity

You are going to produce an article for a popular magazine explaining about one of the following forms of cancer. You must explain what happens to the cells of your body and why in each case the risk of developing cancer is increased. You can add as much extra information about detection, screening, prevention and even treatment as you like – but the key points to focus on are the biology of the cancer. Write about one of the following.

1 Breast cancer resulting from mutations in the BRCA1 and BRCA2 tumour suppressor genes.

2 Cervical cancer caused by the human papilloma virus.

● From an article by Ann Fullick.

1 The photograph below shows a cell in the metaphase stage of mitosis as seen using a light microscope.

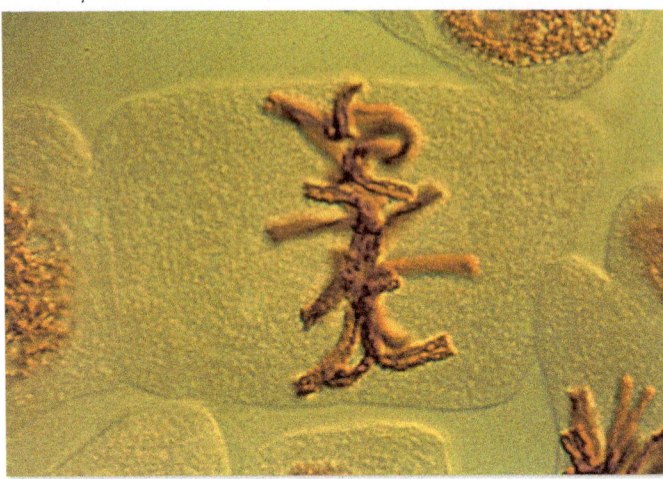

(a) Place a cross (☒) in the box next to the correct statement below.

Metaphase occurs after anaphase and before telophase ☐

Metaphase occurs after prophase and before anaphase ☐

Metaphase occurs after telophase and before prophase ☐

Metapahse occurs after anaphase and before prophase ☐

[1]

(b) Draw and label a diagram to show the appearance of a chromosome in metaphase. [3]

(c) Suggest how the cell shown in the photograph would differ when it is in anaphase. [2]

[Total: 6]

2 (a) The cell cycle includes interphase and mitosis. Mitosis has four phases: prophase, metaphase, anaphase and telophase. The photograph below shows plant root cells at various stages of the cell cycle.

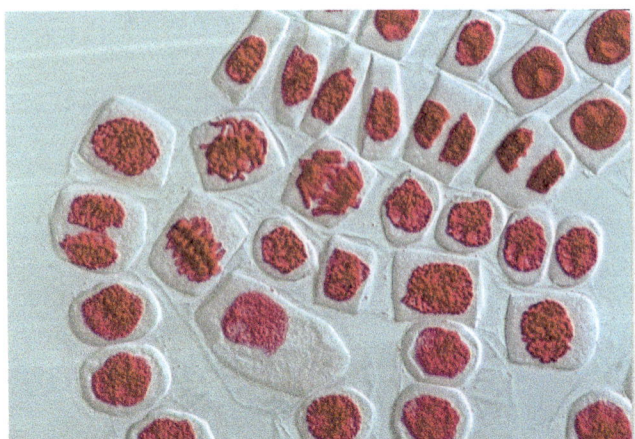

(i) Draw a line to indicate a cell in the photograph that is undergoing **anaphase** and label this line A. [1]

(ii) Draw a line to indicate a cell in the photograph that is undergoing **telophase** and label this line T. [1]

(iii) How many of the cells shown in the photograph are in **telophase**. [1]

(b) Give an account of the events that take place during prophase and metaphase of mitosis. [5]

[Total: 8]

3 The graphs below show changes in the DNA content of cells during the cell cycle in two different plants, **A** and **B**.

Plant A

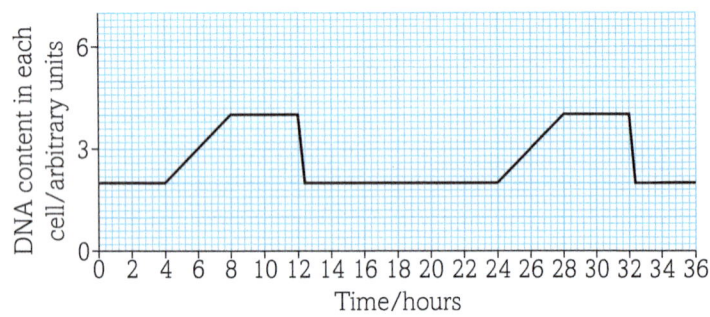

Plant B

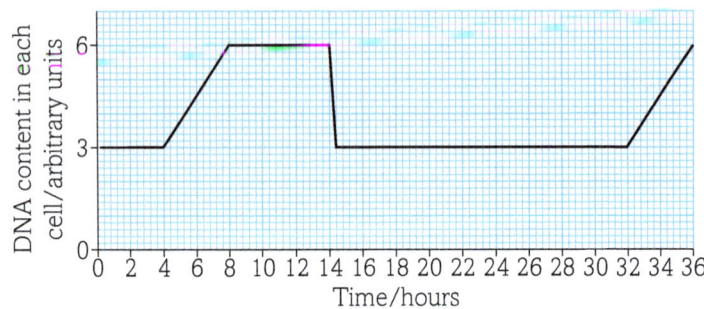

(a) Compare the cell cycle of **plant A** with the cell cycle of plant B. [3]

(b) The DNA content of the cells of **plant A** doubles between 4 and 8 hours. Give an explanation for this change in DNA content. [2]

(c) Describe the events that are occurring inside the cells of plant A between 11 and 13 hours. [2]

[Total: 7]

4 (a) The flow diagram below shows a method for preparing and staining cells in order to observe chromosomes.

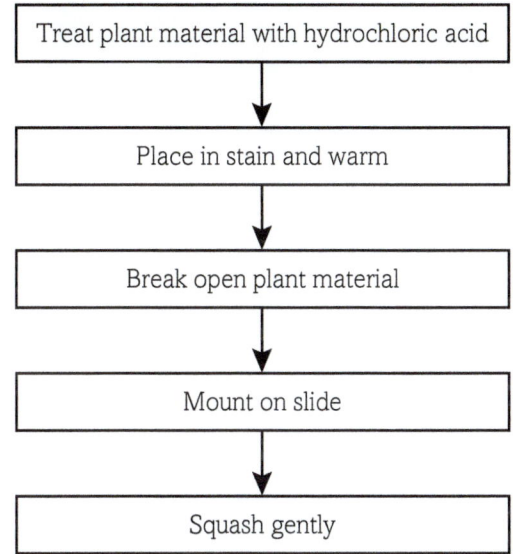

Treat plant material with hydrochloric acid

↓

Place in stain and warm

↓

Break open plant material

↓

Mount on slide

↓

Squash gently

(i) Name a suitable part of a plant to use. Give a reason for your answer. [2]
(ii) Name a suitable stain for this method. [1]
(iii) Explain why it is necessary to squash the plant material at the end of the process. [1]

(b) The diagrams **A**, **B**, **C** and **D** below show four stages of mitosis.

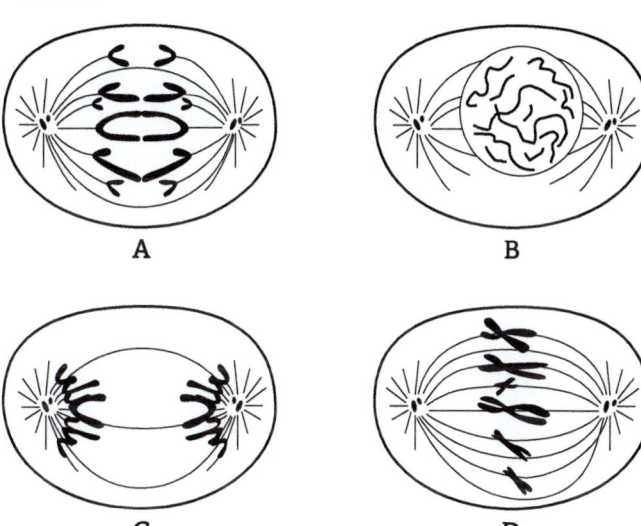

A   B

C   D

(i) Write the letters to show the correct sequence of the stages in mitosis. [1]
(ii) Name the stages shown in diagrams **A** and **C**. [2]

[Total: 7]

CHAPTER 2.4

Meiosis and sexual reproduction

Introduction

When we look at cell division we always talk about eukaryotic cells having two sets of chromosomes, one from each parent. Polysomy – having an extra copy of just one chromosome – is seen as an example of things going wrong. However, many plant species are actually polyploid – they have several complete sets of chromosomes! Increasingly, evidence suggests that polyploidy helps plants survive changing or adverse conditions. This may be why 40% of plant species survived the mass extinction event that wiped out the dinosaurs around 65 million years ago.

The spare genetic information in polyploidy may enable plants to overcome harmful mutations, and they can use the same gene to control different functions. In the Arctic Circle, many of the plants that survive best in the extreme conditions are polyploid – and it is the polyploids that are most successful in colonising bare areas left as glaciers retreat. There is even some evidence that polyploidy may have played a part in human evolution.

In this chapter you will be looking at meiotic cell division in eukaryotic cells. Meiosis is the process by which a reduction in the chromosome number is achieved in the formation of the sex cells or gametes. Meiosis only takes place in the sex organs.

You will learn the process of meiotic cell division and its importance in introducing genetic variation, key to survival in changing conditions, to natural selection and to evolution. You will also consider what happens in chromosome mutations, and in non-disjunction of the chromosomes resulting in either monosomy or polysomy.

You are going to look at gametogenesis – the formation of the sex cells – in both mammals and plants, and then go on to explore how those gametes join at fertilisation to form a new genetic individual. You will follow the mammalian zygote through the early stages of embryonic development until it implants in the uterus of the mother.

All the maths you need

- Carry out calculations using numbers in standard and ordinary form (*e.g. use of magnification*)
- Use scales for measuring (*e.g. measuring sizes of male and female gametes*)
- Find arithmetic means (*e.g. measuring sizes of male and female gametes*)
- Make order of magnitude calculations (*e.g. use and manipulate the magnification formula:*
$magnification = \dfrac{size\ of\ image}{size\ of\ real\ object}$)
- Use and manipulate equations, including changing the subject of an equation (*e.g. magnification*)

What have I studied before?

- The role of meiotic cell division in halving the chromosome number to form gametes
- The ultrastructure of eukaryotic cells including the nucleus, nuclear membrane, chromosomes, centrioles, etc.
- The way in which DNA replicates in the nucleus
- Gene mutations

What will I study later?

- Mutations as a source of new variations
- Classification of plants involves the numbers and role of the cotyledons in the seed
- Post-transcriptional changes in mRNA can result in different products from the same gene (A level)
- Epigenetic modifications and their effect on totipotent stem cells in the embryo (A level)
- The role of random assortment and crossing over in meiosis in giving rise to new combinations of alleles in gametes (A level)
- How random fertilisation during sexual reproduction brings about genetic variation (A level)
- Autosomal and sex linkage of alleles (A level)

What will I study in this chapter?

- The role of meiosis in the production of haploid gametes, including the stages of meiosis
- The replication and division of the genetic material in the main stages of meiosis
- The ways in which meiosis results in genetic variation through recombination of alleles, including independent assortment and crossing over
- Chromosome mutations such as translocations
- Non-disjunction of the chromosomes leading to conditions such as Down's syndrome and Turner's syndrome
- The development of the female and male gametes in mammals – oogenesis and spermatogenesis – and in plants with the formation of the pollen grain and the embryo sac
- Fertilisation in mammals followed by the early development of the embryo
- Fertilisation in plants including the roles of the tube nucleus, the pollen tube and enzymes and the process of double fertilisation inside the embryo sac to form a triploid endosperm and the zygote

By the end of this section, you should be able to...

● describe the role of meiosis in the production of haploid gametes including the stages of meiosis

● explain how meiosis results in genetic variation through recombination of alleles, including independent assortment and crossing over

As you have seen, asexual reproduction can be very successful at producing new individuals, but leaves the population vulnerable to changes in the environment. The offspring are largely identical to their parents. There is very limited genetic variety, although spontaneous mutations do occur.

Relatively few organisms rely solely on asexual reproduction. Most have at the very least a back-up system of sexual reproduction, used when conditions are tough, to introduce the genetic variation that may enable the population or species to survive. In more complex organisms, particularly animals and flowering plants, sexual reproduction is the main way of producing fertile offspring.

Sexual reproduction is the production of a new individual resulting from the joining of two specialised cells known as gametes. The individuals that result from sexual reproduction are not genetically the same as either of their parents, but contain genetic information from both of them. Sexual reproduction relies on two gametes meeting and fusing. It is not always easy to find a mate, particularly if you are a solitary predator. It is also more expensive in terms of bodily resources because it usually involves special sexual organs. But the great advantage of sexual reproduction is that it increases genetic variation as a result of the fusing of gametes from two different individuals. In a changing environment, this gives a greater chance that one or more of the offspring will have a combination of genes that improves their chance of surviving and going on to reproduce.

fig A The genetic variation in offspring produced by sexual reproduction can be very easy to see.

What are gametes?

The nucleus of a cell contains the chromosomes. In most of the cells, the chromosomes occur in pairs. A cell containing two full sets of chromosomes is called **diploid (2n)** and the number of chromosomes in a diploid cell is characteristic for that species. However, if two diploid cells combined to form a new individual in sexual reproduction, the offspring would have four sets of chromosomes, losing the characteristic number for the species. Each new generation would become more heavily loaded with genetic material until eventually the cells would break down and fail to function. To avoid this, **haploid (n)** nuclei are formed with one set of chromosomes (half of the full chromosome number), usually within the specialised cells called gametes. Sexual reproduction occurs when two haploid nuclei fuse to form a new diploid cell called a **zygote** (see **fig B**), a process called **fertilisation**.

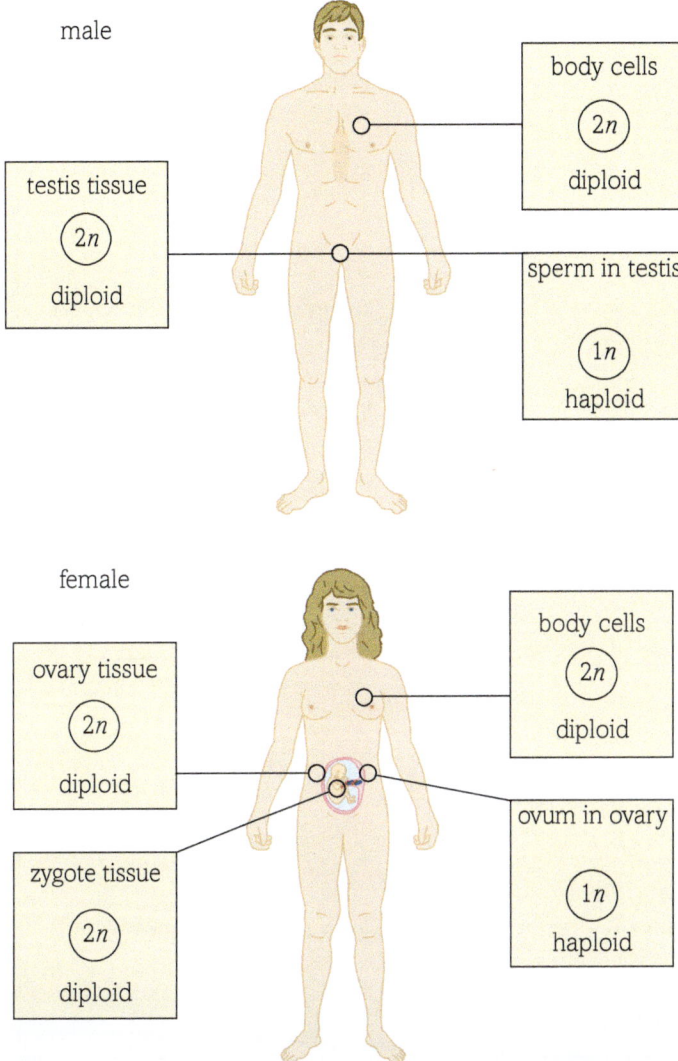

and the female ones are the **ovaries**. The male gametes, **pollen**, are produced in the anthers and the female gametes, **ovules**, are formed in the ovaries. In animals the male gonads are the **testes**, which produce the male gametes known as **spermatozoa**, or more commonly as **sperm**. The female gonads are the ovaries and they produce the female gametes known as **ova** or eggs. The male gametes are often much smaller than the female ones, but they are usually produced in much larger quantities. This can be summarised as:

- male: many, mini, motile
- female: few, fat, fixed

Meiosis

In **Section 2.3.2** you saw that when cells divide by mitosis the number of chromosomes in both the daughter cells is the same as in the original parent cell. However, in the cell divisions that form gametes the chromosome number needs to be halved to give the necessary haploid nuclei. To bring about this reduction in the chromosome number, gametes are formed by a different process of cell division known as meiosis.

Meiosis is a reduction division and it occurs only in the sex organs. In animals the gametes are formed directly from meiosis. In flowering plants meiosis forms special male cells called **microspores** and female cells called **megaspores**, which then develop into the gametes. Meiosis is of great biological significance – it is the basis of the variation that allows species to evolve.

What happens to the chromosomes?

In meiosis two nuclear divisions give rise to four haploid daughter cells, each with its own unique combination of genetic material (see **fig C**). The events of meiosis are continuous although we describe the stages as separate phases. As in mitosis, the contents of the cell, and in particular the DNA, are replicated while the cell is in interphase. Once the cell has all the materials it needs it can enter meiosis.

Many of the stages of meiosis are very similar to those of mitosis, with just a couple of crucial variations. The chromosomes replicate to form chromatids joined by a centromere as in mitosis. However, in meiosis the two chromosomes of each pair, known as **homologous pairs**, stay close together. At this stage **crossing over** or **recombination** takes place, introducing genetic variation as the chromatids break and recombine (see **fig E**). Just as in mitosis, the nuclear membrane and nucleolus break down and the centrioles pull apart to form the spindle. The centromeres do not split in the first division of meiosis, so pairs of chromatids move to the opposite ends of the cell. The cell then immediately goes into a second division without any further replication of the chromosomes. This division is just like mitosis. The centromeres divide and chromatids move to opposite poles of the cell. Finally the nuclear membranes re-form as the chromosomes decondense and become invisible again. Cytokinesis takes place giving four haploid daughter cells, each with half the chromosome number of the original parent cell. These daughter cells later develop into gametes.

fig B The only cells in the human body that are haploid are the gametes.

Did you know?

Polyploidy

Although most eukaryotic organisms are diploid, a number have stable forms of **polyploidy**. Fish can be polyploid; for example salmon have four sets of chromosomes and some fish have as many as 400 chromosomes in total. Some reptiles are also polyploid. Polyploidy is very common in plants, particularly ferns and flowering plants. Potatoes are polyploid and so are the members of the genus *Dendranthema*. The haploid number of the genus is 9, but the polyploid numbers average 18 (2 sets of chromosomes) to 198 (22 sets of chromosomes).

Crop plants are often polyploid. Some wheat varieties are diploid, tetraploid and hexaploid, and the cabbage family is also hexaploid.

The formation of gametes

Gametes are formed in special sex organs. In simpler animals and plants the sex organs are often temporary, formed only when they are needed. In more complex animals the sex organs are usually more permanent structures that are sometimes called the **gonads**. In flowering plants the male sex organs are the **anthers**

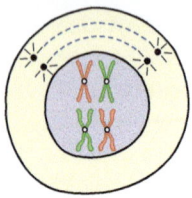

Prophase I – each chromosome appears in the condensed form with two chromatids. Homologous pairs of chromosomes associate with each other. **Crossing over** occurs.

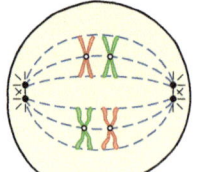

Metaphase I – the spindle forms and the pairs of chromosomes line up on the metaphase plate.

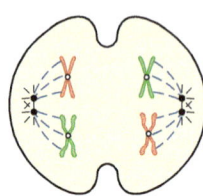

Anaphase I – the centromeres do not divide. One chromosome (pair of chromatids) from each homologous pair moves to each end of the cell. As a result the chromosome number in each cell is half that of the original.

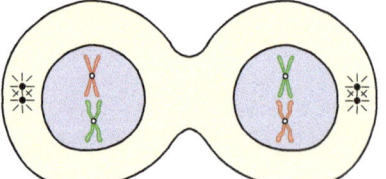

Telophase I – the nuclear membrane re-forms and the cells begin to divide. In some cells this continues to full cytokinesis and there may be a period of brief or prolonged interphase. During this interphase there is *no further replication* of the DNA.

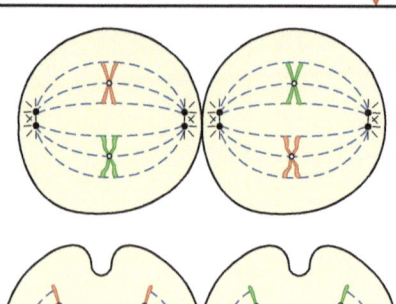

Metaphase II – new spindles are formed and the chromosomes, still made up of pairs of chromatids, line up on the metaphase plate.

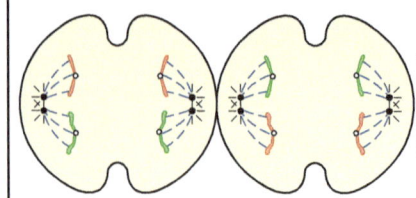

Anaphase II – the centromeres now divide and the chromatids move to the opposite ends of the cell.

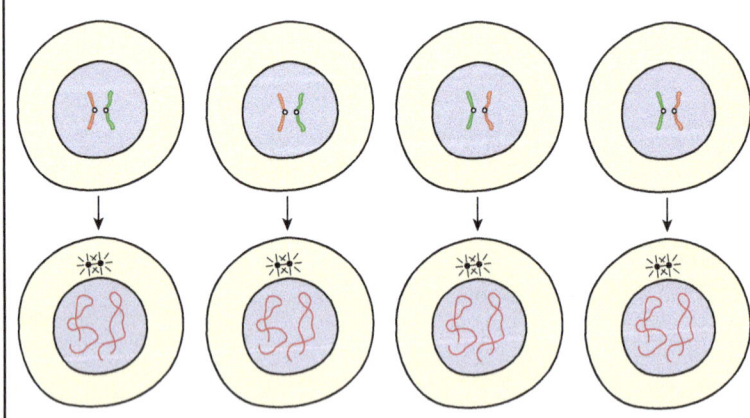

Telophase II – nuclear envelopes re-form, the chromosomes return to their interphase state and cytokinesis occurs, giving four daughter cells each with half the chromosome number of the original diploid cell.

fig C The main steps in the process of meiosis, which results in the formation of haploid gametes. This is a simplified version of meiosis shown in a cell with only two pairs of chromosomes.

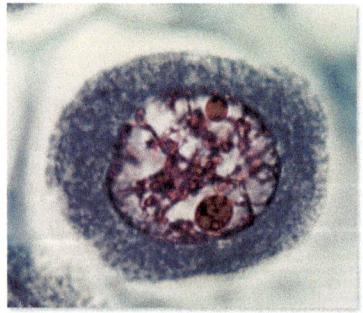
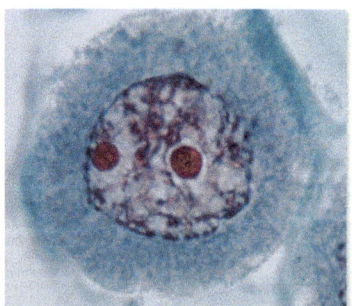
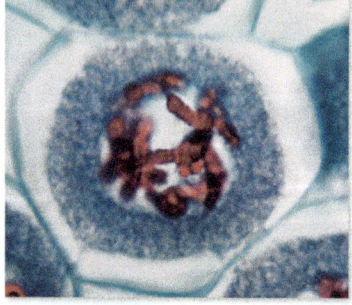
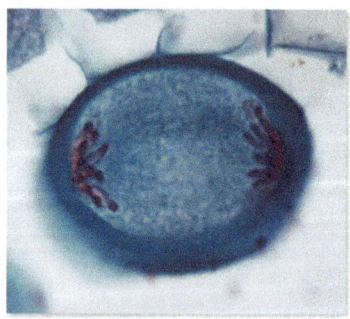
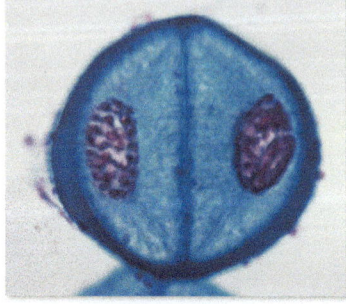
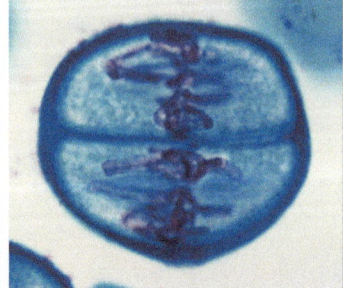
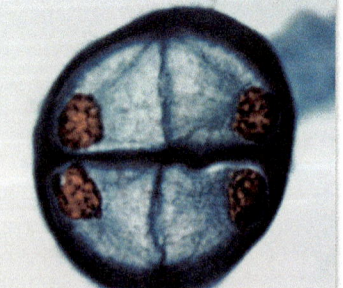

fig D The stages of meiosis are not easy to see in cells, but these images from the testis of a locust and anther of a plant show you many of them.

The importance of meiosis

Meiosis reduces the chromosome number in gametes from diploid to haploid, so that sexual reproduction is possible without each generation carrying an increasing burden of genetic material. It is also the main way in which genetic variation is introduced to a species. This variation is introduced in two main ways:

- **Independent assortment (random assortment)**: the chromosomes that came from the individual's two parents are distributed into the gametes and so into their offspring completely at random. For example, each gamete you produce receives 23 chromosomes. In each new gamete any number from none to all 23 could come from either your maternal or your paternal chromosomes. It has been calculated that there are more than eight million potential genetic combinations within the sperm or the egg. This alone guarantees great variety in the gametes.

- **Crossing over (recombination)**: this process takes place when large, multi-enzyme complexes 'cut and join' bits of the maternal and paternal chromatids together. The points where the chromatids break are called **chiasmata**. These are important in two ways. First, the exchange of genetic material leads to added genetic variation in its own right. Second, errors in the process lead to mutation (see **Sections 1.3.7** and **2.4.2**) and this is a further way of introducing new combinations into the genetic make-up of a species.

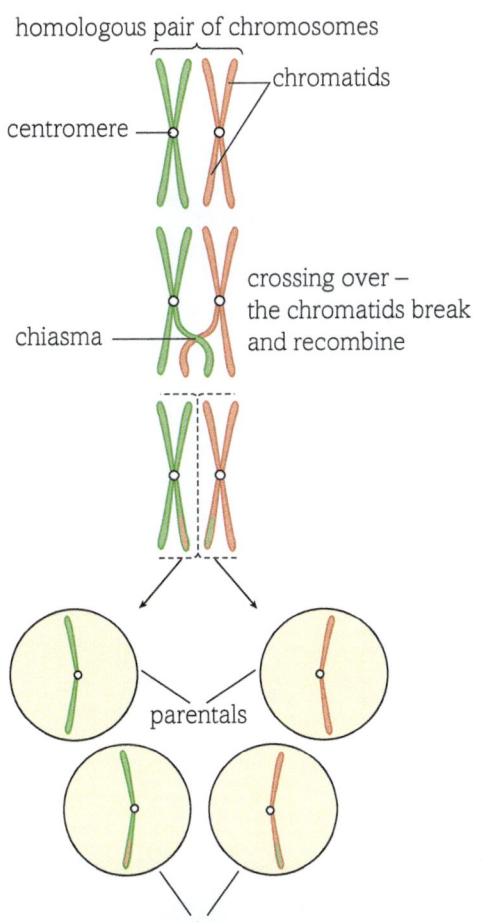

fig E Chromosomes crossing over in meiosis – this process introduces more variation into the gametes.

Questions

1 Give examples of conditions when sexual reproduction would be more advantageous in the production of offspring than asexual reproduction, and vice versa. Explain your choices.

2 Make a table to summarise the main stages in the process of meiosis.

3 Explain how meiosis leads to variation between offspring.

Key definitions

Diploid (2n) signifies a cell with a nucleus containing two full sets of chromosomes.

Haploid (n) signifies a cell with a nucleus containing one complete set of chromosomes.

A **zygote** is the cell formed when two haploid gametes fuse at fertilisation.

Fertilisation is the fusing of the haploid nuclei from two gametes to form a diploid zygote in sexual reproduction.

Polypoidy is when a cell or an organism has more than two sets of chromosomes.

Gonads are the sex organs in animals.

Anthers are male sex organs in plants that produce the male gametes, pollen.

Ovaries are the female sex organs in both animals and plants. They produce the female gametes called ovules in plants and ova in animals.

Pollen is the haploid male gametes in plants.

Ovules are the haploid female gametes in plants.

Testes are the male sex organs that produce the male gametes – sperm.

Spermatozoa (sperm) are the haploid male gametes in animals.

Ova are the haploid female gametes in animals.

Microspores are the result of meiosis in plants that develop into the male gametes, pollen.

Megaspores are the result of meiosis in plants that develop into the female gametes, ovules.

Homologous chromosomes describe a set of one maternal chromosome and one paternal chromosome that pair up during meiotic cell division.

Crossing over (recombination) is the process by which large multi-enzyme complexes cut and rejoin parts of the maternal and paternal chromatids at the end of prophase I.

Independent assortment (random assortment) is the process by which the chromosomes derived from the male and female parent are distributed into the gametes at random.

Chiasmata are the points where the chromatids break during recombination.

2 ▶ Mutations

By the end of this section, you should be able to...

● describe chromosome mutations, illustrated by translocations

● explain how non-disjunction can lead to polysomy and monosomy

In **Section 1.3.7** you learned about gene mutations and the way these can affect the proteins produced by the cells in protein synthesis. However, these are not the only forms of mutations that can affect the genetic material during the process of replication and meiosis.

Chromosome mutations

Sometimes during the process of meiosis, parts of the chromosomes break off and become reattached in the wrong place. These are called chromosome mutations. They are changes on a larger scale than the gene mutations you looked at in **Section 1.3.7**. One of the most common forms of a chromosome mutation is **translocation**. This takes place when a piece from one pair of homologous chromosomes breaks off and reattaches to one of a completely different pair of chromosomes.

Some translocations are balanced – a piece is effectively swapped between two different chromosomes (see **fig A**). People who have balanced translocations are often healthy. However, some translocations are unbalanced – one chromosome loses a piece and another chromosome gains it. These mutations can cause big changes to the phenotype of the individual. They may even be incompatible with life.

If a point or gene mutation is like changing a letter in a word (see **Section 1.3.7**), a chromosome mutation is like changing or removing a word, or completely rearranging the words in the sentence. If we are lucky they will still make sense, but they will probably not mean the same thing as the original sentence.

Translocations can have severe effects. For example, translocations between chromosome 8 and chromosome 21 can result in a type of blood cancer known as core binding factor acute myeloid leukaemia, whilst translocations between chromosome 8 and chromosome 14 can cause Burkitt's lymphoma, a cancer of white blood cells mainly seen in children and young adults.

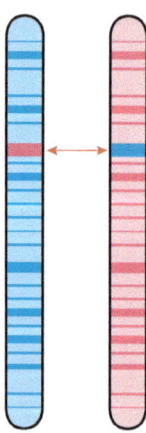

fig A In a balanced chromosome translocation, part of one chromosome is swapped with a section from a completely different pair of chromosomes.

Non-disjunction of the chromosomes

Some mutations affect not single genes or parts of chromosomes, but whole chromosomes. When the cells in the ovaries and testes undergo meiotic division to form the ova and sperm, the chromosome number in the cells is halved. In humans, each gamete should contain 23 chromosomes, including one **sex chromosome**, to pass on to the next generation. However, sometimes an error called **non-disjunction** occurs. During the reduction division of meiosis, the members of one of the homologous pairs of chromosomes fail to separate during anaphase II. As a result, one of the gametes has two copies of that chromosome, and another has no copies. If one of these abnormal gametes joins with a normal gamete and is fertilised, the individual who results will either have **monosomy** with only one member of the homologous pair present from the normal gamete, or **polysomy** with three or more rather than two chromosomes of a particular type. The situation where a cell either lacks a whole chromosome or has more than two of a chromosome is called **aneuploidy**. Most examples of aneuploidy are fatal and affected fetuses do not develop to term. For example, trisomy of chromosome 16 sometimes occurs, but all of the embryos abort spontaneously in the first three months of pregnancy. Trisomy of chromosome 18 causes Edward's syndrome. Most affected babies are stillborn, and those that are born alive almost always die within a year. However, there are some examples of aneuploidy where those affected can and do survive, particularly when the sex chromosomes are affected rather than the autosomes.

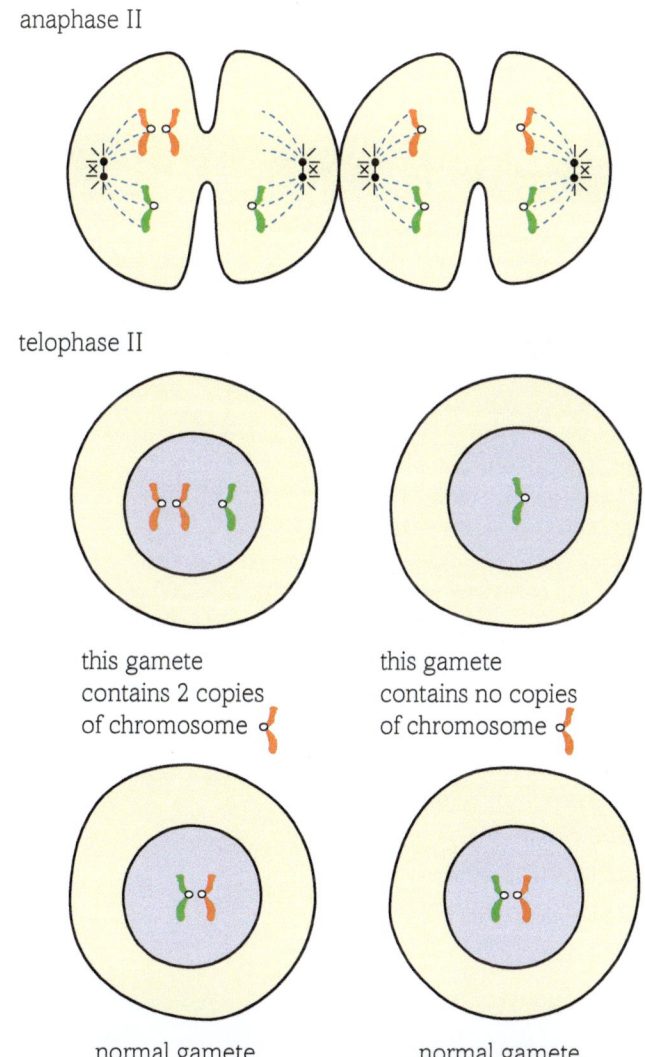

fig B Non-disjunction during meiosis results in gametes with an extra chromosome or a missing chromosome.

Example 1: Down's syndrome (polysomy)

If there is non-disjunction of chromosome 21 in an ovum or sperm, one of the gametes will contain two copies of the chromosome. After fertilisation with a normal gamete, the resulting zygote will have polysomy, with three copies of chromosome 21. Babies born with trisomy of chromosome 21 have Down's syndrome. The extra chromosome affects both mental and physical development. Children born with Down's syndrome will often have problems including heart abnormalities, learning difficulties that can be quite severe, lack of muscle tone and visual problems.

fig C The physical and mental characteristics of Down's syndrome result from one extra copy of chromosome 21.

Did you know?

Factors affecting mutation rates

Mutations are relatively rare events. However, certain factors increase the frequency with which they occur. Ionising radiation is known to increase the **mutation rate**. Certain chemicals (called **mutagens**) also increase the rate of mutation. The rate of non-disjunction increases with the age of the parents – both monosomy and polysomy are more common in children born to older mothers and fathers. The ova are particularly vulnerable. All of the future ova of a woman are suspended in prophase I of meiosis while the woman herself is still an embryo. Meiosis is completed only after ovulation and fertilisation by a sperm (see **Section 2.4.1**). The ova are exposed to all the chemical and radiation hazards of modern life and as a result the rate of non-disjunction increases as the woman gets older.

The effect is clearly seen in the increased incidence in Down's syndrome pregnancies as women get older. The risk of having a baby with Down's syndrome for a 20-year-old woman is 1 in 1667. As she approaches 50 – about the limit of natural child-bearing – the risk has increased. Figures quoted range from 1 in 5 to 1 in 18 births. The same pattern is seen for other conditions involving non-disjunction of the chromosomes.

Example 2: Turner's syndrome (monosomy)

Apart from children affected by Down's syndrome, few fetuses with aneuploidy of the autosomes survive even until birth. However, the absence of a sex chromosome or the presence of extra sex chromosomes is less unusual and less life-threatening, although in most cases it does cause fertility problems. The presence or absence of a Y chromosome determines the route for sexual development in human embryos. Any embryo with at least one Y chromosome will develop male characteristics, whilst any embryo lacking a Y chromosome will develop female characteristics. When there is non-disjunction of the male sex chromosomes, an egg may be fertilised by a sperm that has no sex chromosomes. The resulting embryo will have monosomy – just one X chromosome from the ovum. They will have the genotype XO that results in Turner's syndrome. The affected person is apparently female, but she is infertile and will not undergo puberty without being given extra sex hormones. However, non-disjunction of the male chromosomes will also give sperm carrying both an X and a Y chromosome. If one of these sperm fertilises a normal egg the resulting embryo will be XXY. This is called Kleinfelter's syndrome and affected individuals have small testes and produce little testosterone. They have little facial and body hair, may develop breast tissue, have less muscle development than usual and may be infertile. Around 1 in every 600 live male births is affected by Kleinfelters syndrome, and the individuals may be indistinguishable from their peers or relatively severely affected.

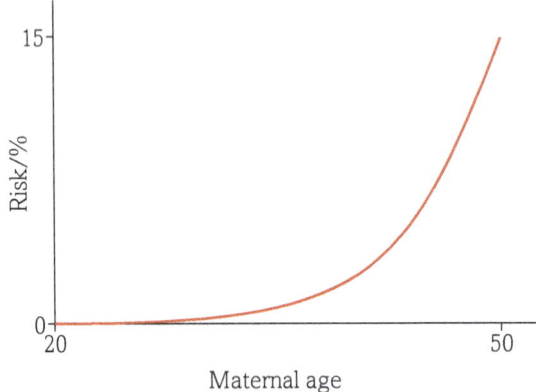

fig D This graph shows how the risk of having a child with Down's syndrome increases with the age of the mother.

Questions

1   Explain the difference between crossing over and translocation in meiosis and the effects they have on the organism involved.

2   What is the difference between polysomy and monosomy and how do they occur?

Key definitions

A **translocation** (noun) is a mutation in which part of one chromosome breaks off and rejoins to another completely different chromosome. It may be balanced, if part of two chromosomes effectively swap, or it may be unbalanced if a piece simply breaks off one chromosome and joins another.

Sex chromosomes are the chromosomes that carry the information that determines the sex of the individual. Human females have two X chromosomes (XX) and males have an X and a Y chromosome (XY).

Non-disjunction is the process that occurs when members of a pair of chromosomes fail to separate during the reduction division of meiosis, resulting in one gamete with two copies of a chromosome and one gamete with no copies of that chromosome.

Monosomy is when only one member of a pair of chromosomes is present in a cell.

Polysomy is when a cell contains three or more rather than two chromosomes of a particular type.

Aneuploidy is when a cell contains too few or too many chromosomes.

The **mutation rate** is the rate at which mutations naturally occur.

Mutagens are chemicals known to increase the rate of mutation.

By the end of this section, you should be able to...

● explain the process of oogenesis and spermatogenesis in mammals

● explain how a pollen grain forms in the anther and the embryo sac forms in the ovule in plants

The gametes that make sexual reproduction possible are formed in a process called **gametogenesis**. Meiosis is just one stage in gamete formation, which produces different male and female sex cells. You are going to consider the way in which sperm and ova are made in the sex organs of mammals, using humans as an example, and also how gametes are formed in flowering plants.

Gametogenesis in mammals

Many millions of sperm are released every time a male mammal ejaculates. The eggs in a sexually mature female are usually numbered in thousands and will eventually run out. Special cells (the **primordial germ cells**) in the gonads divide, grow, divide again and then differentiate into the gametes.

Both mitosis and meiosis play a role in gametogenesis. Mitosis provides the precursor cells. Meiosis brings about the reduction divisions that result in gametes. In human males, the process of gametogenesis continues constantly from puberty. In females, the mitotic divisions take place before birth. The meiotic divisions take place in a few oocytes in each monthly cycle from puberty to menopause and are only completed if the oocyte is fertilised.

Spermatogenesis

Spermatogenesis is the formation of spermatozoa. Each primordial germ cell in the testes results in large numbers of spermatozoa (see **fig A**). There are enormous numbers of primordial germ cells in the testes producing millions of spermatozoa on a regular basis:

• The diploid primordial germ cell divides several times by mitosis to form diploid spermatogonia.

• The spermatogonia then grow without further division until they are big enough to be called primary spermatocytes.

• The spermatocytes undergo meiosis. The first meiotic division results in two haploid cells called secondary spermatocytes.

• The second meiotic division results in four haploid cells called spermatids.

• The spermatids then differentiate in the tubules of the testes to form spermatozoa, the active gametes capable of fertilising an ovum.

Oogenesis

Oogenesis is the formation of ova. Each primordial germ cell in an ovary results in only one ovum (see **fig A**). As a result the number of female ova is always substantially smaller than the number of spermatozoa. Ova contain a much higher proportion of material than sperm so there is a much greater investment of resources in each one. It would not make sense biologically to waste resources by producing too many of them:

• The diploid primordial germ cell divides several times by mitosis to form diploid oogonia. Most of the oogonia do not develop further. They simply degenerate. Only one continues to grow and substantial amounts of storage material go into the cell making it very large compared with the spermatocytes. At this stage the large cell is known as a primary oocyte.

• The oocyte undergoes meiosis. The first meiotic division results in two cells of very unequal size. The larger cell is the secondary oocyte. The other, much smaller, cell sticks to the oocyte and is called the first polar body. At this stage the oocytes do not divide further until after ovulation. What we call ova in the ovary are really secondary oocytes.

- The second meiotic division takes place only after fertilisation occurs. The secondary oocyte divides to form the haploid ovum and another polar body, whilst the first polar body divides to form two more polar bodies. The polar bodies do not seem to have any function except to receive the chromosomes in the meiotic divisions. They degenerate and die as the ovum develops.

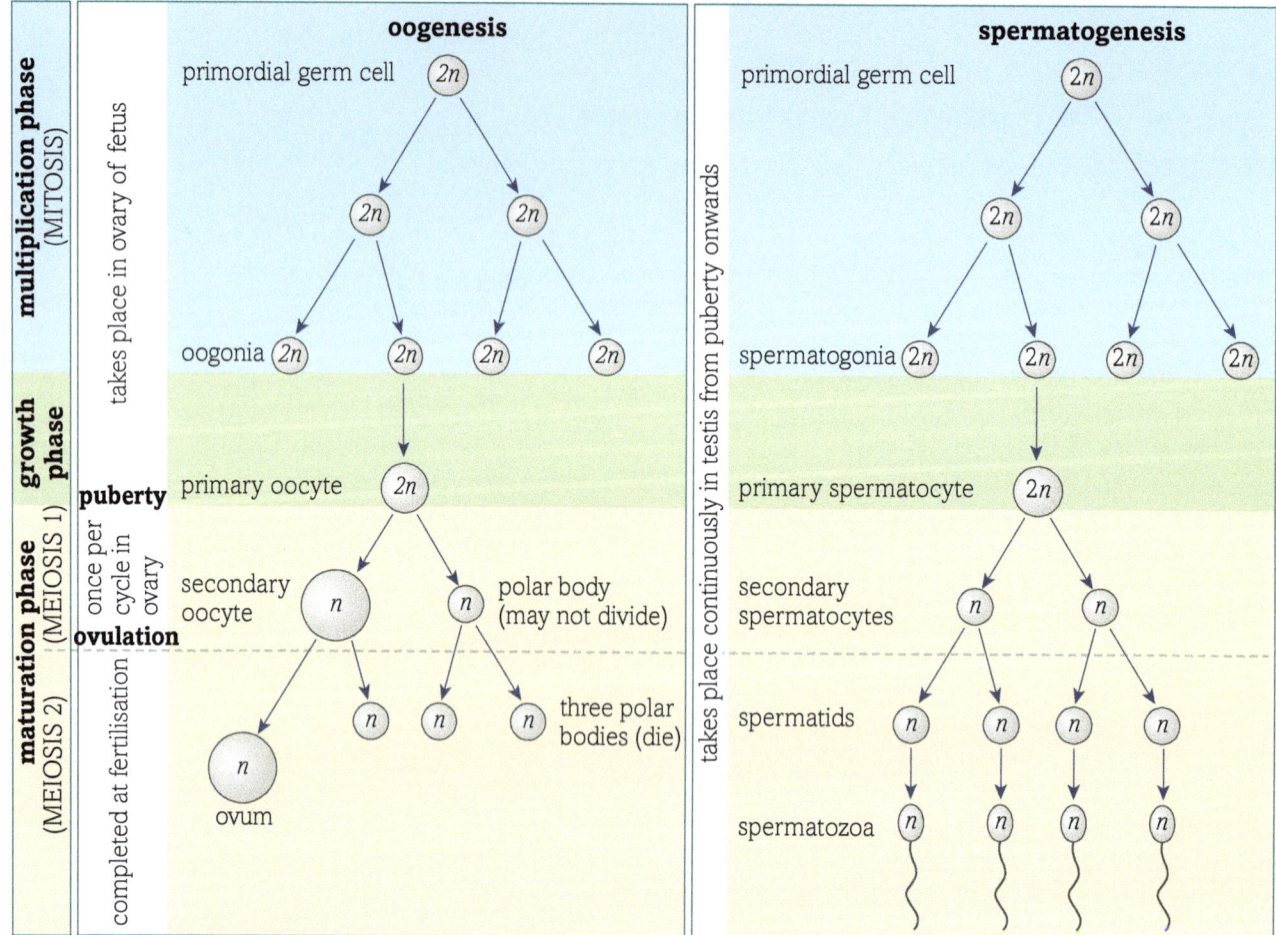

fig A The formation of the sperm in the testis and the ova in the ovary.

Characteristics of the gametes

Spermatozoa: many, mini, motile

The male gametes or spermatozoa of most mammalian species, including humans, are around 50 µm long. They have several tasks to fulfil. They must remain in suspension in the semen so they can be transported through the female reproductive tract, and they must be able to penetrate the protective barrier around the ovum and deliver the male haploid genome safely inside. The close relationship between the human spermatozoan's structure and its functions is shown in **fig B**. Millions upon millions of these motile gametes are produced in the lifetime of a human male. The average size of a UK family is about 1.7 children, with only one spermatozoan needed to fertilise each ovum, and this gives an idea of the scale of biological wastage.

Ova: few, fat, fixed

Although spermatozoa of most animals are very similar in size, the same cannot be said for ova. These vary tremendously in both their diameter and their mass. The human ovum is about 0.1 mm across, whilst the ovum contained in an ostrich egg is around 6 mm in diameter. Eggs do not move on their own, so they do not need contractile proteins, but they usually contain food for the developing embryo. The main difference between eggs of various species is the quantity of stored food they contain. In birds and reptiles a lot of development takes place before the animal hatches, so the egg contains a large food store. In mammals, once the developing fetus has implanted in the uterus it is supplied with nutrients from the blood supply of the mother and so large food stores in the egg are unnecessary.

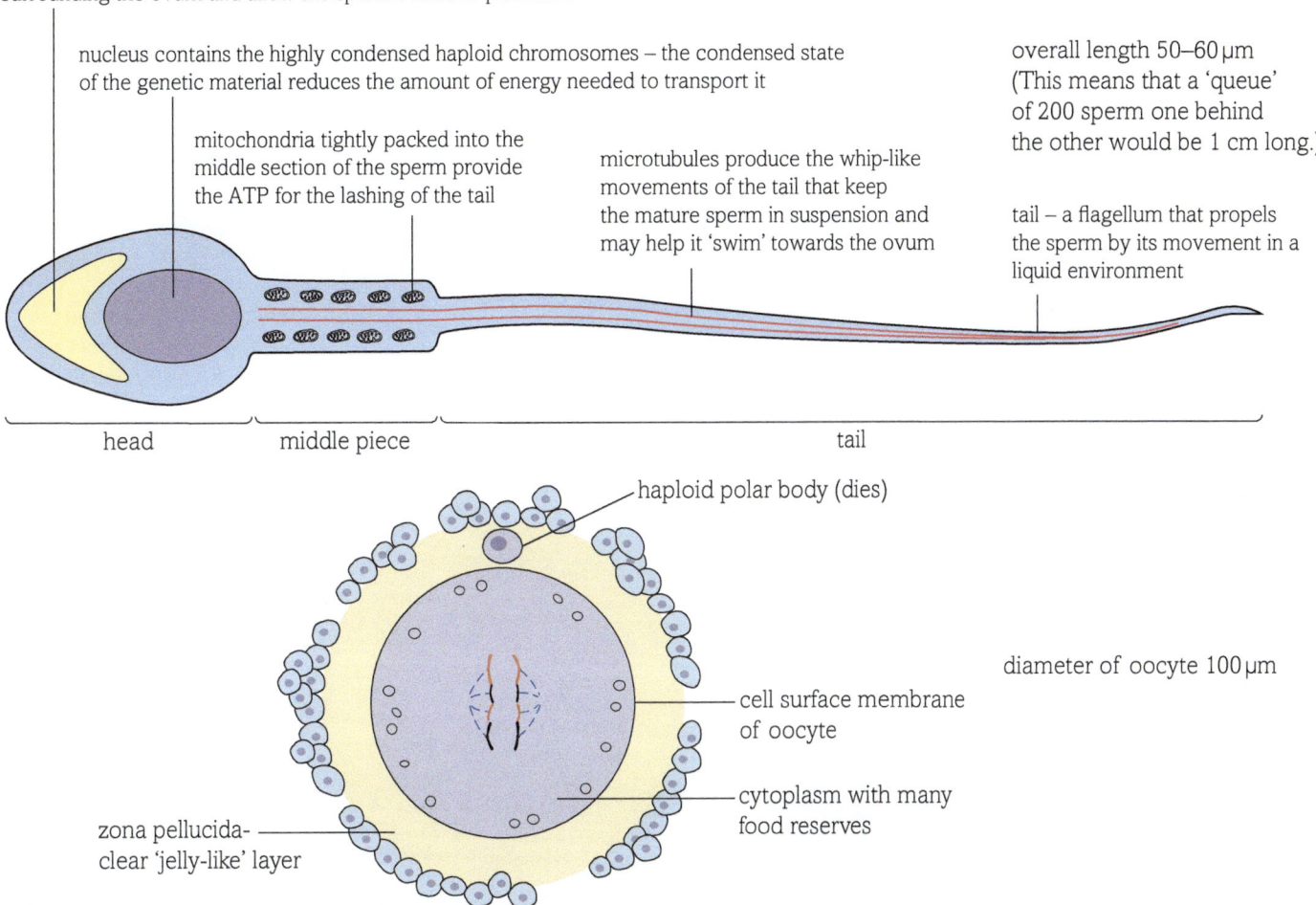

acrosome – membrane-bound storage site for enzymes that digest the layers surrounding the ovum and allow the sperm's head to penetrate

nucleus contains the highly condensed haploid chromosomes – the condensed state of the genetic material reduces the amount of energy needed to transport it

mitochondria tightly packed into the middle section of the sperm provide the ATP for the lashing of the tail

microtubules produce the whip-like movements of the tail that keep the mature sperm in suspension and may help it 'swim' towards the ovum

overall length 50–60 μm (This means that a 'queue' of 200 sperm one behind the other would be 1 cm long.)

tail – a flagellum that propels the sperm by its movement in a liquid environment

head   middle piece   tail

haploid polar body (dies)

cell surface membrane of oocyte

cytoplasm with many food reserves

diameter of oocyte 100 μm

zona pellucida- clear 'jelly-like' layer

fig B Human gametes – the sperm and ova show clear specialisations that fit them for their function. They are not drawn to scale.

Gametogenesis in plants

The formation of gametes in flowering plants is more complex because plants have two phases to their life cycles. The **sporophyte generation** is diploid and produces spores by meiosis. The **gametophyte generation** that results is haploid and gives rise to the gametes by mitosis. In plants such as mosses and ferns, these two phases exist as separate plants. In flowering plants, the two phases have been combined into one plant. The main body of the plant that we see is the diploid **sporophyte**. The haploid gametophytes are reduced to parts of the contents of the anther and the ovary. They are produced by meiosis from spore mother cells.

The formation of pollen, the microgametes

The anthers of flowering plants are analogous to the testes of animals. Meiosis occurs here, resulting in vast numbers of the pollen grains that carry the male gametes. Each anther contains four **pollen sacs** where the pollen grains develop. In each pollen sac there are large numbers of microspore mother cells. These are diploid. They divide by meiosis (see **fig C**) to form haploid microspores, which are the gametophyte generation. The **microgametes** themselves are formed from the microspores by mitosis, with one cell enveloping the other to form a pollen grain containing two haploid nuclei, the **tube nucleus** and the **generative nucleus**. The tube nucleus has the function of producing a **pollen tube** that penetrates through stigma, style and ovary and into the ovule. The generative nucleus then fuses with the nucleus of the ovule to form a new individual.

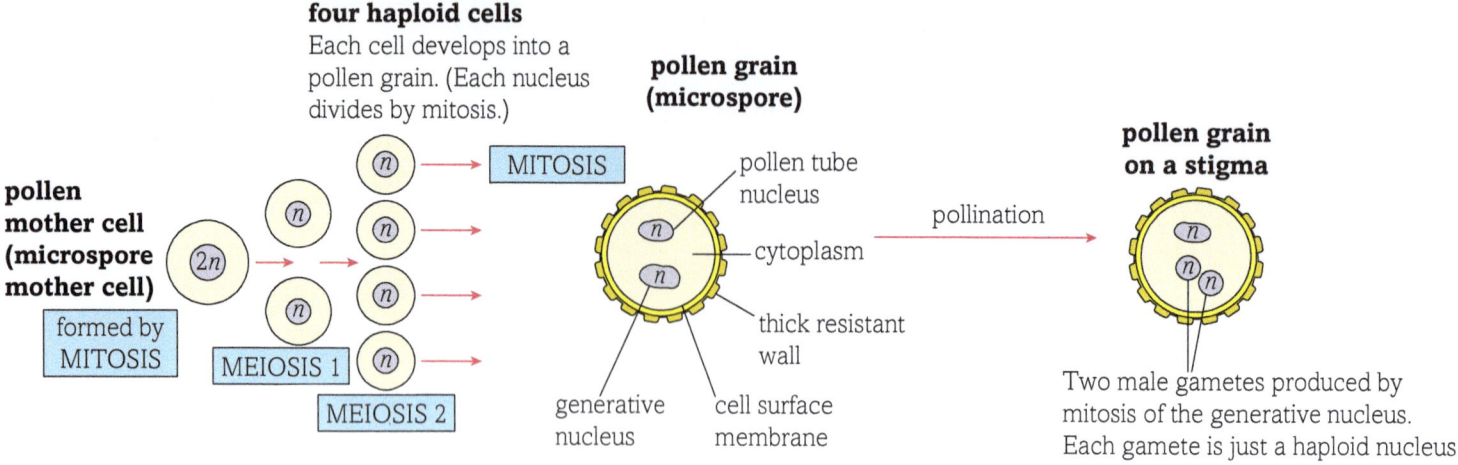

four haploid cells
Each cell develops into a pollen grain. (Each nucleus divides by mitosis.)

pollen mother cell (microspore mother cell)

formed by MITOSIS

MITOSIS

MEIOSIS 1

MEIOSIS 2

pollen grain (microspore)

pollen tube nucleus

cytoplasm

thick resistant wall

generative nucleus

cell surface membrane

pollen grain on a stigma

pollination

Two male gametes produced by mitosis of the generative nucleus. Each gamete is just a haploid nucleus.

fig C Microgametogenesis – the formation of the pollen grain.

Did you know?

The pollen record

The surface patterns of pollen grains are unique and specific to the species. They are extremely tough and resistant to decay and can remain in the soil for thousands of years. Palaeobotanists can tell what plants were growing thousands of years ago and how abundant they were by analysing the pollen they find in archaeological digs.

fig D These amazing looking pollen grains are from plants including sunflowers, morning glory, hollyhocks, lilies, primroses and castor oil plants.

The formation of egg cells, the megagametes

The ovary of the plant is analogous to the animal ovary. Meiosis results in the formation of a relatively small number of ova contained within ovules inside the ovary. Some plants – an example is the nectarine – produce only one ovule (egg chamber), whilst others such as peas produce several. The ovule is attached to the wall of the ovary by a pad of special tissue called the **placenta**. A complex structure of integuments (coverings) forms around tissue known as the nucellus. In the centre the embryo sac forms the gametophyte generation (see **fig E**).

Diploid megaspore mother cells divide by meiosis to give rise to four haploid megaspores, three of which degenerate leaving one to continue to develop. The megaspore undergoes three mitotic divisions that result in an embryo sac containing an egg cell, known as the **megagamete**, two polar nuclei and various other small cells, some of which degenerate.

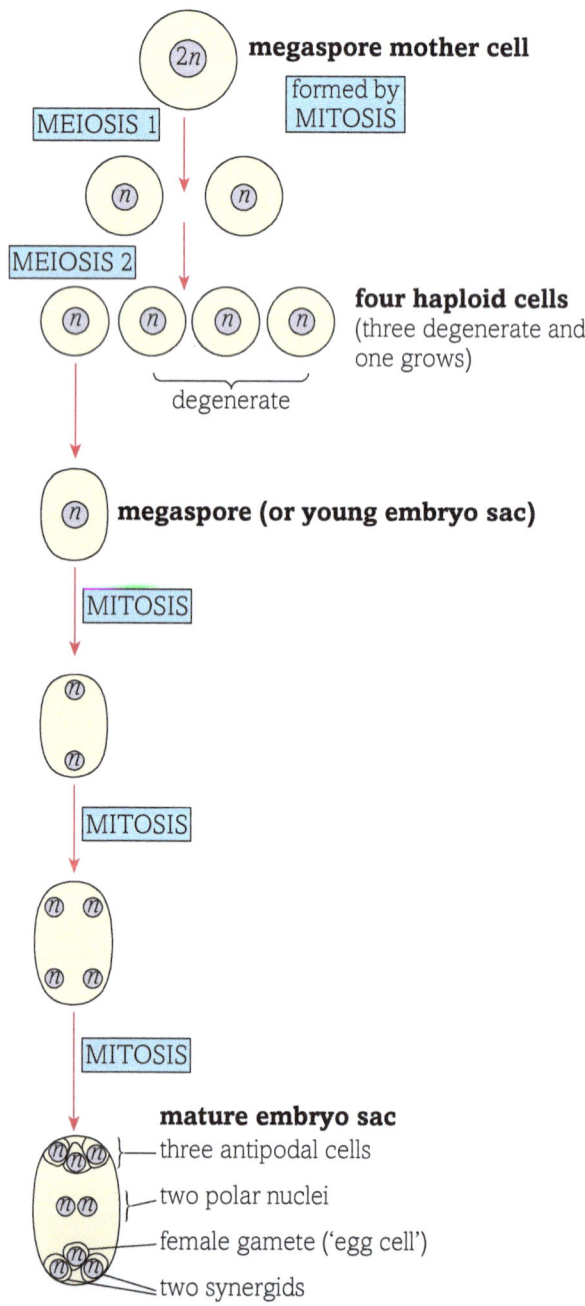

megaspore mother cell

formed by MITOSIS

MEIOSIS 1

MEIOSIS 2

four haploid cells (three degenerate and one grows)

degenerate

megaspore (or young embryo sac)

MITOSIS

MITOSIS

MITOSIS

mature embryo sac
three antipodal cells
two polar nuclei
female gamete ('egg cell')
two synergids

fig E Megagametogenesis – the formation of the egg cell.

Getting together

Asexual reproduction is a guaranteed method of passing on the genes from one individual into the next generation. For sexual reproduction to be successful, the gametes must meet. If these gametes come from two different individuals, the male gamete needs to be transferred to the female gamete. In plants, some flowers attract other organisms such as insects, birds or mammals to transfer the pollen from one plant to another, known as **pollination**. Others rely on the wind to carry their pollen from plant to plant.

Animals use a wide variety of strategies to make sure the gametes meet. They fall into two main categories:

- **External fertilisation** occurs outside the body, with the female and male gametes shed directly into the environment where they meet and fuse. This is common only in aquatic species, because spermatozoa and ova are very vulnerable to drying and are rapidly destroyed in the air. Simpler animals such as jellyfish release copious amounts of male and female gametes into the sea. It is largely a matter of chance whether fertilisation takes place. Many coral colonies have timed spawning events, when several different coral colonies release their eggs and sperm into the water at the same time. The synchronisation is thought to be in response to a series of environmental cues including temperature changes and day length.

- More complex animals such as fish and amphibians have evolved rituals that increase the likelihood of fertilisation by ensuring that the ova and sperm are released at the same time close to each other. In spite of these strategies many of the gametes do not meet. External fertilisation is very wasteful, and is not an option for organisms that live on land.

- **Internal fertilisation** involves the transfer of the male gametes directly to the female. This does not guarantee fertilisation, but makes it much more likely. The way in which the sperm are transferred varies greatly. In many species the male produces packages of sperm for the female to pick up and transfer to her body. More complex animals such as insects and some of the vertebrates have evolved a system whereby the male gametes are released directly into the body of the female during **mating**. This makes sure that the ova and sperm are kept in a moist environment and are placed as close together as possible, which maximises the chances of successful fertilisation.

Key definitions

Gametogenesis is the formation of the gametes by meiosis in the sex organs.

Primordial germ cells are the cells that divide by meiosis to ultimately form the sperm and ova.

Spermatogenesis is the formation of the sperm in the testes.

Oogenesis is the formation of the ova in the ovaries.

The **sporophyte generation** is the diploid generation in plants that produces spores by meiosis.

The **gametophyte generation** is the haploid generation in plants that gives rise to the gametes by mitosis.

The **sporophyte** is the diploid main body of the plant.

Pollen sacs are the parts of the anthers where the pollen grains develop.

A **microgamete** is the male gamete produced in plants, the pollen grain.

A **tube nucleus** is the male nucleus that will control the production of the pollen tube in fertilisation.

A **generative nucleus** is the male nucleus that will fuse with the female nucleus.

A **pollen tube** is a tube that grows out of a pollen grain down the style, into the ovary and through the micropyle of the ovule to carry the two male nuclei to the ovule.

The **placenta (plant)** is the pad of special tissue that attaches the plant ovule to the ovary wall.

A **megagamete** is the female gamete, the egg cell, in plants.

Pollination is the transfer of pollen from the anther to the stigma, often from one flower to another.

External fertilisation is the process of fertilisation in which the female and male gametes are released outside of the parental bodies to meet and fuse in the environment.

Internal fertilisation is the fertilisation of the female gamete by the male gamete, which takes place inside the body of the mother.

Mating is the process by which a male animal transfers sperm from his body directly into the body of the female.

Questions

1 What part do gametes play in sexual reproduction?

2 Explain the role of meiosis in the production of gametes. How does this vary in animals and plants?

3 How are the human male gametes adapted to fit their role?

4 Compare the adaptations of female gametes in plants and mammals.

By the end of this section, you should be able to...

- explain the process of fertilisation in mammals from the first contact between the gametes to the fusion of the nuclei

- explain how the male nuclei formed by division of the generative nucleus in the pollen grain reach the embryo sac, including the roles of the tube nucleus, pollen tube and enzymes

- explain the process of double fertilisation inside the embryo sac to form a triploid endosperm and a zygote

Fertilisation in humans

For sexual reproduction to be successful in humans, as in any other species, the gametes must meet and fuse. The ovum is fully viable for only a few hours. The sperm will survive a day or two in the female reproductive tract. There is little evidence to suggest that the sperm are attracted to the egg in any way – their meeting seems to be entirely a matter of chance. In spite of this, they frequently do meet and fuse. As sperm move through the female reproductive tract the **acrosome** region matures so it is able to release enzymes and penetrate the ovum.

The ovum released at ovulation has not fully completed meiosis and is really a secondary oocyte. It is surrounded by a protective jelly-like layer known as the **zona pellucida** and some of the follicle cells. Many sperm cluster around the ovum, and as soon as the heads of the sperm touch the surface of the ovum the acrosome reaction is triggered (see **fig B**). Enzymes are released from the acrosome, which digest the follicle cells and the zona pellucida. One sperm alone does not produce sufficient enzyme to penetrate the protective layers around the ovum. This seems to be one reason for the very large number of sperm released in ejaculation, providing enough in the oviduct to surround the ovum and digest its defences.

Eventually one sperm will wriggle its way through the weakened protective barriers and touch the surface membrane of the oocyte. This has several almost instantaneous effects. The oocyte undergoes its second meiotic division providing a haploid egg nucleus to fuse with the haploid male nucleus. It is vital that no other sperm enter the egg, as this would result in it being fertilised by too many sperm (**polyspermy**) and would produce a nucleus containing too many sets of chromosomes. The events that follow fertilisation prevent this happening. Ion channels in the cell membrane of the ovum open and close so that the inside of the cell, instead of being electrically negative with respect to the outside, becomes positive. This alteration in charge blocks the entry of any further sperm. It is a temporary measure until a tough **fertilisation membrane** forms around the fertilised ovum. This takes over the job of repelling other sperm as the electrical charge returns to normal.

The head of the sperm enters the oocyte, but the tail region is left outside. Once the head is inside the ovum it absorbs water and swells, releasing its chromosomes to fuse with those of the ovum and forming a diploid zygote. At this point fertilisation has occurred and a new individual has been formed. Fertilisation is also referred to as **conception** in the case of humans.

Fertilisation in plants

The male gamete is contained within the pollen grain. The female gamete is embedded deep in the tissue of the ovary. The pollen grain lands on the surface of the stigma of the flower during pollination. The molecules on the surface of the pollen grain and the stigma interact. If they 'recognise' each other as being from the same species the pollen grain begins to grow or **germinate**. Often the pollen grain will only germinate if it is from the same species, but a different plant. This helps to prevent self-fertilisation, which would reduce variety. Alternatively, pollen grains from the same plant may start to germinate, but be unable to penetrate the carpel.

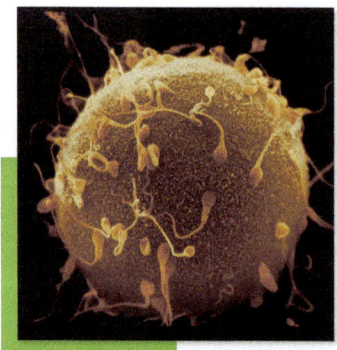

fig A The fertilisation of a human ovum.

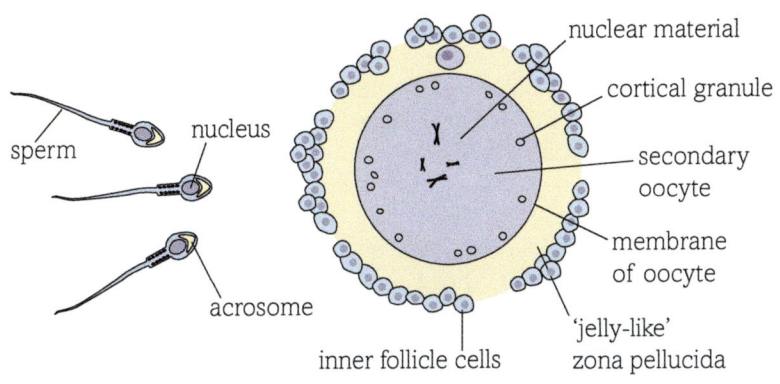

The sperm approach the oocyte in the oviduct. The acrosomes have matured since the sperm left the male reproductive tract.

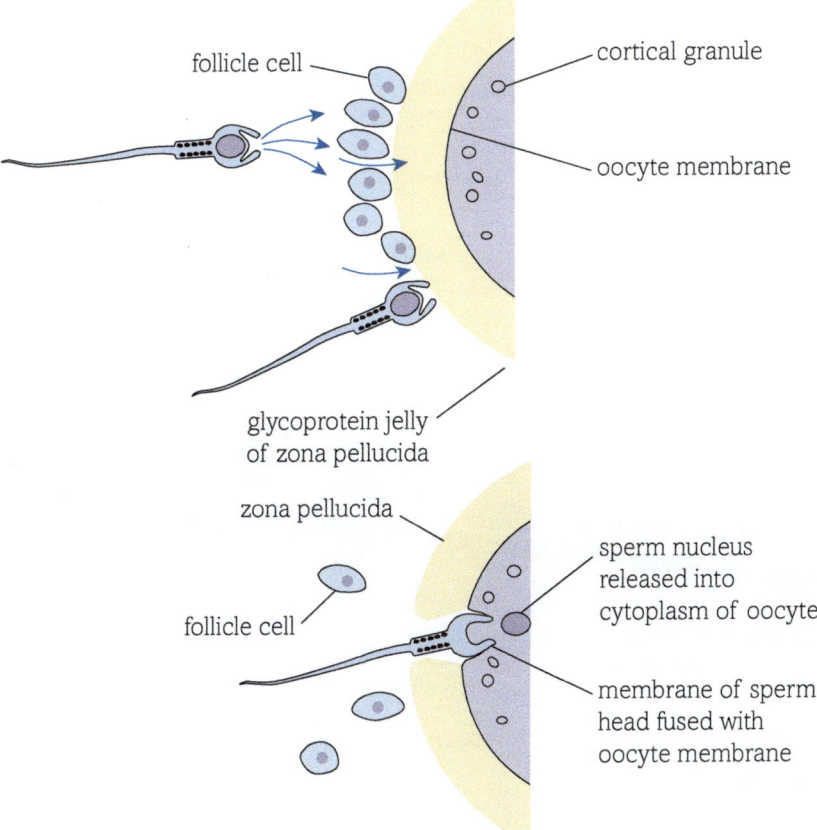

The front of a sperm touches the zona pellucida of the oocyte and the acrosome reaction is triggered. Digestive enzymes pour out of the opened acrosome and begin to digest the zona pellucida. Projections of the sperm surface shoot forward as the result of actin-like proteins.

Fertilisation occurs when one sperm touches the surface of the oocyte and the membranes fuse.

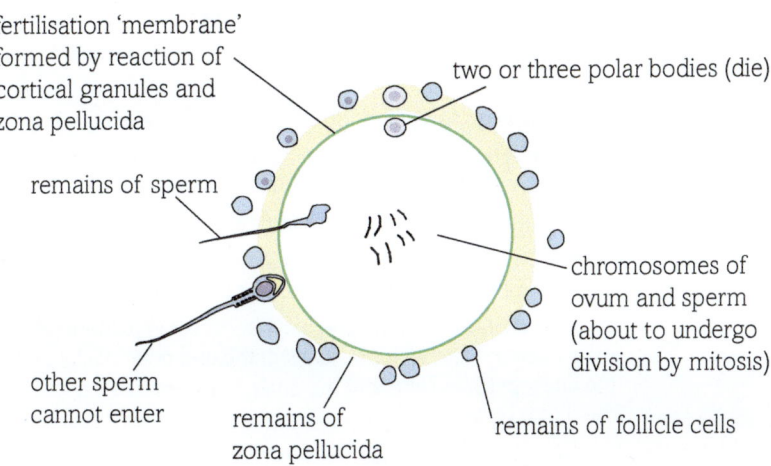

The sperm nucleus rapidly decondenses and releases its chromosomes into the oocyte. Meiosis 2 is completed in the oocyte, forming an ovum. The male and female chromosomes group together. Cortical granules are released from the oocyte that combine with the zona pellucida to form a tough fertilisation membrane.

fig B The acrosome reaction plays a vital role in the successful fertilisation of the egg.

A pollen tube begins to grow out from the tube cell of the pollen grain through the stigma into the style. The tip of the pollen tube produces hydrolytic enzymes to digest the tissue of the style, so the pollen tube can make its way down between the cells. The digested tissue acts as a nutrient source for the pollen tube as it grows. As the pollen tube grows down towards the ovary, the generative cell containing the generative nucleus travels down it. The nucleus of this cell divides by mitosis as it moves down the tube to form two male nuclei. The pollen tube grows through the ovary to reach an ovule and eventually the tip of the pollen tube passes through the micropyle of the ovule. The growth of the pollen tube is very fast due to the rapid elongation of the cell. Once the tube has entered the micropyle, the two male nuclei are passed into the ovule so that fertilisation can occur. Flowering plants undergo what is known as **double fertilisation**. One male nucleus fuses with the two polar nuclei to form the endosperm nucleus, which is triploid. The endosperm is involved in supplying the embryo plant with food when it begins to germinate. The other male nucleus fuses with the egg cell to form the diploid zygote (see **fig C**). At this point fertilisation is complete and the development of the seed and the embryo within can begin.

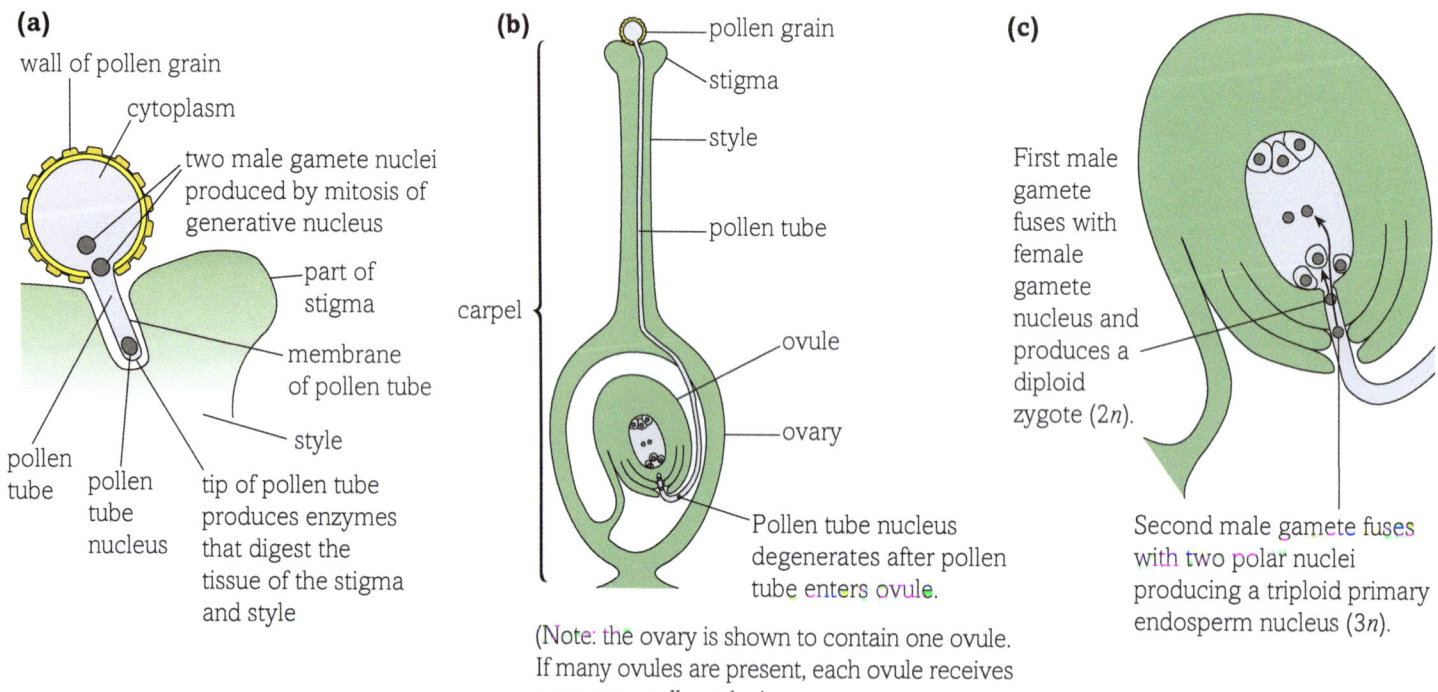

fig C A summary of the events that follow pollination in a flowering plant and lead to the fertilisation of the ovule.

Questions

1 What characteristics would you look for in an organism with:

(a) external fertilisation of the gametes?

(b) internal fertilisation of the gametes?

Explain why these adaptations are important to the method of fertilisation.

2 What is the importance of the reaction between pollen grains and the surface of the stigma in plants?

3 What part is played by enzymes in:

(a) mammalian fertilisation?

(b) fertilisation in flowering plants?

4 How is polyspermy prevented, and why is it important to stop it happening?

Key definitions

The **acrosome** is the region at the head of the sperm that contains enzymes to break down the protective layers around the ovum.

The **zona pellucida** is a layer of protective jelly around the unfertilised ovum.

Polyspermy is the fertilisation of an egg by more than one sperm.

The **fertilisation membrane** is the tough layer that forms around the fertilised ovum to prevent the entry of other sperm.

Conception is the term used for fertilisation of the ovum in humans.

Germination (pollen) is the process by which a pollen tube starts to grow out of the pollen grain to transfer the male nuclei to the ovule.

Double fertilisation is the process that occurs in plants in which one male nucleus fuses with the nuclei of the two polar nuclei to form the endosperm nucleus and the other fuses with the egg cell to form the diploid zygote.

By the end of this section, you should be able to...

● explain the early development of the human embryo to the blastocyst stage

The fusing of the ovum and sperm at fertilisation starts a complex series of events that will eventually lead to the development of a fully formed new individual.

What happens following fertilisation?

In humans, the fertilised egg cell or zygote is said to be **totipotent** – it has the potential to form all of the 216 different cell types needed for an entire new person. The future roles of individual cells are decided relatively early in the development of an embryo.

The first stage of the process is known as **cleavage**. Cleavage involves a special kind of mitosis, in which cells divide repeatedly without the normal interphase for growth between the divisions. This happens as the embryo travels down the oviduct – it takes about 80 hours from ovulation for an ovum to reach the uterus. The result of cleavage is a mass of small, identical and undifferentiated cells forming a hollow sphere known as a **blastocyst** (see **fig A**). In humans this process takes about 5–6 days, occurring as the zygote is moved along the oviduct towards the uterus and as it enters the uterus. One large zygote cell forms a large number of small cells in the early embryo. The tiny cells of the early human embryo are known as **embryonic stem cells**. Stem cells are undifferentiated cells, but have the potential to develop into many different types of specialised cells from the instructions in their DNA. The very earliest cells in an embryo are totipotent like the zygote. In the blastocyst the outer layer of cells goes on to form the placenta, and the inner layer of cells have already lost some of their ability to differentiate. They can form almost all of the cell types needed in the future, but not tissue such as the placenta – we say they are **pluripotent**. These cells are known as **pluripotent embryonic stem cells**. At this stage of development, the woman will not even know she is pregnant.

The blastocyst 'hatches' – breaks free of the outer layer – and begins to implant in the lining of the uterus after about 7 days.

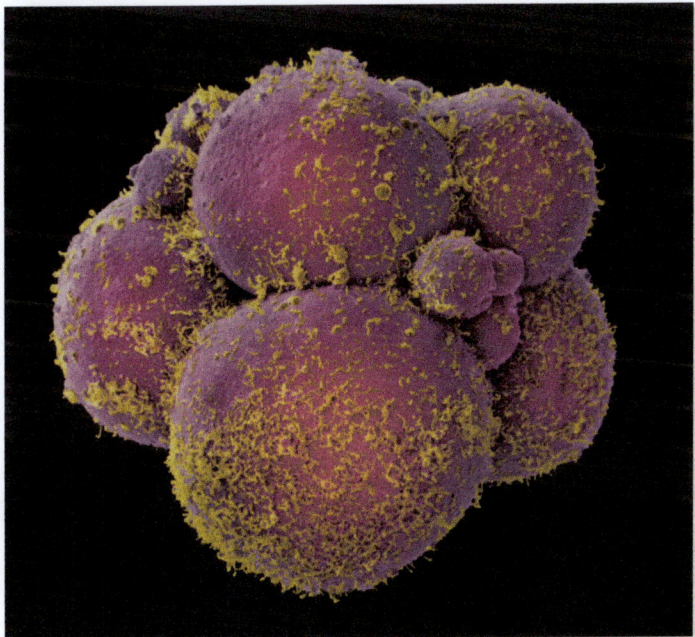

fig A At this early stage of pregnancy any one of these cells has the potential to form another entire human being.

The formation of different cell types

Almost every cell in an organism contains the DNA instructions to make any other type of cell and, in the earliest stages of an embryo, each cell can produce any tissue. However, only days after conception, cells are already predestined or determined to become one type of tissue or another. This **cell determination** is closely linked to the position of the cells in the embryo. If the cells are surgically removed from the embryo and placed somewhere else on the developing body, they will still produce the predetermined cell type – even if it is entirely inappropriate in the new setting. No-one is entirely sure of the mechanism of cell determination. Following this stage the cells can **differentiate** and develop into organs and tissues. If you are studying A level Biology, you will learn more about these amazing processes in **Book 2**.

As cell differentiation takes place to form tissues and organs, different types of cells produce more and more proteins specific to their cell type. The shape of the cell and the arrangement of the organelles will differ. In the nuclei of your cells about 2 metres of DNA is coiled and folded into a space only a few micrometres in diameter. The Human Genome Project found that the human genome consists of around 20 000–25 000 individual genes. In a differentiated cell, between 10 000 and 20 000 of those genes are actively expressed. A different combination will be expressed in different cells, creating the variety of structure and function seen in cells of different tissues.

Did you know?

The transfer dilemma

IVF (*in vitro* fertilisation) is a way of helping infertile couples to have children. Eggs are collected from the ovary of the woman and mixed with sperm. The fertilised eggs begin to develop *in vitro* – in petri dishes in the laboratory, kept warm and supplied with all the nutrients they need.

Originally two or three embryos were replaced in the uterus of the mother after two or three days, when they consisted of four to eight cells with the hope that they would implant. It was very difficult to keep the embryos alive much longer than this in the laboratory.

As techniques for culturing the embryos have improved, more clinics achieve healthy blastocysts and implant them about five days after fertilisation. This is when they would naturally be about to implant in the lining of the uterus and so it increases the chances of successful implantation.

Only the healthiest embryos develop to form a blastocyst so fewer embryos need to be transferred. This reduces the chances of a multiple pregnancy, which in turn increases the chances of a single healthy baby being born at the end of the process. Because, in some cases, no embryos develop to the blastocyst stage, most clinics offer a mixture of techniques depending on the clinical problems of each couple.

fig B It seems amazing that an entire human baby can develop from a simple ball of cells such as the blastocyst in **fig A**.

Questions

1 What are the main differences between totipotency and pluripotency?

2 What is the difference between embryo transfer and blastocyst transfer in the treatment of infertility?

Key definitions

Totipotent means a cell is able able to develop into all different cell types.

Cleavage is a process involving a special type of mitosis with no interphase that results in a mass of small, undifferentiated cells.

A **blastocyst** is a hollow ball of cells formed around five days after fertilisation.

Embryonic stem cells are cells in the early embryo that have the potential to form many other types of cells.

Pluripotent means a cell is able to develop into most different cell types.

Pluripotent embryonic stem cells are embryonic stem cells that can form most, but not all, adult cell types.

Cell determination is the predestination of cells to become particular types of tissue from early in development of the embryo.

To **differentiate** means to develop into specific types of tissues.

THINKING BIGGER

TREATING MALE INFERTILITY

Infertility affects around 1 in 7 couples in the UK and over 30% of those cases are the result of male infertility. Read the following extract taken from a book about *in vitro* fertilisation. The book has been used with patients in infertility clinics and parts have been reproduced on an infertility support website.

…an infertile man

…the lowest number of sperm counted as normal is 20 million sperm per cm³ of semen! Once the sperm count falls below this level it begins to affect fertility. If the sperm count is just a bit below normal there are things which a man can do to increase the numbers – all of them very low tech! If the testes get too warm, the level of sperm production falls, so cool showers or baths, baggy underwear and loose clothing can help to increase the sperm count. Cutting out smoking and drinking less alcohol can also help increase sperm numbers. If the count is really low, things are more difficult.

Numbers aren't everything

…the ability of sperm to fertilise eggs successfully depends on more than numbers. For the man to be fertile his sperm need to have actively lashing tails and around 50% of them must swim forward in straight lines rather than round and round in circles.

It is also very important that the semen does not contain too many abnormal sperm. Every man produces a certain number of sperm with two heads instead of one, or with two tails, or with a break in the small midsection between the head and tail. But if the percentage of these abnormal sperm gets too high, then the chances of a successful pregnancy fall.

Injecting sperm – a major breakthrough

The main technique used to help men with abnormal sperm is known as ICSI (Intra Cytoplasmic Sperm Injection). Eggs are harvested from a woman after treatment with fertility drugs. A single sperm is then injected right into the cytoplasm of each egg cell. Two or three healthy embryos will then be returned to the body of the mother just as in normal IVF treatment.

There are two groups of patients for whom ICSI offers hope … Men who have severe sperm problems, even if they cannot produce semen at all or have very few healthy sperm, can be helped – it only takes one sperm to fertilise each egg. Also, couples who produce healthy eggs and sperm for IVF, but cannot achieve fertilisation can be helped, because whatever the problem with fertilisation it can be overcome by the direct insertion of a sperm into an egg. ICSI is widely used for at least 30% of all the couples who need IVF technology to conceive.

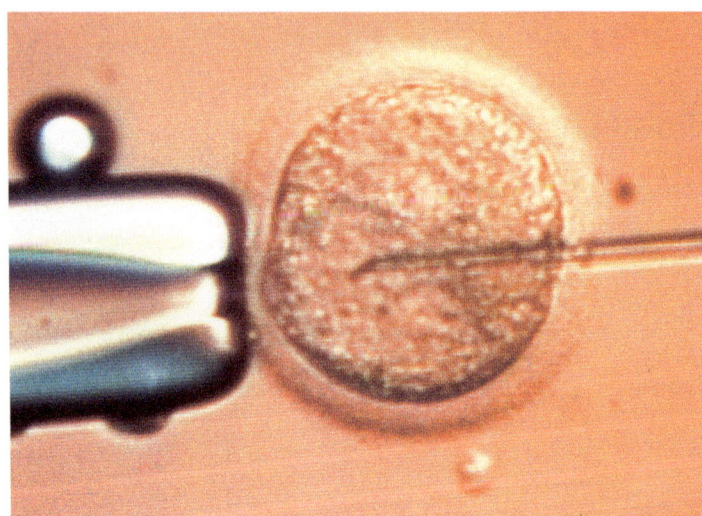

fig A ICSI removes all the normal barriers to conception as the sperm cannot fail to reach the egg.

Where else will I encounter these themes?

1.1   1.2   1.3   1.4   2.1   2.2   2.3   2.4   YO

Let us start by considering the nature of the writing in this article.

1. What are the main problems with writing a book on complex biological issues such as infertility and infertility treatments that aims to inform interested readers and support people who really need the information?

2. The book contains many quotes from doctors, embryologists and research scientists involved in infertility treatments. Do you think this is a useful feature for a book of this type? Justify your response.

3. The book contains a case study of a couple with unexplained infertility who had treatment for many years before conceiving twins by IVF. They went on to have three more children naturally. Discuss the value of a case study like this in a book on infertility treatment.

> **Command word**
> When the word justify is used in this context you are being asked to give clear evidence to support your answer.

Now let us look at the biological basis of the problems of infertility. You have studied gametogenesis, fertilisation and the early stages of the development of the human embryo, which will help you answer these questions.

4. Look at the ways in which a man can try to increase his sperm production and suggest biological reasons for each of the methods suggested.

5. Explain, referring to the normal process of fertilisation of an ovum by a sperm, why the problems described in this excerpt of low sperm numbers, abnormal sperm or inactive sperm would lower or remove the chance of a successful pregnancy.

Activity

Infertility affects around 1 in 7 couples. Around 30% of infertility is due to problems with the female system such as failure to produce ova, or failure of the uterus to support the implantation of the blastocyst. Over 30% of cases are due to male infertility as described above. The rest are because both partners are slightly sub-fertile, or are completely unexplained.

Research the causes and treatment of infertility, bearing in mind what you know about gametogenesis, fertilisation and the implantation of the blastocyst. Produce a visual aid to show the data in the most arresting and informative way possible.

> **Scientific sources**
> As you carry out your research, consider where the sources you find have come from, who wrote them and who they were written for. Established scientific publications are good sources of reliable information, whereas other resources might be less dependable for a number of reasons. Think about what makes a source reliable and why.

● From *Science at the edge. In vitro fertilisation* by Ann Fullick.

Exam-style questions

1 (a) The diagram below shows a section through a *Primula* (primrose) flower.

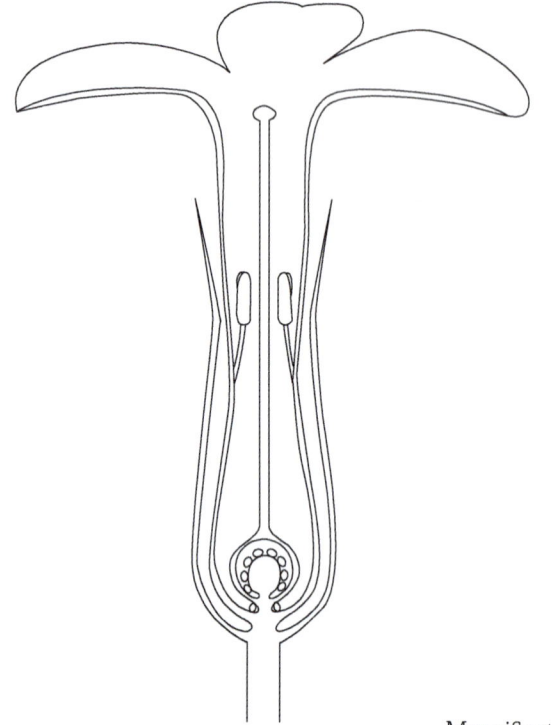

Magnification ×2

Suggest how *Primula* flowers are pollinated. Give an explanation for your answer. [3]

(b) An experiment was carried out to measure the rate of growth of a pollen tube in germinating pollen grains. Fresh pollen grains were placed in a 0.5 mol per dm³ sucrose solution, and kept at a temperature of 20 °C. The pollen tube growth rates were recorded at time intervals of 30 minutes for a period of three hours.
The results of this experiment are shown in the table below.

Time/min	Growth rate of pollen tube/mm per 30 min
30	0.156
60	0.169
90	0.182
120	0.169
150	0.052
180	0.032

Describe how the growth rate of the pollen tube changed during the experiment. [3]

[Total: 6]

2 (a) The diagrams below show some of the stages of meiosis I in an animal cells. The diploid number (2n) of this cell is 4.

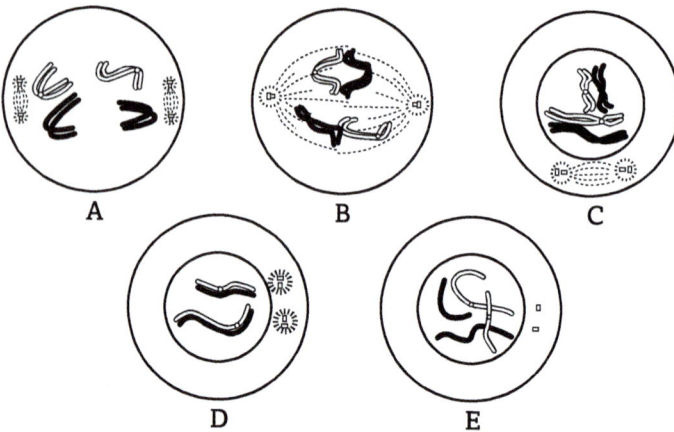

Write the letters in the correct order to show the sequence of stages in meiosis I. [2]

(b) The diploid number of chromosomes in a human cell is 46. State the number of chromosomes present in each of the following:
 (i) A spermatogonium
 (ii) A spermatid [2]

(c) Describe the process of **oogenesis**. [4]

[Total: 8]

3 Read through the following account of meiosis and fertilisation, then write the most appropriate word or words on the dotted lines to complete the account.

During spermatogenesis, a diploid cell called a

.... divides in meiosis 1 to

form two .... Each of

these then divides in meiosis 2, forming four haploid

...., which mature into

spermatozoa.

The random fusion of gametes during fertilisation is one way in

which .... variation is

increased. [4]

[Total: 4]

4 (a) Explain what is meant by the term pollination. [2]
  (b) The substance boron is known to affect the growth of pollen tubes from pollen grains.

  In one experiment, pollen grains were placed in two different media, one containing boron and one without boron.

  The lengths of the pollen tubes were measured every 6 hours, for a total of 36 hours.

  The results are shown in the table below.

Time/hours	Mean length of pollen tubes/μm	
	Medium without boron	Medium with boron
6	20	45
12	45	90
18	70	170
24	90	250
30	100	280
36	100	300

   (i) In the experiment, the mean growth rate of the pollen tubes in the medium without boron from 6 hours until 12 hours is 4.17 μm h^{-1}.
   Calculate the mean growth rate of the pollen tubes in the medium with boron from 6 hours until 12 hours.
   Show your working. [3]
   (ii) Compare the growth of pollen tubes in these two media. [2]
   [Total: 7]

  (b) An investigation into the effect of temperature on pollen tube growth was carried out. Two different varieties of cotton pollen grain were used, variety A and variety B.

  Twenty newly-germinated cotton pollen grains of variety A were placed on growth medium in a Petri dish and incubated in the dark for 24 hours at 15 °C. After this time, the length of each pollen tube was measured and the mean calculated. This was repeated at 5 different temperatures.

  The investigation was then repeated using variety B. The results are shown in the table below.

Incubation temperature/°C	Mean length of pollen tube after 24 hours incubation/mm	
	Variety A	Variety B
15	0.18	0.19
20	0.35	0.48
25	0.53	0.83
30	0.60	0.90
35	0.57	0.60
40	0.10	0.10

   (i) Describe the effect of temperature on the mean length of pollen tubes for variety A. [2]
   (ii) Compare the effect of temperature on the mean length of pollen tubes in variety A with variety B, between 15 °C and 30 °C. [2]
   (iii) Explain the change in the mean length of pollen tubes when the temperature increased from 35 °C to 40 °C. [1]
   [Total: 12]

5 Fertilisation involves the fusion of haploid nuclei.
  (a) The diagram below shows a human sperm cell.

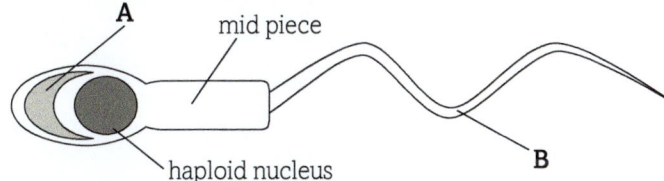

   (i) Name the structures labelled **A** and **B**. [2]
   (ii) Explain why it is important that the sperm has a nucleus that is haploid. [2]
   (iii) Describe the changes in the female gamete from the point when a sperm releases its digestive enzymes to the point when the two nuclei fuse. [3]

TOPIC 3
Classification

3.1 > Classification

Introduction

In 2012, scientists working in Papua New Guinea found the smallest known vertebrate to date – a tiny frog measuring around 7.7 mm in length. *Paedophryne amanuensis* feeds on tiny mites in the leaf litter of its rain forest home – and it can jump up to 30 times its own length. DNA analysis has shown scientists that tiny frogs have evolved 11 times in different areas of the world, all filling a similar niche. In 2014, a new species of dead-leaf toad (*Rhinella yunga*) was discovered in the Peruvian Andes. With its dead-leaf shape, colour and patterning, and the poison it exudes from glands on the back of its head, the toad looks similar to other toads of the same genus. It was only when scientists noticed that these toads lack ear drums that they realised they had discovered another new species. Finding new species is always exciting, but when amphibian species such as the Panamanian golden frog and many of the brightly coloured poison dart frogs are fast becoming extinct due to climate change, human influences and a deadly new fungal disease, finds like these are particularly special.

The scientists used two different methods of identifying the amphibians described above as new species – traditional observation of physical characteristics such as ear drums, and DNA analysis of the genome of the organisms. In this chapter you will be finding out more about how we classify the organisms in the world around us – and why it is important that we do so.

You will learn the main taxonomic groups of the living world including domains, kingdoms and species, and will begin to classify different organisms.

You will consider the problems of defining a species in a way that is useful for all types of organisms, and evaluate the different ones in use. The use of DNA technology is having a major impact on our ability to identify organisms and work out how they are related to other species, and you will look at the changes that are happening. There has been a long-running debate about the numbers of domains and kingdoms which should be used in our classification – see who you think is right!

All the maths you need

- Use scales for measuring (*e.g. measuring size of different organisms and parts of organisms for comparisons when classifying*)

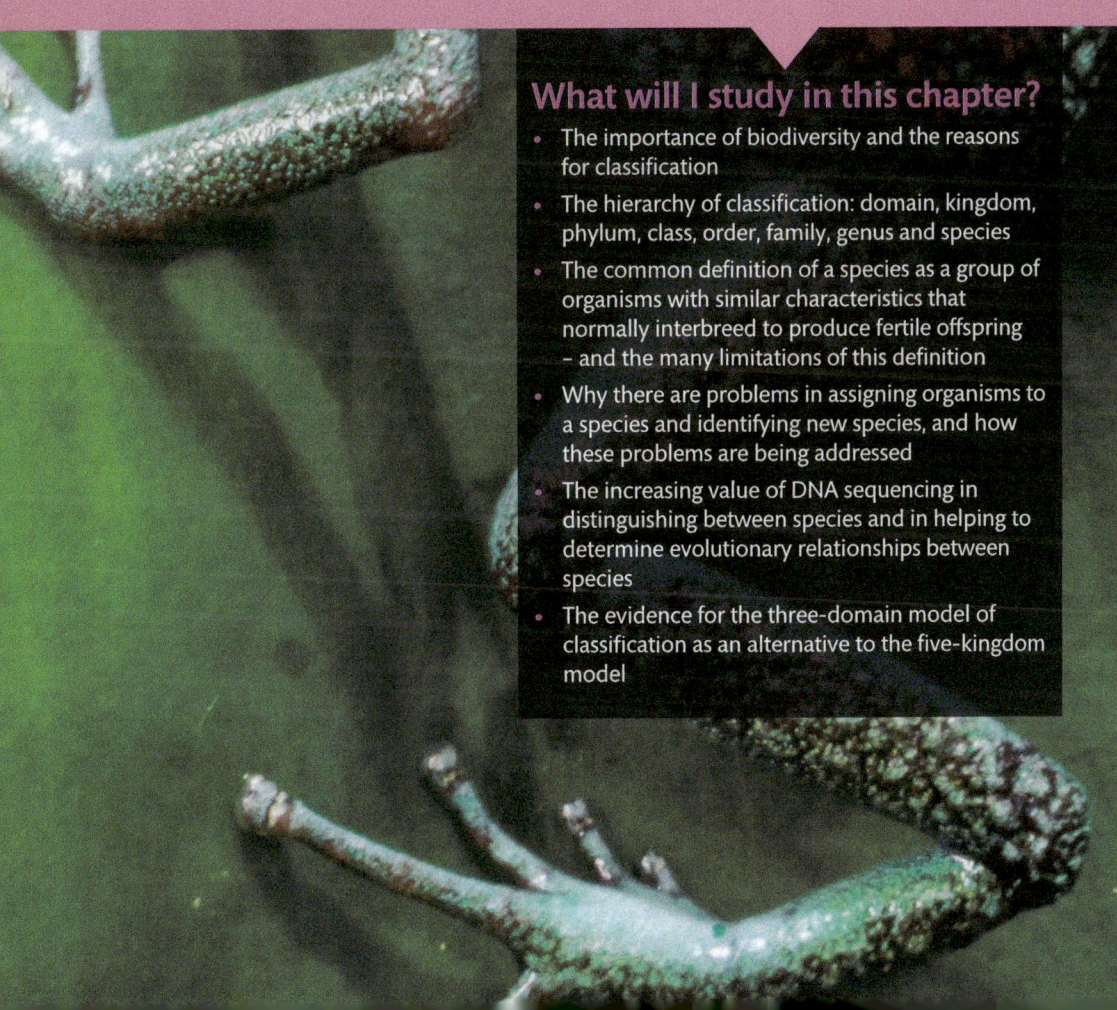

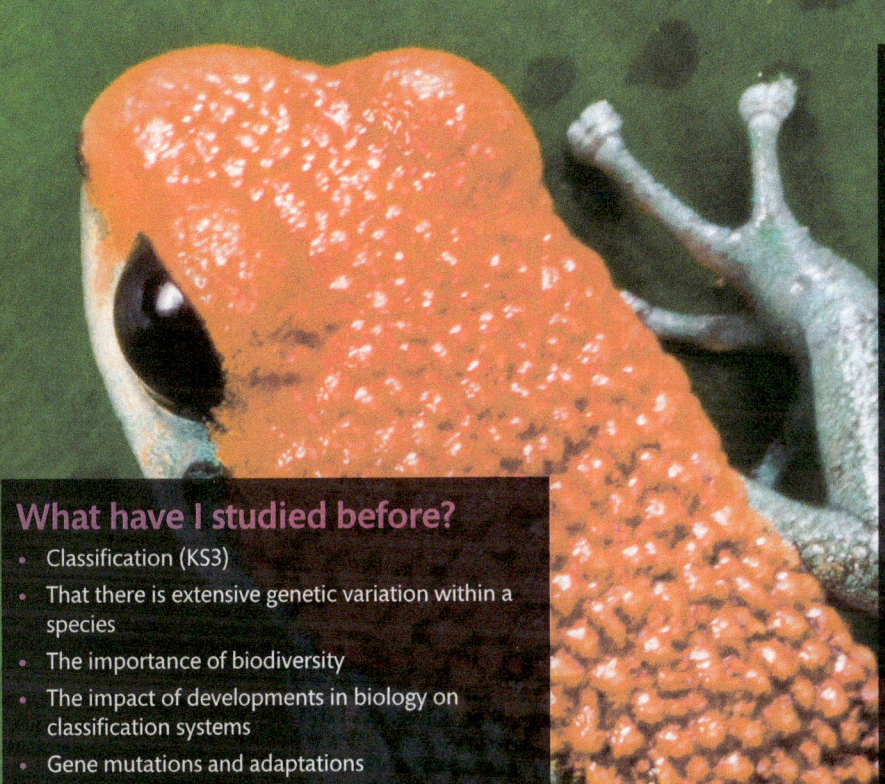

What will I study later?

- That evolution of new species comes about through natural selection acting on variation, which brings about beneficial or damaging adaptations
- How reproductive isolation for different reasons can bring about the formation of new species
- That biodiversity within a habitat (based on species identified in the field) can be assessed using statistical formulae
- Changes in species in an area through colonisation and succession
- Classification of pathogenic bacteria by differences in their cell walls, their mode of infection and their nutritional needs (A level)
- The way in which base sequencing can be used to analyse evolutionary patterns and identify separate species (A level)
- Gene pools and how they can change within a population or a species
- The need to be able to classify organisms for practical investigations of populations in the field

What have I studied before?

- Classification (KS3)
- That there is extensive genetic variation within a species
- The importance of biodiversity
- The impact of developments in biology on classification systems
- Gene mutations and adaptations

What will I study in this chapter?

- The importance of biodiversity and the reasons for classification
- The hierarchy of classification: domain, kingdom, phylum, class, order, family, genus and species
- The common definition of a species as a group of organisms with similar characteristics that normally interbreed to produce fertile offspring – and the many limitations of this definition
- Why there are problems in assigning organisms to a species and identifying new species, and how these problems are being addressed
- The increasing value of DNA sequencing in distinguishing between species and in helping to determine evolutionary relationships between species
- The evidence for the three-domain model of classification as an alternative to the five-kingdom model

By the end of this section, you should be able to...

- describe the classification of organisms into a hierarchy of domain, kingdom, phylum, class, order, family, genus and species

The background to biodiversity

Biodiversity is a measure of the variety of living organisms and their genetic differences. It is a key concept at the moment because the Earth's biodiversity is reducing rapidly. Many scientists think this may affect the future health of the planet. You will find out about biodiversity in more detail later in this topic. Here you will be looking at some of the biology you need in order to understand biodiversity.

Why classify?

The result of millions of years of **evolution** (see **Section 3.2.1**) is an enormous variety of living organisms. Along with this great biodiversity (see **Section 3.3.1**) goes a great variety of names. One organism will have different names not only in different countries, but within different areas of the same country. When biologists from different countries discuss a particular organism they need to be sure they are all referring to the same one. We need an internationally recognised way of referring to any particular living organism. Biodiversity is a very important concept, but to quantify biodiversity we need a way of identifying the different groups of organisms. We solve these problems by classifying the living world, putting organisms in groups based on their similarities and differences. Once scientists know the numbers of different types of organisms that exist in a particular habitat, they can monitor how those populations change. It is also important for biologists to understand how different types of living organisms are related to each other. A good classification system makes these ancestral relationships clear.

fig A This plant is a daisy in the UK, 'la pâquerette' in French, 'margarita europea' or 'coqueta' in Spanish and 'massliebchen' in German. The official classification *Bellis perennis* is used and understood by biologists everywhere.

The history of taxonomy

Taxonomy is the science of describing, classifying and naming living organisms. This includes all of the plants, animals and microorganisms in the world and it is a mammoth task. The aim of a classification system is to group organisms in a way that accurately identifies them and represents their ancestral relationships.

From the time of the Greek philosopher Aristotle onwards, people put organisms into groups based mainly on their physical appearance or **morphology**. People often used **analogous features** to classify organisms – that is, features that look similar or have the same function, but are not in fact of the same biological origin. A moment's thought shows that this system can easily lead to misconceptions. For example, you might put wiggly, legless creatures including snakes, worms, slugs and eels in one classification group and flying animals such as bats, birds and flying insects in another group. A valid classification system must be based on careful observation and the use of **homologous structures** – that is, structures that genuinely show common ancestry.

In the eighteenth century the Swedish botanist Carolus Linnaeus (1707–78) developed the first scientifically devised classification system. We still use many of his principles and his basic naming system today. However, we can now add many more modern techniques to the simple but detailed observation of organisms that he introduced.

The main taxonomic groups

The biggest taxonomic groupings are huge – they divide the whole living world into three domains – the Archaea, the Bacteria (see **Section 2.1.2**) and the Eukaryota (organisms that have eukaryotic cells, see **Section 2.1.3**).

The main taxonomic groups are, from the largest to the smallest: **domain**, **kingdom**, **phylum** (**division** for plants), **class**, **order**, **family**, **genus** and **species**.

The Archaea domain contains just one kingdom:

- **Archaebacteria**: ancient bacteria thought to be early relatives of the eukaryotes. They were thought to be only found in extreme environments, but scientists are increasingly finding them everywhere – particularly in soil.

The Bacteria domain also contains only one kingdom:

- Eubacteria: the true bacteria are what we normally think of when, for example, we are describing the bacteria that cause disease, and which are so useful in the digestive systems of many organisms and in recycling nutrients.

There are four eukaryotic kingdoms:

- Protista: very diverse group of microscopic organisms. Some are heterotrophs – have to eat other organisms – and some autotrophs – they make their own food by photosynthesis. Some are animal-like, some are plant-like and some are more like fungi. Examples include *Amoeba*, *Chlamydomonas*, green and brown algae and slime moulds.
- Fungi: all heterotrophs – most are saprophytic and some are parasitic. They have chitin, not cellulose, in their cell walls.
- Plantae: almost all autotrophs, making their own food by photosynthesis using light captured by the green pigment chlorophyll. These include the mosses, liverworts, ferns, gymnosperms, and angiosperms (flowering plants).

- Animalia: all heterotrophs that move their whole bodies around during at least one stage of their life cycle. These include the invertebrates (e.g. insects, molluscs, worms, echinoderms) and the vertebrates (e.g. fish, amphibians, reptiles, birds, mammals).

The binomial system

The binomial system of naming organisms was originally used by Linnaeus. It is now used universally among biologists. The way different organisms are classified is constantly under review as new data are discovered.

In the binomial system every organism is given two Latin names – binomial literally means two names. The first name is the genus name and the second is the species or specific name which identifies the organism precisely. There are certain rules to writing binomial names:

- use italics
- the genus name has an upper-case letter and the species name a lower-case letter, e.g. *Homo sapiens* – human beings, *Bellis perennis* – common daisy
- after the first use, binomial names are abbreviated to the initial of the genus and then the species name, e.g. *H. sapiens*, *B. perennis*.

A genus is a group of species that all share common characteristics so, for example, the genus *Vanessa* contains both the Red Admiral *Vanessa atalanta* and the Painted Lady *Vanessa cardui*. These lovely butterflies have some very clear similarities, but enough differences for you to see why they are separate species. It is not always so easy to tell species within a genus apart.

fig B These two British butterflies belong to the same genus, but are different species.

Here are a number of different species, with all of their levels of classification shown:

Questions

1 Why is a classification system needed in biology?

2 Draw a diagram to show the main groups of the most commonly used system of classification and how they are related to each other.

3 Discover the classification from domain to species of the following organisms: domestic cat, maize, honey bee and human being.

Key definitions

Biodiversity is a measure of the variety of living organisms and their genetic differences.

Evolution is the process by which natural selection acts on variation to bring about adaptations and eventually speciation.

Taxonomy is the science of describing, classifying and naming living organisms.

Morphology is the study of the form and structure of organisms.

Analogous features are features that look similar or have a similar function, but are not from the same biological origin.

Homologous structures are structures that genuinely show common ancestry.

Domains are the three largest classification categories, including the Eukaryota, the Bacteria and the Archaea.

The **Archaea** domain is made up of bacteria-like prokaryotic organisms found in many places including extreme conditions and the soil. They are thought to be early relatives of the eukaryotes.

A **kingdom** is the classification category smaller than domains. There are six kingdoms: Archaebacteria, Eubacteria, Protista, Fungi, Plantae and Animalia.

A **phylum** (**division** for plants) is a group of classes that all share common characteristics.

A **class** is a group of orders that all share common characteristics.

An **order** is a group of families that all share common characteristics.

A **family** is a group of genera that all share common characteristics.

A **genus** is a group of species that all share common characteristics.

A **species** is a group of closely related organisms that are all potentially capable of interbreeding to produce fertile offspring.

Archaebacteria are ancient bacteria thought to be the oldest form of living organism.

Domain	Bacteria	Eukaryota	Eukaryota	Eukaryota
Kingdom	Eubacteria	Animalia	Fungi	Plantae
Phylum/Division	Proteobacteria	Chordata	Basidomycota	Magnoliophyta
Class	Gammaproteobacteria	Mammalia	Agaricomycetes	Liliopsida
Order	Enterobacteriales	Perissodactyla	Agaricales	Poales
Family	Enterobacteriaceae	Equidae	Amanitaceae	Poaceae
Genus	*Escherichia*	*Equus*	*Amanita*	*Oryza*
Species	*Escherichia coli* *E. coli* common bacterium in the intestines	*Equus caballus* *E. caballus* domestic horse	*Amanita muscaria* *A. muscaria* fly agaric	*Oryza sativa* *O. sativa* rice

What is a species?

By the end of this section, you should be able to...

- explain the limitations of the definition of species as a group of organisms with similar characteristics that interbreed to produce fertile offspring

- explain why it is often difficult to assign organisms to any one species or identify new species

The concept of species

The concept of species is a very important one for biologists. We use species numbers to measure biodiversity (see **Section 3.3.1**). We also look for changes in species to help us monitor the effect of both natural environmental changes and changes that result from human activity. Biologists look for both adaptations within a species and for changes in the numbers or types of species in an environment.

Because species are such an important concept in biology, it makes sense that everyone should be working with the same model. However, this is not as easy as it sounds. Species are defined in many different ways, and the best model changes depending on the circumstances and the type of organism being investigated.

The morphological species concept

The definition of species originally developed by Linnaeus was a **morphological species concept**, based solely on the appearance of the organisms he observed. For many years scientists would look closely at the outer and sometimes inner morphology of the organisms and use the degree of difference or similarity of the physical characteristics to group them into species, genus, etc. Much of the classification we still use is based on morphology. In many cases this approach still works and you can tell just by looking at an organism what it is – for example you would never mistake a lion for a domestic cat. However, the appearance of an organism can be affected by many different things (see **Section 3.2.1**), and there can be a huge amount of variation within a group of closely related organisms. In fact, in organisms that show **sexual dimorphism**, in which there is a great deal of difference between the male and female, the different sexes could well be thought to be different species in a morphological species model.

fig A Even with careful observation, most people would not put these male and female birds in the same species unless you saw them mating.

fig B More examples where careful observation is not always enough. Agaves and aloes look similar – they are adapted to survive in similar conditions, but are not closely related – yet the tiny bog willow is a close cousin of the weeping willow.

The reproductive or biological species concept

For many years a morphological definition of a species was used almost without question. However, in time biologists moved to a basic model of a species based on the reproductive behaviour of the organisms. One widely used definition of a species is:

- a group of organisms with similar characteristics that interbreed to produce fertile offspring.

This definition of species overcomes issues such as sexual dimorphism and is regarded as a good working definition for many animal species, but it has limitations. One obvious limitation is that all of the organisms in a species cannot attempt to interbreed to produce fertile offspring because they do not all live in the same area. So populations of organisms of the same species may not interbreed simply because they are in different places, not because they are different species.

In this species model, if two individuals from different populations mate, they are considered the same species if fertile offspring are produced and genes are combined or 'flow' from the parents to the offspring. So, for example, horses and donkeys look similar, but the offspring produced from a horse and a donkey is a mule, which is sterile. The genes cannot flow to the next generation so they are not the same species. But the offspring produced between a Shire horse and a Shetland pony is fertile – they are extreme variants of the same species. However, this definition is not foolproof. For example, lions and tigers are different species, but if a lion and tiger mate most of the offspring produced are fertile. To help overcome these limitations, two slightly more sophisticated definitions of species based on reproductive capability are:

- a group of organisms with similar characteristics that are all potentially capable of breeding to produce fertile offspring
- a group of organisms in which genes can flow between individuals.

A reproductive concept of species is a good working model for most animals, but it is much less helpful in classifying plants, which frequently interbreed with similar species to form fertile offspring.

Other definitions of species

The definition of a species is constantly developing. Scientists now make decisions about which organisms belong in the same species and how they are related in a number of different ways. Some of these methods are considerably more sophisticated than simple observation. The fundamental chemicals of life such as DNA, RNA and proteins are almost universal. However, whilst these chemicals are broadly similar across all species, differences are revealed when the molecules are broken down to their constituent parts. Scientists use these differences, in the science of **molecular phylogeny**, to build up new models of species and their relationships. But some of the different models of species are no better, and can be even worse than the original morphological model. They include:

- **Ecological species model** – based on the ecological niche occupied by an organism. This is not a very robust way of identifying species, as niche definitions vary and many species occupy more than one niche.
- **Mate-recognition species model** – a concept based on unique fertilisation systems, including mating behaviour. The difficulty with this is that many species will mate with or cross-pollinate other species and may even produce fertile offspring, but are nevertheless different species.

Did you know?

Speciation based on sex organs

fig C Is this moth the Grey Dagger (*Acronicta psi*) or the Dark Dagger (*Acronicta tridens*)? It takes an expert to tell them apart.

The Grey Dagger moth (*Acronicta psi*) looks identical to the Dark Dagger (*Acronicta tridens*) – until you look at the sex organs. The male Grey Dagger has a two-pronged genitalia while the male Dark Dagger has three prongs. The female anatomy has matching differences. These small but vital differences in two otherwise almost identical organisms is enough for them to be classified as two different species.

- **Genetic species model** – based on DNA evidence. This might seem the ultimate, foolproof method of determining species, but people still have to decide how much genetic difference makes two organisms members of different species. Historically, collecting DNA was difficult and it took a long time and cost a lot of money to analyse. As DNA analysis continues to get faster and cheaper, this will ultimately become the main way of classifying organisms (see **Section 3.1.3**).

- **Evolutionary species model** – based on shared evolutionary relationships between species. In this model, members of a species have a shared evolution and are evolving together. This is biologically sound, but it is not always easy to apply. There is not always a clear evolutionary pathway for a particular organism.

Ever-improving DNA analysis means species definitions and evolutionary relationships will become increasingly important in classification. But for now, the biological definition of species combined with basic morphology is still widely used.

Limitations of species models

All of the ways of defining species have limitations, which include:

1 Finding the evidence – many living species have never been observed mating. This is particularly true if a new species is found that is similar to an existing species. Setting up a breeding programme is time-consuming, expensive and may not prove anything.

2 Plants of different but closely related species frequently interbreed and produce fertile hybrids. At what point should the hybrids themselves be regarded as a separate species?

fig D Delicate British bluebells (*Hyacinthoides non-scripta*) have creamy-white pollen, and the sturdier Spanish bluebells (*Hyacinthoides hispanica*) have pale blue pollen. The two interbreed so successfully that the native British bluebell is beginning to disappear from UK woodlands, overtaken by stronger hybrid plants.

3 Many organisms do not reproduce sexually. Any definition involving reproduction or reproductive behaviour is irrelevant for bacteria and the many protists, fungi and others that reproduce mainly asexually.

4 Fossil organisms cannot reproduce and do not, in the great majority of cases, have any accessible DNA, but they still need to be classified.

fig E There are many different species of fossils which need identifying – and breeding experiments or DNA analysis are not much help.

Identifying a species

In spite of all the pitfalls, classifying organisms and identifying the species to which they belong is still a widely used and extremely useful biological tool. Questions about identifying different organisms are absolute, such as is it species X or species Y?, and also comparative, for example, is it a new species that has not been identified before, or just new to a particular scientist or area? Information technology (IT) provides an ideal tool to help scientists answer these questions from simple identification apps to help decide which bird, butterfly or orchid you have just seen, to the prospect of instruments that will be able to identify DNA in the field. We are now at the stage where IT plays a major part in classification.

The Natural History Museum is home to millions of specimens of different organisms from all over the world, collected over several hundred years. The great majority of the species were identified by their external features many years ago, and details were recorded on handwritten and typewritten index cards, which are filed in the museum's vast archives. New specimens are regularly sent to the museum for identification. To reduce the time spent searching the cards, scientists at the museum and the University of Essex are developing VIADOCS to scan and 'read' the card archives, and convert them into an internet-based database and a paper-based catalogue. Once VIADOCS is completed, it will not only make searching for a particular organism much easier, but it will also give scientists around the world access to classification information while working in the field.

Questions

1 Summarise the reasons why biologists need to classify organisms.

2 Compare the advantages and practical difficulties of using classic morphology and reproductive capability to decide if an organism belongs to a particular species.

Key definitions

The **morphological species model** is a species definition based solely on the appearance of the organisms observed.

In **sexual dimorphism** there is a great deal of difference between the appearance of the male and female of a species.

Molecular phylogeny is the analysis of the genetic material of organisms to establish their evolutionary relationships.

The **ecological species model** is a species definition based on the ecological niche occupied by an organism.

The **mate-recognition species model** is a species definition based on unique fertilisation systems, including mating behaviour.

The **genetic species model** is a species model based on DNA evidence.

The **evolutionary species model** is a species model based on shared evolution between groups of organisms.

The importance of DNA

In recent years scientists have developed techniques that allow them to analyse the DNA and proteins of different organisms. In **DNA sequencing** the base sequences of all or part of the genome of an organism is worked out (see **Section 1.3.5**). DNA sequencing leads to **DNA profiling**, which looks at the non-coding areas of DNA to identify patterns. These patterns are unique to individuals, but the similarity of patterns can be used to identify relationships between individuals and even between species.

DNA sequencing and profiling generates so much data that it would be almost impossible for individual scientists to go through it all searching for patterns. This is where the new science of bioinformatics comes into its own. **Bioinformatics** is the development of the software and computing tools needed to organise and analyse raw biological data, including the development of algorithms, mathematical models and statistical tests that help us to make sense of the enormous quantities of data being generated. Using bioinformatics, we can make sense of and use the information generated in DNA sequencing and profiling. You are going to discover some of the ways in which we can use this information to identify species and the relationships between them.

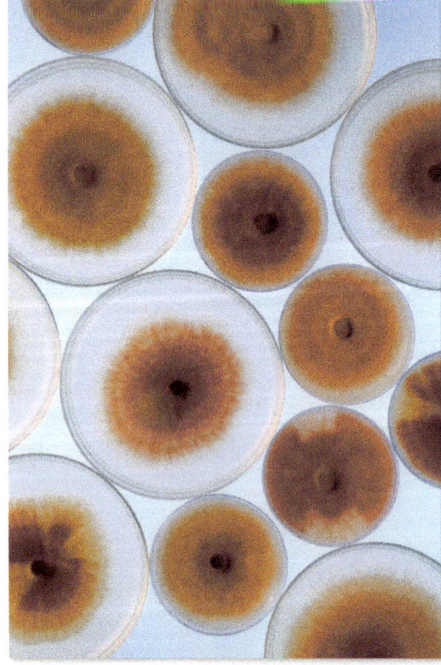

fig A These cultures may all look the same, but DNA evidence shows that they are distinct species of fungi, all of which can cause similar diseases in plants.

The same...

Identifying species from their phenotype can be difficult. External conditions can result in major differences in appearance of individuals of the same species. For example, red deer stags that live in woods and parkland have antlers that are much longer and broader than stags that roam highland mountainsides. They could easily be mistaken for different species, yet DNA evidence shows that they are the same.

...but different

In contrast, for many years the plant disease scab, which can destroy crops such as wheat and barley, was thought to be caused by a single fungus, *Fusarium graminearum*. Molecular geneticists in the United States have investigated the disease to try and help plant breeders and disease control specialists worldwide. DNA evidence, based on the divergence of six different genes, and the proteomic evidence of the proteins they produce, shows that there are at least eight different species of *Fusarium* pathogens, which have a similar effect on crop plants.

In the next few pages you will be looking at how techniques such as DNA sequencing, DNA profiling and protein analysis are changing the way we distinguish between species and determine evolutionary relationships.

The caviar con

Caviar is a luxury food and the very best and most expensive is Beluga caviar, the eggs of the Beluga sturgeon. DNA profiling is used by scientists to identify different but closely related species. Scientists from the American Museum of Natural History developed a series of profiles for different sturgeon species, as some of them are becoming very rare. They then ran DNA profiles on lots of different tins of caviar and discovered that around 25% of the tins claiming to be full of Beluga caviar actually contained the eggs of other, less prestigious species. Similarly, in 2013, DNA analysis identified horse meat in European pork, beef and chicken products – the wrong species by quite a margin. These are examples of bringing cutting-edge classification into the ethics of the marketplace. DNA analysis for species identification is also becoming more and more important in the understanding and treatment of infections (see **Section 2.2.2**).

DNA barcodes

If scientists raise an organism in the lab, they usually know what species it is. But organisms found in the field may not be identified easily. The Consortium for the Barcode of Life (CBOL), the International Barcode of Life project (IBOL) and the European Consortium for the Barcode of Life (ECBOL) are large groups of scientific organisations that are developing DNA barcoding as a global standard for species identification. This involves looking at short genetic sequences from a part of the genome common to particular groups of organisms. For example, a region of the mitochondrial cytochrome oxidase 1 gene (CO1), containing 648 bases, is being used as the standard barcode for most animal species. So far this sequence has been shown to be effective in identifying fish, bird and insect species including butterflies and flies. This region cannot be used to identify plants because it evolves too slowly in these organisms to give sufficient differences between species. However, botanists have identified two gene regions in chloroplasts that have been approved for use to produce a standard barcode for plants, to be used in the same way as the animal barcodes. It is important that every specimen used to produce the definitive bar codes is preserved for reference.

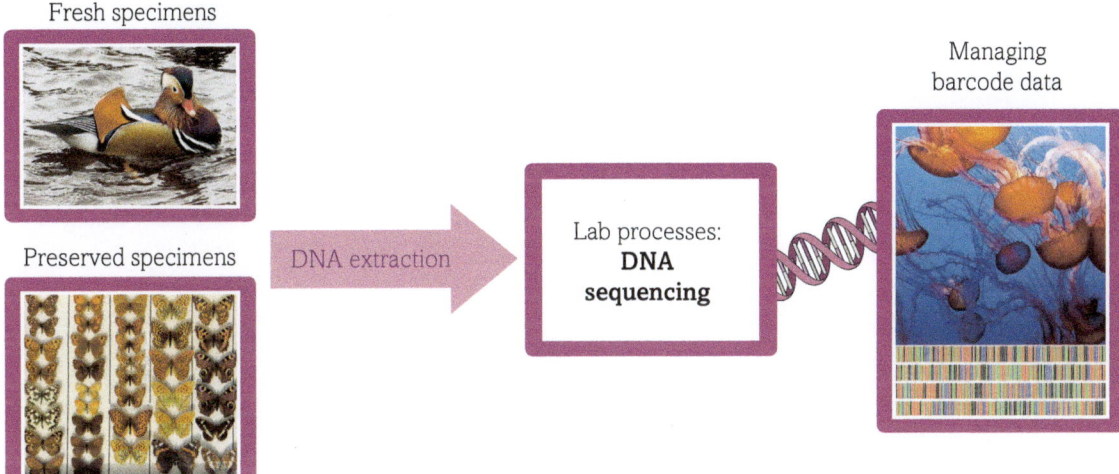

Fresh specimens

Preserved specimens

DNA extraction

Lab processes:
DNA sequencing

Managing barcode data

fig B The production of a DNA barcode.

Bar coding will not replace traditional taxonomy but it will support it. Scientists hope that field instruments will be developed that can be used to analyse genes and identify species as scientists work with organisms in their natural habitats. This will make it so much easier to identify plants with no flowers or fruits, immature animals and larval forms of insects, for example. Quick identification of invasive species, for example, makes it much easier to deal with the threat.

The CBOL project recognises that it will take a long time to barcode all the species of living organisms, but rapid progress is being made. Hundreds of thousands of species are now on the databases, with the numbers increasing all the time. Fortunately the tests needed to get the barcode from the DNA are both fast and relatively cheap. Within the next 20 years it is not unrealistic to think that all known plant and animal species will be identified and barcoded based on DNA analysis. Fungi and bacteria may not be so easy.

Questions

1 Why is it so important to be able to identify individual species?

2 What is bioinformatics?

3 Certain individual proteins such as cytochrome oxidase and haemoglobin, and the genes which code for them, are widely used by scientists to identify both individual species of animals and relationships between them. Why are these particular molecules so useful?

Key definitions

DNA sequencing is the process by which the base sequences of all or part of the genome of an organism is worked out.

DNA profiling is the process by which the non-coding areas of DNA are analysed to identify patterns.

Bioinformatics is the development of the software and computing tools needed to organise and analyse raw biological data, including the development of algorithms, mathematical models and statistical tests that help us to make sense of the enormous quantities of data being generated.

By the end of this section, you should be able to...

- describe how DNA sequencing can be used to distinguish between species and determine evolutionary relationships

- explain the role of scientific journals, the peer review process and scientific conferences in validating new evidence supporting the accepted scientific theory of evolution

Much of the old evidence for evolution relied on similarities in the appearance of living organisms and on fossil evidence. This can cause problems, particularly when two species look similar because they have evolved in response to similar niches, such as moles living underground. Now DNA profiling and proteomics give scientists new insights into evolutionary links.

The more differences (mutations) there are in the DNA from two individuals, the longer the time since they had a common ancestor. Using an agreed mutation rate as a kind of 'molecular clock', you can even estimate how long ago that common ancestor lived. However, problems are caused by the fact that DNA starts to degrade immediately after death. In the most favourable conditions, fragments can survive for 50–100 000 years. Also, the rate at which different parts of the DNA mutates can vary. So the evidence has to be viewed with caution.

Fossil DNA and human evolution

Human evolution has always been difficult to follow owing to limited fossil evidence. Now scientists can extract DNA from suitable fossils under 100 000 years old, and analyse it using the polymerase chain reaction (which amplifies minute traces of DNA) and DNA profiling (see **Book 2**).

At one time it was thought that Neanderthals, who lived 30 000–100 000 years ago, were the ancestors of modern humans. Comparing DNA from a number of Neanderthal fossils with DNA of modern humans has shown both that Neanderthals were not our direct ancestors, but that interbreeding between Neanderthals and modern human ancestors did take place – many people have up to 2% Neanderthal DNA in their genome!

DNA, lice and human evolution

It is not only human DNA that is useful in unravelling human evolution. Modern humans have three types of parasitic lice: head lice, pubic lice and body lice – the latter hang onto clothing but feed off the body. Head and body lice are related to chimpanzee lice. DNA evidence suggests that the common ancestors of humans and chimps diverged about 6 million years ago, and the head lice species diverged at the same time.

DNA analysis of the human head louse and body louse shows that they formed separate species around 72 000 years ago. Body lice live in close-fitting fabric, so this suggests when woven clothing became common. This work was published by R. Kittler, M. Kayser and M. Stoneking in the journal *Current Biology* in 2003.

DNA profiling also shows that the pubic louse is more closely related to the gorilla louse than to other human lice. Work by David Reed and Jessica Light, published in 2007 in the *BMC Biology* journal, shows that human and gorilla ancestors diverged about 6 million years ago, but that pubic and gorilla lice only diverged about 3 million years ago. This suggests that by then humans had lost most of their body hair, and the texture of the pubic hair was already different from the head hair. One explanation is that human ancestors picked up pubic lice from gorillas by using plant bedding that gorillas had slept on. Overall, the DNA evidence from the lice suggests humans had head lice with chimpanzee louse ancestors, and pubic lice with gorilla louse ancestors, long before they began wearing woven clothes and body lice evolved. So the ability to weave clothes was developed long after most of our body hair was lost.

(a)

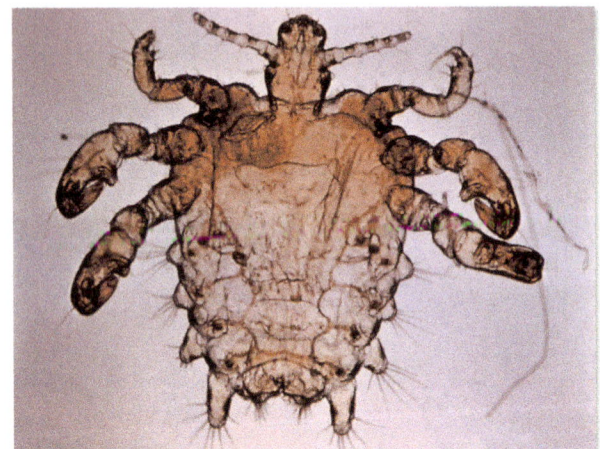

(b)

fig A It is not easy to get hold of gorilla lice for DNA analysis, but this comparison shows that (a) the human pubic louse *Pthirus pubis* is closely related to (b) the gorilla louse *Pthirus gorilla*.

New models support old theories

Scientists use evidence from the DNA analysis of many species to help build diagrams that model the evolution of species from a common ancestor. The process relies heavily on how data are interpreted using statistics and bioinformatics, including computer analysis. It is possible that the same evidence can be interpreted differently. This is why it is important to look for evidence from traditional studies based on appearance, behaviour or fossil data as well as the evidence from modern DNA and protein studies. In an ideal situation, evidence from a variety of sources all agrees. An example is in primate evolution (see **fig B**).

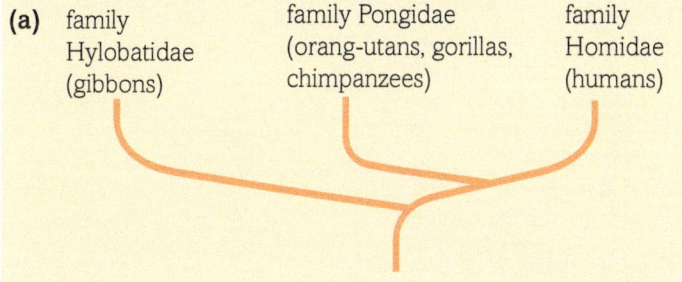

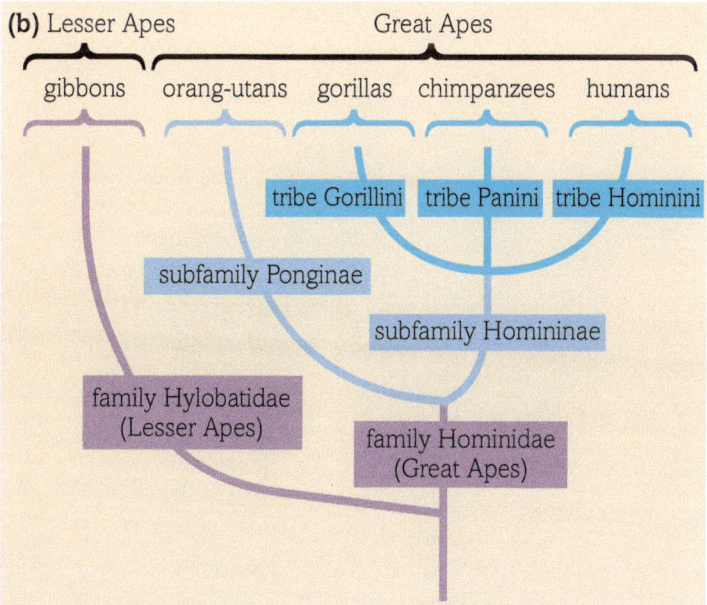

fig B Evolutionary trees showing relationships between the apes based on (a) the fossil evidence available at the time along with morphology and (b) more recent fossil and DNA evidence show how important it is for scientists to keep developing new approaches.

The role of the scientific community

The scientific community plays an important role in validating new evidence, as shown in the example of the evolution of human lice and their relevance to the loss of body hair and the development of clothing. When a piece of research produces useful results and conclusions, it is submitted to a scientific journal. Some of these journals are very famous, others are smaller and very specialised. When an article is submitted, it goes through a process of peer review during which it is read by a number of experts to see if it is reliable. If so, it will be published. A paper should provide enough information for other scientists to carry out the same or similar investigations, so that the conclusions can be validated.

At scientific conferences scientists working in the same field get together to discuss ideas – for example, how accurate is the molecular clock for primates? This helps to promote the development of new techniques in research as well as providing opportunities to challenge the validity of results that are being presented. It facilitates the exchange of ideas, data and techniques between scientists from all over the world.

Questions

1 (a) Describe some of the limitations of using fossils to show evolutionary relationships.
   (b) Explain why DNA profiling is of limited use when building models of evolution.
   (c) Explain why using evidence for evolution from various sources increases its validity.

2 Explain the role of publications and scientific conferences in validating new scientific research.

3 Discuss the value of the peer review process in preparing papers for publication in scientific journals and investigate at least one example of where the process has failed.

By the end of this section, you should be able to...

- explain how gel electrophoresis can be used to distinguish between species and determine evolutionary relationships

- explain how DNA sequencing and bioinformatics can be used to distinguish between species and determine evolutionary relationships

- explain the evidence for the three-domain model of classification as an alternative to the five-kingdom model and the role of the scientific community in validating this evidence

Biochemical relationships

For centuries classification has been based on detailed observations of morphology, such as counting the hairs on the foreleg of a fly or the petals of a flower, or even seeing how similar embryos are, to work out relationships between organisms. Now biochemical relationships are increasingly being used to support or clarify the relationships based on morphology. Scientists need to analyse the structures of many different chemicals as well as the DNA to identify the inter-relationships between groups of organisms. This analysis is known as molecular phylogeny, and not all scientists interpret the results in the same way. Proteins are key molecules in these analyses.

The evidence from biochemical analysis may support or conflict with relationships based on morphology. For example, all green plants have similar complex pathways for making glucose from sunlight using chlorophyll, so it seems safe to assume they all evolved from a common ancestor. In contrast, American porcupines and African porcupines occupy similar niches and look very similar, but biochemical analysis suggests that they are only very distantly related.

Gel electrophoresis – a key technique

The discovery that we can identify patterns in the DNA, RNA and proteins of different individuals, and between different species of organisms has had enormous implications in many areas of science, including species identification and the development of evolutionary models. Increasingly patterns in DNA or RNA fragments are used in species identification (see **Section 3.1.3**). Comparisons between the amino acid sequences of similar proteins in different species or groups of organisms, are also used to help us classify them or trace their evolutionary pathways. A key technique in these processes is **gel electrophoresis**. This is a variation of chromatography which can be used to separate DNA, and RNA fragments, proteins or amino acids according to their size and charge.

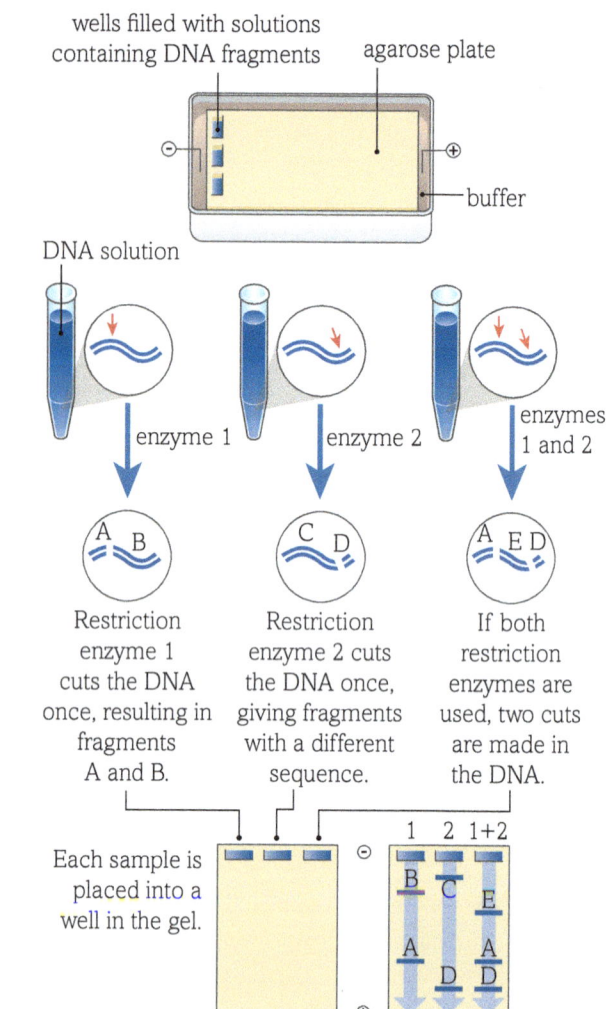

wells filled with solutions containing DNA fragments · agarose plate · buffer · DNA solution

enzyme 1 · enzyme 2 · enzymes 1 and 2

Restriction enzyme 1 cuts the DNA once, resulting in fragments A and B.

Restriction enzyme 2 cuts the DNA once, giving fragments with a different sequence.

If both restriction enzymes are used, two cuts are made in the DNA.

Each sample is placed into a well in the gel.

As fragments of DNA move toward the positive electrode, shorter fragments move faster (and therefore further) than longer fragments.

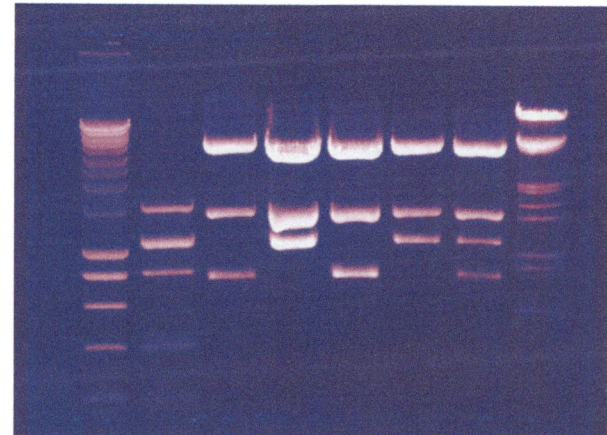

fig A Gel electrophoresis can be used to identify patterns in the DNA, RNA and proteins of different types of organisms to help determine both the species they belong to and their evolutionary links.

The chemicals to be compared are placed in wells in a gel medium in a buffering solution (to maintain a constant pH), with known DNA or RNA fragments, proteins or amino acids to aid identification.

For identifying DNA, the big DNA molecules are cut into fragments by restriction endonucleases – special enzymes that cut DNA into fragments at specific sites. The DNA fragments are added to a gel containing a dye (e.g. EtBr, ethidium bromide) which binds to the fragments in the gel and will fluoresce when placed under ultraviolet (UV) light. A dye is also added to the DNA samples. This does not bind with the DNA but moves through the gel slightly faster than the DNA so that the current can be turned off before all the samples run off the end. An electric current is passed through the apparatus and the DNA fragments move towards the positive anode, because of the negative charge on the phosphate groups in the DNA. The fragments move at different rates depending on their mass and charge. Once the electrophoresis is complete, the plate is placed under UV light. The DNA fluoresces and shows up clearly so the pattern of the different bands can be identified.

When electrophoresis is used to identify and compare the amino acids in a particular protein, dyes are not added to the gel or the mixture. Ninhydrin is added to the gel after electrophoresis has taken place. It reacts with the amino acids so they show up as purple patches and can be compared with known amino acids for identification.

More biochemical relationships

Understanding of the biochemical relationships between organisms is playing an important role in extending our understanding of classification and evolution.

Here are some other examples:

*   The vertebrates and the echinoderms (an invertebrate group of animals that includes starfish and sea urchins) appear, from the evidence of comparative anatomy and embryology, to come from one line of ancestors and the annelid worms, molluscs and arthropods (including insects) from another. Biochemical evidence appears to confirm this unlikely relationship. It shows that phosphagens, molecules that provide the phosphate group for the synthesis of ATP in muscles, are of two different sorts. Phosphocreatine occurs almost exclusively in the muscle tissue of vertebrates and echinoderms whilst phosphoarginine occurs in the other groups.
*   Blood pigments are important in many animal groups. Analysis has shown that any one group contains only one type of blood pigment – all vertebrates and many of the invertebrates have haemoglobin, all polychaete worms have chlorocruorin and all molluscs and crustaceans have haemocyanin. This allows scientists to build up a more detailed picture of the relationships between the different groups.
*   Analysis of the sequence of amino acids in particular proteins can help show the relationships within higher groups, such as a phylum. For example in mammals, the analysis of fibrinogen,

the protein involved in the clotting of the blood, reveals how closely the different mammalian groups are related. Single amino acid changes are used to plot relationships.

*   A combination of DNA analysis, protein analysis and anatomical observations can bring some unlikely relationships to light – for example, the closest living relatives to the hippopotamus appear to be the dolphins and whales.

fig B Starfish and sheep do not seem to have much in common, but biochemical and gel electrophoresis analysis shows that they are more closely related than you might think.

Two domains or three?

For many years biologists divided living organisms into two large groups or domains. They named them the eukaryotes, cells with a complex cell structure (see **Chapter 2.1**), and the prokaryotes which included bacteria (see **Chapter 2.2**). The theory was that eukaryotes had evolved from prokaryotes billions of years ago (see **fig C**). Some scientists think that chloroplasts became part of 'eukaryotic ancestor' cells first, while others think that mitochondria were the first **endosymbionts**. It is possible that both processes happened at the same time as evidence one way or another is almost impossible to obtain.

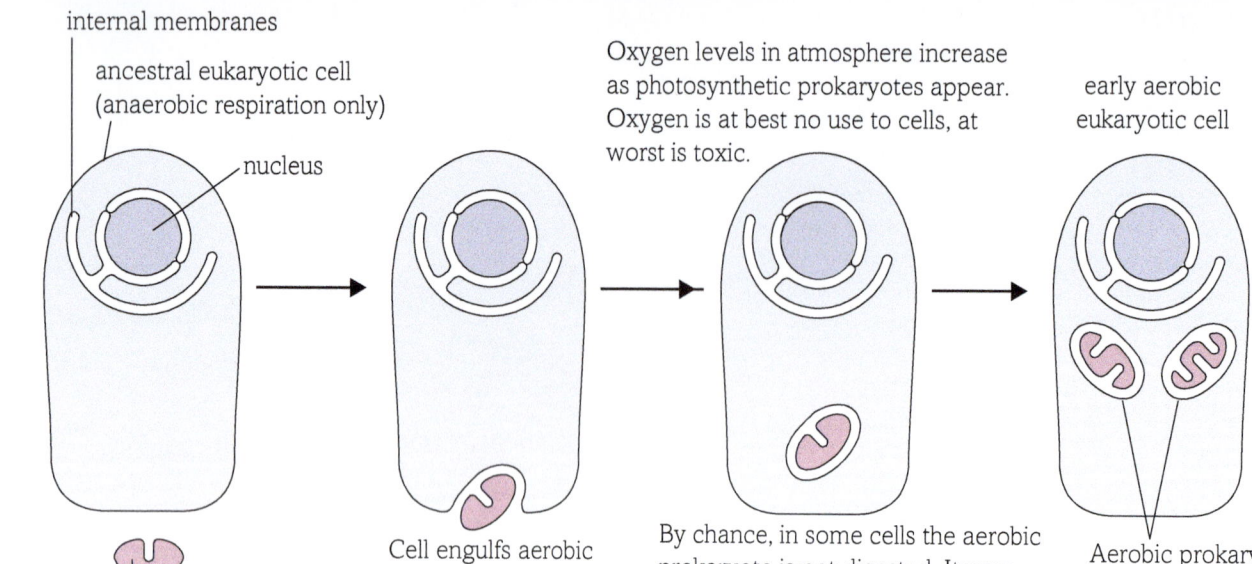

internal membranes
ancestral eukaryotic cell (anaerobic respiration only)
nucleus

Oxygen levels in atmosphere increase as photosynthetic prokaryotes appear. Oxygen is at best no use to cells, at worst is toxic.

early aerobic eukaryotic cell

prokaryotic organism capable of aerobic respiration

Cell engulfs aerobic prokaryote as food.

By chance, in some cells the aerobic prokaryote is not digested. It uses oxygen for cellular respiration. More efficient respiration allows cells to grow and reproduce more rapidly.

Aerobic prokaryotes become permanent feature as mitochondria. Copies are made and passed to daughter cells during reproduction.

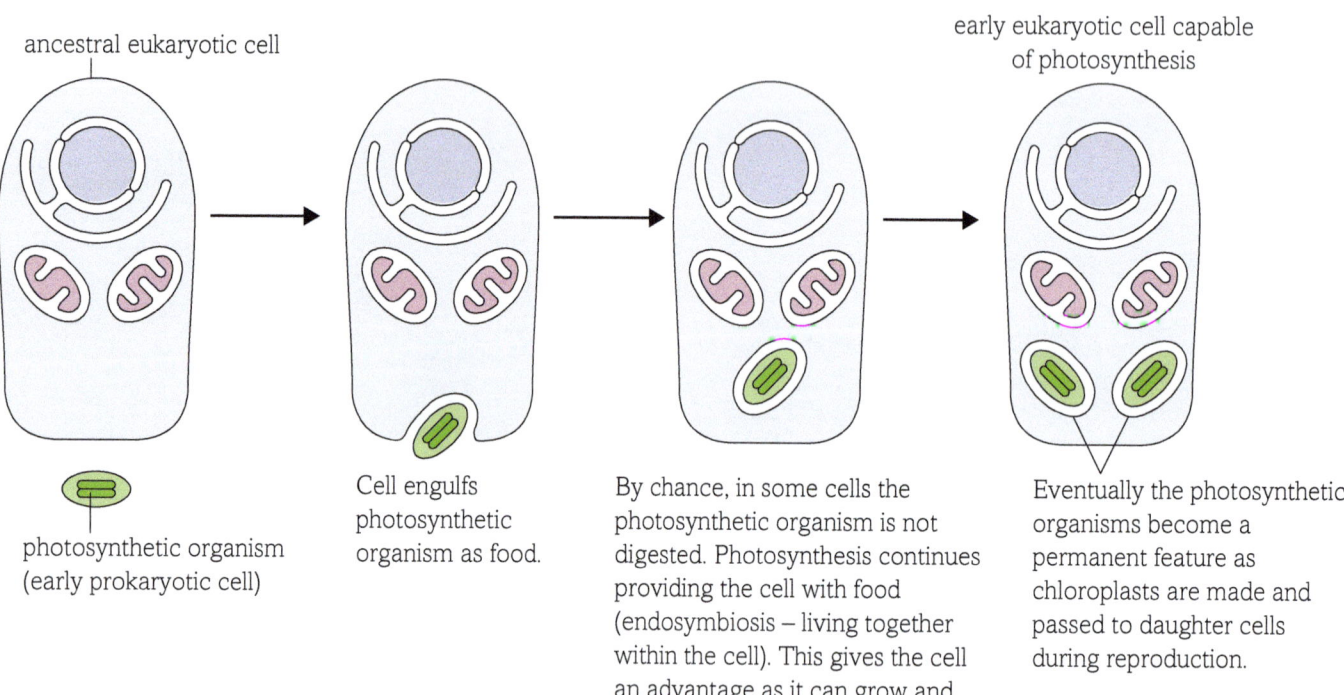

ancestral eukaryotic cell

early eukaryotic cell capable of photosynthesis

photosynthetic organism (early prokaryotic cell)

Cell engulfs photosynthetic organism as food.

By chance, in some cells the photosynthetic organism is not digested. Photosynthesis continues providing the cell with food (endosymbiosis – living together within the cell). This gives the cell an advantage as it can grow and reproduce more rapidly.

Eventually the photosynthetic organisms become a permanent feature as chloroplasts are made and passed to daughter cells during reproduction.

fig C One model of how eukaryotes may have evolved from prokaryotes.

Scientists are using the techniques of molecular phylogeny to investigate this theory further. They have looked at the internal structures of the prokaryotes and eukaryotes, and compared them to the biochemistry of the proteins involved in their ribosomes and enzymes. Much of this analysis involves the use of bioinformatics to process the masses of data generated.

As a result, a new theory developed that there are in fact three domains – two prokaryote domains, the Archaea and Bacteria, and the eukaryotes or Eukaryota. The Archaea and the Bacteria are as different from each other as they are from the eukaryotes (see **table A**). Genetic studies show that all three groups probably had a single common ancestor around three billion years ago. Some evidence suggests the Archaea are more closely related to eukaryotes, including us, than the Bacteria – our last common ancestor was probably around two billion years ago.

Characteristic	Bacteria	Archaea	Eukaryota
membrane-enclosed nucleus	absent	absent	present
membrane-enclosed organelles	absent	absent	present
peptidoglycan in cell wall	present	absent	absent
membrane lipids	ester-linked, unbranched	ester-linked, branched	ester-linked, unbranched
ribosomes	70S	70S	80S
initiator tRNA	formylmethionine	methionine	methionine
operons	yes	yes	no
plasmids	yes	yes	rare
RNA polymerases	1	1	3
ribosomes sensitive to chloramphenicol and streptomycin	yes	no	no
ribosomes sensitive to diphtheria toxin	no	yes	yes
some are methanogens	no	yes	no
some fix nitrogen	yes	yes	no
some conduct chlorophyll-based photosynthesis	yes	no	yes

table A Some of the cellular and molecular characteristics of the three domains of life on Earth.

Apart from the ribosomes, scientists see two key differences in the mass of data that has been generated about the three domains. Archaea replicate by binary fission controlled within a cell cycle, which seems homologous to the cell cycle in eukaryotic cells but is rather different from replication in bacteria. On the other hand, the membrane structure and the membrane proteins of the Archaea are unique – they are different from the bacteria and the eukaryotes, which have homologous structures. The Archaea have an ether link in their lipids, giving branched molecules that may provide extra strength in extreme environments. This adds weight to a model which suggests a different origin for some of the cellular systems of the Archaea and Bacteria, but shows eukaryotes combining features of them both. Work establishing these links has been published in peer-reviewed journals so that other scientists can repeat the procedures to verify the results. Yet in spite of this, there is still debate about the three domains and their origins, although most biologists now accept the three-domain theory. However, evidence showing horizontal gene transfer between groups of organisms in early evolution is increasing, so the idea of a complex interwoven network of ancestry also remains.

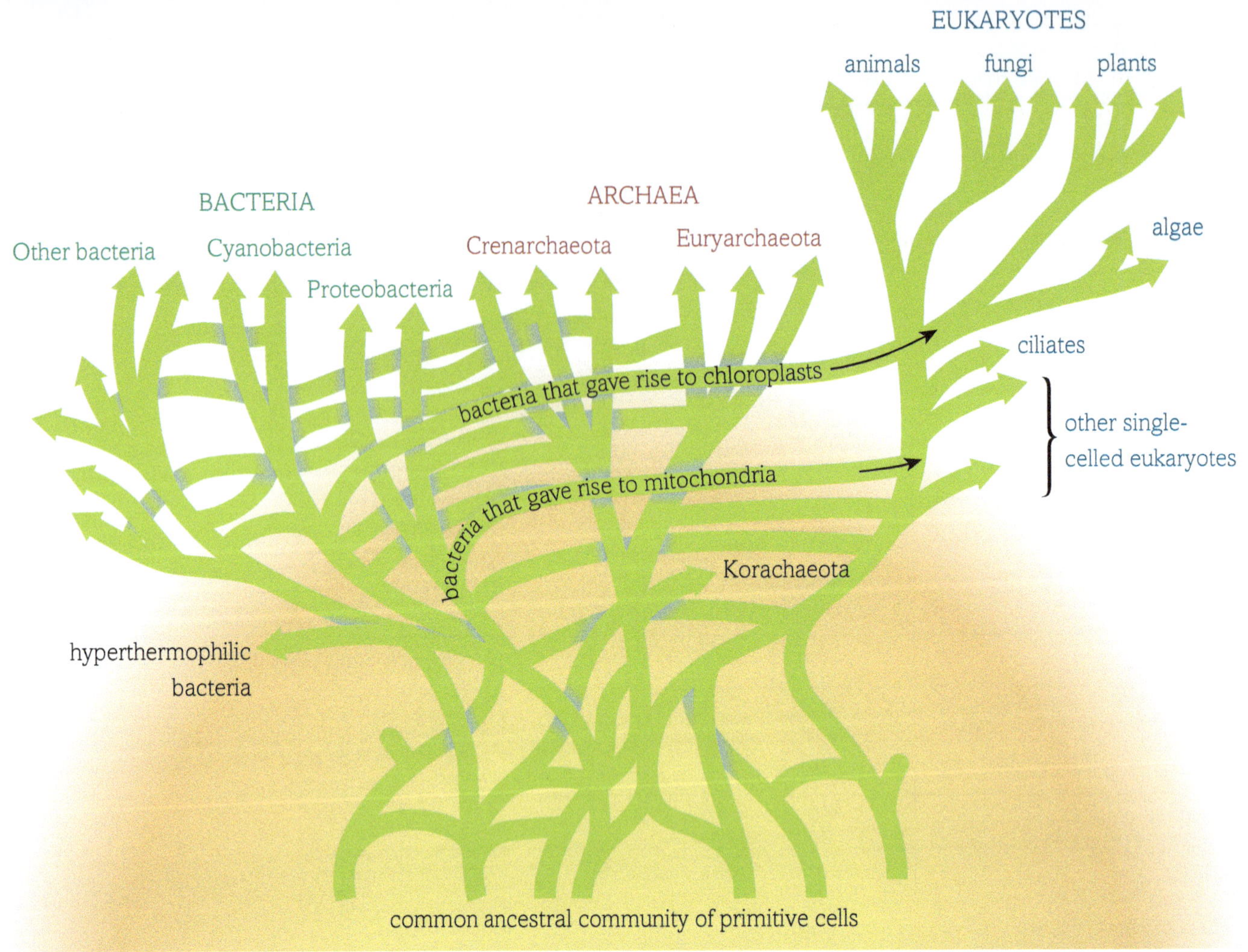

EUKARYOTES

animals   fungi   plants

BACTERIA   ARCHAEA

Other bacteria   Cyanobacteria   Crenarchaeota   Euryarchaeota   algae

Proteobacteria

bacteria that gave rise to chloroplasts   ciliates

bacteria that gave rise to mitochondria   other single-
                                                                 celled eukaryotes

Korachaeota

hyperthermophilic
bacteria

common ancestral community of primitive cells

fig D One of a number of models showing some of the complex relationships that could be the basis for the modern domains of life.

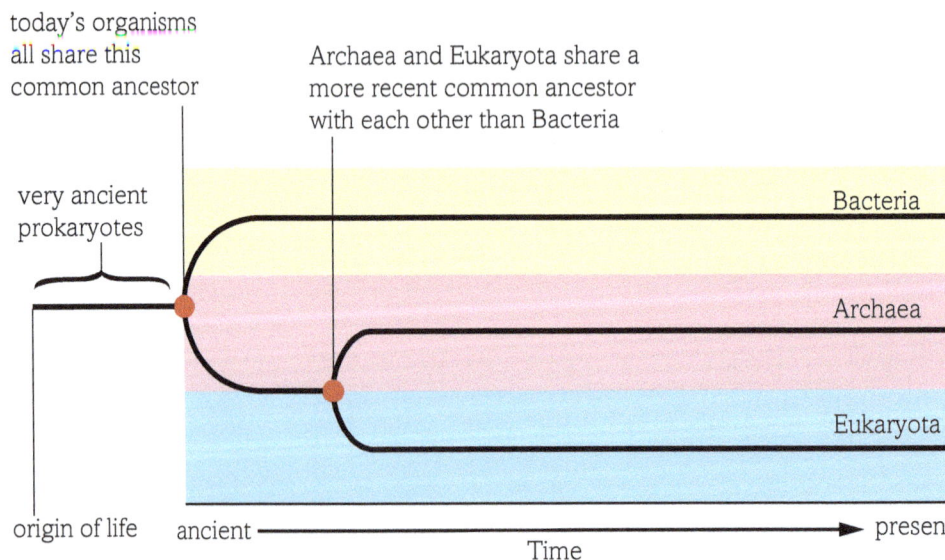

today's organisms
all share this
common ancestor

Archaea and Eukaryota share a
more recent common ancestor
with each other than Bacteria

very ancient
prokaryotes

Bacteria

Archaea

Eukaryota

origin of life   ancient   present
                        Time

fig E The three domains of the living world – most but not all biologists think that all three domains share a common prokaryotic ancestor.

How many kingdoms?

When Linnaeus first worked out his classification system, he proposed all living things fitted into two large groups that he called the plant kingdom and the animal kingdom. Everything including fungi were fitted in – fungi counted as plants because they did not move about.

As technology and scientific knowledge increased, people began to see the world of microscopic organisms, including bacteria and the single-celled organisms we now call the Protista. The structure of fungi, which is very different from the structure of plants, became clear. A new system emerged, based mainly on morphology, resulting in a five-kingdom classification. In this system all of the prokaryotes are put together in one kingdom, the **Monera**. There are the animals, the plants and the fungi. Everything else, such as all of the single-celled organisms and the algae, is grouped together in the **Protista**.

The five-kingdom system is still used, but increasingly biologists now use a six-kingdom classification system. This results from the work done using biochemical and DNA evidence that you have already seen in relation to the three domains. The prokaryotes are no longer regarded as a single mass of bacteria. The Archaea and the Eubacteria are two very different groups of organisms with very different biochemistry.

So modern classification systems include six kingdoms:

Archaebacteria – prokaryotic cells

These are ancient bacteria that have a wide variety of lifestyles and include the **extremophiles** – bacteria that can survive extreme conditions of heat, cold, pH, salinity and pressure. They normally reproduce asexually.

Eubacteria (bacteria) – prokaryotic cells

This kingdom includes the true bacteria and the cyanobacteria, which used to be called the blue-green algae. They normally reproduce asexually.

Protista – eukaryotic cells

This kingdom includes all of the single-celled organisms, the green algae, the brown algae and the slime moulds. It is something of a catch-all for the organisms that do not fit in the other kingdoms. They mainly reproduce asexually.

Fungi – eukaryotic cells

This kingdom includes both unicellular organisms, for example yeasts, and multicellular organisms, for example toadstools and moulds. They are all **heterotrophs**. They reproduce both asexually and sexually.

Plantae – eukaryotic cells

The organisms in this kingdom are all multicellular. They are all **autotrophs** and make their own food by photosynthesis. They include the liverworts and mosses, ferns, gymnosperms and the angiosperms, the true flowering plants. Both asexual and sexual reproduction occurs within this kingdom.

Animalia – eukaryotic cells

The organisms in this kingdom are all multicellular and they are all heterotrophs. They include invertebrates and vertebrates. Sexual reproduction is common but some animals also reproduce asexually.

| kingdom: Monera (prokaryotes) | kingdom: Protista | kingdom: Fungi | kingdom: Plantae | kingdom: Animalia |

five-kingdom classification

| domain: Bacteria | domain: Archaea | domain: Eukaryota (eukaryotes) |

three-domain classification

| kingdom: Eubacteria | kingdom: Archaea | kingdom: Protista | kingdom: Fungi | kingdom: Plantae | kingdom: Animalia |

six-kingdom classification

fig F A six-kingdom classification makes more sense when the three domain system is also in place.

Questions

1 How is molecular phylogeny used to show genetic diversity and similarities between organisms?

2 Discuss the importance of the following in how we distinguish between species and construct evolutionary relationships:
   (a) gel electrophoresis
   (b) bioinformatics.

3 Use the information on these pages along with other details you may find on the internet to help you explain the difficulties of drawing conclusions about how to classify the variety of organisms.

4 Discuss how the development of the three-domain model of classification affects other aspects of classification.

Key definitions

An **endosymbiont** is an organism that lives inside the cells or the body of another organism.

The **Monera** is a kingdom in the five-kingdom classification system that contains the Archaea and Eubacteria.

The **Protista** is a kingdom in the five-kingdom classification system that contains all single-celled organisms, green and brown algae and slime moulds.

Extremophiles are bacteria that can survive extreme conditions of heat, cold, pH, salinity and pressure.

Heterotrophs are organisms that cannot make their own food and have to eat other organisms.

Autotrophs are organisms that can make their own food, either by photosynthesis or chemosynthesis.

Gel electrophoresis is a method of separating fragments of proteins or nucleic acids based on their electrical charge and size.

THINKING BIGGER

TAKING ADVANTAGE OF CHANGE

Changes in the environment can force adaptation on organisms and we often look at how problems resulting from human activity drive natural selection and evolution. But sometimes the changes we make put opportunities in the way of organisms ready to take them. Consider the story of the Oxford ragwort…

OXFORD RAGWORT

by George Short

Stone walls do not a prison make
For plants whose seeds the air can take,
Fluffily floating.
The Oxford gardens could not hold
Your daisy flowers of gleaming gold.
The botanists to crib you tried
In beds beside the Cherwell's side
Where every summer brings along
Its duck and moorhen and a throng
Of students boating.

Disdaining damp luxuriance you
put out in search of pastures new
And took a ride
Down on the railway. Life was fine;
Your seeds were sucked along the line
by passing trains. Soon you were found
on every barren railway ground.
Dry stones and clinkers were no bar
To your swift progress.
Near and far you multiplied.

From vacant lot and mouldering wall
Your florets flash their cheerful call:
Sunshine on earth.
Squalid may be your habitat…
All the more welcome, you, for that.
Your beauty may be brash and cheap…
But winsome, on a rubbish heap
Where there's a dearth.

fig A Oxford ragwort nestling by a railway track in Bristol.

Where else will I encounter these themes?

| 1.1 | 1.2 | 1.3 | 1.4 | 2.1 | 2.2 | 2.3 | 2.4 |

A poem may seem an unlikely source of biological knowledge – but take a look at 'Oxford Ragwort' by George Short, and see what you can learn.

1. Why does poetry seem an unlikely medium for biological knowledge? Would you expect any science expressed in poetry to be accurate and reliable? Explain your responses.

2. George Short studied English at the University of Oxford after the Second World War. He was also a very keen naturalist, and went on to run science broadcasting on the BBC World Service. How does this knowledge alter the way you look at the information in the poem?

3. Using only the information within the poem, identify and summarise the story of the spread of Oxford ragwort.

Extracting information from a poem is not easy – but analyse the content carefully and you should be able to build up a good basic understanding of the spread of Oxford ragwort from the Oxford University Botanic Gardens.

What about the biology of this amazing escape? You have enough information to answer these questions now, but if you are studying A level Biology you might like to look back at this when you learn about the way ecosystems develop over time, and human effects on ecosystems.

4. If you investigate the story of Oxford ragwort online, you will find a wide variety of articles including one from Bristol University School of Biological Sciences and, of course, one on Wikipedia. Read at least two articles on this story. How many of the facts had you worked out based on the poem by George Short?

5. What are the elements of human involvement in the story of the spread of Oxford ragwort?

6. What are the adaptations of Oxford ragwort that enabled it to make such a successful escape from the botanic gardens at Oxford?

Activity

A salty survivor

Oxford ragwort is not the only plant with adaptations that allow it to take advantage of human influences to spread and increase its range. Danish scurvy grass is a plant that thrives in salty, gritty coastal environments and cliff edges. In recent years it has spread across much of the UK – mainly along the motorways, where it thrives in the salty environment provided by the salting and gritting machines that work throughout winter to keep the roads free from ice.

Investigate the on-going story of the spread of Danish scurvy grass and tell the story either as:

• a poem

• a newspaper article for the widest possible audience.

Did you know?

Oxford ragwort is so called because it is said to have escaped from the Oxford botanic gardens a couple of centuries ago. With the construction of the railway it underwent a population explosion which took it to many other places. It is really a Southern European species, and near relative of the groundsel.

● From an unpublished collection of poems by George Short (publication interrupted by the death of the author).

1 The five-kingdom classification system is one of the systems used for classifying living organisms.

(a) Name the kingdom to which bacteria belong. [1]

(b) Give **two** structural features found in a bacterium but not found in a virus. [2]

(c) Some classification systems consider viruses to be living organisms.

Give **two** features of a virus that it shares with living organisms. [2]

(d) There are alternatives to the five-kingdom classification system.

Suggest why there is disagreement about how living organisms should be classified. [2]

[Total: 7]

2 Classification of organisms is important when trying to assess biodiversity.

(a) All organisms can be classified into one of three domains. Name the **three** domains of organisms. [3]

(b) (i) Explain what is meant by a **species**. [2]

(ii) Explain the meaning of the term **genetic diversity** within a species. [2]

(iii) Describe how zoos maintain the genetic diversity of endangered species. [4]

[Total: 11]

3 In the 1990s, a scientist called Woese suggested a new way of grouping organisms into domains.

(a) The table below shows Woese's three domains and gives some of the characteristics of each domain.

Domain	Some characteristics of each domain
P	True nucleus absent Small (70S) ribosomes present Smooth endoplasmic reticulum absent RNA polymerase made up of 14 subunits
Q	True nucleus present Large (80S) ribosomes present Smooth endoplasmic reticulum present RNA polymerase made up of 14 subunits
R	True nucleus absent Small (70S) ribosomes present Smooth endoplasmic reticulum absent RNA polymerase made up of 4 subunits

(i) Place a cross (☒) in the box which shows the two domains which are most **distantly related**.

☐ A P and Q

☐ B P and R

☐ C Q and R [1]

(ii) Place a cross (☒) in the box which shows the domain that represents eukaryotic organisms.

☐ A P

☐ B Q

☐ C R [1]

(iii) The diagram below represents the phylogenetic tree for the three domains.

Place a cross (☒) in the box on the diagram that correctly identifies the eukaryotic domain. [1]

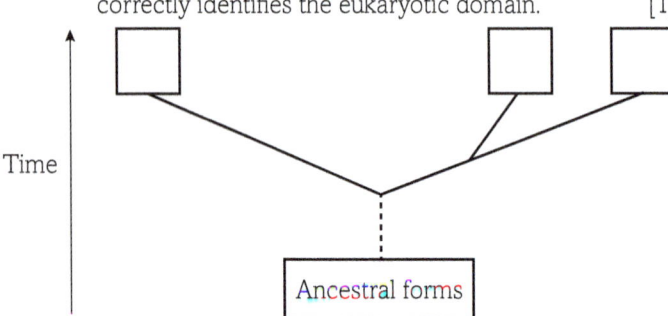

(iv) Give the name of **one** of the other two domains. [1]

(b) One domain includes the plants and these have cells with a cell wall.

(i) Describe the structure of a plant cell wall. [4]

(ii) A student studied the cell wall arrangement between two adjacent plant cells. He noticed several features which he could not name. Two of these are described in the table below.

Complete the table by writing in the name of each feature described.

Feature described	Name of feature
Site where there was no cell wall and the cytoplasm linked the two adjacent cells	
Dark line that is the boundary between one cell and the next cell	

[2]

[Total: 10]

4 Woese was the scientist who proposed a classification of organisms into three domains called the Archaea, Bacteria and Eukaryota (Eucarya).

(a) The table below shows some of the characteristics of the three domains.

Characteristic	Domain		
	A	**B**	**C**
Mitochondria	Absent	Absent	Absent
Cell wall containing peptidoglycan	Yes	No	No
Amino acid carried on tRNA that starts protein synthesis	Formyl-methionine	Methionine	Methionine
Sensitive to antibiotics	Yes	No	No
May contain chlorophyll	Yes	No	Yes

(i) Using the information in the table, suggest which of A, B and C represents the Eukaryota domain. Give a reason for your answer. [2]

(ii) Many scientists believe that the Eukaryota domain is more closely related to the Archaea domain than to the Bacteria domain.

Using the information in the table, suggest which of A, B and C represents the Archaea domain. Give a reason for your answer. [2]

(b) Cells of the Eukaryota domain contain rough endoplasmic reticulum and Golgi apparatus.

Both the rough endoplasmic reticulum and the Golgi apparatus are made up of membrane-bound sacs.

(i) Describe how you would recognise the Golgi apparatus as seen using an electron microscope. [3]

(ii) Explain the roles of rough endoplasmic reticulum and the Golgi apparatus in a cell. [6]

[Total: 13]

5 The scientist Carl Woese suggested that living organisms could be grouped into three domains. There are specific differences between the organisms in the three domains.

(a) Place a cross (☒) in the box that correctly identifies the names of the three domains suggested by Woese. [1]

☐ A Animalia, Archaea and Eukarya

☐ B Animalia, Bacteria and Prokaryotae

☐ C Archaea, Bacteria and Eukarya

☐ D Archaea, Eukarya and Prokaryotae

(b) Carl Woese's ideas were not accepted when he first suggested that every organism could be classified into one of three domains.

(i) Suggest **two** ways in which Woese communicated his findings to the scientific community. [2]

(ii) Describe how the scientific community would have evaluated Woese's theory. [2]

(iii) Woese suggested that organisms could be placed in taxonomic groups based on molecular phylogeny. Explain what is meant by this statement. [4]

[Total: 9]

TOPIC 3
Classification

3.2 ▶ Natural selection

Introduction

The soil in bogs is damp and acidic with few nutrients, but in North America pitcher plants successfully colonise them – because they are carnivores. Pitcher plants are adapted as pitfall traps for insects that visit the pitcher as if it was a flower. But the pitcher contains a deadly liquid. It is a dilute acid, because ants attracted by nectar fall into the pitcher, releasing the formic acid in their body as they decay. The liquid also contains enzymes made by the plant, bacteria acting as decomposers and a chemical that causes waterlogging of insect wings. Inside the pitcher, downwards facing hairs at the top followed by smooth waxy surfaces make it almost impossible for insects to escape once they fall into the liquid, where the acid, enzymes and bacteria digest them. Some of the minerals released are then absorbed into the body of the plant. Amazingly, the larva of a species of blowfly actually develops in the pitcher fluid, escaping to pupate – but returning to the flowers for nectar as an adult, so pollinating them. The strange flowers are self sterile, in spite of dropping pollen and nectar onto their own stigma. The pitcher plant is a prime example of natural selection, with a mass of adaptations for a bizarre but successful way of life.

In this chapter you will consider the theory of evolution by natural selection and how Darwin's ideas have been modified in the light of our current knowledge of genetics. Through a range of examples, you will look at the importance of physiological, anatomical and behavioural adaptations in different organisms, which enable them to survive and reproduce successfully in a wide range of habitats.

If the environment changes, or an individual moves to a different habitat, the adaptations that have been successful may no longer work so well. Some individuals will have extra adaptations that make them better adapted to the new conditions. They are the individuals that are most likely to survive and reproduce successfully, passing on the alleles for the new adaptation. This is natural selection in action and you will be considering a number of examples.

Finally you will be looking at how organisms diverge when they are reproductively separated. As they adapt to slightly different conditions in slightly different ways, new species may eventually emerge. You will be looking at both allopatric and sympatric speciation, considering examples of both.

All the maths you need

- Interpret data from a variety of tables and graphs (*e.g. graph to show effect of adaptations on reproductive success*)

What have I studied before?

- That there is usually extensive genetic variation in a species
- That evolution is a change in the inherited characteristics of a population through time, through a process of natural selection that may result in the formation of new species
- That evolution occurs through natural selection of variants that give rise to different phenotypes
- Gene mutations

What will I study later?

- How natural selection leads to biodiversity
- Why biodiversity is important and how it can be assessed
- Adaptations of mammals, insects and fish for gas exchange
- Adaptations in plants and animals for the mass transport of substances around the body
- How abiotic and biotic factors in an ecosystem can affect population size and exert selection pressures
- Human effects on ecosystems and thus on biodiversity, extinction and speciation (A level)
- The origins of genetic variation (A level)
- The effect of selection pressures on the gene pool of a population or species (A level)

What will I study in this chapter?

- The way evolution comes about through natural selection acting on variation bringing about adaptations
- Examples of the way organisms may occupy a niche as a result of physiological adaptations that involve the way the body of the organism works, making them better adapted for survival and reproduction
- Examples of the way organisms may occupy a niche according to anatomical adaptations that involve their form and structure, making them better adapted for survival and reproduction
- Examples of the way organisms may occupy a niche as a result of behavioural adaptations that involve changes to programmed or instinctive behaviour, making them better adapted for survival and reproduction
- The ways in which reproductive isolation can lead to allopatric and sympatric speciation
- The development of antibiotic resistance in bacteria

By the end of this section, you should be able to...

● explain how evolution can come about through natural selection acting on variation, bringing about adaptations

● explain how organisms occupy niches according to physiological, behavioural and anatomical adaptations

The idea of evolution by natural selection is a key concept in biology. It gives a scientific explanation for the great diversity of life on Earth and for all of the organisms that have existed and become extinct. The idea was put forward by Charles Darwin (1809–82) in his book *On the Origin of Species by Means of Natural Selection*, and a version of his theory still underpins our understanding of biology today.

Observing evolution

Darwin spent five years travelling the world on *HMS Beagle*, observing the natural history of all the countries where they landed. He began to realise that wherever he went, the organisms he saw were adapted to their particular environment. However, he did get some help from people along the way. For example, on the Galapagos Islands where Darwin made some of his most important observations, the giant land tortoises vary from one island to the next. Each island has a different sub-species and they have distinctive shell shapes, which are adapted to the terrain and the way the animals feed. The vice-governor of the islands pointed this out, but initially Darwin ignored it. As he noted in his journal, by the time Darwin realised just how varied the organisms on the different islands were, he had already mixed up the collections from the first two islands!

On his return to England, Darwin spent the next 20 years or so reading, thinking and honing the ideas his observations had triggered. He carried out experiments on organisms including pigeons and plants to support his ideas. Finally, in 1859, he went public with his theory.

fig A Darwin was fascinated and revolted by the marine iguanas of the Galapagos. They are unique among reptiles because they are adapted to absorb heat, which allows them to swim in the sea and feed on the seaweed and algae that is abundant on the rocky coasts of the islands.

The theory of evolution

The main ideas put forward by Darwin to explain the great variety of life include:

- Living organisms that reproduce sexually show great variety in their appearance.
- Organisms produce an excess of offspring – in other words, many of the offspring an organism produces do not survive to reproduce themselves. As a result there is always a struggle for survival, a competition between members of the same species.
- Organisms that inherit characteristics that give them an advantage in this struggle are most likely to survive and pass on the desired feature to their offspring.
- Organisms that inherit characteristics that put them at a disadvantage will be more likely to die out before they can reproduce.

This process is known as the survival of the fittest, where fitness is the ability of an organism to survive and reproduce in the environment in which it is living. Darwin called this process **natural selection**. When the long-term changes in organisms that occur as a result of natural selection produce a new species, this is evolution.

Evolution is a change in the genetic composition of a population of organisms over several generations, as a result of natural selection acting upon variation, bringing about adaptations and in some cases leading to the development of new species. The variation may be the result of sexual reproduction, random mutation, inbreeding or hybridisation.

Did you know?
Friendly competitors

Alfred Wallace (1823–1913), another British naturalist, proposed a similar theory to Darwin – in fact they each published an initial paper at the Linnaean Society at the same time. However, Darwin's ideas were backed up by a much greater bank of research and observations, and he published his book on the subject first – and so it is Darwin who is mainly associated with the theory of evolution. However, Darwin and Wallace always remained in contact and Darwin helped organise a pension for Wallace when he fell on hard times.

Neo-Darwinism – evolution in the twenty-first century

When Charles Darwin developed his theory of evolution by natural selection, no-one had seen chromosomes or knew about DNA, genes and the genetic basis of the variation in living organisms. As our knowledge of genetics, genomics, molecular biology, ecology and palaeontology has grown, our model of evolution has itself evolved.

A modern statement of the theory of evolution might be:

*The evolution of organisms occurs as a result of the differential fertility and survival of organisms with different **genotypes** (genetic variation) leading to different **phenotypes** within a specific environment. Those **alleles** that deliver the adaptations best suited to the environment are most likely to be passed on to the next generation.*

This more modern definition suggests that a disadvantageous trait does not necessarily mean those individuals are wiped out. They may simply be less successful at reproducing. It also recognises that the advantages or disadvantages of a particular trait will differ with the environment. The changes in the frequency of a particular allele in a population of organisms, which may or may not lead to speciation, are almost always driven by a change in the environment or by the organism moving into a slightly different environment.

Learning tip

Make sure you are very clear about the difference between natural selection and evolution.

Natural selection is the *process* by which evolution occurs, but natural selection does not necessarily lead to the evolution of a new species.

Adaptation to a niche

Organisms do not exist in a vacuum. The various species are all part of a complex system of interactions between the physical world and other living organisms that we call **ecology**. Each species exists in a particular **niche**.

The niche occupied by an organism is an important concept that is difficult to define. It describes the role of the organism in the community – rather like a job description or a way of life. You can consider different aspects of a niche, such as the food niche or the habitat niche. Some niches are very large and general, for example organisms that eat grass; some are very small and specific, for example organisms that feed by cleaning the teeth of other, larger organisms.

Successful adaptation

A successful species is well adapted to its niche, meaning that individuals in that species have characteristics that increase their chances of survival and reproduction, and therefore of passing those characteristics on to the next generation. Adaptations may be of many different kinds, including anatomical, physiological and behavioural.

- **Anatomical adaptations** involve the form and structure of an organism, for example the thick layer of blubber in seals and whales and the sticky hairs on the sundew plant, which enable it to capture insects ready to digest.

fig B The sticky hairs of a sundew are anatomical adaptations that enable the plant to capture insects and use their bodies to supplement the nutrients in the poor soil in which it grows.

- **Physiological adaptations** involve the way the body of the organism works and include differences in biochemical pathways or enzymes. For example, diving mammals can stay under water for far longer than non-diving mammals without drowning. Once they are under water their heart rate drops dramatically, so that the blood is pumped around their body less often and the oxygen in their blood is not used as rapidly (see **fig C**). The main body muscles can work more effectively using anaerobic respiration than those of land-living mammals, so the oxygen-carrying blood is directed to the brain and the heart where it is still needed. This is known as the mammalian diving response.

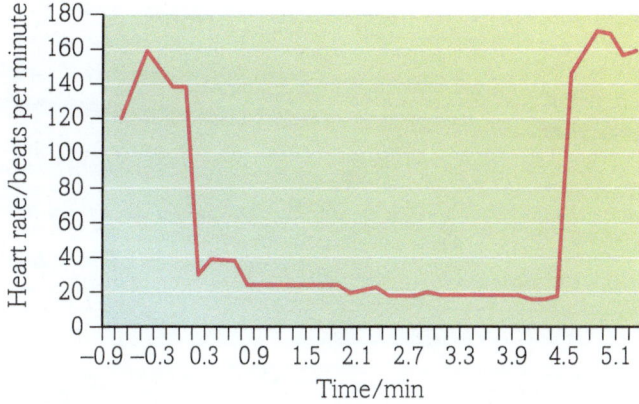

fig C The slowing of the heart rate (bradycardia) as a seal dives.

- **Behavioural adaptations** involve changes to programmed or instinctive behaviour making organisms better adapted for survival. For example, many insects and reptiles orientate themselves to get the maximum sunlight on their bodies when the air temperature is relatively low. This allows them to warm up and move fast enough to feed and to escape predators. When they get hot, they change their orientation to minimise their exposure to the sun, or shelter from it. Social behaviour such as hunting as a team or huddling together for warmth can improve the survival chances of both individuals and a group of organisms. Migrating to avoid harsh conditions, courtship rituals and using tools are other examples of behavioural adaptations.

Successful adaptations enable species to exploit every possible habitat and the different niches within each habitat. Most organisms have a mixture of different types of adaptations that enable them to survive and succeed in their particular environment. Here are just a few examples:

Adaptations for survival

Butterfly brilliance

When grayling butterflies fly, we can see their beautiful wings. However, the underside of the wings is a dull, broken pattern of greys and browns. This anatomical adaptation gives them excellent camouflage against the coastal heathlands where they live because they become almost invisible when they land. However, on sunny days, their shadows could make them visible to predators. They have a behavioural adaptation that allows them to overcome this – they follow the sun, changing their orientation through the day so their shadow is always as small as possible. This also helps to control their body temperature, preventing them from overheating as they absorb as little heat as possible. A variation of the same adaptation means that if they need to warm up to fly, they will risk being seen and angle their wings to absorb more of the heat from the sun.

fig D Variations of the same behavioural adaptation enable grayling butterflies to become almost shadowless and invisible or to warm up ready for flight.

Fungal carnivores

Fungi are often thought of as saprophytes, breaking down dead material in the soil. However, some fungi are active carnivores. Nematode worms are found in huge numbers in the soil. Some groups of fungi have developed adaptations that enable them to capture and feed on these nematodes. Some produce sticky nets or adhesive pads to trap the worms, some live inside the living worms, but *Arthrobotrys anchonia* is a fungus that actively lassoes nematodes. It traps them in hyphal loops called constriction rings as they pass, involving both structural and physiological adaptations (see **fig E**). Three fungal cells form a ring and when a nematode moves into it, a combination of wall changes and the osmotic potential of the cells result in water moving in fast. Within 0.1 seconds the ring inflates and holds the nematode in its grip. Often the nematode puts its tail through another loop as it moves to try and escape. Within hours the fungus grows more hyphae that penetrate the body of the nematode and digest it, absorbing the nutrients and transporting them within the fungus. It is thought that these predatory fungi evolved from saprophytic fungi in a high carbon, low nitrogen environment. The nematodes provide the missing nitrates.

fig E Some carnivorous fungi have adaptations that enable them to lasso their moving prey.

Same species, different environment, different adaptations

David Grémillet (1968–) and his team looked at the relative importance of physiological and behavioural adjustments in the great cormorant (*Phalacrocorax carbo*) in two contrasting environments – Normandy, where the water temperature is 12 °C, and Greenland, where it is 5 °C. Cormorants are not well insulated by fat and have poorly waterproofed feathers so they are easily affected by cold. The team found big differences in the feeding behaviour of birds breeding in the two regions. The birds living in Greenland spent 70% less time in the water than those in Normandy. They spent far less time swimming on the surface of the water between dives, and also returned to the land more often. The total daily energy intake of cormorants was similar in both areas, but prey capture rates in Greenland were 150% higher than those in Normandy because the changes in their behaviour resulted in far greater efficiency at finding food. Behavioural adaptations were more important than physiological ones to their survival in the colder niche.

fig F Handling cormorants needs care and skill, but the work of David Grémillet and his team has given us a better understanding of the behavioural adaptations that enable these birds to survive and breed in the difficult conditions of Greenland.

Natural selection leads to adaptations that give individuals an advantage in a particular niche. If conditions change, those adaptations may not be as successful, and the selection pressure will change. This may lead to changes in the species, and ultimately to the formation of new species.

Questions

1. Compare Darwin's original theory of evolution with the more modern version of the theory and explain the differences.

2. Evolution is based in part on adaptation. Discuss what is meant by adaptation and whether the different types of adaptation can really be looked at in isolation.

3. Explain the importance of the niche concept in understanding the adaptations of organisms.

4. Looking at the data from **fig C**, answer the following questions:
   (a) What do the negative numbers on the *x*-axis represent?
   (b) How long did the recorded dive last?
   (c) What was the percentage depression of the heart rate? Based on this, how long would you predict that the dive might have lasted without the bradycardia?
   (d) Describe the type of adaptation involved in this response and explain why it is so important to the survival of the seal in its ecological niche.

5. How do the behavioural adaptations of cormorants in Greenland help them to survive in their cold-water, fish-eating niche?

Key definitions

Natural selection is the process by which the organisms that are best adapted in a particular environment are most likely to survive and reproduce, passing on their advantageous alleles to their offspring.

The **genotype** is the genetic make-up of an organism with respect to a particular feature.

Phenotypes are the physical traits (including biochemical characteristics) expressed as a result of the interactions of the genotype with the environment.

An **allele** is a version of a gene, a variant.

Ecology is the study of the interactions of organisms with each other and with the environment in which they live.

A **niche** is the role of an organism within the habitat in which it lives.

An **anatomical adaptation** is an adaptation involving the form and structure of an organism.

A **physiological adaptation** is an adaptation involving the way the body of the organism works, including differences in biochemical pathways or enzymes.

A **behavioural adaptation** is an adaptation involving programmed or instinctive behaviour making organisms better adapted for survival.

By the end of this section, you should be able to...

● explain with examples how evolution can come about through natural selection acting on variation bringing about adaptations

Individuals that are not well adapted to their environment may not survive to reproduce or may produce fewer offspring than those that are better adapted, so their characteristics will become less common in the population. This is what Charles Darwin meant by the term 'survival of the fittest' and is what we now call natural selection. If the niche occupied by an organism changes due to changes in the environment, different characteristics may make an individual more successful. Natural selection will favour the survival of individuals with those different characteristics, and we say the **selection pressure** has changed. Changes in selection pressure result in changes (evolution) within the species. Depending on how different the individuals are, they may even be considered a new species.

Oysters adapting to change

Malpeque Bay on Prince Edward Island in Canada is home to massive oyster populations (*Crassostrea virginica*). In 1915 the oyster fishermen of Malpeque Bay began to notice that amongst their usually healthy catches there were a few diseased oysters that were small and flabby with pus-filled blisters. Before long the oyster beds had been all but wiped out by this new Malpeque disease. However, a small number of the millions of offspring produced by each oyster in a year carried an allele giving them resistance to the disease. Because only individuals that had this allele were able to survive and reproduce, the frequency of this gene in the population increased rapidly. By 1935 a small oyster harvest was again possible and by 1940 the beds were as prolific as ever, but with a rather different gene pool – now containing a high frequency of disease-resistance alleles (see **fig A**). These oysters were not a different species – they were still *C. virginica* – but they had adapted by natural selection to a changed environment.

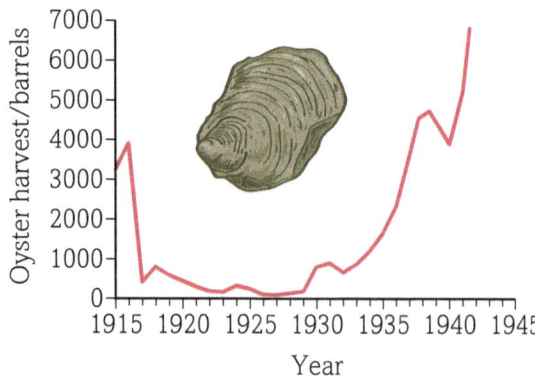

fig A Oyster yields from Malpeque Bay 1915–40. Disease devastated the populations, but as a result of the increased selection of the disease-resistance allele within the population, large and healthy oyster beds returned.

Natural selection in moths

Bernard Kettlewell (1907–79) at Oxford University in the 1950s studied the peppered moth *Biston betularia* and proposed that the moth had undergone natural selection in response to environmental changes. The normal, or *typical*, form of *B. betularia* is a creamy speckled moth found in British woodlands. In the eighteenth century, black specimens or melanics resulting from a random dominant mutation were captured occasionally. They were easily visible against the pale bark of the trees, both for human collectors and for birds looking for a meal. This selection pressure meant that the frequency of the dark allele in the population remained low.

Then in the mid-nineteenth century soot and smoke from the factory chimneys of the Industrial Revolution darkened the bark of the trees and the surfaces of buildings. As a result the melanic form of *betularia* was at a selective advantage and the frequency of the allele within the population began to increase as more and more of the light-coloured moths fell prey to predators. This process became known as **industrial melanism**.

Kettlewell set up several experiments to examine the selection of the two forms. In an unpolluted area with clean trees in Dorset, he released equal numbers of light and dark moths; 12.5% of the light moths were subsequently recaptured, but only 6% of the dark ones. In other observations birds were seen to eat 26 light moths and 164 dark ones as they rested on a light tree trunk. In Birmingham, at the time a highly polluted industrial area, 40% of the dark moths were recaptured, but only 19% of the light ones. Of the moths picked off the blackened trees from equal releases, 43 were light and 15 were dark. Kettlewell concluded that the change in frequency of melanic moths was due to the selection melanic moths had become the dominant form.

fig B The camouflage effects of the two main forms of *B. betularia* on the bark of a polluted tree and a clean tree.

Reversing the trend

Anti-pollution legislation that was passed in the 1960s resulted in cleaner, paler buildings and trees again, so the selection pressure has moved back in favour of the paler moth (see **fig C**). The frequency of the *typical* or pale allele in the population has increased again.

There is evidence for similar industrial melanism in over 70 different species of British moths. For example, the melanic form of the marbled beauty moth *Cryphia domestica* was dominant in London in the 1970s and 1980s but, since the 1990s, the pale form has reappeared in strength.

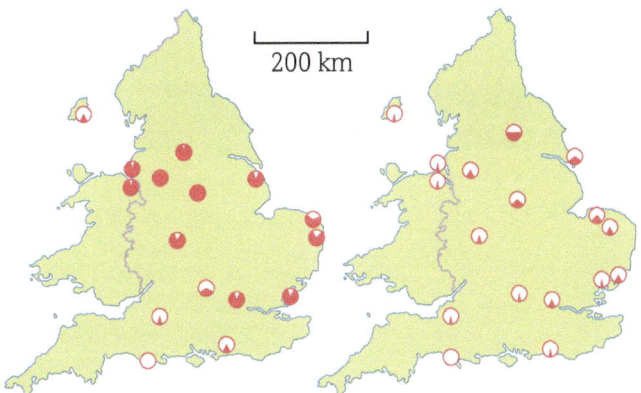

fig C The map on the left shows the proportions of the melanic moths in the population from Kettlewell's 1956 data. The map on the right shows data on moths collected in similar areas by Bruce Grant and his colleagues in 1996.

Selecting for reproductive success

Sometimes natural selection operates on adaptations that are less to do with survival in a particular habitat than with attracting a mate. For example, male African long-tailed widow birds have very long tails that appear to have very little use except in the mating season to attract mates. To investigate this, male birds were captured and their tails artificially lengthened (by gluing extra feathers on) or shortened (by cutting them). Once released, their reproductive success was measured by the number of nests with eggs and/or young in the territory of each male. The birds with artificially lengthened tails were clearly the most successful (see **fig D**).

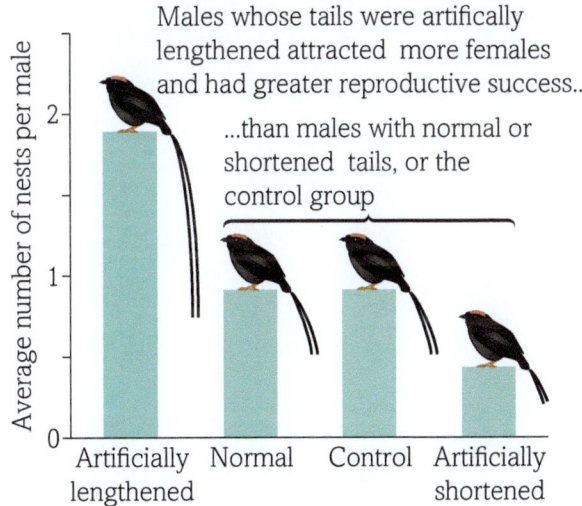

Males whose tails were artifically lengthened attracted more females and had greater reproductive success...

...than males with normal or shortened tails, or the control group

fig D The selection of an anatomical adaptation that gives an advantage in getting a mate can be seen in African long-tailed widow birds.

Many animals show similar selection for features that give no advantage within their niche, but are linked to reproductive success, for example the thick skulls and enormous horns of some antelopes, and the spectacular tail of the peacock. Plants also show some amazing adaptations. All insect-pollinated flowers have adaptations to bring pollinators to the plant, but they have not all evolved to the level of the bee orchid, which is adapted to look like a bee.

fig E The amazing patterns and structures of the bee orchid have no benefit to the plant except to attract bees to pollinate the flower and so ensure reproductive success.

Directional selection

The oysters of Malpeque Bay and the moths of the British woodlands are good examples of **directional selection**, 'classic' natural selection showing a change from one phenotypic property to a new one more advantageous in the circumstances. Directional selection occurs anywhere that environmental pressure is applied to a population.

The introduction of the rabbit disease myxomatosis into Britain in 1953 is another example. It almost wiped out the rabbit population over the following ten years, but rabbits are now common once more. Many of them carry an allele that renders them immune to myxomatosis – the frequency of that allele in the rabbit **gene pool** has increased enormously.

Questions

1 How can changes in a niche or habitat lead to changes in a species?

2 Using the data from **fig A**, how long did it take for Malpeque disease to virtually wipe out the oyster population in the bay, and how long did it take for the resistance allele to become sufficiently dominant for the population to recover? Why is this an example of the adaptation of an organism to its niche?

3 Relatively recently, some scientists challenged Kettlewell's elegant ideas about industrial melanism. Even more recently, many more scientists have supported his work and the story of *B. betularia* is still regarded as one of the best documented examples of industrial melanism. Using this section, the support material and any other resources you have, investigate the evidence for and against industrial melanism based on *B. betularia*.

Key definitions

Selection pressure is the pressure exerted by a changed environment or niche on individuals in a population, causing changes in the population as a result of natural selection.

Industrial melanism is the evolution of dark-coloured individuals in a habitat that has been made darker by industrial pollution, e.g. soot.

Directional selection is natural selection showing a change from one dominant phenotype to another in response to a change in the environment – one phenotype is selected for over all the others.

A **gene pool** is all of the variants of all of the genes in a population.

The evolutionary race between pathogens and medicines

By the end of this section, you should be able to...

- Explain the evolutionary race between pathogens and the development of medicines to treat the diseases they cause

Mutation, adaptation and natural selection are not only seen in eukaryotes such as the cormorants, fungi and moths you have considered in the previous chapter. They are also seen in pathogens (disease-causing organisms) such as bacteria and viruses. As a result the medicines that we develop to destroy pathogens can quickly become ineffective. The pathogens are in an evolutionary race against us and the medicines we develop to treat the diseases that they cause.

Beating bacteria, step 1

Bacteria cause a wide range of diseases, from throat infections to tuberculosis and septicaemia. We had no medicines to cure bacterial diseases on a large scale until the mid1940s. Before then, millions of people died globally every year as a result of bacterial infections. When the work of Alexander Fleming, Howard Florey and Ernst Chain resulted in the mass production of penicillin, the effect of the drug seemed almost miraculous. People recovered from infections that would have meant certain death and everyone thought that the battle against bacterial disease was won.

However, the discovery and development of penicillin was the beginning of the antibiotic story, not the end. Penicillin did not kill all of the different types of bacteria. As you saw in **Section 2.2.1**, the penicillin family of antibiotics affect Gram-positive bacteria, with their thick peptidoglycan walls, but do not affect other bacteria. A range of antibiotics were developed that targeted different types of bacteria and for a short time it appeared that bacterial diseases would become almost a thing of the past. Antibiotics were prescribed freely and everyone came to expect an antibiotic prescription when they went to the doctors with an

infection. It is estimated that, on average, antibiotics add 20 years to each person's life in countries like the UK and USA.

Bacteria fight back

A sign of the problems to come emerged rapidly. Within a year of penicillin first being used as a medicine there were reports of penicillin resistant *Staphylococcus aureus*. By the 1960s, many bacteria had become **resistant** to penicillin. A tiny percentage of the original bacterial populations

(estimated at about 1%) must have carried a random mutation (see **Section 1.3.6**) giving them resistance to damage by penicillin. In many cases the adaptation was the presence of an enzyme called penicillinase that splits the penicillin molecule so it no longer works. This adaptation gave the bacteria a great advantage and so, as a result of natural selection, resistance to penicillin became more and more prevalent in bacterial populations. Once again, people became seriously ill and died from bacterial infections that were no longer affected by penicillin.

Beating bacteria, step 2

As penicillin resistance spread, scientists produced a new antibiotic called methicillin. This put people ahead again in the evolutionary race against pathogenic bacteria as common pathogens had no resistance to methicillin. However, this didn't last long and methicillin resistance spread rapidly through bacterial populations, and now the antibiotic has few uses. Resistant pathogens such as methicillin-resistant *Staphylococcus aureus* (MRSA) are a serious problem in hospitals and care homes in many different countries (see **Book 2 Section 6.1.6**). The bacteria were ahead again.

At the moment there are growing numbers of bacteria that are resistant to not just one but many antibiotics. These multi-resistant strains of bacteria are almost untreatable. Some are even resistant to vancomycin, one of the most powerful antibiotics we have, which still cures the majority of serious bacterial infections.

original population includes some mutations   bacteria with resistant mutation more likely to survive   antibiotic taken   new population almost entirely resistant to antibiotic apart from new mutants

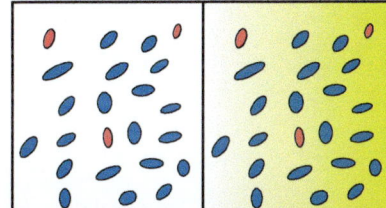 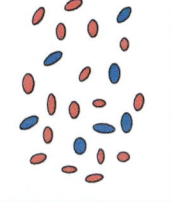 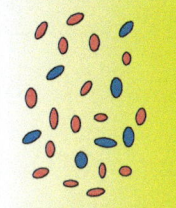 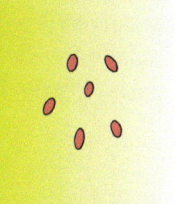 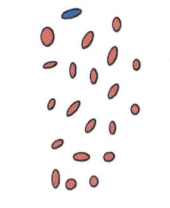

antibiotic taken   new population has higher proportion containing the advantageous resistant mutation   only bacteria with resistant mutation survive

fig A The evolution of antibiotic resistance by natural selection.

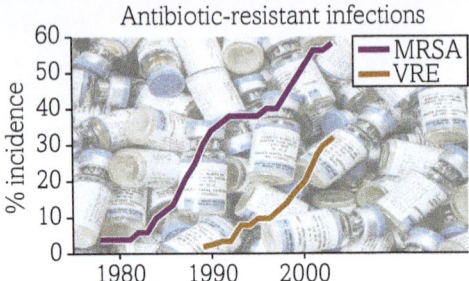

Antibiotic-resistant infections

fig B The rise in MRSA- and vancomycin-resistant enterococcus (VRE) shows how fast bacteria are getting ahead in the *Evolutionary race* against the medicines we have to treat the diseases they cause.

Antibiotic	Antibiotic first given to patients	Antibiotic resistance reported
penicillin	1941	1942
vancomycin	1956	2002 (partial resistance, 1997)
methicillin	1960	1961
linezolid	2000	2001
daptomycin	2003	2005
tigecycline	2005	2012

table A The speed at which bacterial resistance has developed to a range of antibiotics.

What does the future hold?

Unless things change, the future could look very bleak. There are serious fears that we could return to a time when bacterial diseases are one of the biggest killers in the UK. Not only is antibiotic resistance increasing, but the number of new antibiotics being developed and brought onto the market has been steadily falling (see **fig C**).

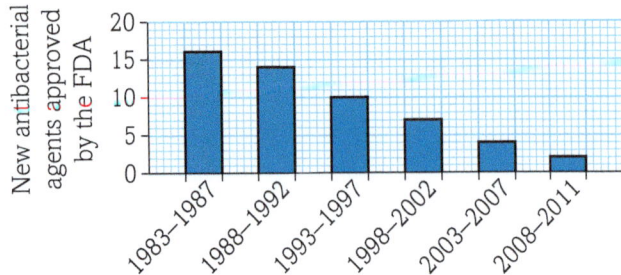

fig C The fall in the number of new antibiotics approved for use by the FDA in the USA.

Factors that contribute to the problem include:

- Antibiotics are too widely prescribed and used.
- Wide-spectrum antibiotics are often used to make sure they have an effect, rather than testing to determine if an infection is bacterial and if so which bacteria are involved.
- People do not complete courses of antibiotics, which makes it easier for resistance to develop.
- In some countries antibiotics are widely used in the food chain.
- A lack of basic hygiene in hospitals and care homes has encouraged the spread of antibiotic resistant organisms such as MRSA.
- There is no big financial incentive for pharmaceutical companies to develop new antibiotics as new antibiotics will be used sparingly to prevent the development of resistance.

Although bacteria appear to be getting ahead in the evolutionary race, we humans still have a few tricks up our sleeve. Here are some of the ways in which we hope to overcome the problems of antibiotic resistance and maintain effective antibacterial therapies for the future:

- Reducing the use of antibiotics.
- Better education so people understand that they do not always need antibiotics.
- Reducing the use of antibiotics in farm animals.
- The Longitude Prize 2014 set the challenge to develop a quick test confirming bacterial infections in the doctor's surgery, making it easier to prescribe the best antibiotic for maximum effect.
- DNA sequencing will help identify bacteria and find new ways of targeting them with antibiotic drugs, and genetic engineering will enable large amounts of new drugs to be produced.
- Development of new antibiotics: Scientists are looking at a wide range of substances that show antimicrobial properties in the natural world and designing new molecules using computer modelling. In recent years scientists have looked at chemicals in places including the blood of crocodiles, honey, soil, fish slime and the deep ocean abyss to try and find novel antibiotics.

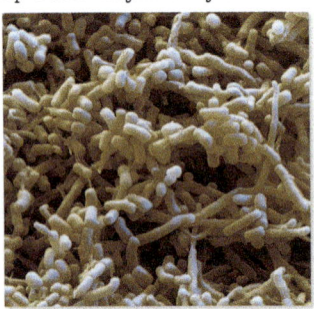

fig D These *Streptomyces* bacteria, collected in Pacific Ocean sediments, produce compounds that may give us a completely new class of antibiotics – the anthracimycins. Scientists think the new drugs may be effective against anthrax and possibly MRSA.

If you continue to study Biology A level you will learn more about the evolutionary race between pathogens and the development of medicines to treat the diseases they cause, including the mechanisms of resistance.

Questions

1 Explain how a mutation can give one bacterium an advantage over others of the same type.

2 Produce a flow diagram to explain how antibiotic resistance develops in a population of bacteria.

3 Explain how the following lead to the development of antibiotic resistance:
   (a) overprescribing antibiotics
   (b) individuals who do not complete their course of antibiotics.

4 If someone wins the Longitude Prize, discuss how this invention could reduce the development of antibiotic resistance in bacteria.

Key definition

A **resistant** bacterium is not affected by an antibiotic.

By the end of this section, you should be able to...

- explain how reproductive isolation can lead to allopatric and sympatric speciation

Changes in populations of organisms are very interesting and they can tell us a great deal about the process of natural selection. But a species is usually a much bigger entity than a single population, and often consists of a large number of populations spread across a country or even countries. So how is a new species formed?

Isolation and speciation

A species is a group of organisms sharing a number of structural and evolutionary features, which are capable of interbreeding to produce fertile offspring. **Speciation**, also known as the formation of a new species, happens as a result of the isolation of parts of a population. The key factor in the process is reproductive isolation. The two isolated populations experience different conditions, and this in turn means that natural selection acts in different directions on the two populations. As a result, over time both the genotype and the phenotype of the isolated groups will change. This can continue to the point where, even if members of the split population are reunited, they can no longer successfully interbreed.

Speciation can also take place as a result of **hybridisation** and this is particularly common in plants. Sometimes two closely related species can breed and form fertile hybrids that are successful in their own right and may be better adapted to the niche. In some cases these hybrids do not produce fertile offspring if they are crossed back to their parent plants so a new species is formed, which may out-compete the parent plants, for example, the hybrid formed between native and Spanish bluebells.

Isolating mechanisms

For different species to evolve from an original species, different populations of the species usually have to become reproductively isolated from each other, so that mating and therefore gene flow between them is restricted. There are a number of ways in which this may happen:

- **Geographical isolation**: a physical barrier such as a river or a mountain range separates individuals from an original population.
- **Ecological isolation**: two populations inhabit the same region, but develop preferences for different parts of the habitat.
- **Seasonal isolation** (also known as **temporal isolation**): the timing of flowering or sexual receptiveness in some parts of a population drifts away from the norm for the group. This can eventually lead to the two groups reproducing several months apart.
- **Behavioural isolation**: changes occur in the courtship ritual, display or mating pattern so that some animals do not recognise others as being potential mates. This might be due to a mutation that changes the colour or pattern of markings.
- **Mechanical isolation**: a mutation occurs that changes the genitalia of animals, making it physically possible for them to mate successfully with only some members of the group, or it changes the relationship between the stigma and stamens in flowers, making pollination between some individuals unsuccessful.

Reproductive isolation is the key factor in speciation.

Allopatric speciation

Allopatric speciation takes place when populations are physically or geographically separated in some way. Scientists recognise allopatric speciation as the main evolutionary process. Allopatric speciation is of enormous importance in the history of evolution, as great land masses moved and separated. The physical isolation of populations continues to occur as a result of natural changes, for example as islands form and disappear, as ice floes melt, rivers change course and lakes either dry up or appear – and so allopatric speciation continues to be very important. Some of the changes that result in allopatric speciation are the result of human interventions such as dams, roads and cities.

There are many examples of allopatric speciation. Some of the most striking are when organisms become completely isolated – for example when islands are formed. When a species evolves in geographical isolation and is found in only one place it is said to be **endemic**.

Endemic species of Madagascar

Madagascar, a large island off the coast of East Africa, provides good examples of endemism as a result of allopatric speciation. Almost all of the species found there are endemic to the island. These range from the amazing giant baobab trees to ring-tailed lemurs and from the bizarre elephant's foot plant to a small, prolifically breeding mammal, the tailless tenrec which can have over 30 babies at a time. The only species that are not endemic have been taken to the island by people in relatively recent times. Imported species can cause many problems to the endemic species.

fig A The ring-tailed lemur is just one of the unique species endemic to Madagascar as a result of allopatric speciation.

Desert pupfish

In the Nevada desert, close to Death Valley, is the Ash Meadows National Wildlife Refuge. At one stage this area had many springs, streams and rivers, but around 50 000 years ago the climate changed and the area became dry and arid. Most of the water dried up, but small individual ponds and springs remained. The fish trapped in each area could no longer interbreed and evolved independently. Now in this oasis there are four completely separate species of desert pupfish, each with different colouring and different courtship displays, each endemic to one place. They are the Ash Meadows Armagosa pupfish, the tiny Devils Hole pupfish, the Warm Springs pupfish and the Ash Meadows speckled dace. The Devils Hole is perhaps the most extreme of the tiny ecosystems – the fish there have adapted to survive in warm water that rises and falls when there are earthquakes in Mexico.

(a)

(b)

fig B (a) Devils Hole; and (b) Devils Hole pupfish.

Adaptive radiation

Allopatric speciation is frequently followed by **adaptive radiation**. Adaptive radiation takes place when one species evolves rapidly to form a number of different species, which all fill different ecological niches. There are a number of well-known examples of adaptive radiation:

- **Australian marsupials and monotremes**: Australia is well-known for its unusual fauna and flora. Perhaps most unusual are two groups of mammals, the **marsupials**, which protect their young in pouches, and the even rarer egg-laying **monotremes**. In the rest of the world, the **placental mammals** dominate. Until about 5.5 million years ago, Australia was joined to the rest of the world, when the only mammals were marsupials and monotremes. After Australia separated from the other continents, the marsupials evolved to fill an enormous range of niches, from the large herbivorous kangaroo with its wide-ranging niche to the koala with its eucalyptus tree niche, and others with different carnivorous niches, for example the quoll and the Tasmanian devil. On other continents placental mammals evolved and mostly replaced the marsupials and monotremes but they did not reach Australia until humans arrived, eventually bringing other mammals with them.

- Darwin's finches: these birds provide a classic example of how the availability of a range of different niches produces different selection pressures and results in adaptive radiation. The finches were discovered by the great nineteenth century naturalist Charles Darwin on his voyage on *HMS Beagle*. On the Galapagos Islands near the equator there are a number of feeding niches

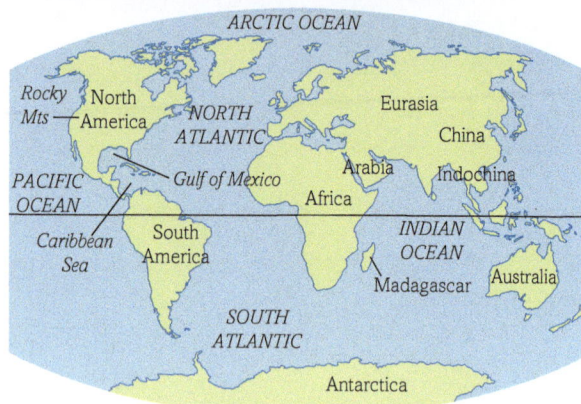

fig C The unique fauna of Australia is all the result of allopatric speciation and adaptive radiation as a result of its geographical isolation.

for birds, for example small seeds, large nuts and insects living in rotten bark. The original finches that arrived on the islands were of a single species. No-one is quite sure how they got there because the islands are 500 miles from land, but a small flock was probably carried there by a storm or a hurricane.

Within the birds that arrived at the islands, there would have been variation in alleles and characteristics, and different niches on the islands would have favoured individuals with different variations. For example, a bird with a slightly smaller, stronger bill would get more food by eating mainly seeds. This would enable it to thrive, reproduce and pass on its beak characteristics to its offspring. Over generations, natural selection resulted in individuals with small strong beaks ideally adapted to eating seeds. Similarly, a finch with a longer, thinner beak might well be more successful in probing dead wood for insects and so begin to feed relatively exclusively in that way. By exploiting different niches the finches avoided competing for the same relatively scarce food resources. As a result, at least 14 different species of finch have evolved on the Galapagos Islands over several million years from one common ancestral species.

Because food was such an important selection pressure, it was important to mate with a finch with a similarly shaped beak to pass on the advantageous characteristic. Mating with a finch that had a differently shaped beak would produce a variety of offspring that were less well adapted to feeding, so there was a selective pressure on choosing the right kind of mate. As a result any phenotypic and behavioural changes that made choosing the right mate easier were also selected, so the different species look different. Although the finches specialise and feed on particular types of food, and vary considerably in size and appearance, DNA analysis has shown that genetically they are remarkably similar.

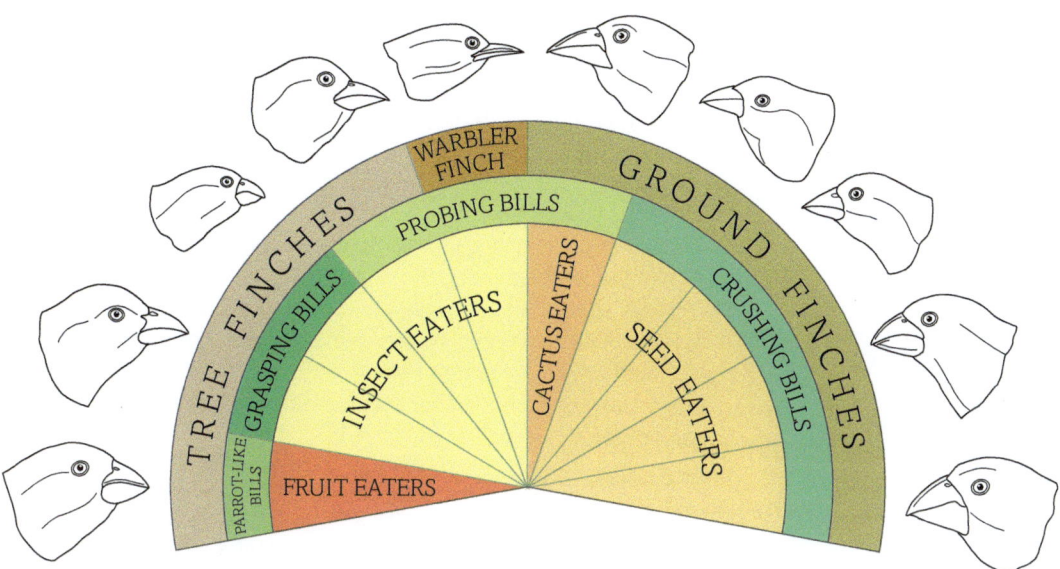

fig D Over several million years, at least 14 species of finches evolved from the original ancestor species. The anatomy of the beaks and adaptations for feeding of 10 of these finches is shown in this diagram.

Sympatric speciation

Sympatric speciation takes place between populations of a species living in the same place that become reproductively isolated by mechanical, behavioural or seasonal changes. Gene flow continues to some extent as speciation takes place – a very different model to allopatric speciation. Sympatric species are closely related and occupy overlapping ranges. Many scientists are reluctant to classify examples of sympatric speciation, suggesting that when speciation occurs there is always an adaptive pressure, including the presence of unrecognised microhabitats, which drive the formation of two species and produce a barrier between the populations. DNA evidence often shows that species originally thought to be the result of sympatric speciation actually show evidence of cross-breeding or geographical or niche separation at some point.

Sympatric plants

There are two species of palm trees endemic to Lord Howe Island in New South Wales, which appear to be an example of pure sympatric evolution. *Howea forsteriana* and *Howea belmoreana* diverged from each other a very long time after the island was formed 6.9 million years ago. The island is so tiny that geographical isolation is not possible. The trees are wind pollinated and produce lots of pollen so they could interbreed if they were not separate species. The type of soil they grow on seems to influence the timing of flowering and could be the driver for the speciation. DNA evidence supports the theory of sympatric speciation for these trees.

Sympatric speciation in progress

There appears to be an example of sympatric speciation in progress in the United States. Until the mid-1800s the tiny fruit fly *Rhagoletis pomonella* lived only on hawthorn bushes, laying eggs on the fruits. The larvae respond to the smell of the hawthorns and return as adults to hawthorns to reproduce. However, over 150 years ago along the Hudson River Valley many huge apple orchards were planted. Genetically, apple trees are quite closely related to hawthorns. Some female *R. pomonella* laid their eggs on the apples, either by mistake or because they could not find hawthorns. The larvae did not do particularly well, but some survived to adulthood. These flies responded to the smell of apples, not hawthorns, and a breeding group of apple-dwelling flies evolved. Now there are two breeding groups of *R. pomonella* in the Hudson River Valley. One feeds on hawthorns, the other on apples. The two populations show increasing reproductive isolation because they mate only with flies on the same food source. The apple-dwelling flies have adapted to life on apple trees so they now emerge from their pupae at a different time of the year (see **fig E**). Apples provide more food and better protection for the maggots from parasitic wasps. Scientists have analysed the frequency of a number of alleles in the flies and have found they are becoming increasingly different. It seems likely that two entirely different species will evolve, which will no longer interbreed as their reproductive cycles will be completely out of synchronisation.

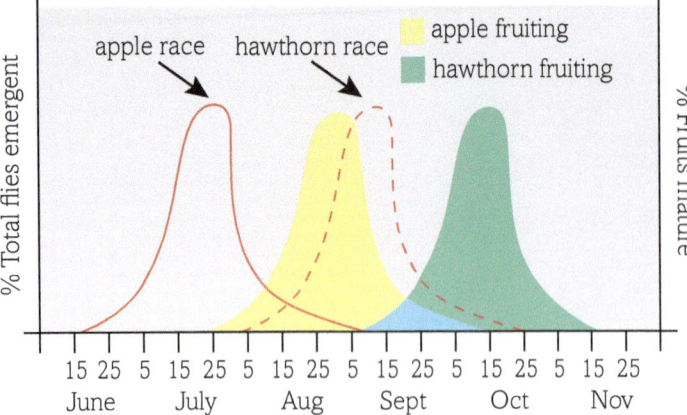

fig E The apple race of flies now emerges earlier than the hawthorn race, so there will be less chance for the flies to interbreed and a greater chance that this will lead to sympatric speciation (data from Reissig 1991, Feder and Filchak 1999 and others).

Cichlid fish in the African lakes

The cichlid fish of the African lakes are an example of speciation on a grand scale, illustrating allopatric speciation, adaptive radiation and sympatric speciation.

The family Cichlidae contains an enormous variety of fish, including tilapia, an increasingly important source of protein for many people in Africa and around the world. Cichlids show immense variation in shape, size, colour, feeding habits, courtship displays and breeding habits. They include a number of species that are mouth-brooders, which means they carry their eggs in their mouths until they hatch and then allow the tiny fish to retreat back into their mouths when danger threatens.

A lake is a largely enclosed environment, so within a lake there will be many different habitats and microhabitats, but no readily available way of moving on to another lake except under exceptional circumstances such as major flooding. As a result, speciation takes place within each lake independently. Within the individual great lakes of Africa, the cichlid fish have undergone speciation and adaptive radiation to produce an amazing variety of species over time.

Scientists have not yet found and identified all of the species of cichlids that live within these enormous bodies of water. There are

fig F Examples of the diversity of the species of cichlids in Lake Malawi.

probably around 1500 species of cichlids within Lakes Victoria, Malawi and Tanganyika alone, and the evidence suggests that in Lake Victoria at least, much of this speciation has taken place over the last 15 000 years.

Cichlid speciation

Molecular phylogeny suggests that the great majority of the cichlid fish evolved much more recently than the lakes were formed. The fish in each lake are more closely related to the other species of fish endemic in their own lake than they are to the fish in the other lakes. This suggests that the fish have diversified in the individual lake systems after they have become separated from each other:

- For example, in Lake Malawi, DNA evidence suggests that all of the species have evolved from a common ancestor within the last 5 million years. They have less than 6% difference in their mitochondrial DNA (see **Section 2.1.3**). The original ancestor is thought to resemble the swamp-dwelling fish *Astatotilapia calliptera*, which is the only one of the cichlid species found in Lake Malawi that is also found in other lakes.

- This ancestral species fed by sifting algae from the mud sediments at the bottom of a swamp or lake. The cichlids that now fill the lake have evolved to fill almost every available feeding niche in a living example of adaptive radiation on a grand scale. There is a species that removes and eats the parasites from the skin of catfish, which also live in the lake. There are cichlids that eat scales, that bite fins and that eat the eggs and young of other cichlids. Some live at great depths and have very large eyes to see in the dimmest light. Others feed on insects that land on the rocks at the edge of the lake and have eyes and other sense organs that work in the air as well as the water. Some species eat crabs, some snails, some rasp algae off rocks and some eat plants. The evidence suggests that these different species rarely interbreed in spite of living in the same lake.

- Lake Victoria is much younger than the other two big lakes, and scientists know it has dried out three times during its 400 000-year history. The last time it refilled was only 15 000 years ago. The few cichlid species that survived the dry period, hidden in the mud or in tiny remaining pools of water, have evolved to produce the 500 or more species scientists have so far recorded in the lake today. By identifying micro-environments and looking at the alleles that have changed, scientists are unravelling the selection pressures that have driven this vast and rapid evolution. They have discovered, for example, that the cloudiness of the water can drive the evolution of species with very similar feeding habits. The water gets cloudier with depth, and this affects the wavelength of light that penetrates. Research shows that species living in deeper and cloudier water have different optical pigments in their eyes than fish with similar feeding habits living in the clearer, shallower water. The females are less affected by colour in their choice of mates, and the males tend to have red and yellow display colours – wavelengths of light that penetrate the cloudy water.

Cichlids and sympatric speciation

The best example of sympatric speciation in cichlid fish comes not from the biggest lakes, but from Lake Barombi Mbo in Cameroon. This is a small lake, only 2.5 km wide, but 110 m deep, formed in the crater of a volcano. Only the top 40 m contains enough oxygen to sustain vertebrate life.

The lake contains 11 species of cichlids. Molecular phylogenetic studies based on both nuclear and mitochondrial DNA show that the species are more closely related to each other than to any other cichlids. What is more, the closest relative of these 11 species is the only cichlid fish found in the streams surrounding the lake. This suggests these fish entered the lake from the streams, perhaps during a rainy season flood, and they have then evolved into 11 different species all in the same small volume of water, without any geographical barrier between them. They have evolved to fill different niches, including:

- *Pungu maclareni*, a yellow, blue and black fish that eats a form of sponge endemic to the lake.
- *Konia dikume*, a deep water fish that can tolerate low oxygen and feeds on invertebrates.
- *Stomatepia pindu*, which range from black to purple and feed on other fish.
- *Stomatepia mongo*, which has a long snout – it may be a detritivore.

Lake Ejagham is another small lake in Cameroon that shows a clear example of sympatric speciation. The lake has a surface area of only 0.5 km² and is 18 m deep, but it has a number of different habitats. The bottom of the lake is sandy near the shore, covered with leaves and twigs from the surrounding trees. The lake bottom in the middle of the lake is muddy. Lake Ejagham contains two closely related species of cichlids. The smaller species feeds on microscopic animals in the deeper areas of the lake. The larger species feeds near the edges on invertebrates, including insects that feed on the leaves that fall into the water near the shore. They breed as well as feed in different habitats and so, although they are in the same tiny lake, they are reproductively isolated and have formed separate species.

Fear for the future

The cichlids of the great lakes are valuable in many ways, both as a food source for local people and as a tremendous resource for scientists studying the mechanisms of speciation. However, the diversity of the lakes is under threat. In Lake Victoria, Nile perch and water hyacinths were introduced. These changed the ecology – the Nile perch is a voracious predator and water hyacinths grow and cover the surface. Deforestation leads to soil erosion so the water has become full of silt and cloudy. As a result, the cichlids are under threat. Their numbers have been substantially reduced and scientists fear many species will have disappeared before they have been fully identified. Up to two-thirds of the species are now classified as endangered or extinct. Fortunately, so far the other lakes have fewer problems – but for the cichlids of Lake Victoria, it is a race against time.

Questions

1 Give examples of isolating mechanisms that lead to:
 (a) allopatric speciation (b) sympatric speciation.

2 Discuss the effect of adaptive radiation on the species richness of an ecosystem.

3 Look at the drawings of the Darwin's finches. They look very different yet are genetically quite similar. Explain how this may have come about.

4 Explain how both allopatic and sympathetic speciation are involved in the adaptive radiation of the cichlid fish of African lakes.

5 Why are the great lakes of Africa such an ideal place to study speciation?

6 Explain how both allopatric and sympatric speciation are involved in the adaptive radiation of the cichlid fish of the African lakes.

Key definitions

Speciation is the formation of a new species.

Hybridisation is the production of offspring as a result of sexual reproduction between individuals from two different species.

Geographical isolation occurs when a physical barrier such as a river or a mountain range separates individuals from an original population.

Ecological isolation occurs when two populations inhabit the same region, but develop preferences for different parts of the habitat.

Seasonal isolation occurs when the timing of flowering or sexual receptiveness in some parts of a population drifts away from the norm for the group. This can eventually lead to the two groups reproducing several months apart.

Behavioural isolation happens when changes occur in the courtship ritual, display or mating pattern so that some animals do not recognise others as being potential mates. This might be due to a mutation that changes the colour or pattern of markings.

Mechanical isolation happens when a mutation occurs that changes the genitalia of animals, making it physically possible for them to mate successfully with only some members of the group, or it changes the relationship between the stigma and stamens in flowers, making pollination between some individuals unsuccessful.

Allopatric speciation is speciation that takes place when populations are physically or geographically separated and there can be no interbreeding or gene flow between the populations.

An **endemic** species is a species that evolves in geographical isolation and is found in only one place.

Adaptive radiation is a process by which one species evolves rapidly to form a number of different species that all fill different ecological niches.

Marsupials are mammals that give birth to very immature young and then protect them in pouches.

Monotremes are primitive mammals that lay eggs and feed their offspring with milk from mammary glands.

Placental mammals are mammals that provide for the developing fetus during gestation through a placenta.

Sympatric speciation is speciation that takes place between populations of a species living in the same place. They become reproductively isolated by mechanical, behavioural or seasonal mechanisms and gene flow continues between the populations to some extent as speciation takes place.

THINKING BIGGER

REVIVING THE QUAGGA

Until recently it was thought that the last quagga, a species similar to the plains zebra, had died in Amsterdam zoo in 1883. In recent years DNA evidence suggested that the quagga was in fact a sub-species of the plains zebra, and a rebreeding programme in South Africa set out to restore the quagga to the African plains where it belongs.

QUAGGA REBREEDING: A SUCCESS STORY

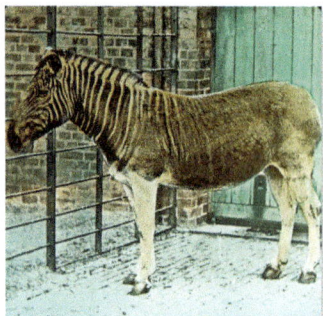 

fig A This stripe pattern on this restored quagga is approaching the pattern seen in the only existing photo of a quagga, taken in London Zoo in 1870.

12:05 (GMT+2), Tuesday, April 15, 2014

Until recently, it was believed that the last quagga died in Amsterdam Zoo in 1883. Today, however, this iconic animal is alive and back in the Western Cape. How was it possible to revive an animal from extinction? Keri Harvey speaks to the Quagga Project's Craig Lardner.

Contrary to popular belief, the quagga (*Equus quagga quagga*) is not a species in its own right. DNA analysis of quagga kept as museum specimens has proven that the extinct quagga was in fact a Burchell's or plains zebra with a colour variation, in which some of its leg and rump stripes disappeared. This also means that Burchell's or plains zebra still carry genes from the extinct quagga, though these may be more diluted now than before.

Vanishing stripes

Why exactly the Burchell's or plains zebra lost some of its stripes is unclear, but … differing colouration seems to provide optimal camouflage: the quagga in each area blend better into their specific surroundings. Another purported reason for the quagga's vanishing stripes, apart from camouflage and hence protection from predators, is tsetse flies. It has been suggested that the zebra's stripes repel tsetse flies and so too the diseases they carry. Because the quagga lived outside the tsetse fly areas, the distinct stripes became obsolete.

…When it was discovered that the Burchell's or plains zebra is a DNA match for the extinct quagga, the project set about attempting to 'rebreed' the quagga. This was done by selecting brownish zebra with reduced stripes and white tail bushes. In this way, the quagga genes could be concentrated to produce an animal that looks precisely like the 'extinct' quagga.

Only mitochondrial DNA was available from museum specimens and not nuclear and living DNA. For this reason, it was impossible to compare the rebred quagga to the original ones that became extinct. Nonetheless, the quagga in the Western Cape are believed to be the 'real thing', as it was in fact only coat pattern that distinguished a quagga from a Burchell's or plains zebra. Thus the Quagga Project seems to have succeeded in rectifying the tragedy that saw them being hunted to extinction.

Where else will I encounter these themes?

1.1   1.2   1.3   1.4   2.1   2.2   2.3   2.4

Let us start by looking at the nature of the writing in this article.

The article opposite is from *Farmer's Weekly*, which is published both in print and online in South Africa and aimed at farmers across Southern Africa. Consider the article and think about the type of writing being used. Try and answer the following questions:

1. What aspects of this article lead you to think it is not a scientific piece of writing?
2. Using the information from this article, make a summary of what you now know about quaggas and the way they have been rebred.

> Consider the format – it is a story told by a journalist based on speaking to someone from the Quagga Project. What about this makes you feel the story is reliable? What might make you wonder if the details are correct?

Now let us look at the biology of this amazing story. You know about DNA in cells, classification, natural selection and adaptation. Use all of these ideas to help you answer the following questions:

3. What is the importance of the classification of quaggas as a separate species or a sub-species of Burchell's zebras?
4. Explain the importance of DNA sequencing in convincing people that rebreeding the quagga might be worth attempting.

> **Writing scientifically**
> As you read these articles identify everything that contributes to writing a scientific article. For instance, the vocabulary, sources and the way information is presented. Make sure your own answers are written scientifically.

Activity

Now read this extract, which is an abstract from a peer-reviewed scientific journal.

Abstract

Twenty years ago, the field of ancient DNA was launched with the publication of two short mitochondrial (mt) DNA sequences from a single quagga (*Equus quagga*) museum skin....(Higuchi et al. 1984, Nature 312, 282–284). This was the first extinct species from which genetic information was retrieved. The DNA sequences of the quagga showed that it was more closely related to zebras than to horses. However, quagga evolutionary history is far from clear. We have isolated DNA from eight quaggas and a plains zebra (subspecies or phenotype *Equus burchelli burchelli*). We show that the quagga displayed little genetic diversity and very recently diverged from the plains zebra...

....However, our results could be consistent with the quagga and the plains zebra being synonymized, as suggested earlier (e.g. Rau 1978; Groves & Bell 2004). Owing to priority, the correct name for plains zebras would thus be *E. quagga*, with, according to Groves & Bell (2004) five living and one extinct subspecies, the quagga (*E. quagga quagga*)...

....We estimate that this divergence took place in the Pleistocene, about 120 000 to 290 000 years ago... (Dawson 1992). Therefore, the distinct coat colour of the quagga (Bennett 1980) must have evolved quite rapidly. Existing plains zebras show a geographical gradient in coloration with progressive reduction in striping from north to south, which has been explained as an adaptation to open country and for which the quagga represented the extreme limit of the trend (Rau 1974, 1978). Thus, the rapid evolution of coat colour in the quagga may be explained by either of two factors, or a combination of them: the disruption of gene flow owing to geographical isolation and/or an adaptive response to a drier habitat.

1 Compare and contrast the writing styles of the two pieces about the quagga.

2 Summarise the information about quaggas and the rebreeding programme you get from this paper and compare it to the information you got from the first article. How does the information differ? Which gave you the most information? Which was easiest to extract information from? Which did you find the most interesting?

3 From the information on quaggas in the above articles, put together a presentation for potential sponsors to support a fund-raising effort towards the reintroduction of the quagga onto the South African plains.

> **Command word**
> Note that when the words compare and contrast are used, you are expected to look for both similarities and differences between two or more things.

• From South African *Farmer's Weekly* magazine and the following journal article: Leonard, Jennifer A., Nadin Rohland, Scott Glaberman, Robert C. Fleischer, Adalgisa Caccone and Michael Hofreiter. 'A rapid loss of stripes: the evolutionary history of the extinct quagga.' *Biology letters* 1, no. 3 (2005): 291–295.

1 If a horse, *Equus caballus*, is mated with a donkey, *Equus asinus*, a hybrid known as a mule is produced.

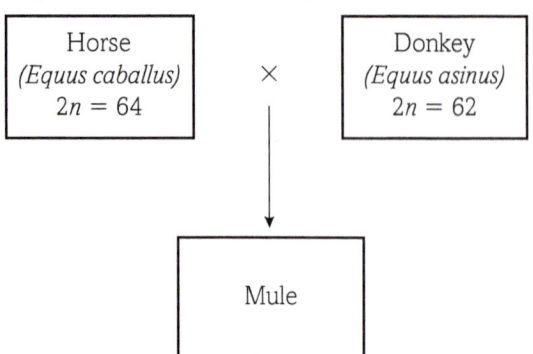

Mules are almost always sterile and produce no offspring. This phenomenon is an example of a post-zygotic isolating mechanism.

(a) State the diploid number of chromosomes in a mule and suggest why mules are unable to produce offspring. [3]

(b) State what is meant by the term **isolating mechanism**. Suggest why the production of a mule by mating a horse with a donkey is described as a post-zygotic isolating mechanism. [3]

(c) It has been suggested that the mule should be named as a new species, *Equus mulus*.

Suggest why this might not be acceptable to some biologists. [2]

[Total: 8]

2 The white tobacco fly (*Bemesia tabaci*) is a pest of many plants, especially glasshouse crops. It is found in Southern Europe, although there have been hundreds of outbreaks in England in recent years. Biologists studying this pest believe the species is undergoing evolutionary change. They have identified two strains, Biotype A and Biotype B. Biotype B flies grow more quickly than Biotype A flies. Biotype B flies are becoming more resistant to insecticides.

(a) Using the information provided above, state which of the two biotypes, A or B, is a more serious pest. [1]

(b) The two biotypes of white tobacco fly live in the same areas. Explain how the white tobacco fly could evolve into two species. [4]

(c) The white tobacco fly is found in glasshouses. Explain why biological control would be more suitable for controlling this pest than the use of insectides. [3]

[Total: 8]

3 An investigation was carried out into the mating preferences of cichlid fish from three populations (A, B and C) taken from Lake Malawi. The fish were all the same species, but the males of each population showed distinct physical differences. Male fish were separated into different areas of a tank by transparent plastic sheets. The plastic sheets had holes which allowed any female to enter, but prevented the males from leaving. The diagram below shows the arrangement of the tank.

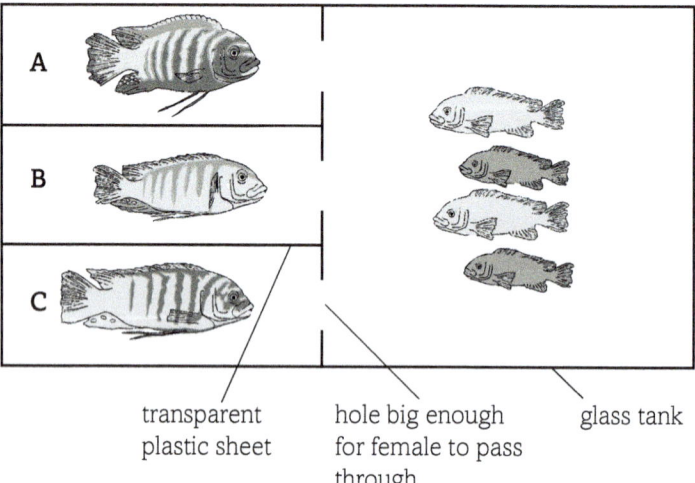

transparent plastic sheet   hole big enough for female to pass through   glass tank

Females from each population were allowed to choose one mate, and their offspring were collected. The male parent of the offspring was determined using DNA analysis.

The table below shows the number of times mating occurred between individuals of the different populations in a range of trials.

	Female from population		
Male from population	A	B	C
A	29	0	0
B	0	26	4
C	0	1	8

(a) Explain how the DNA analysis provides reliable evidence for the identity of male parents. [3]

(b) (i) Calculate the percentage of the matings that were between individuals of the same population. Show your working. [2]

(ii) Describe the mating preferences shown by the female fish in this investigation. [2]

(c) Suggest how the data support the hypothesis that population A is the most likely to become a separate species. [4]

[Total: 11]

4 Rhododendrons are shrubby plants that are widely distributed throughout the northern hemisphere. The flowering periods and habitats of two species of rhododendron, found on Yakushima Island in Japan, are shown in the table below.

Species	Flowering period	Main flowering period	Habitat
Rhododendron eriocarpum	April to July	May	Rocky areas in lowland regions
Rhododendrum indicum	May to July	June	High mountainous regions

Where these populations overlap, hybrid plants are found that have arisen as a result of cross-fertilisation between these two species. These hybrid plants are capable of flowering and producing viable seeds.

(a) Suggest why some scientists might prefer to classify *Rhododendron eriocarpum* and *Rhododendron indicum* as varieties within the same species rather than as two separate species. [3]

(b) (i) Explain what is meant by the term **genetic diversity** in a species. [2]

   (ii) Explain why there is likely to be a greater genetic diversity in the hybrid plants than in either of the two separate species. [2]

(c) Explain how the two different species of Rhododendron on Yakushima Island may have evolved from a single population of an ancestral species. [6]

[Total: 13]

CHAPTER 3.3 > Biodiversity

Introduction

Biodiversity is basically the variety of life – all of the different types of organisms, the variation between organisms of the same species and the ecosystems of which they are a part. In some places there is tremendous biodiversity in a very small area – this is well-known in tropical rain forests and coral reefs, but the immense biodiversity of the microorganisms in the human body or microorganisms, fungi and invertebrate fauna of the soil is not always as well recognised. In regions with more extreme climates, such as the lava fields in Iceland or deserts around the world, there is less biodiversity – but it is still amazing how many different types of organisms can be observed.

In this chapter you will be defining biodiversity and considering why it is so important. You will see how biodiversity can be assessed at very different scales, from species in a habitat to genes in a population. There are a number of different ways of assessing the biodiversity of a habitat – you can look at the number of different species in an area or the relative numbers of the members of the different species in an area. It is possible to measure biodiversity taking both of these factors into account by calculating an index of diversity.

Biodiversity within a species is also an important concept and you will be looking at examples that demonstrate this importance. You will also consider ways of measuring variation within the gene pool of a population using modern techniques of DNA sequencing.

All the maths you need

- Understand the principles of sampling as applied to scientific data (*e.g. use an index of biodiversity to assess the biodiversity of a habitat*)
- Use a given equation and substitute numerical values into algebraic equations (*e.g. calculate an index of diversity*)

What have I studied before?

- How to investigate the distribution and abundance of organisms in the field
- How to determine the numbers of an organism in a given area
- The way human interactions with ecosystems can have both positive and negative effects on biodiversity
- Some of the benefits and challenges of maintaining local and global biodiversity
- That meiosis and mutations are a source of variation between individuals of the same species

What will I study later?

- How random fertilisation during sexual reproduction brings about genetic variation
- How abiotic and biotic factors in an ecosystem can affect population size and exert selection pressures
- Human effects on ecosystems and thus on biodiversity, extinction and speciation
- That mutations, along with the processes of random assortment and crossing over during meiosis, give rise to genetic variations within a species (A level)
- The development of antibiotic resistance in bacteria (A level)
- The effect of selection pressures on the gene pool of a population or species (A level)

What will I study in this chapter?

- What biodiversity is and why it is important
- How biodiversity can be measured as the number of species in a habitat or the abundance of the individuals of the different species in a habitat
- How these biodiversity measures can be combined in the following index of diversity:

$$D = \frac{N(N-1)}{\sum n(n-1)}$$

- How biodiversity can be assessed by looking at the variety of alleles in the gene pool of a population

By the end of this section, you should be able to...

● describe what is meant by biodiversity

● recognise that biodiversity can be assessed within a habitat at the species level using a formula to calculate an index of diversity

Biological diversity is decreasing at an alarming rate. Most scientists are agreed that this is not a good thing, even if they disagree about the causes. But what is biodiversity? Why should we preserve it – and how can we do this?

Defining biodiversity

Most people have some idea what the term biodiversity means, but defining it clearly is not easy. The Convention on Biological Diversity, the largest international organisation working on the subject, uses biodiversity as a term to describe the variety of life on Earth, from the smallest microbes to the largest animals and plants. They suggest the concept of biodiversity includes genetic diversity both between individuals within a species and between different species, as well as the variety of different ecosystems.

The number of different species is a useful basic measure, but the concept of biodiversity is more complex than this. The differences between individuals in a species, between populations of the same species, between communities and between ecosystems are all examples of biodiversity. Biodiversity can be assessed on different scales, from species level in a habitat to the genetic level within a population.

fig A Biodiversity in the harsh environment of an Icelandic lava field, where mosses and lichens are the dominant species, is very different from the biodiversity of a tropical rainforest or a coral reef.

Why is biodiversity important?

Does it really matter if there are fewer species of snails or beetles in the world, if an unknown plant species ceases to exist or if the genetic variation between the members of a rare population gets less and less? All the evidence suggests that it does.

In your work at GCSE and earlier, you learned how all the organisms in an ecosystem are interdependent, and how they can affect the physical conditions around them. Rich biodiversity allows large scale ecosystems to function and self-regulate. These ecosystems are also interlinked on a larger scale across the Earth. If biodiversity is reduced in one area, the natural balance may be destroyed elsewhere. The air and water of the planet are purified by the action of a wide range of organisms. Waste is decomposed and rendered non-toxic by many organisms, including bacteria and fungi. For example, microorganisms in soil and water convert ammonia into nitrate ions, which are then taken up and used by plants.

Photosynthesis by plants plays an important part in stabilising the atmosphere and the world climate. Plants absorb vast amounts of water from the soil, which then evaporates into the atmosphere through transpiration, producing clouds that in turn produce rain. Therefore plants help to determine where rain will fall. Plant roots, along with fungal mycelia, also hold the soil together, affecting how water runs off the soil surface and reducing the risk of flooding. Plant pollination, seed dispersal, soil fertility and nutrient recycling in systems such as the nitrogen and carbon cycles are vital for natural ecosystems and farming, and they all depend on a rich biodiversity.

Biodiversity also provides the genetic variation that has allowed us to develop the production of crops, livestock, fisheries and forests, and enables further improvement by cross-breeding and genetic engineering. This variation will help us to cope with problems arising from climate change and disease. Plant biodiversity also provides the potential of plants to produce chemicals that are important in many areas of human life, including new medicines.

The benefits of a biodiverse and healthy ecosystem are being increasingly assessed and valued as ecosystem services. You will consider this idea further in **Section 3.3.3**.

Assessing biodiversity at the species level

Biodiversity can be measured in a number of ways. There are two main factors which need to be considered when measuring biodiversity at the species level. One is the number of different species in an area – the **species richness**. The other is the evenness of distributions of the different species – the relative abundance of the different types of organism that make up the species richness.

Species richness

Biodiversity varies enormously around the world in terms of numbers of species. The wet tropics are generally the areas of highest biodiversity. For example, in 0.1 hectares (less than four tennis courts) of Amazon rainforest you would expect to find 150–280 tree species. Almost every tree is a different species. Imagine the numbers of other plants and animals associated with each type of tree and you can begin to appreciate the species richness of these areas.

As you move away from the wet tropics, the species diversity tends to fall. In temperate rainforests, tree species richness drops to 20–25 species in the same size of sample area. Further north again in the boreal forest in Scandinavia and Northern Canada, it falls to 1–3 species. To highlight this, scientists have identified a number of **biodiversity hotspots (fig B)** of unusual biodiversity and endemism. They occupy only 15.7% of the Earth's land surface, but are home to 77% of the Earth's terrestrial vertebrate species. Unfortunately, these areas often coincide with areas with resources that people want to use. For example, the Latin American rainforests have huge biodiversity, but are also a rich source of wood, gas, oil and minerals, and people are rapidly destroying the rainforests to access these resources.

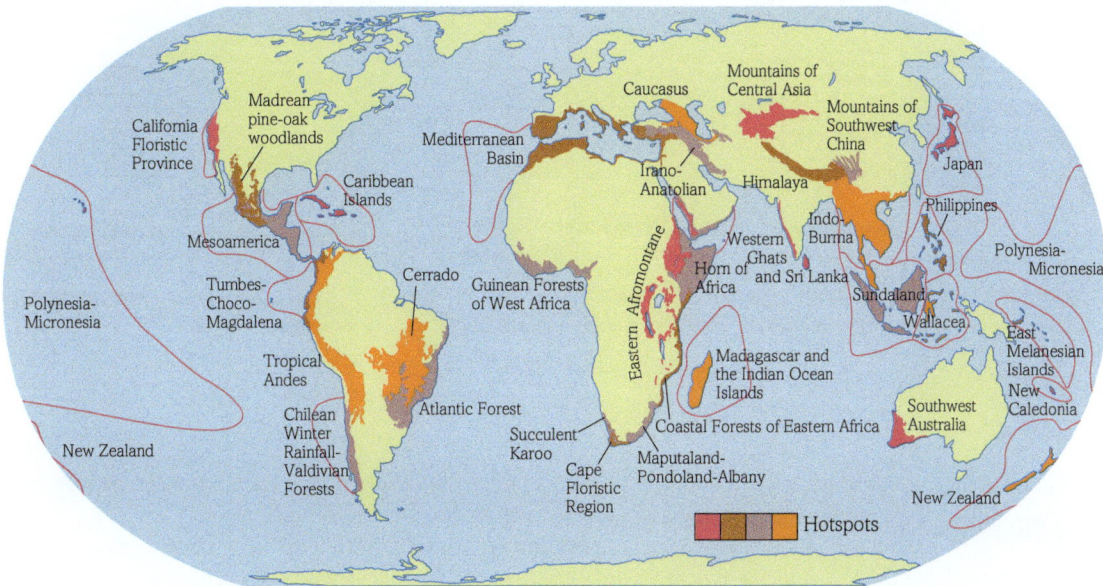

fig B Known biodiversity hotspots around the world.

Tropical areas also show areas of high marine and freshwater species richness, and include of all the different types of organisms. Coral reefs are the marine equivalent of the tropical rainforests, and they are the key areas of marine biodiversity. However, species richness is not the only important factor in a biodiversity hotspot.

Endemism

Another important criterion is the number of endemic species in an area – species that are found nowhere else (see **Section 3.2.3**). If you look at **fig C** you can see that the areas of greatest biodiversity are not always the same as the areas with the biggest number of endemic species. This is why it is so difficult to prioritise areas for conservation. There have been many ideas about why some areas have particularly rich biodiversity – in fact around 125 different theories have been published. Most have been eliminated because they do not apply to all organisms or they are not supported by the evidence. The best current model suggests that a very stable ecosystem allows many complex relationships to develop between species. High levels of productivity (when photosynthesis rates are very high) can support more niches. In areas where organisms can grow and reproduce rapidly, it is more likely that more mutations occur, leading to adaptations which allow organisms to exploit more niches.

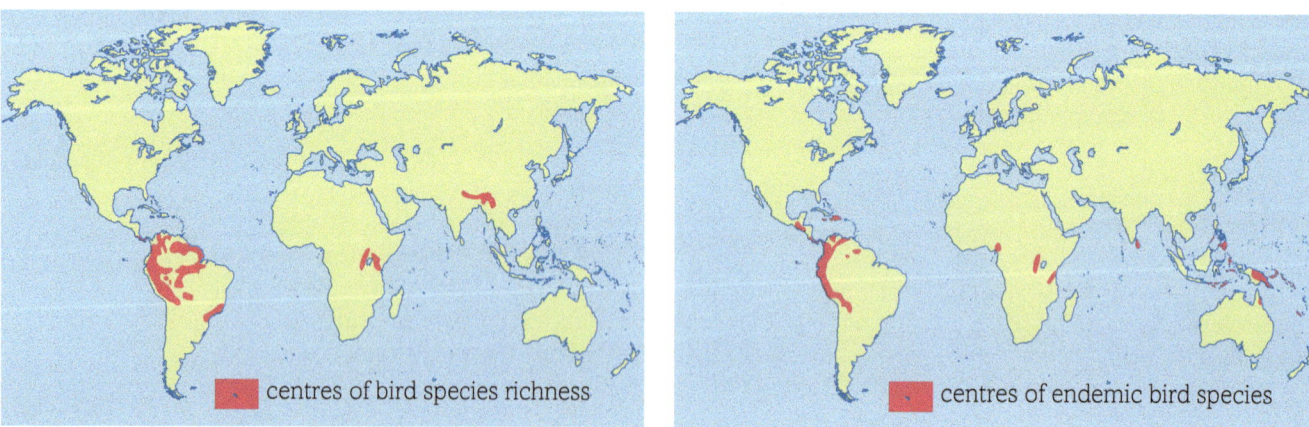

fig C Hotspots of bird biodiversity measured by species richness and by endemic species.

Species abundance

The absolute number of species is not the only important factor in biodiversity. **Relative species abundance** is also key – the relative numbers of the different types of organisms.

Picture two areas of land, **plot A** and **plot B**. Both have five species of plant growing on them – grass, daisies, buttercups, dandelions and lady's bedstraw, in the proportions shown in **table A**.

plot A   plot B

	Grass	Daisies	Dandelion	Buttercups	Lady's bedstraw
plot A	95%	2%	1%	1%	1%
plot B	30%	20%	15%	15%	20%

table A

Plot B is more diverse, because the numbers of the different species are more evenly spaced. **Plot A** might well be a well-tended garden lawn or city park, while **plot B** could simply be a less well-cared for version of the same site. An area showing an even abundance of different species is considered to be more biodiverse than one containing the same number of different species but dominated by one or two of those species.

The risks to biodiversity are not evenly spread around the world. Certain areas are much more vulnerable to damage and loss, particularly small isolated ecosystems such as islands, rainforests, coral reefs, bogs and wetlands. Many of these areas are also biodiversity hotspots, so if they are damaged many species will be lost. Every time a species becomes extinct, the biodiversity of the world decreases. On the other hand, every time a new species evolves biodiversity increases.

Unless people understand the level of biodiversity in an area, it is impossible for them to recognise how important its loss might be, and so decide what to do about it. How can biodiversity be measured?

Measuring biodiversity

Scientists have developed many different ways of measuring the biodiversity of an ecosystem. Some are more effective than others, and all have limitations. In this example, both the species richness and the species abundance of an area are taken into account in a formula that gives a diversity index at the species level within a habitat:

$$D = \frac{N(N-1)}{\sum n(n-1)}$$

where

D = diversity index

N = the total number of organisms of all species

n = the total number of organisms of each individual species – the abundance of the different species

$\sum$ = the sum of all the values that follow. (You need to calculate $n(n-1)$ for each species and then add them together).

A simple worked example is given below based on some samples collected from a local pond.

Species	Number of organisms collected (n)	($n-1$)	$n(n-1)$
stickleback	4	3	4 × 3 = 12
mosquito larvae	12	11	12 × 11 = 132
water boatman	3	2	3 × 2 = 6
pond skater	4	3	4 × 3 = 12
cyclops	17	16	17 × 16 = 272
Total number of organisms (N)	**40**	**39**	**40 × 39 = 1560**

fig D A diversity index can be used to calculate the biodiversity of a pond such as this.

In a sample from a small pond, the species shown in the above table were found. You can calculate the diversity index for this pond as follows:

$$D = \frac{N(N-1)}{\sum n(n-1)}$$

Using the data from the table

$$D = \frac{1560}{12 + 132 + 6 + 12 + 272}$$

$$D = \frac{1560}{434}$$

Diversity index = 3.60

The higher the value of the diversity index, the greater the variety of living organisms found in the area.

How biodiversity varies

There have been many ideas about why some areas have particularly rich biodiversity and around 125 different theories for this have been published. Most of these theories have been eliminated because they do not apply to all organisms or they are not supported by the evidence. The best current model suggests that high biodiversity is seen in:

- very stable ecosystems as this allows many complex relationships to develop between species

- area where there are high levels of productivity (when photosynthesis rates are very high) as this can support more niches

- areas where organisms can grow and reproduce rapidly, as it is more likely that more mutations occur, leading to adaptations that allow organisms to exploit more niches.

In general, when an environment has relatively extreme environmental conditions, the biodiversity is low. Any change in this extreme environment has a big impact on population numbers. This type of ecosystem tends to be unstable and very susceptible to change. A particularly severe frost, a flood or a new pathogen can devastate or even wipe out one or more populations. This type of environment also has a number of unfilled niches so an incoming organism can become established very rapidly and overpower an existing species if they are competing for food or territories.

In less hostile environments, biodiversity can be very high. This results in a very stable ecosystem, because a new species moving in or out will not have a lot of effect.

As a result of these factors, the risk of loss of biodiversity is not evenly spread around the world. Biodiversity can be lost due to natural events such as a volcanic eruption or flooding, and also as a result of human activities. Certain habitats are much more vulnerable to damage and loss, particularly small isolated ecosystems such as islands, coral reefs, bogs and wetlands. Many of these areas are also biodiversity hotspots, so if they are damaged many species will be lost. Every time a species becomes extinct, the biodiversity of the world decreases. On the other hand, every time a new species evolves, biodiversity increases.

When to measure biodiversity?

Biodiversity is not constant. For example, the animal species in an area can vary with the time of day. Many bat species flying on a warm summer evening in the UK will not be visible the next morning. What is more, in the temperate and alpine areas of the world there are distinct seasons. This means that the picture of biodiversity in an area will change considerably through the year. The number of plant species in the same area of a UK woodland floor or meadow measured during the summer will differ to that measured in winter, and show different biodiversity. Similarly, wetland feeding sites for migrating birds are alive with aquatic and wading birds in the winter months, but during the summer they are relatively empty.

fig E The great migration of millions of wildebeest and zebra across the Serengeti plains of East Africa means that the biodiversity of the region changes dramatically as the migrating animals arrive and then move on.

Questions

1 The term biodiversity is often used in the media simply to indicate the number of species of living organisms. Why does this give a limited picture?

2 In a small area of woodland, a group of students counted both the number of species present in a measured area and the numbers of individuals of the different species.

Species	Number of organisms collected (n)
holly	9
bramble	3
oak	3
butcher's broom	5
ivy	3
yew	1
Total number of organisms (N)	**24**

Using this information, calculate the index of diversity for this area of woodland and decide whether you think it has high or low biodiversity.

3 Using **fig C** explain how the areas of high bird biodiversity and high bird endemism differ, and why this might be.

Key definitions

Species richness refers to the number of different species in an area.

A **biodiversity hotspot** is an area with a particularly high level of biodiversity.

Relative species abundance refers to the relative numbers of species in an area.

By the end of this section, you should be able to...

● recognise that biodiversity can be assessed within a species at the genetic level by looking at the variety of alleles in the gene pool of the population

Biodiversity is not only about the numbers of different species and how they are distributed. Biodiversity within an individual species is also a very important concept. The gene pool of a species is all of the genes in the genome, including all the different variants of each gene. Modern DNA analysis allows us to measure biodiversity on a different scale, at a genetic level – and we are discovering that genetic diversity within a species is very important.

Gene and allele frequency

Mutations are changes in the DNA structure (see **Section 1.3.7**). These changes may be tiny, such as a change in a single base pair, or they may involve the obliteration or duplication of an entire chromosome. Many mutations have no effect at all on the phenotype, while others have severe or even lethal effects.

Mutations can increase the gene pool of a population by increasing the number of different alleles available. The relative frequency of a particular allele in a population is known as the **allele frequency**. If a mutation results in an advantageous feature, the frequency of that allele in the population will be selected for and so increase in frequency. If the mutation is disadvantageous, natural selection will sometimes result in its removal from the gene pool. More frequently it will be retained at a very low frequency unless it also transfers some benefits. A disadvantageous allele in one set of environmental conditions may become an advantageous allele if conditions change. The changes in allele frequency due to natural selection may lead to the evolution of new species. Many mutations are neutral, conferring neither advantage nor disadvantage. For example, on average a baby has 100 new mutations in its DNA compared with its parents' chromosomes, but quite probably none will have an effect on the phenotype.

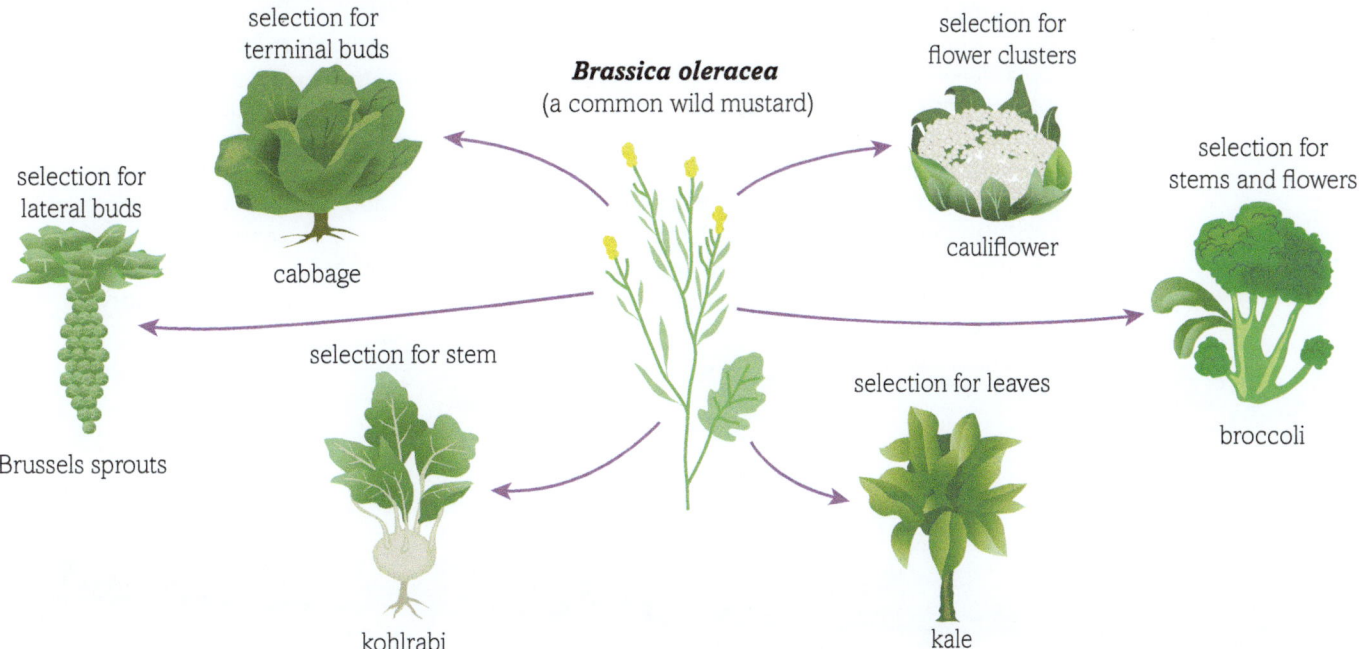

fig A The wide variety of crop plants that have resulted from selective breeding of *Brassica oleracea* shows the huge amount of genetic variety present in that gene pool. It is this variety that both natural and artificial selection act on.

Measuring genetic biodiversity

The genetic variety within a population is also an important measure of biological health and wellbeing – without variety a population is vulnerable. Modern technology has made it possible to build up a clear model of **genetic diversity** within a population by analysing the DNA and comparing particular regions for similarities and differences.

For example, scientists have discovered that cheetahs have very little genetic diversity. These beautiful members of the cat family, the fastest land animals on Earth, are not only low in numbers, but have so little genetic biodiversity that they are in danger of being wiped out by a single disease or a small change in their environment. Scientists think that they must have been almost destroyed 10 000–20 000 years ago. As a result of this bottleneck the modern populations are related to just a few founder members (see **Book 2 Section 8.2**). Now that numbers are dwindling as their habitat disappears, there are serious worries about the survival of the species.

fig B The genetic diversity of these beautiful members of the cat family is dangerously low.

Models of the molecular phylogenetic relationships (see **Section 3.1.2**) between related organisms based on DNA and other evidence have proved to be a very useful tool for measuring biodiversity. For example, scientists at the Natural History Museum have built up contrasting maps of biodiversity based on both numbers of species and DNA similarities. The ones shown in **fig C** show bee populations – the genetic variation map changes the most biodiverse area for bees from Ecuador (highest species richness) to Kashmir (highest genetic diversity). This type of study can have huge importance for conservation work. If you are trying to conserve biodiversity with limited funding – and funding is always limited – you need to be confident that you are choosing the area with the highest biodiversity. Maps like these can be generated for overall biodiversity or for the diversity of particular groups of animals and plants. They can be produced for the whole world, for individual countries or for smaller local areas. The value of this type of data is that it can be used to highlight areas that need protection and, with regular updating, provides a way of monitoring changes in biodiversity anywhere.

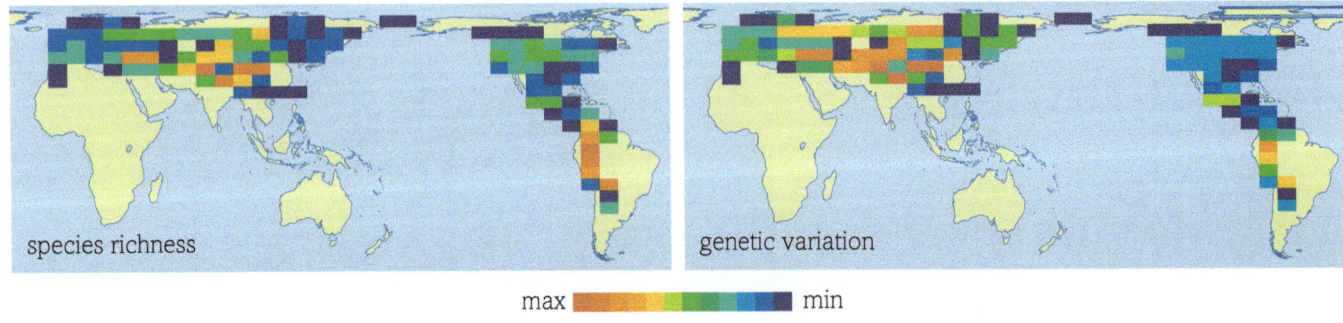

fig C These maps show the biodiversity of bees around the world measured by species richness and by genetic diversity.

The isolated islands of Hawaii

The Hawaiian island populations show clearly how living organisms adapt to a particular niche or role in the community. They also demonstrate why it is important for scientists to look at biodiversity at both the level of species numbers and abundance and also at the level of genetic diversity.

The Hawaiian Islands are very isolated – 4000 km from the nearest continental land mass and 1600 km from the nearest other islands. They also have a great deal of biodiversity in terms of species numbers – 1000 species of native flowers, 10 000 species of insects, 1000 species of land snails and around 100 species of birds. But before people introduced them, there were no reptiles and only one species of mammal – a bat. Analysis of the DNA of the native populations shows that they are very closely related, even though some of them look very different. All those insect species seem to have evolved from only around 400 original species, while there appear to have been only seven founder species of land birds. So in these isolated circumstances, a small group of founder organisms have adapted and evolved to take advantage of the different ecological niches that were available to them. Places where endemism is common often have a rich biodiversity in terms of species numbers, but relatively low genetic diversity. This is one reason why areas with many endemic populations are very vulnerable to the introduction of disease.

Questions

1 How does **fig A** illustrate the concept of genetic diversity in a species?

2 Using the information supplied here and other sources, discuss why it is important to measure both species richness and genetic diversity to give a full picture of biodiversity.

Key definitions

Allele frequency is the frequency with which a particular allele appears within a population.

Genetic diversity is a measure of the level of difference in the genetic make-up of a population.

Ecosystem services

By the end of this section, you should be able to...

● Explain the ethical and economic reasons (ecosystem services) for the maintenance of biodiversity

In recent years people have become increasingly aware of biodiversity and its importance on both a local and a global level. But increasingly, not only scientists but also politicians are becoming aware of the economic importance of healthy, biodiverse ecosystems.

Ethical reasons for maintaining biodiversity

There are many ethical reasons for maintaining the environment and supporting biodiversity. For example:

- If we destroy the biodiversity of an ecosystem, we are denying future generations the opportunity to use these renewable natural resources.
- The natural world and the biodiversity within it is a great source of pleasure for many people and so should be protected and maintained.
- If biodiversity is lost when a species becomes extinct, unique combinations of DNA are lost. Many people would argue that extinction and loss of biodiversity due to human activities or competition for habitat is unethical.
- Human activities also have the potential to cause mass extinctions through global climate change and this interference with biodiversity on a massive scale is regarded as unethical.

fig A Beautiful places such as this are valuable in many ways and it seems unethical to destroy them.

Unfortunately ethical arguments do not always carry a lot of weight with governments. On the other hand, if economic arguments can be made for the maintenance of high biodiversity, then governments and politicians take notice.

Ecosystems services

The idea of **ecosystems services** is a relatively new one, but one that is rapidly gaining ground and making politicians in countries all around the world take notice. Ecosystems services are services provided by the natural environment that are of benefit to people. The main categories of ecosystems services defined by the Millennium Ecosystem Assessment (or MA) are fairly widely accepted, although definitions do vary. They include the following.

Provisioning services

Ecosystems provide us with all sorts of provisions we need, including food, fibres for clothing, building materials, fuel, genetic resources for crop improvements, fresh water and medicines. The greater the biodiversity, the more potential sources of all these services there are available to us.

Regulating services

Ecosystem processes help to maintain and regulate our environment. They are involved in processes including water purification and sewage treatment, maintaining air quality, disease regulation, pest control and pollination. People are also increasingly aware of the importance of biodiverse ecosystems in maintaining the climate of the planet.

Supporting services

Biodiverse ecosystems provide support for other ecosystems services that we need. For example, soil formation and nutrient cycling in the environment are both ecosystems services without which we could not grow food to eat.

Cultural services

A biodiverse and healthy ecosystem is important for human health and well-being, and is also used in recreation and education. In some areas, the economy of a country can depend on people visiting to observe its wild animals and plant life.

Taking an ecosystems approach

Once biodiversity is lost from an area it is not easily regained. The growing tendency to treat ecosystems as part of a financial equation is an encouraging one, as economic arguments for the maintenance of biodiversity are likely to have more effect than purely ethical ones.

People are increasingly taking factors that would have a negative or positive impact on ecosystems and their biodiversity into account in policy and decision making in governments. Not only do ecosystem services produce economic benefits, but if biodiversity is neglected or destroyed, it can be very expensive. When human activities cause environmental disasters such as hurricanes, floods or forest fires, either directly or directly, it costs a lot of money to put things right. If areas become deserts and soil is eroded away as a result of deforestation, there is both an economic cost and a human cost as people cannot farm and therefore suffer malnutrition. If non-sustainable resources are harvested and not replaced, that industry will collapse along with the ecosystem.

Loss of biodiversity can also cause us potential financial losses for the future. Any loss of biodiversity reduces the chances of finding a new drug, a new food or some genes that can be used in a crop plant to increase yield or increase resistance to disease.

When a government or community takes an approach to making planning decisions based on ecosystem services, they will recognise and respect the limits of the biological systems and conserve the structure of the ecosystem as far as possible. It is

an approach that involves optimising productivity for long-term benefits, rather than simply taking the maximum short-term gain. Economic value has to be placed on all aspects of ecosystems services for this approach to work, but there are encouraging signs that more and more people in positions of power and influence recognise the need for action. For example DEFRA (the UK Department for Environment, Food and Rural Affairs) has

fig B Extreme weather events and changes in climate can cause millions of pounds worth of damage and lost income. Healthy biodiverse global ecosystems can help prevent such events, and this is something politicians can use in policy development.

produced a number of publications to support other government departments as they undertake sustainable developments with ecosystem services in mind.

fig C The UK Government appears to be taking ecosystems services seriously.

Questions

1 Suggest two ethical arguments for the maintenance of biodiversity.

2 (a) What is meant by ecosystem services?
   (b) Use one clear example to explain exactly what is mean by the terms provisioning services, regulating services, supporting services and cultural services.

3 Discuss the potential importance of the concept of ecosystem services in maintaining local and global biodiversity.

Key definition

Ecosystems services are services provided by the natural environment that are of benefit to people.

By the end of this section, you should be able to...

● Explain the principles of ex-situ (zoos and seed banks) and in-situ (protected habitats) conservation, and the issues surrounding each method

Conservation means keeping and protecting a living and changing environment. It is an active process involving an enormous range of projects from reclaiming land after industrial use to helping set up sustainable agriculture systems in the developed world, and from the protection of a single threatened species to global legislation on pollution levels and greenhouse gas emissions. Around the world there are many animals and plants that are threatened with extinction. This is often, but not always, as a result of human activities causing habitat loss or climate change.

There are two main ways of conserving animals and plants. **Ex-situ conservation** takes place outside of their natural habitat, for example, in zoos or seed banks. **In-situ conservation** takes place in the natural habitat of the organism.

Ex-situ conservation

The United Nations Convention on Biological Diversity in 1992 defined ex-situ conservation as the conservation of components of biological diversity (living organisms) outside their natural habitats. Sometimes when an organism is threatened with extinction there is not time to conserve their habitat or protect them in situ (on site). By removing some of the animals or plants from their natural habitat it is sometimes possible to conserve the species. At worst this enables their genetic material to be conserved and at best a breeding population can eventually be returned to their natural habitat. Ex-situ conservation is always seen as a complementary approach to in-situ conservation, and in an ideal situation it takes place in the country where the threatened species originates.

Ex-situ conservation of plants

It has been estimated that 25% of the world's flowering plant species could disappear within the next 50 years. There are thought to be about 242 000 species of flowering plants now, so this would mean 60 500 species disappearing in less than one human lifetime!

Plants are of vital importance to all our lives. The genetic material of these extinct species would be lost forever, which would be a disaster not only for the plants but also possibly for human survival. Crossbreeding crop plants back to original wild plants, or using wild plants to supply genes for genetic engineering, are ways in which the long-term health of our crop plants can be maintained.

Botanic gardens (zoos for plants) maintain collections of many of the world's most interesting and unusual plants. In the 1960s, with a view to conservation, the Royal Botanic Gardens at Kew set up a seed bank that is now home to the seeds of nearly 4000 wild plants (**fig B**). Called the Millennium Seed Bank, it had two main aims. The first was to collect and conserve the seeds of the entire UK native flora by the year 2000. Of the 1442 native plants, over 300 are already threatened with extinction. The second aim was to conserve the seeds of an additional 10% of the flora of the whole world by the year 2010, concentrating particularly on the drylands (arid, subarid and subhumid regions), which are experiencing some of the most rapid loss of habitat.

fig A Rare plants like these British fly orchids are not just beautiful – who knows how we might use them in the future?

A seed bank can preserve many plants in a state of effective suspended animation. Live seeds are collected from the wild, removed from the fruits and cleaned. They are screened using X-rays to make sure that they contain fully developed embryos. Then they are dried, put into jars and stored at between −20 and −40°C where many will survive and remain capable of germinating for up to 200 years. In general, the lifespan of a seed doubles for every 5°C drop in temperature or 2% fall in relative humidity. Some of the seeds stored may even germinate in several thousand years' time.

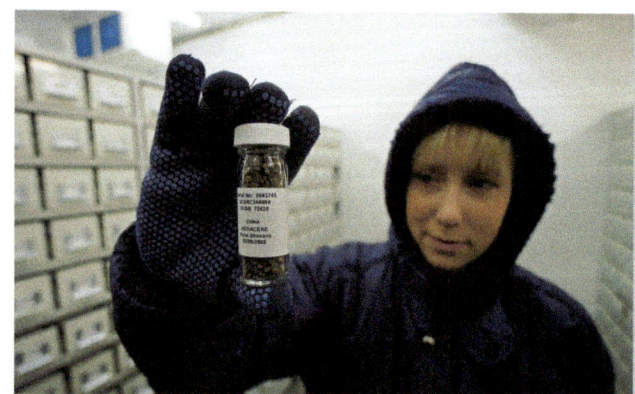

fig B Stored at low temperatures, seeds like these will still be able to germinate in hundreds of years – by which time, hopefully, their habitats have been restored and conserved.

Most plants make huge numbers of seeds, so they can be collected without damaging the natural population. Seeds are usually small, so large numbers of them can be stored relatively cheaply in a small space. They contain all the genetic material of the plant so they are a record of the genetic make-up of the species as well as a potential new plant for the future. Other countries have set up similar projects and there are now over 1000 seed banks around the world.

About 80% of the known species of plants could be stored in seed banks, however, the seeds of some species do not store well. Unfortunately these include many crop plants such as mango, rubber, oak, avocado, cacao and coconut. These plants have to be conserved differently. They may be grown where they are found naturally, in field gene banks such as plantations, orchards and arboretums or as tissue cultures. In this way the species is grown on, year after year. One problem is that field gene banks take up a lot of room and a lot of work. For example, the world potato collection at the International Potato Centre in Peru contains around 4100 different clones of potato, all of which have to be planted annually. Using tissue cultures to conserve plants and growing plants on as needed takes up a lot less space and time, but allows more variety to be conserved.

Ex-situ conservation of animals

It is not always possible to conserve animal species in the wild because the conditions that have put them under threat of extinction continue. Zoos and wildlife parks used to exist just for people to look at the animals, but today they play an important role in animal conservation. In captive breeding programmes individuals of an endangered species are bred in zoos and parks in an attempt to save the species from extinction. Usually the ultimate aim is to reintroduce the captive-bred animals into the wild to restore the original populations.

Reintroduction does not always work but can be more successful in national parks or other protected areas. Species that have been saved by captive breeding and successfully reintroduced into protected areas in their own countries include the Californian condors and Przewalski's horses. Captive breeding programmes for the white and black rhino, along with large amounts of conservation work in the field in east Africa, lead to hopes that these amazing creatures will also be saved from extinction.

fig C The Mkomazi Game Reserve Rhino Sanctuary in Tanzania has been established to help build up the population of black rhinos and protect them from the poachers who have hunted them almost to extinction.

There are several problems with captive breeding and reintroduction:

- There is not enough space or sufficient resources in zoos and parks for all the endangered species.

- It is often difficult to provide the right conditions for breeding, even if scientists know what those conditions are. For example, the giant panda is notoriously difficult to breed even when conditions are ideal.

- Reintroduction to the wild will be unsuccessful unless the original reason for the species being pushed to the edge of extinction is removed.

- Animals that have been bred in captivity may have great problems in adjusting to unsupported life in the wild.

- When the population is small, the gene pool is reduced and this can cause serious problems. Zoos try to overcome this by keeping detailed records of the genetic data of their breeding individuals. Sperm can be swapped with other zoos (for artificial insemination) to maximise genetic variation in the offspring.

- Reintroduction programmes can be very expensive and time-consuming, and they may fail.

There has been interest in saving endangered species by cross-species cloning. This is the cloning of animals using closely related species as surrogate mothers. In 2001 scientists cloned a gaur, an endangered species of ox. DNA from the cells of a male animal that had died 8 years earlier was fused with cows' ova from which the nucleus had been removed. Of the 692 ova used, only one produced a healthy embryo and this was implanted in an ordinary cow. The gaur bull calf was born strong and healthy but he died within 48 hours from a common gut infection. In the same year, scientists cloned a mouflon, a rare breed of wild sheep, using a domestic sheep as the surrogate mother. The lamb lived a few months. Although there have been few real successes, some scientists believe this technology can eventually be used to bring back species that have recently become extinct. Conversely, there are scientists who feel that this cutting-edge research is a waste of valuable resources, at least until the conditions which drove the organism to extinction have been addressed.

Did you know?

The Frozen Ark

The Frozen Ark project, set up in 1996, plans to conserve the genetic resources of animals in the world by conserving their DNA and viable cells. It is the animal equivalent of the Millenium Seed Bank for plants. The Frozen Ark aims to save DNA samples from endangered species. DNA is very stable and can be stored frozen for hundreds of years, and samples take up very little room – 10 million samples could be stored in the volume of an average house. This gives scientists the possibility of increasing genetic variation in critically endangered species or even cloning extinct species once their habitat has been restored and conserved. The Frozen Ark does not replace in-situ and ex-situ conservation – it supports them.

In-situ conservation

The UN Convention on Biological Diversity (1992) defined in-situ conservation as the conservation of ecosystems and natural habitats, and the maintenance and recovery of viable populations of species in their natural surroundings. It is the internationally accepted primary conservation strategy.

For a species to be conserved, it needs a long-term habitat. Often it is not single species but large areas of countryside that are threatened, which may include whole habitats and even ecosystems. Plants and animals are conserved together within their natural relationships. To conserve and protect endangered species their whole habitat has to be conserved. Governments need to be involved to protect large areas such as these.

The biggest units of protected land are the National Parks, which have been set up around the world to protect and conserve native species. These parks work to protect the animals and plants within them. In the UK National Park land is protected from development and allowed to sustain a rich diversity of animals and plants in a number of different ways. In England alone there are over 200 National Nature Reserves, covering around 90 000 hectares of countryside, and Sites of Special Scientific Interest (SSSIs) covering another 7% of land. These SSSIs are usually quite small areas that contain a particularly rich diversity of life, or a group of endangered plants or animals which thrive there. SSSIs play an important role in the conservation of many habitats in the British countryside.

The biggest units of protected land in the UK are the 14 National Parks, which represent about 8% of the land area of the country. These areas are all home to farming communities, villages and towns, but they are managed with a view to the best possible compromise between economic demands and the conservation of the countryside and the wildlife. Examples include Dartmoor, the New Forest, the Peak District, the North York Moors, and Lundy Island, a marine nature reserve.

fig D The New Forest is home to many rare species, from the silver studded blue butterfly and sand lizards, to the long-leaved sundew and the New Forest ponies.

Similarly, National Parks are set up around the world to protect and conserve native species. Different countries and parks have different levels of protection for the plants and animals that are found within them. Unfortunately, poachers, who will kill protected species to make money from the skin, horns and other products from protected species, can be a problem in some countries.

Examples of National Parks include:

- The Great Barrier Reef Marine Park (Australia): This protects more than 2900 coral reefs, which are home to some of the greatest marine biodiversity in the world including 6 out of the 7 species of marine turtles and around 1500 species of fish.

fig E The biodiversity of the Great Barrier Reef is almost unique in the oceans.

- Volcanoes National Park (Rwanda): This contains 5 volcanoes that are covered in rainforest and bamboo. It is home to much biodiversity including 178 recorded species of birds and the extremely rare mountain gorilla.

fig F Rare mountain gorillas in the Volcanoes National Park in Rwanda.

- Serengeti National Park (Tanzania): One of the most famous of all Africa's National Parks, the Serengeti is known the world over for the great migrations of wildebeest and zebra that take place every year, as well as for its populations of leopards, lions, cheetahs, elephants, rhinos and buffaloes.

- Semuliki National Park (Uganda): This National Park is relatively new and covers part of the only lowland tropical rainforest in east Africa. The area has a particularly high level of biodiversity of both plants and animals, including over 400 bird species and the extremely rare forest ground thrush. There are also more than 60 mammal species including pygmy hippos and eight primate species. It is also very rich in insects, with around 300 species of butterflies alone.

- Kanha National Park (India): This large National Park has saal and bamboo forests, grasslands, lakes and streams and is home to many endangered species including tigers.

- Everglades (USA): A large subtropical wilderness, this mangrove rich National Park is home to 36 protected species including the West Indian manatee, the Florida panther and the American crocodile.

In-situ conservation strategies

There are a number of different strategies that are used as part of in-situ conservation, both within and outside of National Parks and other conservation areas. These include habitat restoration and recovery, strategies for the sustainable use and management of biological resources, and managed recovery programmes for threatened or endangered species, which might involve ex-situ breeding programmes. The genetic diversity of ecosystems is monitored within specific areas, as are any threats to biodiversity. These threats range from the invasion of alien species to over-exploitation of a resource. The preservation of traditional knowledge and traditional land management practices is also very important for the maintenance of biodiverse habitats. The formulation and implementation of appropriate legislation is also key to successful in-situ conservation.

Conflicts in conservation

When land is set aside for conservation there are often conflicts between the needs of people living there and the needs of the animals and plants that are being conserved. It costs money to maintain and conserve the area, which could be spent on health and education instead. People need the land to live and to earn money, but their activities can lead to pollution, erosion and many other problems. Which is more important – the long-term biodiversity of the local ecosystem or the local people? And who makes the decisions about which gets priority treatment? These are difficult questions to answer everywhere in the world, but perhaps particularly so in countries where many people may struggle to survive and need the land to grow food to feed themselves and their families.

Sustainability

Habitats and ecosystems can be conserved with less conflict by encouraging sustainable methods of land use. For example, illegal logging operations in rainforests practise 'slash and burn' (cutting down all the trees and burning the ground afterwards) to harvest wood to sell and clear the soil for farming. The soil is soon spent and biodiversity lost. However, if we harvest the trees selectively and replant for the future, biodiversity can be maintained while people continue to use the forest for income. This is sustainable forestry.

Sustainable agriculture includes farming methods that minimise damage to the environment and avoid monoculture. These are becoming increasingly important around the world. They involve using organic fertilisers where possible, minimising the use of artificial fertilisers and chemical pesticides, using biological pest control, maintaining hedgerows and planting in rotation to avoid the soil becoming exhausted. Large-scale farming is vital to provide the food we need but often sustainable methods such as using biological pest control can increase yields and improve profits while being cheaper in the long-term than using chemicals.

It is important to get our priorities right. People and politicians often have limited sympathy for ecology and so it is important to target spending accurately. Around the world there is a growing understanding of the need for sustainable agriculture and sustainable tourism to conserve biodiversity whilst still providing the food and income that people need. Research continues into how food and other resources can be produced in a way that minimises loss of biodiversity, and even increases it again. Similarly, tourism can be developed in a way that is sustainable, does minimal damage to the environment, provides jobs and money for local people and at the same time conserves the environment and thus maintains biodiversity.

Maintaining biodiversity is a major issue for the twenty-first century. Success or failure will affect the whole planet, and the potential consequences of failure could be devastating for us all. In a global economy, we can all play a vital role in maintaining biodiversity by the choices that we make now.

Did you know?

Restoring the forests in Ethiopia

Professor Legesse Negash is a pioneer in the propagation of Ethiopia's indigenous trees. Like so many African countries, Ethiopia has tremendous natural biodiversity that is threatened both by possible climate change and by deforestation. By finding ways to propagate some of the indigenous trees, and storing the germlines, Legresse Negash aims to protect the soil against erosion and help to save the biodiversity of the country. He has set up the Center for Indigenous Trees Propagation and Biodiversity Development in Ethiopia. The centre not only produces thousands of young trees, it also aims to educate farmers and train new young biologists to work in this important area so that new trees can be planted when the older ones are harvested.

fig G Professor Legesse Negash, seen here planting a keystone indigenous tree species (*Ficus vasta Forssk*) at his Center for Indigenous Trees Propagation and Biodiversity Development in Ethiopia.

Questions

1 Preservation means preserving something exactly as it is now. How does conservation differ from preservation?

2 (a) What is the difference between ex-situ and in-situ conservation?

(b) How can ex-situ conservation be used to support in-situ conservation?

3 Why is it much easier to conserve plant biodiversity than animal biodiversity?

4 Discuss the main advantages and disadvantages of captive breeding and reintroduction programmes.

Key definitions

Conservation refers to maintaining and protecting a living and changing environment.

Ex-situ conservation is the conservation of components of biological diversity (living organisms) outside their natural habitats.

In-situ conservation is the conservation of ecosystems and natural habitats, and the maintenance and recovery of viable populations of species in their natural surroundings.

1 Biodiversity is an important concept in conservation.
   The diagrams below show four identically sized areas A, B, C and D.
   Different shapes represent different species.

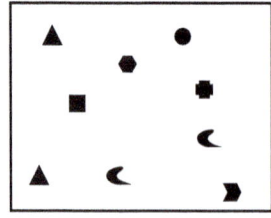

Area A

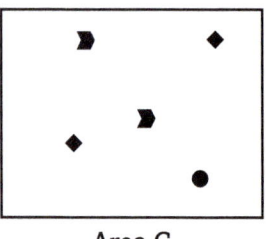

Area B

Area C

Area D

Place a cross (☒) in the box next to the correct letter to complete each of the following statements.

(a) (i) The area with highest species richness is

 A

☐ B

☐ C

☐ D   [1]

(ii) The area with lowest species richness is

 A

☐ B

☐ C

☐ D   [1]

(iii) State which area contains an endemic species, giving reasons for your answer.   [3]

(b) Zoos help to conserve species through captive breeding programmes and reintroduction programmes.

   Describe how zoos use these programmes to help conserve rare species.   [5]

   [Total: 10]

2 The yarrow plant, *Achillea* sp., shown in the photograph below, has been extensively studied to try to assess the influence of genes and the environment on its adaptation.

(a) A large number of seeds from wild yarrow plants were grown in controlled environmental conditions. The resulting population contained plants of different heights.

   (i) Suggest which of the graphs, A, B, C or D shown below, would represent the height distribution in this plant population and give an explanation for this height distribution.

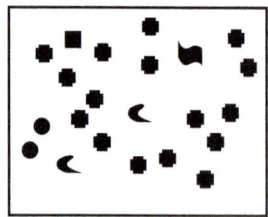

[3]

   (ii) Suggest **two** environmental conditions that were controlled in this investigation.   [2]

(b) The mean height of yarrow plants, growing at various altitudes (height above sea level), was recorded. The results are shown in the table and graph below.

Altitude/m	0	1000	2000	2500	3000
Mean height of yarrow plant/cm	130	75	50	30	20

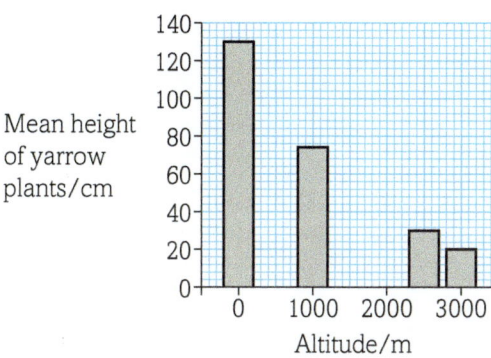

(i) Using the information in the table, complete the graph above. [1]

(ii) Describe the effect of increasing altitude on the mean height of yarrow plants. [2]

(c) In a further investigation, plants with a height of 50 cm, growing at 2000 m above sea level, were cloned to produce a large number of plants. These plants were then grown at three different altitudes and their mean height at each altitude was recorded.

The results are shown in the graph below.

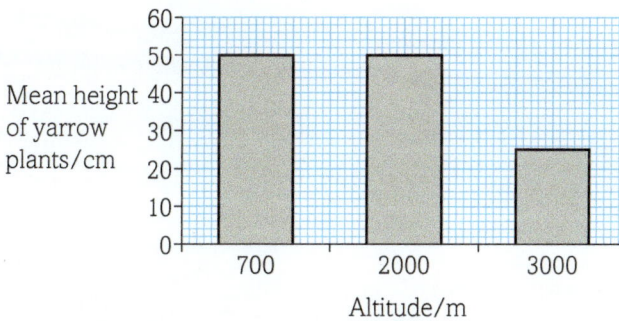

The evidence from the graph suggests that a combination of genetic and environmental factors influence the height of yarrow plants at different altitudes above sea level.

(i) Give **one** piece of evidence from the graph which suggests that genetic factors influence the height of yarrow plants at different altitudes. [1]

(ii) Give **one** piece of evidence from the graph which suggests that environmental factors influence the height of yarrow plants at different altitudes. [1]

(iii) Suggest a reason for using cloned plants rather than using seeds in this investigation. [1]

(iv) Suggest a reason for growing cloned plants at 2000 m in this investigation. [1]

[Total: 12]

3 Plants like daisies, dandelions, clover and yarrow have broad leaves. They can be regarded as weeds on school playing fields as they compete with the narrower-leaved grasses like *Agrostis* and *Festuca*. Selective herbicides can be used to kill off the broad-leaved plants to allow the grass to grow better. These herbicides can take up to a week to work.

A scientist surveyed the biodiversity of plants growing on the school playing field. One hundred quadrats (sampling areas) were used and the results aggregated.

These aggregated results were then put into a table and used to calculate an index of diversity.

Species	Number of plants growing in 0.25 m² quadrat (n)	$n-1$	$n(n-1)$
Daisy (*Bellis perennis*)	2000		
Dandelion (*Taraxacum officinale*)	3000		
Red clover (*Trifolium pratense*)	3500		
Yarrow (*Achillea millefolium*)	2000		
Grass (*Agrostis spp*)	4000		
Grass (*Festuca spp*)	4500		
Bee Orchid (*Ophrys apifera*)	1		
Totals			

(a) (i) Complete the table by filling in the missing boxes. [5]

(ii) Use the information in the table to calculate the index of diversity given by the following formula:
$$D = \frac{N(N-1)}{\Sigma n(n-1)}$$
[3]

(b) A rare plant called a Bee Orchid (*Ophrys apifera*) was found in one part of the field.

(i) Why was it important for the scientist to take 100 samples of the field? [2]

(ii) Explain why the occurrence of this one Bee Orchid made no difference to the biodiversity as measured by the index of diversity. [1]

(c) Selective herbicides were sprayed on the playing field, and the scientist came back a week later to see how this treatment affected the biodiversity.

(i) Describe and explain what happened to the numbers of dandelion plants and the numbers of grass plants over the week. [4]

(ii) Explain why the recalculated index of diversity was found to be 2.26. [2]

[Total: 17]

TOPIC 4
Exchange and transport

Cell transport mechanisms

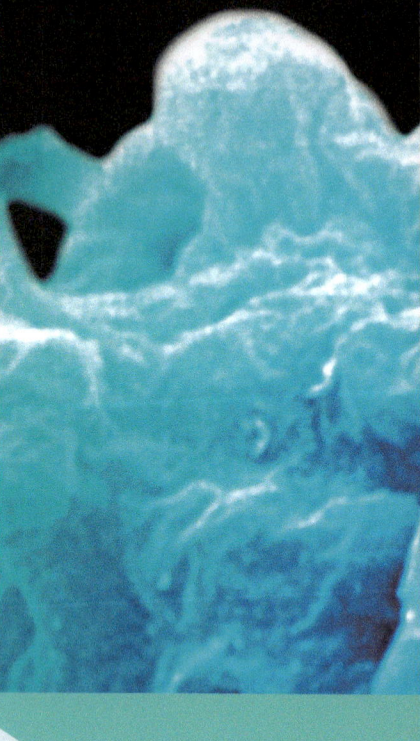

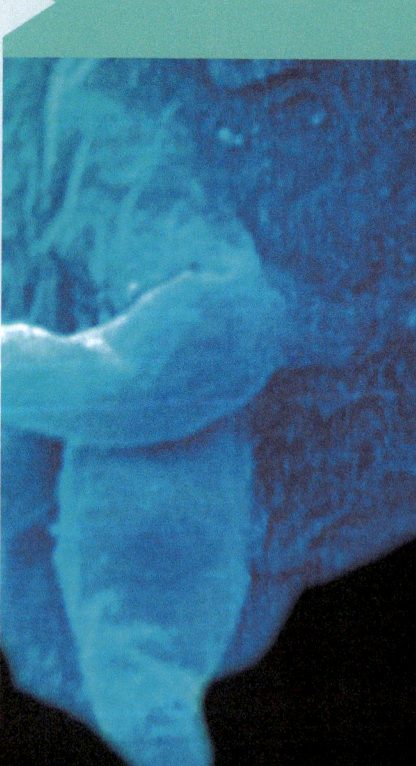

Introduction

The single-celled *Amoeba proteus* has a problem. It lives in fresh water. The cytoplasm of the cell contains lots of dissolved minerals and sugars, and the outer membrane is partially permeable. As a result, water moves in by osmosis, a passive process common to all living things. The *Amoeba* also has a contractile vacuole, a specialised organelle for getting rid of the excess water. As water moves into the cytoplasm passively, it is moved into the contractile vacuole by a process that involves the active transport of ions into the vacuole. The water flows down a concentration gradient. Once the vacuole is full it contracts, emptying the water out of the cell. Scientists have been studying *Amoeba* for many years, but they are still not sure exactly how this happens. However, they do know that osmosis and active transport all take place within this single-celled organism.

In this chapter you will be looking at how substances are transported into and out of cells. Diffusion is a key concept – the net movement of particles down a concentration gradient from an area of high concentration to an area of lower concentration as a result of the random movement of particles. It is an important form of transport in all organisms.

You will also consider osmosis – a specialised form of diffusion that involves the movement of water down a concentration gradient through a partially permeable membrane. Osmosis is important in the formation of tissue fluid in animals, and in plants it is vital for maintaining turgor and supporting structures from the leaves to the whole plant.

However, sometimes things need to be moved against concentration or electrochemical gradients – and this is where active transport comes in. Active transport, along with endocytosis and exocytosis, uses ATP to move substances against a concentration or electrochemical gradient in many different circumstances in the cell.

All the maths you need

- Recognise and make use of appropriate units in calculations (*e.g. nm^2, nm^3, molar quantities for serial dilutions, etc.*)
- Recognise and use expressions in standard and decimal form (*e.g. use of magnification*)
- Substitute numerical values into algebraic equations using appropriate units for physical quantities (*e.g. water potential calculations*)
- Solve algebraic equations in a biological context (*e.g. water potential equations*)

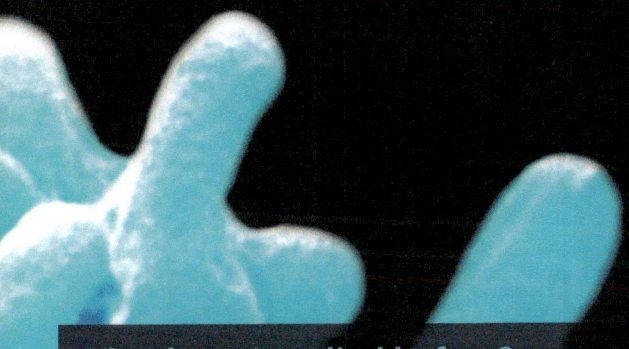

What will I study later?

- That larger multicellular organisms need a transport system, because their surface area : volume ratio makes direct diffusion of substances in and out of all cells impossible on a realistic timescale
- The importance of diffusion in gas exchange in arthropods, fish and mammals
- The importance of osmosis in the formation of tissue fluid
- The role of osmosis in moving water into a plant through the root hair cells and through the plant to the xylem and phloem
- The importance of osmosis in support (turgor), growth, movement, tropisms and photosynthesis in plants
- Homeostatic mechanisms in animals that maintain a relatively constant internal environment to avoid osmotic damage to the cells (A level)

What have I studied before?

- How substances are moved into and out of cells by diffusion, osmosis and active transport
- The concept of surface area : volume ratio in determining if a transport system is needed
- The role of phospholipids in the structure of the cell membrane
- The nature of ATP

What will I study in this chapter?

- Why a cell does not need a specific transport system
- The fluid mosaic model of the cell surface membrane, including protein pores and active pumps
- The role of passive transport in the cell, including diffusion, facilitated diffusion and osmosis
- The importance of osmosis in animal and plant cells
- Water potential in plant cells, including the equation:

$$\psi = P + \pi$$

- The transport of large molecules into and out of the cell through the formation of vesicles in the processes of endocytosis and exocytosis
- The process of active transport into and out of the cell, including the hydrolysis of ATP

By the end of this section, you should be able to...

● recall the structure of the cell surface membrane with reference to the fluid mosaic model and explain how this structure relates to the different types of transport in cells

The need for transport

Cells are microscopic, but as you saw in **Chapter 2.1**, they contain many different organelles and a wide range of chemical reactions take place within them. Cells require a supply of chemicals, such as glucose and oxygen for cellular respiration. These must be transported from outside the organism into the cells. Respiration supplies energy for the other reactions of life, but it also produces the toxic waste product carbon dioxide. This and other waste products need to be removed from the cells before they cause damage to them. Cells need to transport substances made in one area to another or out of the cell completely. Cells also have to transport the raw materials they need across the surface cell membrane into the cell. The way substances move into, out of and within cells often depends on the structure of the cell membrane.

The fluid mosaic model of the cell membrane

The cell surface membrane acts as the gatekeeper to the cell, controlling the transport of materials into and out of the cell. Membranes also control transport within the cells – the endoplasmic reticulum and the Golgi apparatus are both membrane-bound structures that move substances about inside the cell (see **Section 2.1.4**). In **Section 2.1.2** you also studied the fluid mosaic model of cell membranes and how the properties of the polar lipids and proteins determine its structure. Look at **fig A** to remind yourself of the key points of this structure.

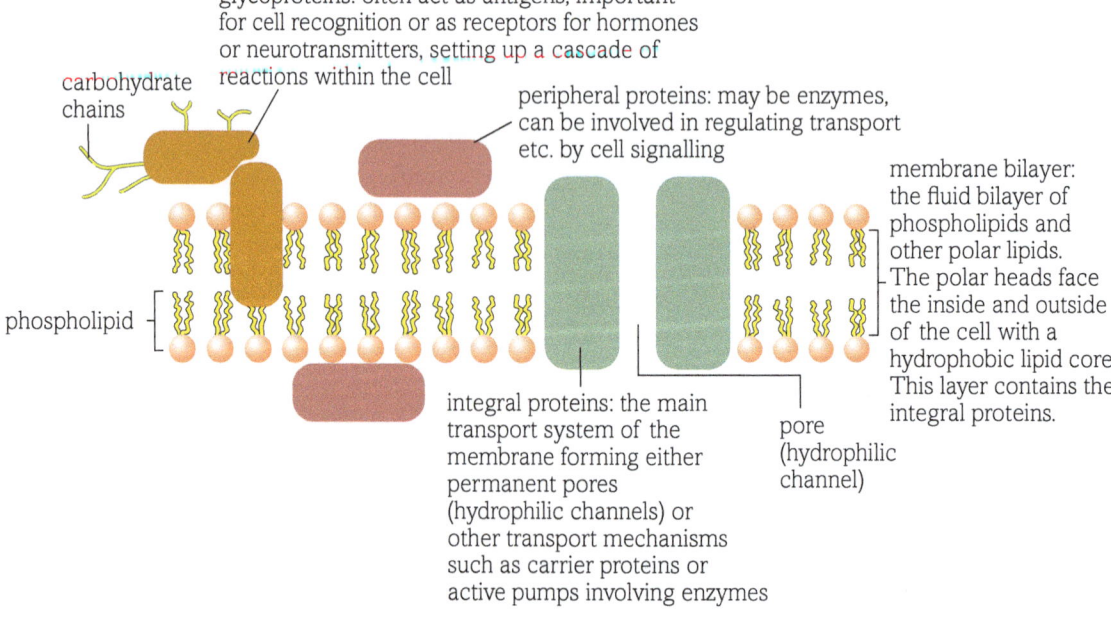

glycoproteins: often act as antigens, important for cell recognition or as receptors for hormones or neurotransmitters, setting up a cascade of reactions within the cell

carbohydrate chains

peripheral proteins: may be enzymes, can be involved in regulating transport etc. by cell signalling

membrane bilayer: the fluid bilayer of phospholipids and other polar lipids. The polar heads face the inside and outside of the cell with a hydrophobic lipid core. This layer contains the integral proteins.

phospholipid

integral proteins: the main transport system of the membrane forming either permanent pores (hydrophilic channels) or other transport mechanisms such as carrier proteins or active pumps involving enzymes

pore (hydrophilic channel)

fig A The fluid mosaic model of the cell membrane – a phospholipid sea with associated proteins, which may be floating or anchored within the membrane.

Learning tip

Be clear about the different roles of glycoproteins, peripheral proteins and integral proteins in the fluid mosaic model of a cell membrane.

The main types of transport in cells

Substances are transported into, out of and around cells by a variety of different mechanisms. **Passive transport** takes place as a result of concentration, pressure or electrochemical gradients and involves no energy from the cell. **Active transport** involves moving substances into or out of the cell, by using adenosine triphosphate (ATP) produced during cellular respiration.

Passive transport mechanisms

There are three main types of passive transport in cells:

- **Diffusion** – the movement of particles in a liquid or gas down a concentration gradient. They move from an area where they are at a relatively high concentration to an area where they are at a relatively low concentration as a result of random movements. Cell membranes are no barrier to diffusion of small particles such as the gases oxygen and carbon dioxide.
- **Facilitated diffusion** – diffusion that takes place through carrier proteins or protein channels. The protein-lined pores of the cell membrane make facilitated diffusion possible.
- **Osmosis** – a specialised form of diffusion that involves the movement of solvent molecules (usually water) down a concentration gradient through a partially permeable membrane. The partially permeable nature of the cell membrane means solutes can be accumulated either side of the membrane and results in the movement of water by osmosis across the membrane.

Active transport mechanisms

- **Endocytosis** – the movement of large molecules into cells through vesicle formation. The fluid nature of the cell membrane makes it possible to form vesicles.
- **Exocytosis** – the movement of large molecules out of cells through vesicle formation.
- **Active transport** – the movement of substances across the membrane of cells directly using ATP. The proteins in the membrane act as carriers or enzymes, making ATP energy available to move ions or molecules through the membrane.

You will learn more about the ways in which the membrane is adapted to its transport functions as you look at the different transport mechanisms in more detail.

fig B The rapid movements of the Venus fly trap as it catches its prey are the result of different types of transport within the cells of the plant.

Questions

1 Explain why transport systems are needed in cells.

2 Summarise the structure of the cell membrane and explain how this is related to its transport roles within the cell.

Key definitions

Passive transport is transport that takes place as a result of concentration, pressure or electrochemical gradients and involves no energy from a cell.

Active transport is the movement of substances into or out of the cell using ATP produced during cellular respiration.

Diffusion is the movement of the particles in a liquid or gas down a concentration gradient from an area where they are at a relatively high concentration to an area where they are at a relatively low concentration.

Facilitated diffusion is diffusion that takes place through carrier proteins or protein channels.

Osmosis is a specialised form of diffusion that involves the movement of solvent molecules down a concentration gradient.

Endocytosis is the movement of large molecules into cells through vesicle formation.

Exocytosis is the movement of large molecules out of cells by the fusing of a vesicle containing the molecules with the surface cell membrane. The process requires ATP.

By the end of this section, you should be able to...

- explain how passive transport is brought about in the cell by diffusion and facilitated diffusion

- explain how the properties of molecules affect how they are transported, including solubility, size and change

We know from scientific analysis that the concentration of substances either side of a membrane can be very different. This suggests that a membrane exercises control over the passage of substances across it. The properties of the membrane affect the transport of substances into and out of the cell, but the properties of the molecules to be transported also have an effect. The size of a molecule is important to how it is transported through cell membranes and so is its solubility in lipids and water. The presence or absence of charge on a molecule also affects how it is transported. For example, some substances, particularly those that dissolve very easily in lipids, simply pass through the membrane in a process of diffusion. Other very small molecules, such as the gases oxygen and carbon dioxide also pass freely in and out of cells through the membrane. Some large molecules, such as steroid hormones, are not transported through the membrane. Many charged particles, such as sodium ions, need specific carriers and pores to make it through from one side to the other.

Diffusion

In physical terms, diffusion is the movement of the molecules of a liquid or gas from an area where they are highly concentrated to an area where they are at a lower concentration. We say that they move down their concentration gradient. This occurs because of the random motion of molecules due to the energy they have, which is dependent on the temperature. If you have a large number of molecules tightly packed together, random motion will result in their spreading out and eventually reaching a uniform distribution. The molecules do not stop moving once they reach a uniform distribution; however, the movement no longer causes a net change in concentration because equal numbers are moving in all directions.

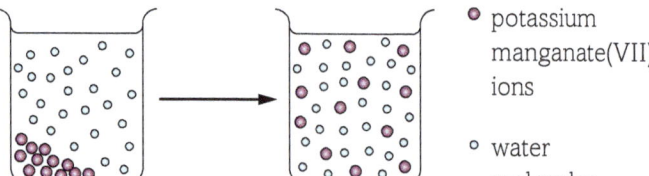

- potassium manganate(VII) ions

- water molecules

fig A If the beaker is left to stand, diffusion takes place as the random movement of both the water and the potassium manganate(VII) ions ensures that they are eventually evenly mixed.

For many small molecules, like oxygen and carbon dioxide, the membrane is no barrier and they can diffuse freely across it. This movement, by diffusion alone, is a form of passive transport.

However, larger hydrophilic molecules and ions larger than carbon dioxide molecules cannot move across the membrane by simple diffusion.

Facilitated diffusion

Substances with a strong positive or negative charge and large molecules cannot cross cell membranes by simple diffusion. Nevertheless, they may move into and out of the cell down a concentration gradient by a specialised form of diffusion. Facilitated diffusion involves proteins in the membrane that allow only specific substances to move through passively down their concentration gradient (see **fig B**). They may simply be channel proteins that form pores through the membrane. Each type of

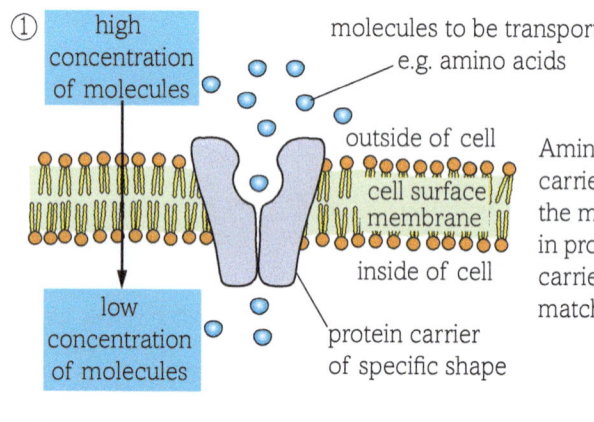

① high concentration of molecules

molecules to be transported e.g. amino acids

outside of cell

cell surface membrane

inside of cell

low concentration of molecules

protein carrier of specific shape

Amino acids are carried across the membrane in protein carriers of matching shape.

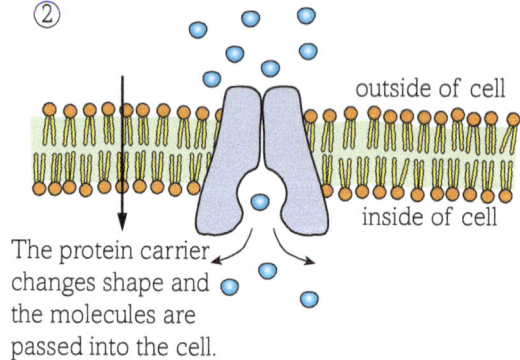

②

outside of cell

inside of cell

The protein carrier changes shape and the molecules are passed into the cell.

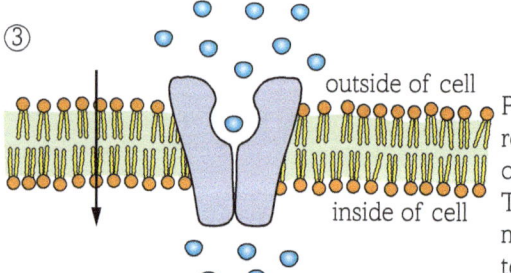

③

outside of cell

inside of cell

Protein carrier returns to its original shape. This allows more molecules to enter.

fig B Facilitated diffusion acts as a ferry across the lipid membrane sea. It is not an active process, so it can only work when the concentration gradient is in the right direction.

channel protein allows one particular type of molecule through, dependent on its shape and charge – for example, some are sodium ion channels and others form potassium ion channels. Some channels open only when a specific molecule is present or there is an electrical change across the membrane, such as during the passage of nerve impulses along neurones. These are called gated channels.

Another form of facilitated diffusion depends on carrier molecules floating on the surface of the membrane. The carriers will be found on the *outside* surface of the membrane structure when a substance is to be moved *into* the cell or organelle, and on the *inside* for transport *out* of the cell or organelle. The protein carriers are specific for particular molecules or groups of molecules, depending on the shape of the protein carrier and the substance to be carried. Once a carrier has picked up a molecule it rotates through the membrane to the other side, carrying the molecule with it, and then releases the molecule. The movement through the membrane takes place because the carrier changes shape once it is actually carrying something. The process can only take place down a concentration gradient – from a high concentration of a molecule to a low one. It does not use energy, so is considered a form of diffusion. For example, red blood cells have a carrier to help glucose move into the cells rapidly.

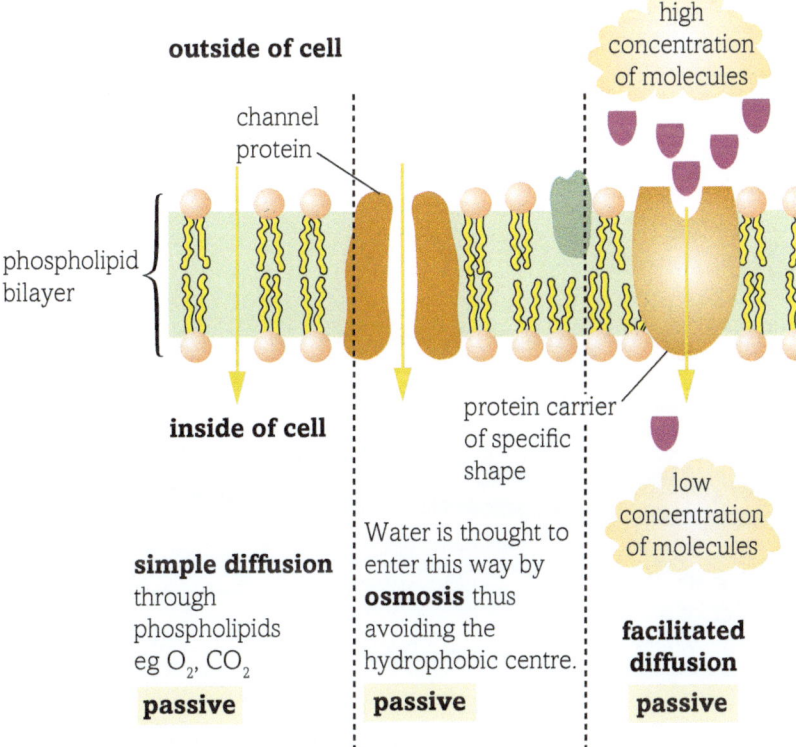

fig C Three of the main transport routes through a cell surface membrane.

Questions

1 Describe the conditions needed for the passive transport of molecules into a cell.

2 Water and ions often enter the cell through protein pores but they cannot pass through the lipid layer in the same way as oxygen and carbon dioxide. Why not?

3 Explain the differences between simple and facilitated diffusion.

By the end of this section, you should be able to...

● explain how osmosis brings about passive transport of water into and out of the cell

● explain the water potential of plant tissue

As you have seen, diffusion takes place when molecules or ions can move freely. Free water molecules – water molecules that are not involved in hydrogen bonding or any other type of bonding – can move easily, even through partially permeable membranes. So, as a result of random motion, water molecules will tend to move across a partially permeable membrane down their concentration gradient.

What is osmosis?

Osmosis can be defined as the net movement of solvent molecules from a region where they are at a high concentration to a region where they are at a lower concentration through a partially permeable membrane. In living organisms the solvent is always water, and membranes in cells are generally partially permeable, in that they let some molecules through, but not others. So osmosis in cells involves the movement of water from a region of high concentration of water molecules (in other words, a dilute solution of the solute) to a region of lower concentration of water molecules (a more concentrated solution of the solute) across a partially permeable membrane such as the cell surface membrane or nuclear membrane. It may help you to think of osmosis in terms of water molecules moving from the side of a membrane where there is a more dilute solution to the side where there is a more concentrated solution.

If the solution bathing the outside of a cell has a lower concentration of dissolved substances (solutes) than the solution inside the cell, there will be a concentration gradient that encourages water molecules to move into the cell. If the opposite is true and the solution bathing the cell has a higher concentration of solute than the cell contents, water will move out of the cell. The **osmotic concentration** of a solution concerns only those solutes that have an osmotic effect. Many large insoluble molecules found in the cytoplasm, for example starch and lipids, do not affect the movement of water and so are ignored when considering osmotic concentration. Only soluble particles are considered, including the big plasma proteins such as albumin and fibrinogen.

In the context of living cells the movement of water by osmosis, and the control of this process, is very important. In animal cells in particular it is vital that water does not simply move continuously into the cells from a dilute external solution, because the end result would be that the cells would swell up and burst.

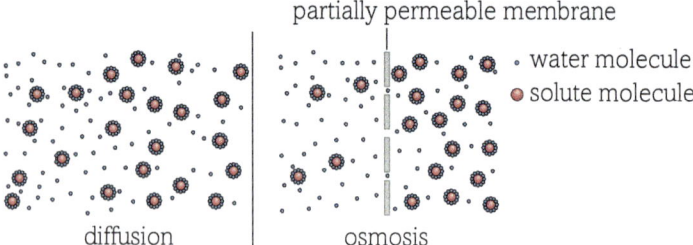

partially permeable membrane

• water molecule
● solute molecule

diffusion | osmosis

fig A In diffusion the random movement of particles results in an even distribution of both solute and solvent particles. In osmosis a partially permeable membrane means only solvent molecules and very small solute particles can move freely.

Modelling osmosis in cells

You can make a model cell using an artificial membrane that is permeable to some molecules – in particular water – and impermeable to others, such as sucrose. There are many experiments showing the movement of water in these circumstances, and one of the simplest is illustrated in **fig B**. The presence or absence of sucrose in the different regions of the model can be shown by carrying out Benedict's test for non-reducing sugars on the solutions (see **Section 1.2.1**).

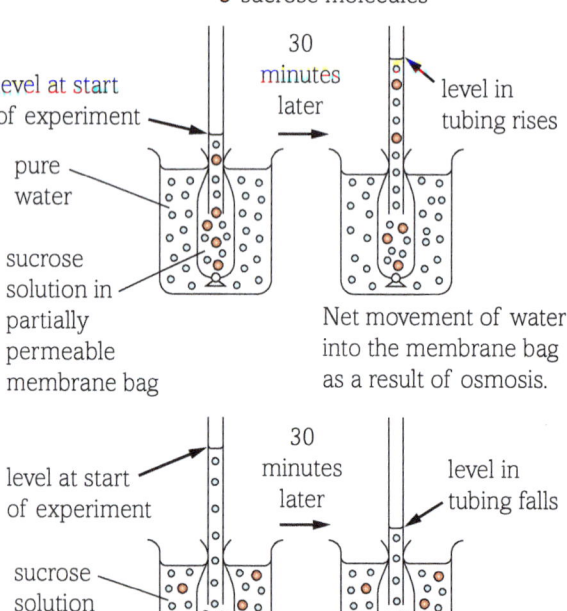

○ water molecules
● sucrose molecules

level at start of experiment

30 minutes later

level in tubing rises

pure water

sucrose solution in partially permeable membrane bag

Net movement of water into the membrane bag as a result of osmosis.

level at start of experiment

30 minutes later

level in tubing falls

sucrose solution

pure water

Net movement of water out of bag as a result of osmosis.

fig B The artificial partially permeable membrane in this experiment models the cell surface membrane. It allows water to pass through freely, but not the solute molecules.

Osmotic concentrations

During osmotic experiments cells are often immersed in solutions of varying osmotic concentrations. The osmotic concentration of a solution only takes into account those dissolved substances that have an osmotic effect. This is particularly important in living things because many of the large molecules found in the cytoplasm of a cell do not affect the movement of water into or out of the cell and so we ignore them when calculating osmotic concentration:

- In an **isotonic** solution, the osmotic concentration of the solutes in the solution is the same as that in the cells.

- In a **hypotonic** solution, the osmotic concentration of solutes in the solution is lower than that in the cytoplasm of the cells.

- In a **hypertonic** solution, the osmotic concentration of solutes in the solution is higher than that in the cytoplasm.

Osmosis in animal cells

Osmosis needs to be carefully controlled in animal cells, where the net movement of water in or out needs to be kept to a minimum. Animal cells are effectively like fragile balloons filled with jelly. When too much water moves in, the cells burst; when too much moves out, the cells shrivel as the concentrated cytoplasm loses its internal structure and the chemical reactions that normally take place in the cell stop working.

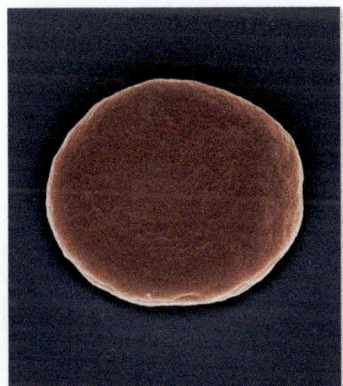 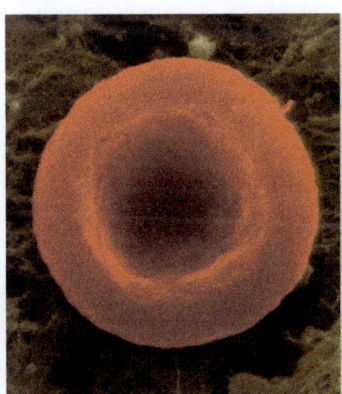 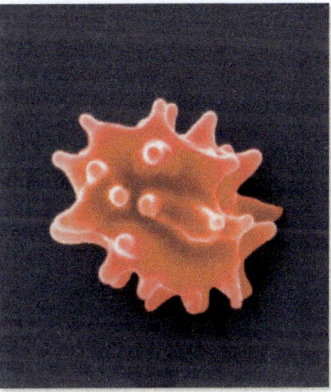

fig C The effects of osmosis on red blood cells show why the systems of the body that maintain solute concentrations and water balance are so important.

Osmosis in plant cells

Plant cells are also like fragile balloons filled with jelly – but the balloon is inside the rigid box of the cellulose cell wall. In plant cells the cellulose cell wall prevents cells bursting. If the surrounding fluid is hypotonic to the cytoplasm of a plant cell, water will enter the cell by osmosis – but not indefinitely. As the cytoplasm swells and presses on the cell walls, it generates hydrostatic pressure. The inward pressure of the cell wall on the cytoplasm increases until it cancels out the tendency for water molecules to move in. This inward pressure is called the **pressure potential**. When the osmotic force moving water into the plant cell is balanced by the pressure potential forcing it out, the plant cell is rigid, in a state known as **turgor**. Most plant cells are in a state of turgor most of the time – the rigid structure supports the stems and leaves of the plant.

If plant cells are put in a solution which is just slightly hypertonic, water moves out of the cell by osmosis and turgor is lost. The cell membrane begins to pull away from the cell wall as the protoplasm shrinks. This is called **incipient plasmolysis**. We measure incipient plasmolysis using serial dilutions, looking for the point at which 50% of the cells are plasmolysed and 50% are not. This is the concentration that is equivalent to the solute potential of the cell sap (see **page 218**). If the cell is placed in a hypertonic solution, so much water will leave the cell that the vacuole will be reduced and the protoplasm will shrink away from the cell walls – the cells suffer **plasmolysis**. However, the size and shape of plant cells does not change much whether they are fully turgid or fully plasmolysed, because of the cell wall. They do not swell and burst, or become very small. It is only the contents that change.

(a) 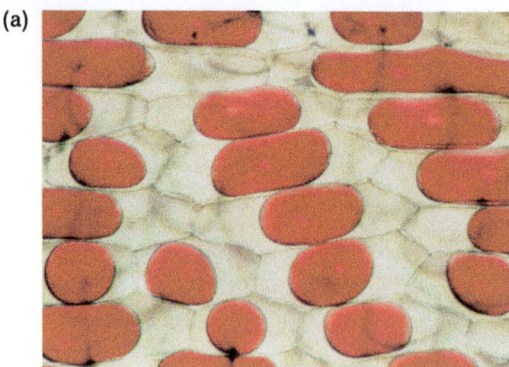 (b)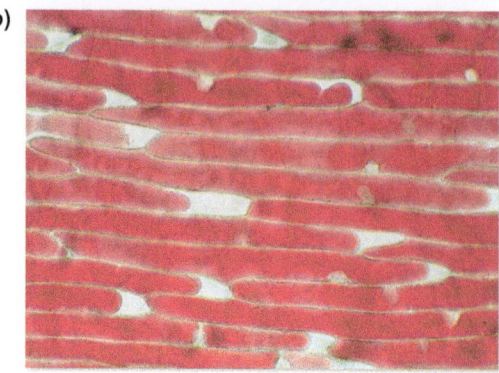

fig D Plant cells showing (a) plasmolysis; and (b) turgor.

Water potential in a plant cell

Most of the work on osmosis is done on plant cells. They are generally bigger and easier to see with the light microscope than animal cells, and the changes are easier to see and measure than those in animal cells. When studying osmosis, there are some important terms that help us calculate what is happening inside the cells, and what the impact of any changes we make is likely to be.

- **Water potential (Ψ)** – a measure of the potential for water to move *out* of a solution by osmosis. Pure water has the highest water potential, because water molecules will always move from pure water into any solution on the other side of a partially permeable membrane. This maximum water potential is always given as zero. All other solutions will have a lower water potential than pure water, because their concentration of water molecules must be lower than pure water. As a result the water potential Ψ for any solution other than water will be negative. Water always moves from an area of high water potential to an area of low water potential.

This gives us an alternative definition of osmosis as:

- The net movement of water molecules from an area of higher water potential to an area of lower water potential through a partially permeable membrane.

Two variables interact inside a plant cell: the **turgor pressure** exerted by the cell wall and the osmotic potential of the cell contents.

- **Turgor pressure (P)**. As water moves into a plant cell by osmosis, the protoplasm swells. It cannot keep expanding because the rigid cell wall limits it. As water continues to move into the cell, hydrostatic pressure is generated as the swelling protoplasm pushes against the cell wall. An inward opposing pressure is exerted by the cell wall against the expanding cell contents. (Think of a balloon in a shoebox!) The hydrostatic pressure generated as the cell contents push against the cell wall of a plant cell as a result of the movement of water into the cell by osmosis is known as turgor pressure. Turgor pressure will rise until the osmotic force pulling water into the cell is balanced by the turgor pressure opposing further entry of water. At this balancing point the turgor pressure equals the osmotic potential of the cell and turgor remains constant. Turgor pressure usually has a positive value. It can be measured directly in an individual cell using a pressure probe.

- **Osmotic potential (π)**: This is the potential of water to move across a partially permeable membrane from a solution of low concentration (hypotonic) of dissolved solutes to one of high concentration (hypertonic). Pure water has the highest (least negative) osmotic potential; a solution containing dissolved solutes has a lower (more negative) osmotic potential than pure water. The greater the solute concentration of a solution, the lower (more negative) the osmotic potential. Osmotic potential is always negative.

This gives us another alternative definition of osmosis as:

- The net movement of water molecules from an area of higher osmotic potential to an area of lower osmotic potential through a partially permeable membrane.

The combination of the osmotic potential of the protoplasm and the turgor pressure exerted by the protoplasm against the plant cell walls gives us the **water potential** of the cell.

The water potential (Ψ) is a measure of the energy state of the water at any point. It is the combined effect of osmotic potential pulling water into a cell and turgor pressure forcing water out. When the turgor pressure and osmotic potential are equal, the cell is at full turgor and the water potential of the cell is zero. If the water potential is different between two regions that are separated by a partially permeable membrane, then water will diffuse from the area of higher water potential (less negative) to that where water potential is lower (more negative). This diffusion of water is the basis of osmosis.

water potential of cell	=	turgor pressure	+	osmotic potential
(usually negative)		(usually positive)		(always negative)
Ψ	=	P	+	π

In animal cells, the water potential of the cell is simply the osmotic potential of the cytoplasm as there is no cell wall to exert a hydrostatic pressure.

The movement of water in the right direction is vital for the healthy functioning of all cells. This movement depends largely on the concentrations of various solutes both in the cytoplasm and in the surrounding fluids. Cells control the movement of water by selecting which solute molecules move into and out of the cell. Many molecules move passively by diffusion, but cells have another way of selecting which molecules cross the membranes, as you will see in the next section.

Questions

1 Write a definition of osmosis in your own words and give two examples of where osmosis is important in living organisms.

2 In an experiment, human cheek cells were placed in three solutions: an isotonic solution, a hypertonic solution and a hypotonic solution. Describe what you would expect to happen in each case. Explain your answers in terms of osmosis.

3 Without osmosis, plants as we know them would not survive. True or false? Discuss.

Key definitions

Osmotic concentration is a measure of the concentration of the solutes in a solution that have an osmotic effect.

An **isotonic** solution is a solution in which the osmotic concentration of the solutes is the same as that in the cells.

A **hypotonic** solution is a solution in which the osmotic concentration of solutes is lower than the cell contents.

A **hypertonic** solution is a solution in which the osmotic concentration of solutes is higher than that in the cell contents.

Pressure potential is a measure of the inward pressure exerted by the plant cell wall on the protoplasm of a cell, opposing the entry of water by osmosis. It usually has a positive value.

Turgor is the state of a plant cell when the solute potential causing water to be moved into the cell by osmosis is balanced by the force of the cell wall pressing on the protoplasm.

Incipient plasmolysis is the point at which so much water has moved out of the cell by osmosis that turgor is lost and the cell membrane begins to pull away from the cell wall as the protoplasm shrinks.

Plasmolysis is the situation when a plant cell is placed in hypertonic solution when so much water leaves the cell by osmosis that the vacuole is reduced and the protoplasm is concentrated and shrinks away from the cell walls.

Water potential (Ψ) is a measure of the potential for water to move out of a solution by osmosis.

Turgor pressure (P) is a measure of the inward pressure exerted by the plant cell wall on the protoplasm of the cell as the cell contents expand and press outwards, a force which opposes the entry of water by osmosis.

Osmotic potential (π) is a measure of the potential of a solution to cause water to move into the cell across a partially permeable membrane as a result of dissolved solutes.

By the end of this section, you should be able to...

● explain the process of active transport, including the role of ATP

● describe how the hydrolysis of ATP provides an accessible supply of energy for biological processes

● describe how large molecules can be transported into and out of cells through the formation of vesicles, in the processes of endocytosis and exocytosis

Diffusion and facilitated diffusion are passive processes that allow small molecules to move across membranes. Cells can maintain steep concentration gradients by simply 'mopping up' the substance as soon as it arrives inside the cell. They can do this by immediately starting to metabolise the substance – by chemically changing it to something else – or by using a carrier molecule on the surface of an organelle to take it into the organelle.

Both diffusion and facilitated diffusion rely on a concentration gradient in the right direction to move a substance into the cell. However, cells have another system called active transport that enables them to move substances across membranes against a concentration or electrochemical gradient using energy supplied by the cell.

How does active transport work?

Active transport involves a **carrier protein**, which often spans the whole membrane (see **fig A**). It may be very specific, picking up only one type of ion or molecule, or it may work for several relatively similar substances that have to compete with each other for a place on the carrier.

The energy needed for active transport is provided by molecules of adenosine triphosphate (ATP). You looked at the relationship between the structure of ATP and its functions as the universal energy supplier for biological processes in **Section 1.3.1**. Cells that carry out a lot of active transport generally have many mitochondria to supply the ATP they need. The active transport carrier system in the membrane involves the enzyme ATPase. This enzyme catalyses the hydrolosis of ATP, breaking one bond and forming two more (see **Chapter 3.1**) to provide the energy needed to move carrier system in the membrane or to release the transported substances and return the system to normal.

Active transport is a one-way system for each specific substance – the carriers will not transport a substance back through the membrane. An active transport system moves substances only in the direction required by the cell. In some cases they will move out again through open channels, down the concentration or electrochemical gradient that has just been overcome, but active transport can move substances in faster than they can move out by diffusion.

In active transport the movement of a substance is often linked with that of another particle, such as a sodium ion. One of the best known examples of active transport is the sodium pump that actively moves potassium ions into the cell and sodium ions out. This pump is vital for the working of the nervous system – each nerve impulse depends on an influx of sodium ions through the axon membrane. These ions have to be actively pumped out of the neurone again so that another impulse can pass.

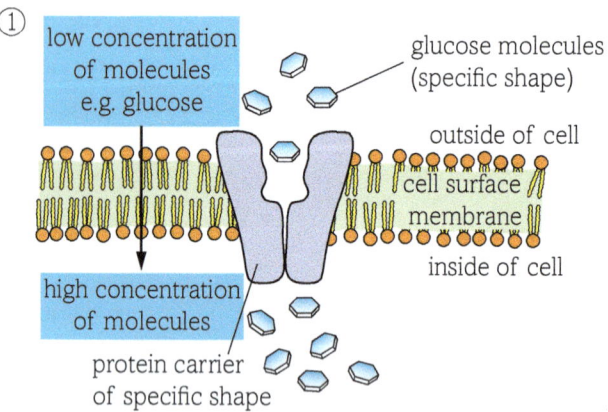

① low concentration of molecules e.g. glucose — glucose molecules (specific shape)

outside of cell

cell surface membrane

inside of cell

high concentration of molecules

protein carrier of specific shape

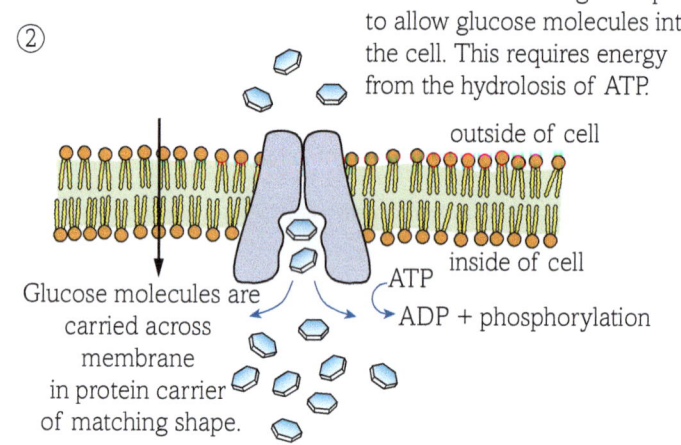

Protein carrier changes shape to allow glucose molecules into the cell. This requires energy from the hydrolosis of ATP.

② outside of cell

inside of cell

ATP

ADP + phosphorylation

Glucose molecules are carried across membrane in protein carrier of matching shape.

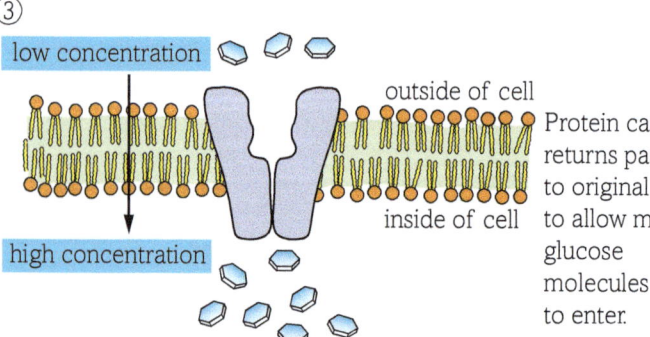

③ low concentration

outside of cell

inside of cell

high concentration

Protein carrier returns passively to original shape to allow more glucose molecules to enter.

fig A Using active transport, cells can move selected substances into or out of the cell, even when the concentration gradient is in the wrong direction.

The combination of diffusion, facilitated diffusion and active transport means that the cell surface membrane provides control over what moves into or out of the cell. The concentration of ions and molecules within the cell can be maintained at very different levels from those of the external fluids. In a similar way the membranes around the organelles and in the cytoplasm provide a range of microenvironments within the cell itself, each suited to different functions, such as the protein-packaging systems in the Golgi body.

Evidence for active transport

Active transport requires energy in the form of ATP produced during cellular respiration. Much of the evidence for active transport comes from linking these two processes together, showing that without ATP active transport cannot take place:

1 Active transport takes place only in living, respiring cells.

2 The rate of active transport depends on temperature and oxygen concentration. These affect the rate of respiration and so the rate of production of ATP.

3 Many cells that are known to carry out a lot of active transport contain very large numbers of mitochondria – the site of aerobic cellular respiration and ATP production.

4 Poisons that stop respiration or prevent ATPase from working also stop active transport. For example, **cyanide** prevents the synthesis of ATP during cellular respiration. It also stops active transport. However, if ATP is added artificially, active transport starts again.

Endocytosis and exocytosis

Diffusion and active transport allow the movement of small particles across membranes. However, there are times when larger particles need to enter or leave a cell, for example when white blood cells ingest bacteria or gland cells secrete large steroid hormones. Membrane transport systems cannot do this job, but the membrane has properties that make it possible to move larger particles into or out of the cell.

Materials can be surrounded by and taken up into membrane-bound vesicles in a process known as endocytosis (see **fig B**). This can occur at a relatively large scale, for example during the ingestion of bacteria during **phagocytosis** (cell eating). It also happens at a microscopic level, when tiny amounts of the surrounding fluid are taken into minute vacuoles. This is known as **pinocytosis** (cell drinking). Electron microscope studies have shown that pinocytosis is very common as cells take in the extracellular fluid as a source of minerals and nutrients. Exocytosis is the term for the emptying of a membrane-bound vesicle at the surface of the cell or elsewhere (see **fig B**). For example, in cells producing hormones, vesicles containing the hormone fuse with the cell surface membrane to release their contents. These processes are made possible by the fluid mosaic nature of the membrane. The formation of vesicles and the fusing of vesicles with the surface cell membrane are both active processes, requiring energy supplied by ATP.

intake of materials – endocytosis
large particles – phagocytosis
liquids – pinocytosis

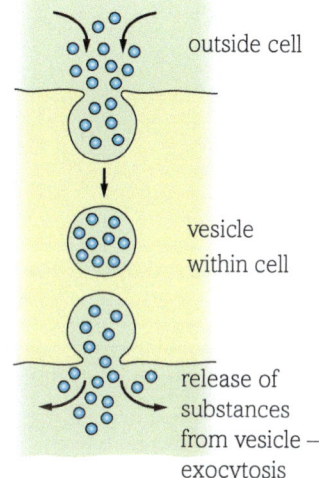

outside cell

vesicle within cell

release of substances from vesicle – exocytosis

fig B The properties of the cell membrane allow cells to take in large particles or release secretions.

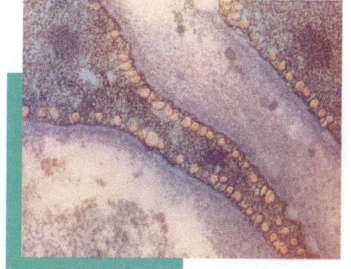

fig C The mass of tiny vesicles along these cell membranes show pinocytosis.

Questions

1 Explain the importance of active transport in cells.

2 Discuss the role of ATP in active transport in the cell.

3 Suggest how endocytosis and exocytosis provide evidence for the fluid mosaic model of membranes.

Key definitions

Carrier proteins are proteins that move a substance through the membrane in active transport – usually linked to an ATPase to break down ATP.

Cyanide is a metabolic poison that stops mitochondria working.

Phagocytosis is the active process when a cell engulfs something relatively large such as a bacterium and encloses it in a vesicle.

Pinocytosis is the active process by which cells take in tiny amounts of extracellular fluid by tiny vesicles.

1 (a) Describe the structure of the cell surface (plasma) membrane. [5]

(b) The ability of a substance to pass into a cell depends on its solubility in oil and water. The oil-water partition coefficient is a measure of the solubility of a substance in oil compared to water. The equation below shows how it is calculated.

$$\text{Oil-water partition coefficient} = \frac{\text{Solubility in oil}}{\text{Solubility in water}}$$

The graph below shows the relationship between membrane permeability and the oil-water partition coefficient for four different substances A, B, C and D.

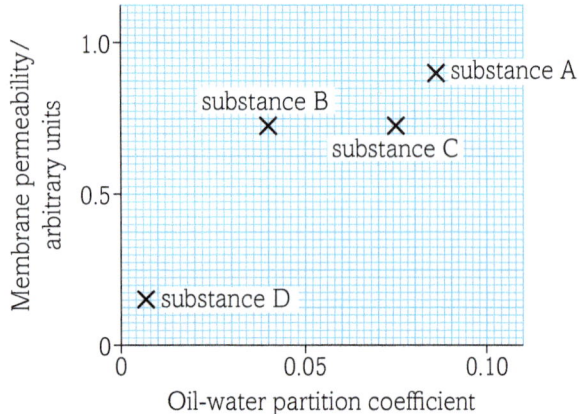

(i) Compare the ability of substances A, B, C and D to cross a cell surface membrane. [3]

(ii) Using the information shown in the graph and your knowledge of the cell surface membrane, suggest how substance A crosses a membrane. Give an explanation for your answer. [3]

[Total: 11]

2 The diagram below represents the structure of the cell surface membrane.

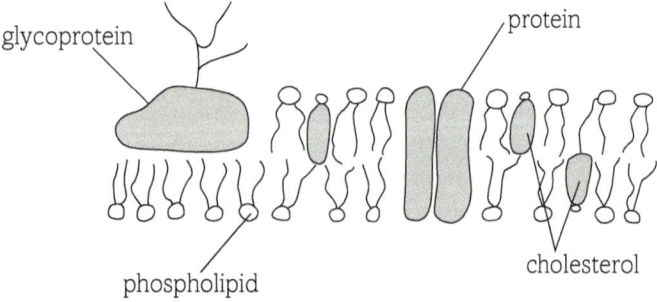

(a) Explain why the phospholipid molecules form a bilayer. [3]

(b) A student carried out an experiment to investigate the effect of alcohol concentration on the permeability of beetroot membranes.

Beetroots are root vegetables that appear red because the vacuoles in their cells contain a water-soluble red pigment. This pigment cannot pass through membranes. Eight pieces of beetroot were cut. One piece of beetroot was placed into a tube containing 15 cm³ of water and left for 15 minutes. The procedure was repeated for seven different concentrations of ethanol. After 15 minutes, each piece of beetroot was removed from the tubes and a sample of the fluid removed and placed in a colorimeter. The colorimeter was used to determine the intensity of red coloration of the fluid.

The results of the investigation are shown in the graph below.

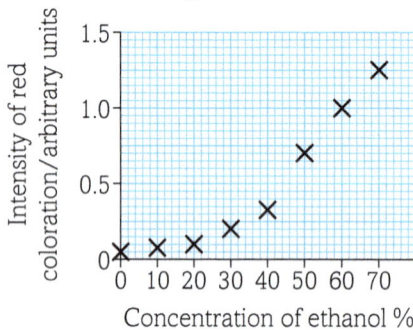

(i) Suggest **two** variables, other than those stated above, which should be kept constant during this experiment. [2]

(ii) There was some red coloration in the tube containing only water. Suggest an explanation for this. [2]

(iii) Describe what the student should have done to reduce the red coloration in the tube containing only water. [1]

(c) The graph above shows that ethanol has an effect on the permeability of beetroot.

(i) State the effect that the ethanol concentration has on the intensity of the red coloration. [1]

(ii) Suggest an explanation for this effect. [2]

[Total: 11]

3 Molecules are transported into and out of cells by several mechanisms.

(a) Read through the following passage that describes some of these mechanisms, then write on the dotted lines the most appropriate word or words to complete the passage. [4]

Some molecules move across a cell surface membrane by passing down a concentration gradient, through the phospholipid bilayer. The movement of some polar molecules across the membrane involves carrier and channel ……………………………… molecules. When this movement occurs down a concentration gradient,

the process is called ....
and when it occurs against a concentration gradient the
process is called ....
Energy in the form of ....
is used in the movement of molecules against a
concentration gradient.

(b) A student wanted to sweeten some strawberries, so she
sprinkled some sugar on top of them, one hour before
eating them. The student noticed that the sugar that she
had sprinkled on them was no longer visible and that there
was some juice at the bottom of the bowl.

The student thought that the juice was the sugar dissolved
in water and that the water had come from the fruit.

In order to test this hypothesis, she weighed some fresh
strawberries and sprinkled them with sugar. One hour later
she rinsed off the juice and reweighed the strawberries. The
mass of the strawberries before adding the sugar was 77 g.

The mass after rinsing off the juice was 70 g.

  (i) Calculate the percentage decrease in the mass of the
  strawberries. Show your working. [2]

  (ii) Suggest **one** possible source of error in the student's
  procedure that could make this value for the percentage
  decrease in the mass of the strawberries inaccurate.
  Explain how this source of error would affect the value
  for the percentage decrease in the mass of the
  strawberries. [3]

  (iii) Using your knowledge of cell transport mechanisms
  and the properties of water, explain how the juice
  is formed from the water that came from the fruit. [3]
  [Total: 12]

4 *Amoeba* is a single-celled aquatic organism. Substances in the
water can enter the cell by a variety of mechanisms.
An experiment was carried out to compare the uptake into
Amoeba of substance A and substance B.
Some of these organisms were placed in a solution containing
equal concentrations of both substances and kept at 25 °C. The
concentration of substances A and B, in the cytoplasm of these
organisms, was measured every 30 minutes over a period of 5 hours.
The results of this experiment are shown in the graph at the
top of the next column.

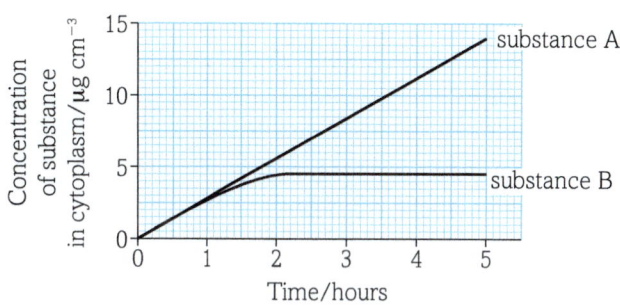

(a) Using the information in the graph, compare the uptake
of substance A with the uptake of substance B during this
period of 5 hours. [3]

(b) Substance B enters the cells by diffusion. Describe and
explain how the results of this experiment support this
statement. [4]

(c) Substance A enters the cells by active transport. Give **two**
differences between active transport and diffusion. [2]
[Total: 9]

5 Substances are moved in and out of cells by diffusion and
active transport.

(a) Explain how molecules move by **diffusion**. [2]

(b) The graph below shows the changes in concentration
of substance A on the inside and outside of a partially
permeable membrane, during a 50 minute period.

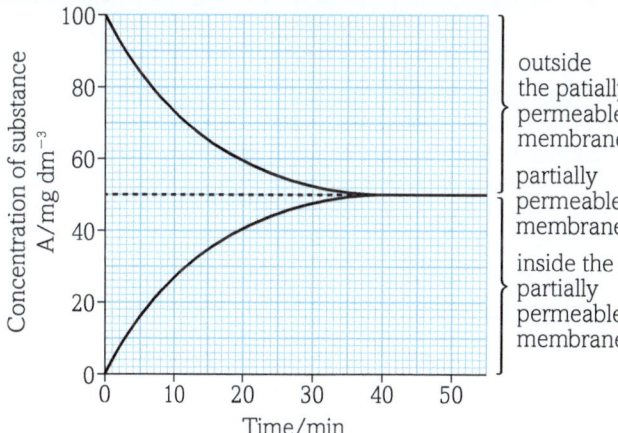

Substance A crosses from one side of this membrane to the
other by diffusion.

Describe how the information given in the graph supports
this statement. [3]

(c) (i) Give **two** differences between diffusion and active
  transport. [2]

  (ii) Suggest how the changes in concentration of
  substance A would differ if active transport were used
  to transfer substance A across the membrane. [2]
  [Total: 9]

TOPIC 4
Exchange and transport

4.2   Gas exchange

Introduction

The gills of fish are perfectly adapted for efficient gas exchange between the blood of the fish and the water – but other animals have evolved some perhaps less elegant, but equally effective, ways to survive underwater. Dragonfly larvae have gills in their rectum – water is pumped in and out of the anus by contractions of the abdominal muscles, giving them a 'jet-propelled' escape mechanism as well. Other insects such as diving beetles carry a bubble of air around with them, covering one or more of their spiracles so they can 'breathe' underwater. The red, worm-like larvae of some midges have haemoglobin in their body fluids – the oxygen-carrying molecule also found in your blood. This allows them to survive, even in relatively polluted water where the oxygen levels are low.

In this chapter you will be considering why organisms need specialised gas exchange surfaces and transport systems as they get bigger and their metabolic rate increases, including calculations of surface area to volume ratios.

All successful gas exchange surfaces have certain features in common. You will be discovering these features and looking for them in the gas exchange systems of mammals (using humans as an example), insects and fish.

Gas exchange is also very important in plants, and you will be looking at how it takes place in the leaves. You will also consider the roles of the stomata and the lenticels in making gas exchange possible in different parts of the plant.

All the maths you need

- Recognise and make use of appropriate units in calculations (*e.g. work out the units for breathing rate*)
- Recognise and use expressions in standard and decimal form (*e.g. use of magnification*)
- Use ratios (*e.g. calculate surface area to volume ratio*)
- Estimate results (*e.g. numbers of stomata in a given area of leaf or lenticels in an area of tree trunk*)
- Find the arithmetic mean of a range of data (*e.g. the mean number of stomata on the leaves of a plant, or lenticels in an area of tree trunk*)
- Calculate or compare the mean, median and mode of a set of data (*e.g. the numbers of stomata or lenticels in different types of plants or different areas of the same plant*)
- Calculate the surface areas and volumes of regular shapes (*e.g. work out the approximate surface area and volume of model organisms*)

What have I studied before?

- How substances are moved into and out of cells by diffusion
- The need for exchange surfaces in multicellular organisms in terms of surface area : volume ratio
- Why oxygen and carbon dioxide need to be moved into and out of organisms

What will I study later?

- How oxygen is carried in the blood from the lungs to the cells in mammals
- How carbon dioxide is carried from the cells to the lungs in the blood of mammals
- How gas exchange in plants is linked to water movements in transpiration
- Why oxygen is needed in cells and how it is used in the biochemistry of cellular respiration, and how carbon dioxide is produced in the same process (A level)
- Why it is important for plants to take in carbon dioxide for the biochemistry of photosynthesis and how oxygen is formed as a waste product of the process (A level)
- The effect of *Mycobacterium tuberculosis* on the tissue of the lungs (A level)
- Control of the breathing rate in mammals (A level)

What will I study in this chapter?

- Why organisms need a mass transport system and specialised gas exchange surfaces as they increase in size and metabolic rate
- Calculating the approximate surface area : volume of a number of model organisms and comparing the diffusion distances
- The main features of a successful gas exchange surface
- The structure and function of the gas exchange system in humans, including the adaptations that make it effective
- The structure and function of the gas exchange system in insects including a dissection of an insect to show the structures involved in gas exchange
- The structure and function of the gas exchange system in fish
- Gas exchange mechanisms in flowering plants, that will include gas exchange in the leaves, the role of the stomata in controlling gas exchange and the importance of the lenticels in allowing gas exchange in other parts of the plant

By the end of this section, you should be able to...

● explain how surface area to volume ratio affects transport of molecules in living organisms

● explain why organisms need specialised gas exchange surfaces as they increase in size

Organisms respire – and for aerobic respiration they need oxygen and produce waste carbon dioxide. They exchange these gases with the environment in which they live. One of the main ways substances move in and out of cells is by diffusion, the free movement of particles in a liquid or a gas down a concentration gradient from an area where they are at a relatively high concentration to an area where they are at a relatively low concentration.

Gas exchange in small organisms

For a single-celled organism, such as an *Amoeba,* and very small multicellular organisms including many marine larvae, the nutrients and oxygen they need can diffuse directly into the cell from the external environment and waste substances can diffuse directly out. This works because:

- The diffusion distances from the outside to the innermost areas are very small.

- The surface area in contact with the outside is very large relative to the volume of the inside of the organism. That is, its **surface area to volume ratio (sa : vol)** is large, so there is a relatively big surface area over which substances can diffuse into or out of the organism (see **fig A**).

- The metabolic demands are low – the organisms do not regulate their own temperature and the cells do not use much oxygen and food or produce much carbon dioxide.

Single-celled organisms and very small multicellular organisms do not need specialised gas exchange or transport systems because diffusion is enough to supply their needs.

Modelling surface area : volume ratios

The surface area to volume ratio of an organism is the key factor that determines whether diffusion alone will allow substances to move into and out of all of the cells. However, it is not easy to calculate the surface area to volume ratio of organisms such as elephants, people and oak trees. It is difficult even for an *Amoeba* because of its irregular shape.

So scientists use models to help show what happens in the real situation. A simple cube makes surface area to volume calculations easy. The bigger the organism gets, the smaller the surface area to volume ratio becomes. The distance from the outside of the organism to the inside gets longer, and there is

proportionately less surface for substances to enter through. So it takes longer for substances to diffuse in, and they may not reach the individual cells quickly enough to supply all their needs.

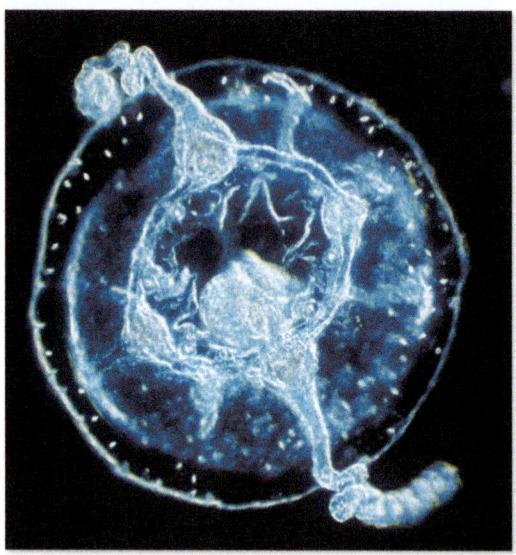

fig A The surface area : volume ratio of this tiny jellyfish larva is relatively large and so simple diffusion can supply all its needs.

 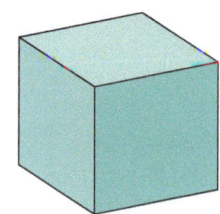

Larger surface area compared to volume.

Smaller surface area compared to volume.

	1 mm cube	2 mm cube	4 mm cube
Surface area	6 sides × 1² = 6 mm²	6 sides × 2² = 24 mm²	6 sides × 4² = 96 mm²
Volume	1³ = 1 mm³	2³ = 8 mm³	4³ = 64 mm³
Surface area to volume ratio	6 : 1	3 : 1	3 : 2

radius

	0.5 µm	1 µm	2 µm
Surface area $4\pi r^2$	3.14 µm²	12.57 µm²	50.27 µm²
Volume $\frac{4}{3}\pi r^3$	0.52 µm³	4.19 µm³	33.51 µm³
Surface area to volume ratio	6 : 1	3 : 1	3 : 2

fig B In this diagram the cubes and spheres represent models of organisms.

Gas exchange in large organisms

In contrast to unicellular organisms such as an *Amoeba*, larger organisms are made up of billions of cells, often organised into specialised tissues and organs. Substances need to travel long distances from the outside to reach the cytoplasm of all the cells. Nutrients and oxygen would eventually reach the inner cells of the body by simple diffusion, but not fast enough to sustain the processes of life.

The metabolic rate of larger organisms, especially larger animals, tends to be higher than that in smaller animals. Mammals and birds, which control their own body temperatures and are very active, have very high metabolic rates. The demands of each individual cell for oxygen and food, and the amount of carbon dioxide and other wastes produced, is much higher in each individual mammal cell than in, for example, an *Amoeba*.

Complex organisms have evolved specialised systems to exchange the gases they need, taking oxygen in and removing carbon dioxide. In humans, gas exchange takes place in the lungs, in fish it is the gills, in insects the tracheal system and in plants most gas exchange takes place in the leaves.

Factors affecting the rate of diffusion

Gas exchange systems are specialised for the exchange of oxygen and carbon dioxide between the body of the organism and the environment. These gases are exchanged by simple diffusion. The rate of diffusion across a membrane is controlled by a number of factors:

- the surface area – the bigger the surface area the more particles can be exchanged at the same time

- the concentration gradient of the particles diffusing – particles diffuse from an area where they are at a relatively high concentration to an area where they are at a relatively low concentration, so the more particles there are on one side of a membrane compared with the other, the faster they move across. Maintaining the gradient, e.g. by transporting substances away once they have diffused, makes diffusion faster

- the distance over which diffusion is taking place – the shorter the diffusion distance the faster diffusion can take place.

Learning tip

When you look at gas exchange systems in different organisms, take note of the adaptations that enable diffusion to take place as fast as possible and see which factors increase the rate of diffusion.

Questions

1 Explain why large animals cannot take in all the substances they need from outside the body through their skin.

2 Here are three facts about gas exchange in humans. Oxygen enters the body and carbon dioxide leaves it through the lungs. The lungs are made of thousands of tiny air sacs surrounded by blood vessels. The surface area of the lungs is approximately $50\,m^2$. Explain how this helps the two gases to diffuse quickly into and out of the blood.

Key definition

Surface area to volume ratio (sa : vol) is the relationship between the surface area of an organism and its volume.

By the end of this section, you should be able to...

● explain how mammals are adapted for gas exchange

For organisms that live on land there is a perpetual conflict between the need for oxygen and the need for water. The conditions that favour the diffusion of oxygen into an organism also favour the diffusion of water out of that organism. This conflict of needs illuminates the way in which many respiratory systems have evolved. Animals and plants need a large, moist surface area for successful gas exchange, and yet they need to limit the water loss from this same surface as much as possible. The evolution of lungs has solved the problem for air-breathing vertebrates. Some lungs are quite simple and merely add a little extra to the area for gas exchange already provided by the body surface – frogs are an example of this. Other animals cannot use their outer surface for gas exchange at all and so are much more dependent on efficient lungs, including all the birds and mammals.

Effective gas exchange

Effective gas exchange systems have a number of features in common:

- A large surface area giving sufficient gaseous exchange to supply all the needs of the organism – it has to compensate for the relatively small surface area : volume ratio of the organism as a whole.

- Thin layers to minimise the diffusion distances from one side to the other.

- In animals, a rich blood supply to the respiratory surfaces. The blood is involved in the transport of the respiratory gases to and from the site of gaseous exchange, helping to maintain a steep concentration gradient.

- Moist surfaces as diffusion takes place with the gases in solution.

- Permeable surfaces that will allow free passage of the respiratory gases.

Mammals, including humans, have very efficient gas exchange systems. They are well adapted to enable the maximum volume of oxygen to move into the body, and the maximum volume of carbon dioxide to be removed. You will be looking at the human gas exchange system as a mammalian example.

The human gas exchange system

Most of the human gas exchange system is found within the chest. It is linked with the outside world through the mouth and nose (see **fig A**). The passages of the nasal cavity have a relatively large surface area, but no gaseous exchange takes place here. The passages have a good blood supply, and the lining secretes mucus and is covered in hairs. This means that the external air is prepared before entering the rest of the system. The hairs and mucus filter out and 'clean up' much of the dust, small particles and pathogens such as bacteria that you breathe in. The moist surfaces increase the level of water vapour in the air and the rich blood supply raises the temperature of the air if this is necessary. This means that the air entering the lungs has as little effect as possible on the internal environment.

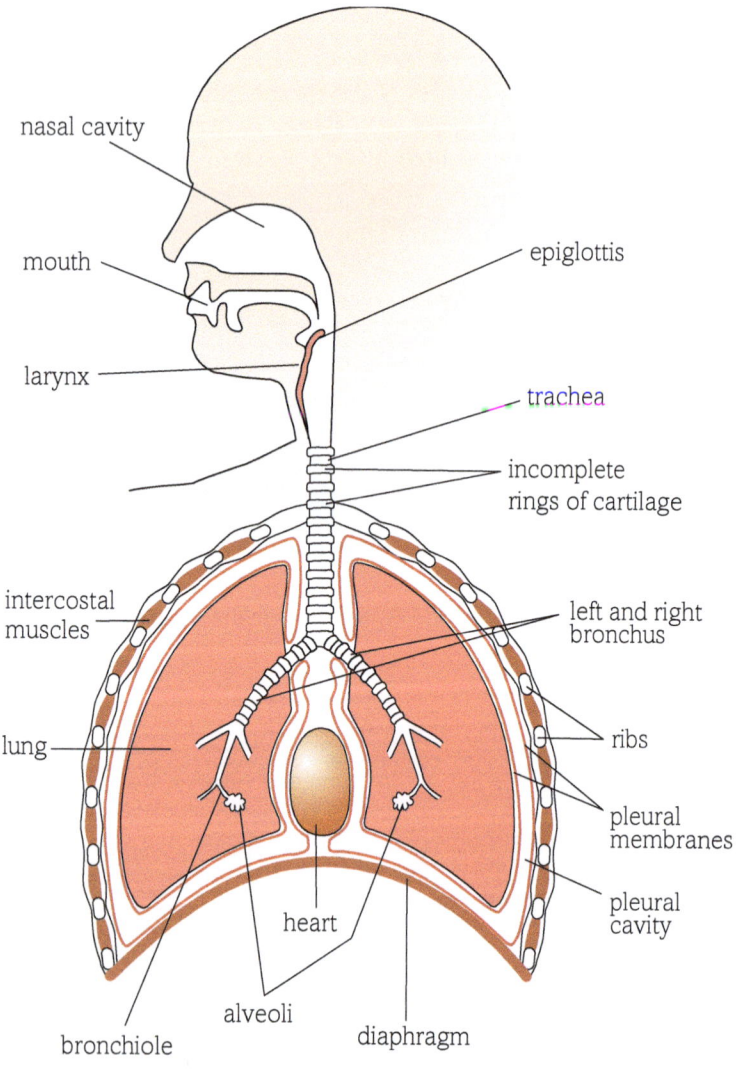

fig A The human respiratory system.

Part of the gas exchange system	Function
Nasal cavity	The main route by which air enters the gas exchange system.
Mouth	Air can enter the respiratory system here, but misses out on the cleaning, warming and moistening effects of the nasal route.
Epiglottis	Flap of tissue that closes over the glottis in a reflex action when food is swallowed. This prevents food from entering the gas exchange system.
Larynx	The voice box, which uses the flow of air across it to produce sounds.
Trachea	Major airway to the bronchi, lined with cells including mucus-secreting cells. Cilia on the surface of the trachea move mucus and any trapped microorganisms and dust away from the lungs.
Incomplete rings of cartilage	Prevent the trachea and bronchi from collapsing but allow food to be swallowed and moved down the oesophagus.
Left and right bronchus	Tubes leading to the lungs are similar in structure to the trachea but narrower, and divide to form bronchioles.
Lung	The organ where gas exchange takes place.
Bronchioles	Small tubes that spread through the lungs and end in alveoli. The larger tubes have cartilage rings but once the diameter is 1 mm or less, there is no cartilage and they collapse quite easily. Their main function is still as an airway but a little gas exchange may occur.
Alveoli	The main site of gas exchange in the lungs.
Ribs	Protective bony cage around the gas exchange system.
Intercostal muscles	Found between the ribs and important in breathing.
Pleural membranes	Surround the lungs and line the chest cavity.
Pleural cavity	Space between the pleural membranes, usually filled with a thin layer of lubricating fluid that allows the membranes to slide easily with breathing movements.
Diaphragm	Broad sheet of tissue that forms the floor of the chest cavity, also important in breathing movements.

Gas exchange in the alveoli

In the lungs most of the gas exchange occurs in tiny air sacs known as alveoli (singular: alveolus) (see **fig B**). An alveolus is made of a single layer of flattened epithelial cells. The capillaries

that run close to the alveoli also have a wall that is only one cell thick. Between the two is a layer of elastic connective tissue holding everything together. The elastic tissue helps to force air out of the lungs, which are stretched when you breathe in. This is known as the elastic recoil of the lungs. The alveoli have a natural tendency to collapse, but this is prevented by a special phospholipid known as **lung surfactant** that coats the alveoli and makes breathing easier.

Gaseous exchange occurs by a process of simple diffusion between the alveolar air and the deoxygenated blood in the capillaries. This blood has a relatively low oxygen content and a relatively high carbon dioxide content.

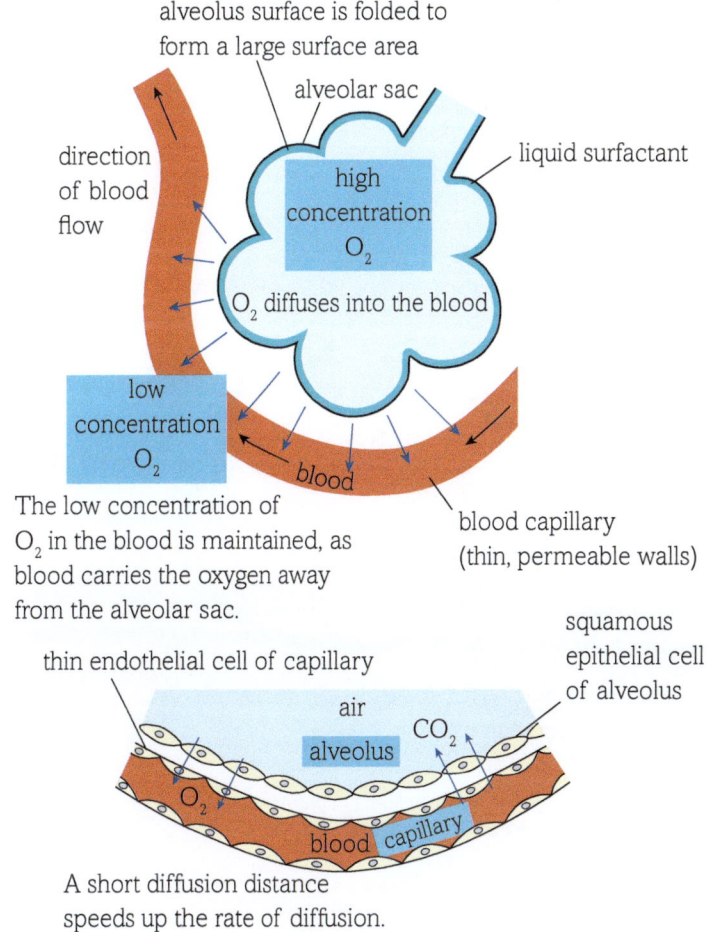

The low concentration of O_2 in the blood is maintained, as blood carries the oxygen away from the alveolar sac.

A short diffusion distance speeds up the rate of diffusion.

fig B The alveoli are the main gas exchange surfaces of the lungs.

Large surface area

The alveoli provide an enormous surface area for the exchange of gases in the human body. Recent calculations have shown that an average adult human has around 480–500 million alveoli in their lungs, which gives a surface area for gas exchange of around 40–75 m^2 packed into your chest – that is the surface area of between 10 and 18 table tennis tables.

Short diffusion distance

The walls of the alveoli are only one cell thick, as are the walls of the capillaries that run beside them. This means the distance that diffusing gases have to travel between them is only around 0.5–1.5 μm (micrometres, microns, 10^{-6} m).

Steep concentration gradient

Blood is continuously flowing through the capillaries past the alveoli, exchanging gases. The continuous flow of the blood maintains the concentration gradient on the capillary side. The air within the alveoli is constantly being refreshed with air from outside by breathing (see **table A**). Movement of gases into and out of the alveoli is mainly by diffusion, but movement of air into and out of the lungs is by a **mass transport system** (see **Section 4.3.1**).

	Percentage of gas in:		
	Inspired air	**Alveolar air**	**Expired air**
oxygen	20.70	13.20	14.50
carbon dioxide	0.04	5.00	3.90
nitrogen	78.00	75.60	75.40
water vapour	1.24	6.20	6.20

table A The composition of the gases in the human gas exchange system.

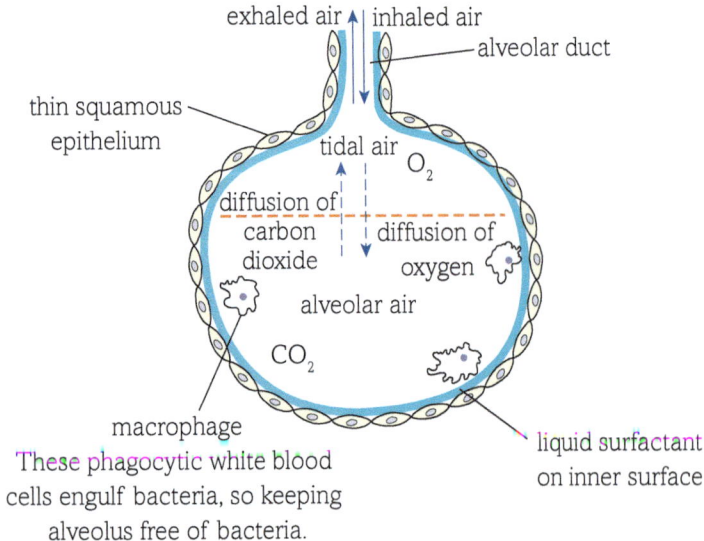

fig C Diffusion across the alveolar surfaces provides the blood with oxygen and disposes of carbon dioxide.

Breathing

Although the exchange of gases at the alveolar surfaces in the lungs happens by passive diffusion alone, moving air between the lungs and the external environment is an active process known as **breathing** or **ventilation**. There are two parts to the process of breathing – taking air into the chest, known as **inhalation** and breathing air out again, called **exhalation**. The chest cavity is effectively a sealed unit for air, with only one way in or out – through the trachea. Breathing involves a series of pressure changes in the chest cavity that in turn bring about movements of the air.

Inhalation is an active, energy-using process. The muscles around the diaphragm contract and as a result it is lowered and flattened. The intercostal muscles between the ribs also contract, raising the rib cage upwards and outwards. These movements result in the volume of the chest cavity increasing, which reduces the pressure

in the cavity. The pressure within the chest cavity is now lower than the pressure of the atmospheric air outside, so air moves in through the trachea, bronchi and bronchioles into the lungs to equalise the pressure inside and out.

Normal exhalation is a passive process. The muscles surrounding the diaphragm relax so that it moves up into its resting domed shape. The intercostal muscles also relax so that the ribs move down and in, and the elastic fibres around the alveoli of the lungs return to their normal length. As a result, the volume of the chest cavity decreases, causing an increase in pressure. The pressure in the chest cavity is now greater than that of the outside air, so air moves out of the lungs, through the bronchioles, bronchi and trachea to the outside air (see **fig D**).

If you need to, you can force air out of your lungs more rapidly than passive exhalation allows. The internal intercostal muscles contract, pulling the ribs down and in, and the abdominal muscles contract forcing the diaphragm upwards. This increases the pressure in the chest cavity, causing exhalation. This is known as forced exhalation. Singers use this to achieve a powerful voice and to maintain long notes, and free divers do it before a dive so they can fill their lungs with as much air as possible. Coughing is an exaggerated form of forced exhalation which is used to force mucus out from the respiratory system.

shape of thorax when breathing in

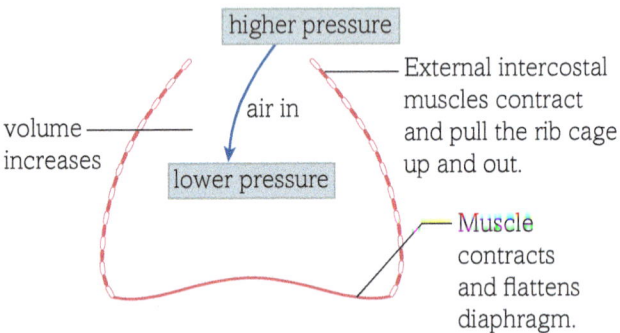

shape of thorax when breathing out

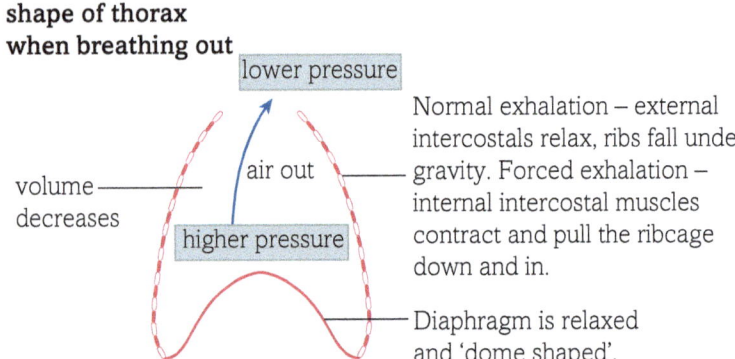

fig D You can feel the movements of your ribs during inhalation and exhalation, but the movements of your diaphragm are less obvious.

Learning tips

Remember that inhalation is active and uses ATP for muscle contraction.

Normal exhalation is passive and does not use energy.

Protecting the lungs

Your gas exchange system carries out the exchange of oxygen and carbon dioxide. As well as gases, the air you breathe in also carries lots of tiny particles such as dust, pollen grains and smoke particles which could block the tiny alveoli. It also carries microscopic organisms like bacteria and viruses. Some of these microscopic organisms are **pathogens** – they cause diseases – and the respiratory system provides a potential route inside the body. To reduce the chances of damage happening to your lungs and of infection, your respiratory system produces lots of mucus that lines your airways and traps these tiny particles and organisms. The mucus is usually very runny, so that it is easily moved up the airways by cilia that sweep upwards to the back of your throat (see **fig E**). Here the majority of the mucus is swallowed without you even noticing it. The acid in your stomach and your digestive enzymes digest the mucus and everything carried with it.

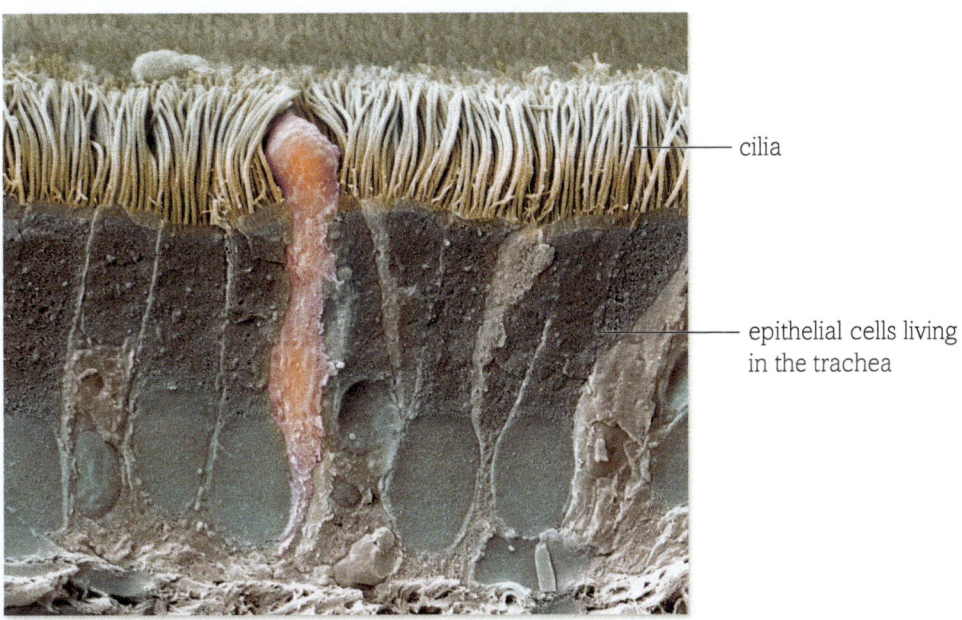

cilia

epithelial cells living in the trachea

fig E The cilia in your trachea and bronchi beat constantly to move mucus with its load of pathogens and dirt out of your gas exchange system.

Questions

1 Explain carefully why humans need a complex internal gas exchange system.

2 Why is breathing through your nose better for your body than breathing through your mouth?

3 Explain why breathing is important in maintaining concentration gradients in the alveoli air.

4 Compare the last two columns in **table A**. Explain the difference in oxygen and carbon dioxide percentages in expired air compared with alveolar air.

Key definitions

Lung surfactant is a special phospholipid that coats the alveoli and makes breathing easier.

A **mass transport system** is an arrangement of structures by which substances are transported in the flow of a fluid with a mechanism for moving it around the body.

Breathing or **ventilation** is the process in which physical movements of the chest change the pressure so that air is moved in or out.

Inhalation is breathing in.

Exhalation is breathing out.

A **pathogen** is a microorganism that causes disease.

● explain how insects are adapted for gas exchange

The human gas exchange system has evolved to cope admirably with the problems of gaseous exchange a large, complex, land-dwelling animal. All mammals and most land vertebrates have developed a similar system. But internal lungs are not the only way to solve these problems. Arthropods, the invertebrate phylum which includes insects, crustaceans, millipedes, centipedes and many others, have a very different gas exchange system that nevertheless enables some of them – such as the flying insects – to live very active lives.

Gas exchange in insects

Insects, like mammals, are complex and largely land-dwelling animals, with relatively high oxygen requirements and an external surface through which little or no gaseous exchange can take place. However, insects have evolved an approach to gas exchange that is very different from other complex animals. The respiratory system of insects has evolved to deliver the oxygen directly to the cells and to remove the carbon dioxide in the same way (see **fig A**).

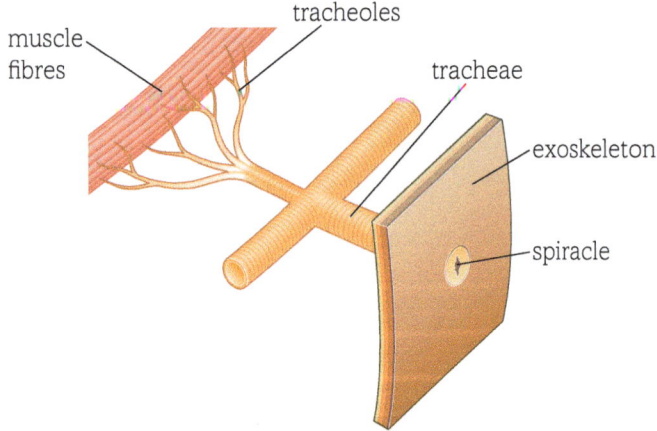

fig A The gas exchange system of an insect has to fulfil the same requirements as the human gas exchange system. In spite of its very different design, there are many strikingly similar features.

The gas exchange system in insects

- **Spiracles**: found along the thorax and abdomen of most insects. They are the site of the entry and exit of the respiratory gases. In many insects the spiracles can be opened or closed by sphincters, which is of great value in the control of water loss.

- **Tracheae** (singular: **trachea**): these are the largest tubes of the insect respiratory system and may be up to 1 mm in diameter. They carry air directly into the body for gas exchange with the cells, running both into the body of the insect and along it (see **fig A**). The tubes are supported by spirals of chitin, the same material that makes up the insect cuticle. The chitin spirals hold the tracheae open if they are squashed or deformed as the insect moves – rather like the rings of cartilage in the mammalian trachea. However, the chitin makes the tracheae relatively impermeable to gases, and so little gas exchange takes place in these vessels. The tracheae divide to form narrower and narrower tubes until they branch out into the tracheoles.

- **Tracheoles**: these are minute tubes of diameter 0.6–0.8 μm. Each one is a single elongated cell and they have no chitin lining. As a result they are freely permeable to gases. The tracheoles spread throughout the tissues of the insect. They are so small they run between and even penetrate into individual cells. It is in the tracheoles that most of the gas exchange takes place in insects.

How does the insect respiratory system work?

Air enters the system through the spiracles, but they are also the major site of water loss from the surface of the insect. To minimise the amount of water lost, the spiracle sphincters are kept closed as much as possible. So, for example, an adult flea which has spiracle sphincters is much more resistant to drying out than a larval flea that has no sphincters.

The spiracles of an inactive insect may all be closed. One or two pairs will open occasionally and this provides enough air for gas exchange. When the insect becomes active, the oxygen demand is higher and more of the spiracles open.

Air moves along the tracheae and tracheoles by diffusion alone and research shows this is sufficient for it to reach all the tissues. The huge network of tiny tracheoles gives a very large surface area and this is where most of the gas exchange takes place in the insect.

The tracheoles may contain water towards the end of their length. This limits the penetration of the gases for diffusion. However, when the insect is very active and needs more oxygen, lactic acid builds up in the muscle tissues. This affects the osmotic concentration of the cells and so water moves out of the tracheoles into the cells by osmosis (see **Section 4.1.1**). This exposes additional surface area in the tracheoles for gaseous exchange.

All of the oxygen needed by the cells is supplied to them by the gas exchange system. However, up to 25% of the carbon dioxide produced by the cells is lost directly through the cuticle.

The opening and closing of the spiracles normally controls the rate of gas exchange in an insect. This is coordinated by respiratory centres in the nervous system, which are stimulated by increasing carbon dioxide levels and by the lactic acid that builds up in active tissues when there is a lack of oxygen. It seems that a combination of the two factors – lack of oxygen and carbon dioxide build-up – work together to provide the insect with a flexible and responsive gas exchange system more than capable of supplying its needs.

Very active insects

Insects with very active lifestyles, including dragonflies, bees and wasps, large beetles, flies, moths and butterflies, have high energy demands. To supply the extra oxygen needed they have evolved ways of ventilating their gas exchange system:

- Mechanical ventilation: air is actively pumped into the tracheal system. The spiracles open and the insect makes muscular pumping movements of the thorax, abdomen or both. These 'ventilating movements' change the volume and therefore the pressure inside the body, drawing air in and out of the tracheae and tracheoles.

- Collapsible tracheae or air sacs that act as air reservoirs: these increase the volume of air moved through the respiratory system. The ventilating movements of the thorax and abdomen inflate and deflate them. In some insects they can be ventilated by the general body movements. For example, when a locust is in flight, automatic ventilation of the air sacs within the muscles plays a vital role in the supply of oxygen to these very active tissues.

fig B Active insects such as this dragonfly need to pump extra air into the body to meet all the energy needs.

Questions

1 How does gas exchange take place in an insect?

2 Suggest three ways in which an insect can increase the amount of gaseous exchange that takes place in its body.

Key definitions

Spiracles are openings along the side of the thorax and abdomen of an insect that are the site of the entry and exit of the respiratory gases. They may be opened or closed by sphincters.

Tracheae (singular: **trachea**) are the largest tubes of the insect respiratory system, carrying air directly into the body for gas exchange with the cells. They run both into and along the body of the insect.

Tracheoles are minute tubes of diameter 0.6–0.8 μm that are the site of gaseous exchange in insects.

Gas exchange in fish

By the end of this section, you should be able to...

● explain how fish are adapted for gas exchange

Water is 1000 times denser than air and 100 times thicker – more viscous. Air is 20.9% oxygen – it contains 209 cm³ of oxygen per litre, but water at the same temperature and pressure is only 0.5% oxygen – around 5.0 cm³ per litre. Lungs would not work as gas exchange organs for water-dwelling animals such as fish, because it would use up enormous amounts of energy to move the water in and out of them. The gas exchange organs of fish are called **gills** and water flows over them in one direction only. This is much more effective and efficient in energy terms for fast-moving active animals living in water.

The gas exchange system in fish

Bony fish, such as cod, salmon and sticklebacks, have a high oxygen demand because they are very active. However, they cannot undergo gas exchange through their scaly external covering because it is not very permeable to gases. They have evolved a gas exchange system that works very well in the high viscosity and slow rate of oxygen diffusion found in the water. Gills have a large surface area for diffusion, a good blood supply to maintain concentration gradients and thin walls giving short diffusion distances, so they are well adapted for successful gaseous exchange. The gills of bony fish are contained in a gill cavity and covered by a protective bony flap called the **operculum**. The operculum is important in maintaining a flow of water over the gills, even when the fish is stationary (see **fig A**).

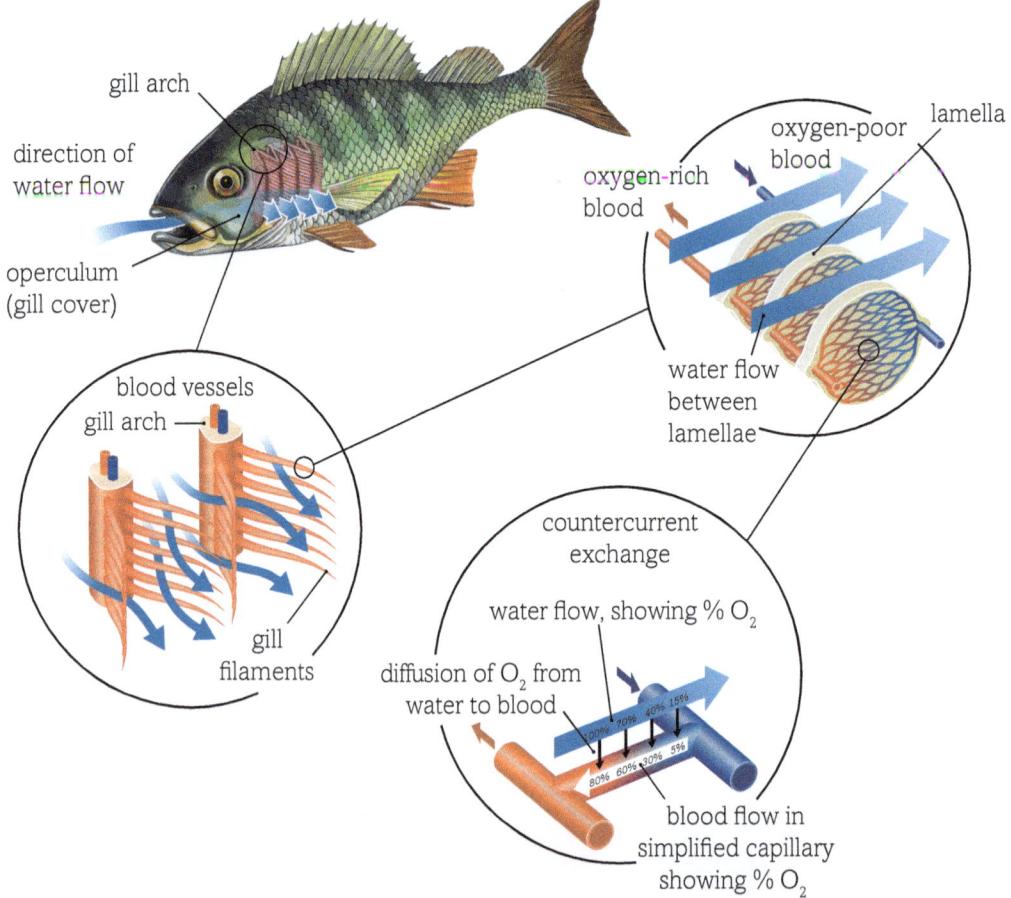

fig A Gills are the gas exchange surfaces of fish. Although they have many common features with the gas exchange surfaces of land animals, they are adapted to their particular environment.

The structure of the gills

The fragile gill filaments occur in large stacks. They need water to keep them apart and so to expose the large surface area needed for gas exchange. The gill lamellae are the main site of gas exchange. They have a very rich blood supply and give the gill filaments their large surface area. Blood leaving the gills flows in the opposite direction to the incoming water, therefore ensuring the most effective possible exchange of gases.

Diffusion in water tends to be slow, so to get the oxygen they need for cellular respiration, fish require a constant flow of water over the gills and the most efficient possible exchange system between the blood and the water. When a fish is out of water it cannot survive long because the gill filaments all stick together. The remaining exposed surface area is not big enough for effective gas exchange to take place.

Ventilating the gills

Sharks and rays, the cartilaginous fish, do not have an operculum. They have to swim all the time to keep the water flowing in through their mouths and out over their gills. Most bony fish have evolved a system using the operculum that means they can ventilate their gills even when they are not moving (see **fig B**).

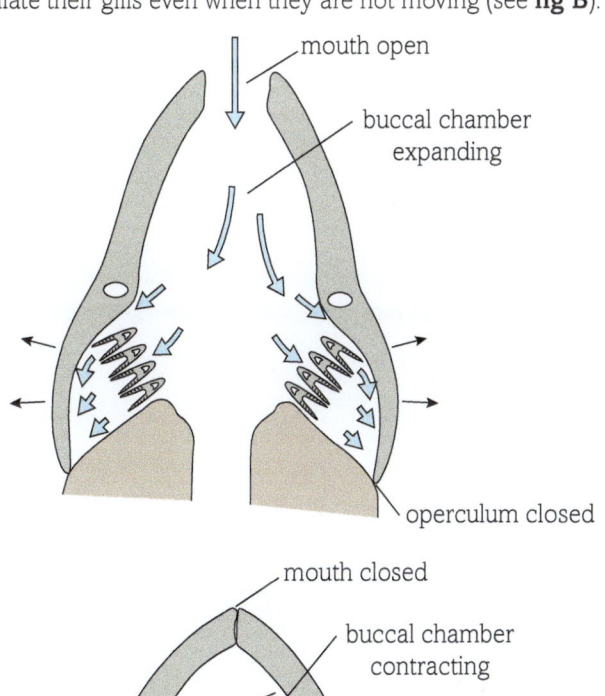

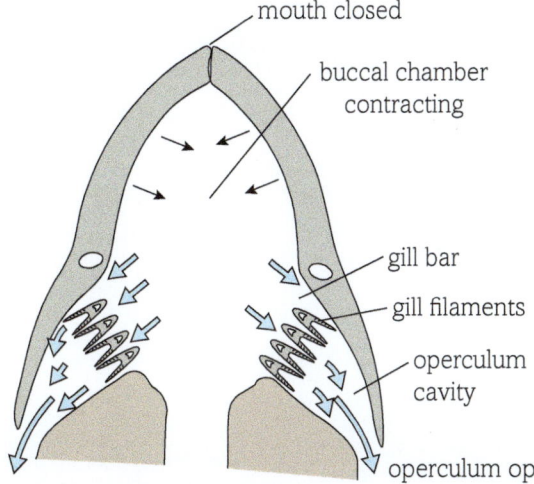

fig B This process ensures there is a continuous flow of water over the gills, even if the fish is not moving.

Maximising gas exchange

Gills have all the usual features of a successful gas exchange – a large surface area, rich blood supply and thin walls. However, the very efficient gas exchange that enables fish to be so active in water is the result of two extra adaptations:

- The blood in the gill filaments and the water moving over the gills flow in different directions. This is called a **countercurrent exchange system**. As you know, diffusion occurs down a concentration gradient and the steeper the concentration gradient, the more effectively diffusion occurs. A countercurrent system maintains steeper concentration gradients than if blood and water flowed in the same direction (see **fig A**). As a result more gas exchange can take place – bony fish extract about 80% of the oxygen from the water flowing over their gills.

- Overlapping gill filaments: diffusion in water tends to be slow, so if the water passes over the gills too quickly it limits the amount of oxygen and carbon dioxide that can be exchanged. The tips of adjacent gill filaments actually overlap, increasing the resistance to the flow of water. This slows down the flow of water over the gill surfaces, giving more time for the exchange of gases to take place.

From these examples it can be seen that gas exchange systems are very highly adapted to enable larger and more complex organisms to carry out gas exchange. Mammalian lungs, insect tracheal systems and the gills of fish are not the only ways of carrying out gas exchange. However, they demonstrate clearly the type of features that must be in place to enable any animal with an unfavourable surface area : volume ratio to obtain the oxygen it needs for life, and remove the waste carbon dioxide it produces.

Questions

1 Which features of the gill system of a fish make it a successful gas exchange system?

2 Make a flow chart to show how bony fish maintain a constant flow of water over their gills.

3 Compare the gas exchange system of a fish with that of an arthropod and a mammal, commenting on both similarities and differences.

Key definitions

Gills are the organs of gas exchange in a fish.

The **operculum** is the bony protective flap that covers the gills of bony fish.

A **countercurrent exchange system** is a system in which two fluid components flow in opposite directions and some properties are exchanged between the two fluids.

By the end of this section, you should be able to...

● describe gas exchange in flowering plants and explain the roles of stomata, the gas exchange surfaces in the leaf and lenticels

Plants need oxygen for respiration and produce waste carbon dioxide. Plant cells respire day and night. However, in plants the situation is complicated because they also photosynthesise, taking in carbon dioxide and producing oxygen, so the net movements of gases in plant tissues vary. During the day the photosynthesising tissues – the green leaves and stems – need to take more carbon dioxide into their cells than they produce by respiration. They also make more oxygen than they use in respiration so they release oxygen into the surrounding air. At night, they simply take in oxygen and release carbon dioxide in respiration. The non-photosynthesising parts of the plant take in oxygen and release carbon dioxide for respiration all the time.

The gas exchange surfaces in plants

The main site of gas exchange in a plant is the leaves.

waxy cuticle
upper epidermis
palisade mesophyll
xylem vessel
phloem tissue
air spaces
lower epidermis
spongy mesophyll
stoma   guard cell

fig A The leaves of a plant are the main site of gas exchange.

Externally the leaves provide a large surface area. This is increased by the gas exchange surfaces of the **spongy mesophyll cells** inside the leaf. These cells have irregular shapes, increasing their surface area, and they are arranged with large air spaces between them. The surfaces of the spongy mesophyll cells are also moist. As a result, gas exchange occurs freely between the cells of the leaf and the air spaces by diffusion. During the day carbon dioxide, required for photosynthesis, moves by diffusion into the cells and oxygen, the main waste product of photosynthesis, moves out. Water also passes by evaporation from the cells into the air spaces.

The impermeable waxy cuticle on the top surface of the leaf acts as a barrier to the diffusion of gases, and particularly the evaporation of water, through the surface of the leaf. A particularly thick waxy cuticle can virtually eliminate evaporation. Yet in spite of this, gases including water vapour move into and out of the leaf, maintaining a concentration gradient so that gas exchange continues within the leaf. Gases move in and out of the leaf itself by diffusion through the **stomata (**singular **stoma)**, specialised pores found mainly in the epidermis on the underside of the leaf (see **figs A** and **B**).

fig B The ability to open and close means that the stomata offer much more than simply a route from the inside to the outside of the leaf. They control gaseous exchange, including water loss.

Controlling gas exchange

There is a constant state of conflict in the needs of a plant. For photosynthesis to occur successfully at its maximum rate, carbon dioxide must move into the leaf and oxygen must move out. To maintain a steady flow of water all the way up from the soil into the aerial parts of the plant involves the evaporation of water from the cells in the leaf (see **Section 4.4.2**). These processes require the stomata to be open. On the other hand, water is often in relatively short supply. In dry climates or in drying conditions, an enormous amount of water would be lost from the leaves if free evaporation were possible – to such an extent that the plant could suffer damage or death. So how is this conflict resolved?

The stomata are not simple pores. Each stoma is bordered by two **guard cells**, specialised examples of epidermal cells. They are sausage-shaped and like other epidermal cells contain a sap vacuole. However, unlike other epidermal cells. they contain chloroplasts and the cellulose of their walls is unevenly distributed.

Opening and closing the stomata

The opening and closing of the stomata is a turgor-driven process. The guard cells respond to carbon dioxide levels in the leaf. When turgor in the guard cells is low, the asymmetric thickening of the cellulose on the cell walls closes the pore. When conditions are favourable for photosynthesis and carbon dioxide is needed by the cells of the leaf, solutes, particularly potassium ions, are moved into the guard cells by active transport. Water then moves into the guard cells from the surrounding epidermal cells by osmosis and the guard cells swell so the turgor pressure increases (see **Section 4.1.3** for details of turgor in plants). The stomatal pores open, again because of the uneven bending resulting from the arrangement of cellulose in the cell walls.

We know that stomatal opening involves active transport because metabolic poisons stop both stomatal opening and the accumulation of potassium ions. When conditions are less favourable for photosynthesis or when it is dark, the active pumping of potassium ions into the cell stops and potassium ions are excreted. As a result, water leaves the cell by osmosis, turgor is reduced and the guard cells become flaccid, closing the stomatal pore.

Although most stomata are found on the underside of leaves, they are also present in stems to allow gas exchange to take place.

Stoma closed

thickened inner wall

thin outer wall

chloroplast

nucleus

small vacuole

guard cell

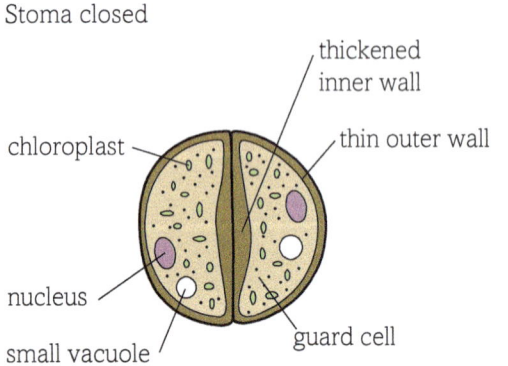

Stoma open (surface view)

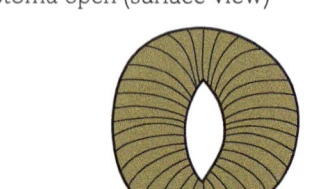

When guard cells become turgid, spiral thickenings of cellulose in the cell walls mean that only the outer walls stretch. The cells become semicircular and a pore opens between them.

Stoma open

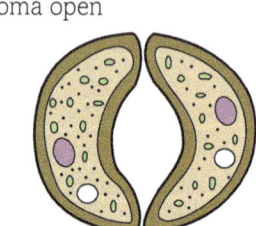

fig C The mechanics of stomatal opening and closing.

Lenticels and gas exchange

The leaves of a plant are not the only tissues that respire and so need to carry out gas exchange. Stomata are found on green stems as well as leaves. However, once plants become thickened and woody, there are no stomata on the surface. When impermeable layers of cork and bark form on a tree, the tissues underneath still need to take up oxygen for cellular respiration and remove carbon dioxide. Special spongy areas called **lenticels** develop, which are made up of loosely arranged cells with many air spaces. They link the inner tissues of the trunk or woody stem with the outside world so gas exchange can take place. Lenticels also form on roots so they can exchange gases with the air in the soil.

(a)

lenticel

(b)

lenticels

fig D Lenticels enable gas exchange to take place in the woody tissues of plants.

Did you know?

Mangroves and lenticels

The term mangrove is used to describe a number of different species of trees that all live in tropical and subtropical coastal regions. Their roots are usually submerged in brackish water and the soil is very low in oxygen. To overcome the lack of oxygen in the soil, red mangroves have evolved specialised prop roots that develop from the lower branches. They only penetrate a few centimetres into the soil and most of the root is above the water. They are covered in lenticels through which they exchange gases. The gases then move into and out of the submerged root tissue by diffusion. The surfaces of the lenticels are so hydrophobic that they keep water out, even when the aerial roots are submerged at high tide.

fig E The typical roots that give mangroves their weird appearance are an adaptation for gas exchange.

Questions

1 Why is gas exchange so important to plants?

2 Summarise the main adaptations for gas exchange in plants.

Key definitions

Spongy mesophyll cells are the cells inside the leaf of a plant where gas exchange takes place.

Stomata (singular **stoma**) are specialised pores found mainly in the epidermis on the underside of the leaf through which gases diffuse into and out of the cell.

Guard cells are the cells that open and close the stomatal pores, controlling the rate of gas exchange.

Lenticels are spongy areas with loosely packed cells that are the site of gas exchange in woody stems and roots.

ASThINKING BIGGER

ASTHMA

When you breathe, air moves easily in and out of your lungs so efficient gas exchange can take place. If someone has asthma, everything changes. In this activity you are going to look at some information about asthma from an educational web resource and consider the information in the light of what you have learned about the human gas exchange system.

Asthma causes the airways of the lungs to narrow so people have difficulty in breathing. People with asthma have over-sensitive airways that become irritated by triggers such as pollen, house dust mites, pet hairs, exercise, smoke or even cold air. Asthma can also be triggered by stress. Someone who has asthma isn't affected all the time. They may have attacks several times a day or only a few times a year.

During an asthma attack the cells lining the bronchioles release chemicals called histamines. The histamines cause the lining cells to become inflamed, produce large amounts of mucus and swell. Histamines also make the muscles in the walls of the bronchioles contract. As a result of all these changes the airways narrow, making it very difficult to move air into and out of the lungs.

air enters the breathing system through the nose and mouth

healthy airways are relaxed and open, with relatively little mucus, so it is easy to inhale and exhale

in patients with asthma, the bronchial muscles contract so the tubes are constricted, the linings of the bronchi and the bronchioles are inflamed and extra mucus is produced, all of which make it more difficult to breathe in and out

inflamed bronchial tube of an asthmatic

normal bronchial tube

fig A The changes in the bronchioles during an asthma attack can make it difficult to breathe.

Measuring the effect of an asthma attack

Doctors can measure the effects of asthma attacks using a spirometer. The patient being tested breathes out as quickly as they can through a mouthpiece. The instrument produces a graph of the amount of air they breathe out and how quickly they do it (the Forced Expiratory Volume in one second or FEV1). A healthy person will be able to exhale over 75% of

their total lung volume in less than a second. Someone affected by asthma or other lung conditions won't be able to do this.

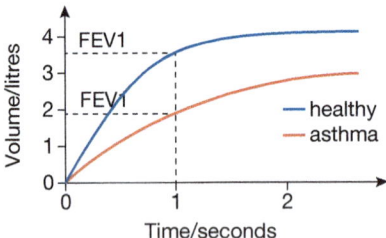

fig B A graph comparing the forced expiratory volume in 1 second (FEV1) of a healthy person and a person with asthma.

People with asthma can also monitor their asthma using a simple instrument called a peak flow meter. This helps them to see how well their breathing system is working and adjust their medication if they need to.

There are two main ways of treating asthma:

1 Relievers are chemical compounds which give immediate relief of the symptoms of asthma. They are used when someone has an asthma attack. Relievers are drugs which are similar to the natural hormone adrenaline. Adrenaline is released in the body when you need to run away or fight. It attaches to active sites in the muscles of your airways and makes the muscles relax. This opens up (dilates) your airways so you can get more air in and out of your lungs. The relievers used by people with asthma work in a very similar way. When they are used during an asthma attack, they relax the bronchial muscles and open up the airways making it much easier to breathe.

2 Preventers are medicines which are taken regularly every day. They reduce the sensitivity of the lining of the airways and so make asthma attacks much less likely. Most preventers are steroids which are taken by inhalers.

If people with asthma use their medication and make sensible choices about their lifestyles, they can often control their asthma well. This means they can enjoy an active, healthy lifestyle with very few asthma attacks.

Where else will I encounter these themes?

1.1　1.2　1.3　1.4　2.1　2.2　2.3　2.4

Let us start by looking at the nature of the writing in this webpage extract:

1. The information here comes from a much larger online resource on breathing and asthma produced by the Association of the British Pharmaceutical Industry (ABPI) to support biology education in UK schools. Aimed at GCSE students, it pushes the levels of knowledge towards the beginning of A level studies. The online resource includes a number of animations including one showing the effects of asthma and asthma-relieving medication on the human gas exchange system.

   a. Compare the advantages and disadvantages of an online resource for delivering biological knowledge.

   b. How can a resource describing asthma help students get a better understanding of the structure and functions of the gas exchange system?

   c. How do you think animations might help develop understanding of processes such as inspiration and expiration, the changes in the gas exchange system during an asthma attack and the impact of asthma-relieving drugs?

You may like to visit the original site of these extracts to help you answer the questions.

Now apply your biological knowledge to this information on the effects of asthma. You should be able to answer all of these questions now, although you may wish to return to the final question if you go on to study hormones at A level:

2. How do the features of an asthma attack as described here interfere with what you know is needed for efficient gas exchange in the lungs?

3. Look at the graph showing the effect of asthma on the forced expiratory volume and use it to answer the following questions:

   a. What is the percentage reduction in the forced expiratory volume of a healthy individual compared to someone affected by asthma:

   i after 1 second　　ii after 2 seconds　　iii after 2.5 seconds?

   b. Using the data to support your decision, evaluate reasons why FEV1 is measured rather than the total volume of air breathed out or the volume of air breathed in when assessing the severity of an asthma attack.

4. In the UK, 1 in 11 children and 1 in 12 adults are affected by asthma. Many more of them take relieving medications rather than preventative medications. Find out more about the action of one common medication used to relieve asthma and one used to prevent asthma on the cells lining the bronchi and bronchioles of the gas exchange system. Produce clear diagrams explaining the action of each at a cellular level, showing how they return the body to normal function and help to maintain normal function respectively.

Command word
An evaluation should review all the information to form a conclusion. You should think about the strengths and weaknesses of the evidence and information and come to a supported judgement.

Activity

Understanding gas exchange

An understanding of gas exchange in organisms as diverse as *Amoeba*, oak trees, dragonflies and human beings is a vital part of biology.

Look through your work on this topic and come up with a single resource that you think would catch the imagination of other AS/A level students and help develop understanding of gas exchange, either in a specific organism or in all organisms. It might be an infographic or a poster – if so, produce it! It could be an animation or an app – if so, produce a story board showing how it would work and the information it will carry. It could be the case study of a particular disease that affects the gas exchange system. If so – develop it. The choice is yours.

● From a website produced by the Association of the British Pharmaceutical Industry (ABPI), www.abpischools.org.uk/page/modules/breathingandasthma

1 The diagram below shows a section through an alveolus and the surrounding tissue.

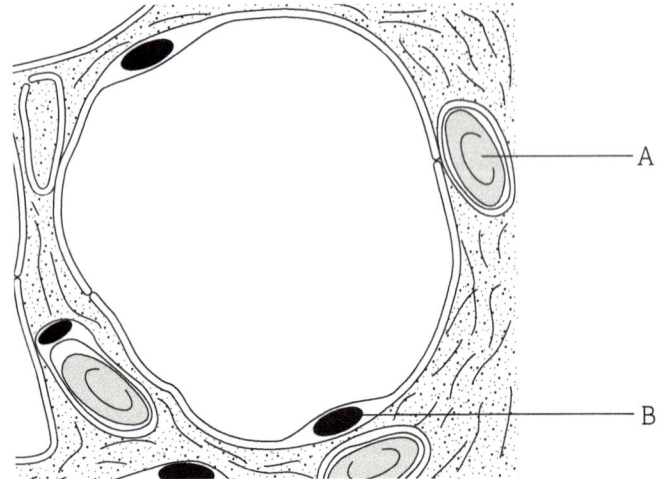

(a) Name the cells labelled A and B. [2]

(b) Describe and explain how alveoli are adapted for the function of gas exchange. [4]

(c) The pulmonary ventilation rate is found by multiplying the tidal volume by the number of breaths taken per minute. Calculate the pulmonary ventilation rate for a person breathing a tidal volume of $0.45\,dm^3$, 18 times per minute. [1]

[Total: 7]

2 A study was carried out into the number of cigarettes smoked by men per year and the number of deaths from lung cancer. The graph below shows the results of this study.

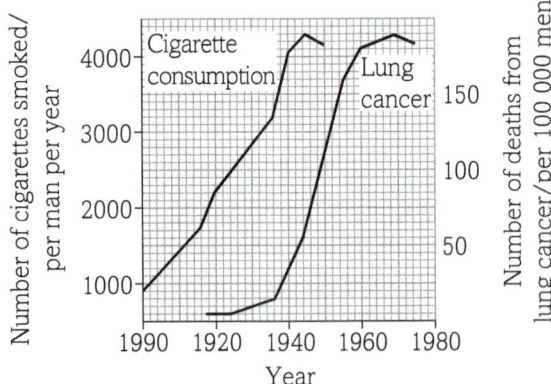

(a) Describe the changes in the number of deaths from lung cancer between 1920 and 1975. [3]

(b) The results of this study indicate that there is a correlation between cigarette smoking and lung cancer.

(i) Describe the meaning of the term **correlation**. [1]

(ii) Describe the evidence shown in this graph that suggests there is a correlation between cigarette smoking and the number of deaths from lung cancer. [2]

(iii) Give **two** additional pieces of information that would increase the validity of any conclusions made from this study. [2]

(c) Emphysema is another lung disease associated with cigarette smoking. One symptom of emphysema is shortness of breath. This is due to the damage to the alveoli and destruction of capillaries surrounding the alveoli. The diagram below show alveoli from a lung of a person with emphysema and some alveoli from a healthy person.

alveoli from a lung of a person with emphysema

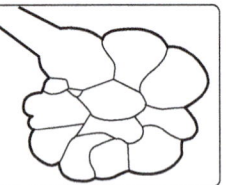

alveoli from a healthy person

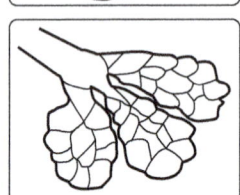

Use your knowledge of the structure of the lung and its adaptations for gas exchange to explain why a person with emphysema has problems with gas exchange. [4]

[Total: 12]

3 (a) Describe a method to determine the mean number of stomata per unit area in the lower epidermis of a leaf. [4]

(b) The table below shows the mean numbers of stomata per mm^2 in the upper epidermis and in the lower epidermis of leaves from three different species of plants.

Type of plant	Mean number of stomata per mm²	
	Upper epidermis	Lower epidermis
Sunflower	120	175
Tobacco	50	190
Castor oil	182	270

(i) Compare the mean numbers of stomata in the upper epidermis of these three plants. [2]

(ii) Using the data in the table, suggest which type of plant is least well-adapted to growing in dry conditions, Give an explanation for your answer. [3]

[Total: 9]

4 (a) Some invertebrates, such as insect larvae, show various adaptations to living in freshwater with a low concentration of dissolved oxygen. Some other invertebrates are adapted to living in fast-flowing water.

   (i) Describe and explain **two** ways in which invertebrates are adapted to living in fresh water with a low concentration of dissolved oxygen [4]

   (ii) Suggest and explain **one** way in which an invertebrate might be adapted to living in fast-flowing water. [2]

(b) An experiment was carried out to investigate the relationship between temperature and the concentration of dissolved oxygen in freshwater. The results are shown in the table below.

Temperature/°C	Concentration of dissolved oxygen/mg dm^{-3}
5	12.8
10	11.3
15	10.2
20	9.2
25	8.2
30	7.5

Describe the relationship between temperature and the concentration of dissolved oxygen, as shown by the data. [2]

[Total: 8]

5 (a) The photographs below show a larva of *Chironomus* and a mosquito larva. These insect larvae are adapted to living in freshwater with very low concentrations of dissolved oxygen.

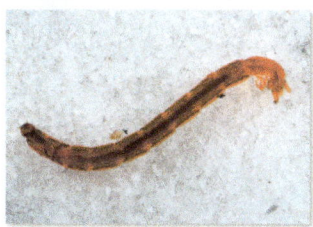 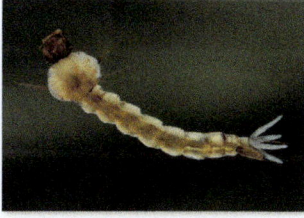

Describe and explain **one** way in which each of these insect larvae is adapted to living in water with a very low concentration of dissolved oxygen. [4]

(b) Some other freshwater invertebrates, such as stonefly nymphs, are adapted to living in fast-flowing water.

Suggest and explain **two** ways in which stonefly nymphs are adapted to living in fast flowing water. [4]

[Total: 8]

6 The picture below shows a number of stomata from the leaf of *Lilium sp.*

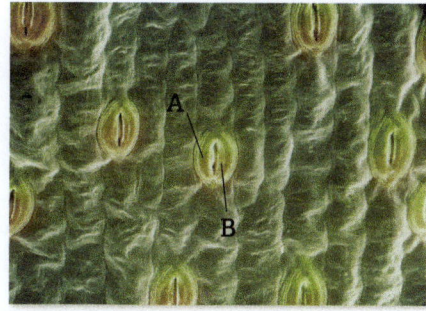

(a) Name the cell labelled **A**. [1]

(b) Name the organelle labelled **B**. [1]

(c) Outline the processes that cause the stomata to open. [4]

(d) An investigation was carried out into the transpiration rate of plants under a number of conditions. The results are shown below. For each treatment, suggest the effect on transpiration and explain your reasoning. [6]

Treatment of plant	Effect on transpiration
Darkness	
Addition of phenylmercuric acetate (a metabolic poison)	
Plant left without water for ten days	

[Total: 12]

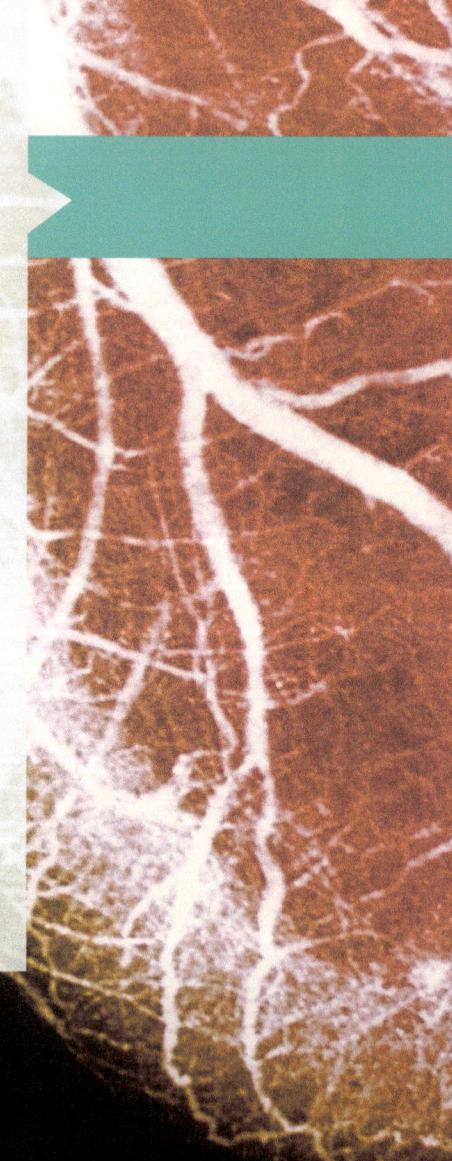

TOPIC 4
Exchange and transport

CHAPTER
4.3 › Circulation

Introduction

If your car breaks down, mechanics can replace worn-out parts, put in new oil and transmission fluid, change perished or worn-out pipes... but we do not expect doctors to be able to do the same for our bodies. Of course they cannot – but a lot can be done to replace or repair the various parts of the circulatory system. The heart can have new valves, new blood vessels to supply the muscle and can even be replaced in a transplant. The blood vessels can be opened up, unblocked or replaced with grafts from other healthy areas of the body. Blood can be replaced by transfusions, and even the bone marrow that makes the blood cells can be replaced by transplants. Doctors have even developed techniques by which they can operate on the circulatory system of a fetus in the uterus, to give blood transfusions or even repair some heart conditions long before birth.

In this chapter you will be looking at mammalian transport systems. This involves studying the general principles of circulatory systems and the details of the human blood, blood vessels and heart. The heart is a complex organ and you will learn not only the events of the heartbeat but also how the heartbeat is controlled from within the heart itself, and how to interpret ECG traces of the heart.

You will consider the transport of gases in the blood, and the role of the oxygen-carrying pigments including haemoglobin in adults and the different pigments found in the developing fetus and in the muscles. You will also be looking at the way materials are transferred between the circulatory system and the cells of the body. This includes understanding the formation of tissue fluid and the differences between tissue fluid and lymph.

All the maths you need

- Recognise and make use of appropriate units in calculations (*e.g. work out the unit for the heart rate*)
- Find the mean for a range of data (*e.g. measuring mean heart rate*)
- Interpret data for a variety of graphs (*e.g. explain electrocardiogram (ECG) traces*)
- Substitute numerical values into algebraic equations using appropriate units for physical quantities (*e.g. water potential calculations*)
- Solve algebraic equations in a biological context (*e.g. cardiac output + stroke volume × heart rate, water potential equations*)
- Calculate the surface areas and volumes of regular shapes (*e.g. work out the approximate surface area and volume ratio of a single cell*)
- Use ratios (*e.g. calculate surface area to volume ratio of a single cell*)

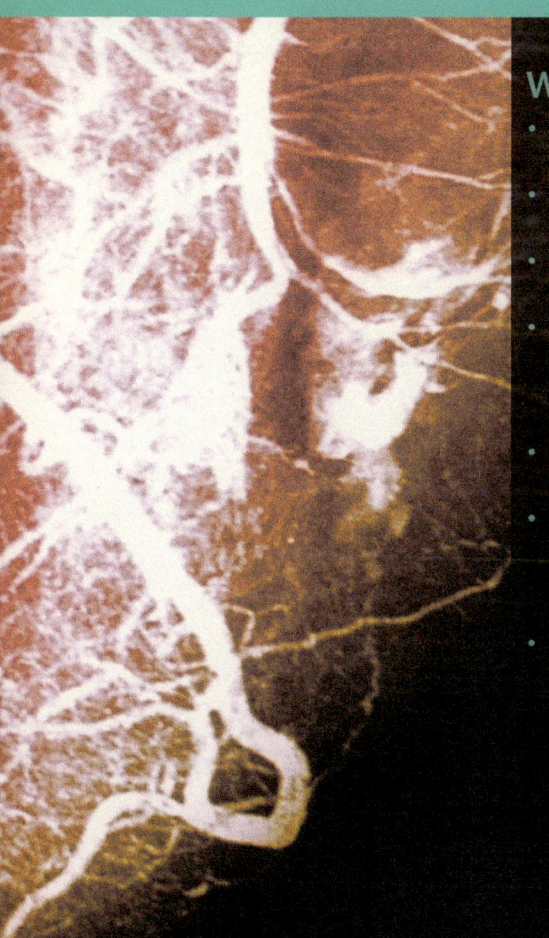

What have I studied before?

- The human circulatory system and how it is related to the gas exchange system
- How the structure of the heart and the blood vessels are adapted to their functions
- How red blood cells, white blood cells, platelets and plasma are adapted to their functions in the blood
- The movement of oxygen and carbon dioxide between the air in the lungs and the blood
- The need for specific transport systems as organisms get larger and have higher metabolic rates relating to the surface area : volume ratio

What will I study later?

- Transport systems in plants (A level)
- The link between the oxygen, dissolved food molecules and carbon dioxide transported in the blood and the events of cellular respiration (A level)
- The links between the sucrose and dissolved mineral ions transported with water around a plant and the events of photosynthesis and cellular respiration (A level)
- The effect of hormones such as adrenaline on the circulatory system (A level)
- The movement of substances out of and into the blood in the nephrons of the kidney (A level)
- The control of the heart rate by the nervous system (A level)

What will I study in this chapter?

- The advantages of a double circulation in mammals over the single circulatory system in fish
- The structure of the heart, arteries, veins and capillaries related to their functions
- The sequence of events of the cardiac cycle and the myogenic stimulation of the heart
- The structure of the blood including the different cells, plasma, plasma proteins and platelets and functions including transport, defence and the formation of lymph and tissue fluid
- The role of platelets and plasma proteins in blood clotting
- The transport of oxygen in the blood, the roles of haemoglobin, fetal haemoglobin and myoglobin and the effect of carbon dioxide concentrations on the carriage of oxygen
- The transfer of materials between the circulatory system and the cells, involving the formation of tissue fluid as a result of oncotic and hydrostatic pressure, and the formation of lymph from the tissue fluid that is not reabsorbed into the blood

By the end of this section, you should be able to...

● explain why organisms need a mass transport system

● explain the advantages of a double circulatory system in mammals over the single circulatory systems in in bony fish, including the facility for the blood to be pumped to the body at higher pressure and the splitting of the oxygenated and deoxygenated blood

Within any organism, substances need to be moved from one place to another. In single-celled organisms and microscopic multicellular organisms, diffusion is sufficient to supply all their needs. However, as you saw in **Section 4.2.1**, the surface area : volume ratio is key. When organisms reach a certain size, diffusion alone is not enough.

The need for transport

Large multicellular organisms have internal transport systems that carry substances to every cell in the body, delivering oxygen and nutrients and removing waste so that cells can carry out their functions efficiently. In large complex organisms such as humans, chemicals made in a cell in one part of the body such as a hormone like insulin or adrenaline may have an effect on a different type of cell elsewhere in the body. So substances made internally need to be moved around the body as well.

In many animals, including all vertebrates, this transport system is the heart and circulatory system and the fluid that flows through it. This is an example of a mass transport system – substances are transported in the flow of a fluid with a mechanism for moving it around the body. All large complex organisms have some form of mass transport system. Substances are delivered over short distances from the mass transport system to individual cells deep in the body by processes such as diffusion, osmosis and active transport (see **Chapter 4.1**).

Features of mass transport systems

Mass transport systems are very effective for moving substances around the body. Most mass transport systems have certain features in common. They have:

• a system of vessels that carry substances – these are usually tubes, sometimes following a very specific route, sometimes widespread and branching

• a way of making sure that substances are moved in the right direction, e.g. nutrients in and waste out

• a means of moving materials fast enough to supply the needs of the organism – this may involve mechanical methods such as the pumping of the heart or ways of maintaining a concentration gradient so that substances move quickly from one place to another, e.g. using active transport

• a suitable transport medium.

Circulation systems

Many animals have a circulatory system in which a heart pumps blood around the body. Insects have an open circulatory system, with the blood circulating in large open spaces. However, most larger animals including the mammals have a closed circulatory system with the blood contained within tubes. The blood makes a continuous journey out to the most distant parts of the body and back to the heart.

Animals such as fish have a **single circulation system**. The heart pumps deoxygenated blood to the gills, the organs of gas exchange where the blood takes in oxygen and becomes oxygenated, giving up carbon dioxide at the same time. The blood then travels on around the rest of the body of the fish, giving up oxygen to the body cells before returning to the heart.

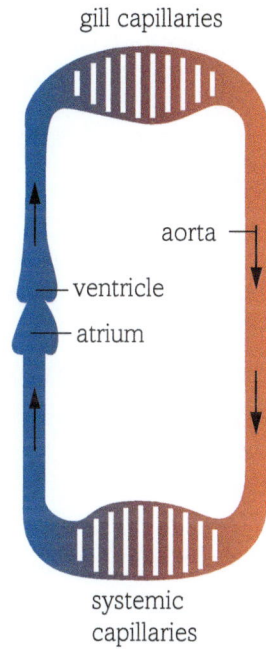

fig A The single circulation of a fish.

Birds and mammals need far more oxygen than fish. Not only do they have to move around without the support of water, but they also maintain a constant body temperature that is usually higher than their surroundings. This takes a lot of resources, so their cells need plenty of oxygen and glucose and produce a lot of waste products that need to be removed quickly. Birds and mammals have evolved the most complex type of transport system, known as a **double circulation** because it involves two circulatory systems. The **systemic circulation** carries **oxygenated**, also known as oxygen-rich, blood from the heart to the cells of the body where the oxygen is used, and carries the **deoxygenated** blood, blood that has given up its oxygen to the body cells, back to the heart. The **pulmonary circulation** carries deoxygenated blood from the heart to the lungs to be oxygenated, and carries the oxygenated blood back to the heart (see **fig B**).

The separate circuits of a double circulatory system make sure that the oxygenated and deoxygenated blood cannot mix, so the tissues receive as much oxygen as possible. Another big advantage is that the fully oxygenated blood can be delivered quickly to the body tissues at high pressure. The blood going through the tiny blood vessels in the lungs is at relatively low pressure so it does not damage the vessels and allows gas exchange to take place. If this oxygenated blood at low pressure went straight into the big vessels that carry it around the body it would move very slowly. Because it returns to the heart, the oxygenated blood can be pumped hard and sent around the body at high pressure. This means it reaches all the tiny capillaries between the body cells quickly, supplying oxygen for an active way of life.

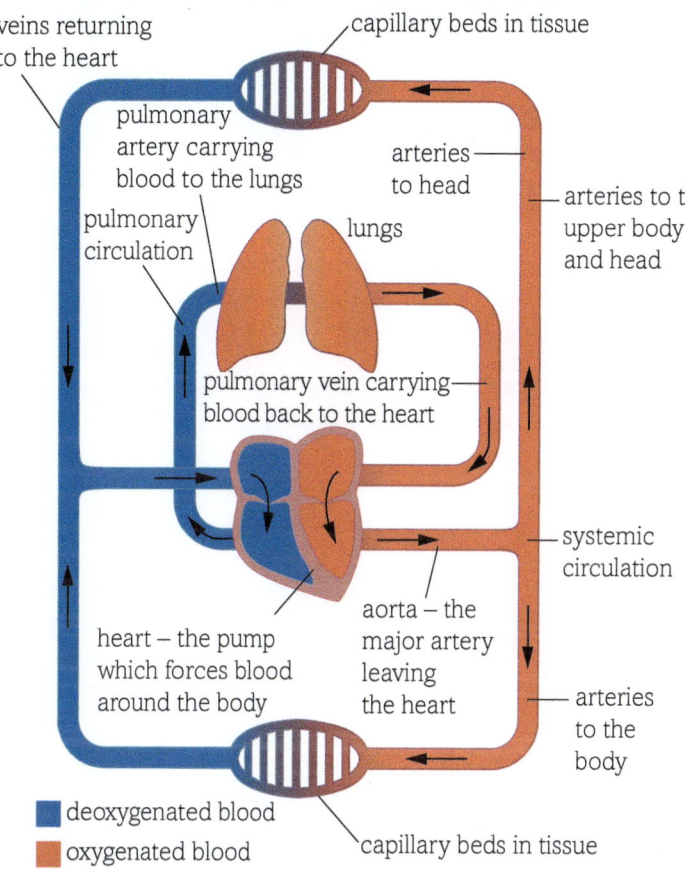

fig B A double circulation sends blood at high pressure, carrying lots of oxygen to the active cells of the body.

Questions

1 What are the main characteristics of a mass transport system?

2 In fish, the blood cannot be supplied to the body tissues at high pressure. Why not – and why does this not matter?

3 Why is a double circulation ideal for an active animal that maintains its own body temperature independently of the environment?

Key definitions

Single circulation system is a circulation in which the heart pumps the blood to the organs of gas exchange and the blood then travels on around the body before returning to the heart.

Double circulation system is a circulation that involves two circulatory systems, one of deoxygenated blood flowing from the heart to the gas exchange organs and back oxygenated to the heart, and one of oxygenated blood leaving the heart and flowing around the body, returning deoxygenated to the heart.

The **systemic circulation** carries oxygenated blood from the heart to the cells of the body where the oxygen is used, and carries the deoxygenated blood back to the heart.

Oxygenated blood is blood that is carrying oxygen.

Deoxygenated blood is blood that has given up its oxygen to the body's cells.

The **pulmonary circulation** carries deoxygenated blood to the lungs and oxygenated blood back to the heart.

By the end of this section, you should be able to...

- describe the structure of blood as plasma and blood cells, including erythrocytes and leucocytes (neutrophils, eosinophils, monocytes and lymphocytes)
- describe the function of blood as transport, defence, and formation of lymph and tissue fluid

In mammals the mass transport system is the **cardiovascular system**. This is made up of a series of vessels with the heart as a pump to move blood through the vessels. The blood is the transport medium and its passage through the vessels is called the **circulation**. The system delivers the materials needed by the cells of the body, and carries away the waste products of their metabolism.

It also carries out other functions, such as:

- carrying hormones (chemical messages) from one part of the body to another
- forming part of the defence system of the body
- distributing heat.

You are going to study all three parts of the cardiovascular system, starting with the transport medium – blood.

The components of the blood and their main functions

Your blood is a complex mixture carrying a wide variety of cells and substances to all areas of your body (see **fig A**).

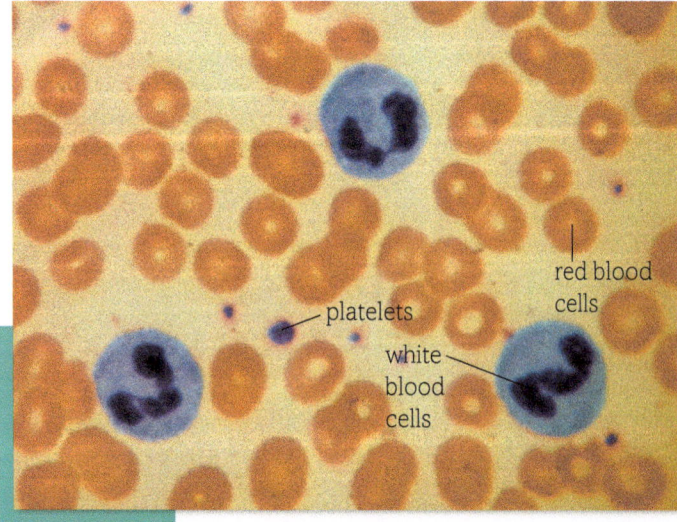

fig A This light micrograph shows red blood cells, white blood cells and platelets.

Plasma

Your blood is a complex mixture carrying a wide variety of cells and substances to all areas of your body. Blood carries out a wide variety of functions, but they fall into three main categories – transport, defence and the formation of tissue fluid and lymph.

Each of the different components of the blood has particular functions.

- digested food products (e.g. glucose and amino acids) from the small intestine to all the parts of the body where they are needed either for immediate use or storage
- nutrient molecules from storage areas to the cells that need them
- excretory products (e.g. carbon dioxide and urea) from cells to the organs such as the lungs or kidneys that excrete them from the body
- chemical messages (hormones) from where they are made to where they cause changes in the body.

The plasma also helps to maintain a steady body temperature by transfering heat around the system from deep-seated organs (e.g. the gut) or very active tissues (e.g. leg muscles in someone running). It acts as a buffer to pH changes.

Substances move between the plasma or red blood cells and the body cells by diffusion or active transport.

Erythrocytes

There are approximately 5 million erythrocytes per mm^3 of blood (4–5 million per mm^3 in women, 5–6 million per mm^3 in men). They contain **haemoglobin**, a red pigment that carries oxygen and gives them their colour. They are formed in the bone marrow. Mature erythrocytes do not contain a nucleus and have a limited life of about 120 days.

The erythrocytes transport oxygen from the lungs to all the cells (see **Section 4.3.3**). They are well adapted for their function. The biconcave disc shape of the cells means that they have a large surface area to volume ratio, so oxygen can diffuse into and out of them rapidly. Having no nucleus leaves much more space inside the cells for the haemoglobin molecules that carry the oxygen. In fact, each red blood cell contains around 250–300 million molecules of haemoglobin and can carry approximately 1000 million molecules of oxygen. Haemoglobin also carries some of the carbon dioxide produced in respiration back to the lungs. The rest is transported in the plasma.

Leucocytes

Leucocytes are much larger than erythrocytes, but can also squeeze through tiny blood vessels as they can change their shape. There are around 4000–11 000 per mm^3 of blood and there are several different types. They are formed in the bone marrow although some mature in the thymus gland. Their main function is to defend the body against infection. They all contain a nucleus and have colourless cytoplasm, although some types contain granules that can be stained. There are a number of different types of erythrocytes, which you will study further in **Book 2 Chapter 6.3**.

- **Granulocytes:** These leucocytes have granules in the cytoplasm of the cells that take up stain and are obvious under the microscope. They have lobed nuclei. Granulocytes include:

    Neutrophils – part of the non-specific immune system, they engulf and digest pathogens by phagocytosis. They have multi-lobed nuclei. Up to 70% of all leucocytes are neutrophils.

    Eosinophils – part of the non-specific immune system. They are stained red by eosin stain. They are important in the non-specific immune response of the body against parasites, in allergic reactions and inflammation, and in developing immunity to disease.

    Basophils – part of the non-specific immune system. They have a two-lobed nucleus. They produce histamines involved in inflammation and allergic reactions.

- **Agranulocytes:** These leucocytes do not have granules to take up stain in their cytoplasm. They have unlobed nuclei. Agranulocytes include:

    Monocytes – part of the specific immune system. They are largest of the leucocytes. They can move out of the blood into the tissues to form macrophages, that also play an important part in the specific immune system. They engulf pathogens by phagocytosis.

    Lymphocytes – small leucocytes with very large nuclei that are vitally important in the specific immune response of the body.

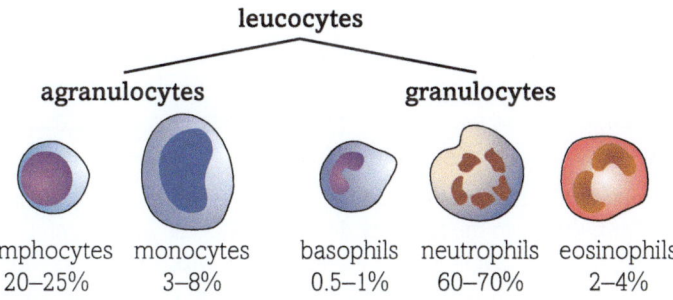

leucocytes

agranulocytes   **granulocytes**

lymphocytes   monocytes   basophils   neutrophils   eosinophils
20–25%   3–8%   0.5–1%   60–70%   2–4%

fig B The main types of leucocytes in human blood.

Platelets

Platelets are tiny fragments of large cells called **megakaryocytes**, which are found in the bone marrow. There are about 150 000–400 000 platelets per mm^3 of blood. They are involved in the clotting of the blood (see **Section 4.3.4**).

Substances move between the plasma or red blood cells and the body cells by diffusion or active transport. The tiniest blood vessels have walls only one cell thick so diffusion distances are short and substances pass easily across these into other cells. Every cell in the body is close to one of these small vessels.

Questions

1 What are the main functions of human blood?

2 Explain the role of the blood transport system in the defence against infection.

3 Red blood cells are unusual in not having a nucleus. Explain how this is an adaptation for their role in carrying oxygen, and why they have a limited life.

Key definitions

The **cardiovascular system** is the mass transport system of the body made up of a series of vessels with a pump (the heart) to move blood through the vessels.

Circulation is the passage of blood through the blood vessels.

Haemoglobin is a red pigment that carries oxygen and gives the erythrocytes their colour.

Granulocytes are leucocytes that have granules in the cytoplasm of the cells that take up stain and are obvious under the microscope. They have lobed nuclei and include neutrophils, eosinophils and basophils.

Neutrophils are part of the non-specific immune system. They have multi-lobed nuclei and engulf and digest pathogens by phagocytosis. Up to 70% of all leucocytes are neutrophils.

Eosinophils are part of the non-specific immune system. They are stained red by eosin stain and are important in the response of the body against parasites, in allergic reactions and inflammation, and in developing immunity to disease.

Basophils are part of the non-specific immune system. They have a two-lobed nucleus and produce histamines in inflammation and allergic reactions.

Agranulocytes are leucocytes that do not have granules to take up stain in their cytoplasm. They have unlobed nuclei and include monocytes and lymphocytes.

Monocytes are part of the specific immune system. They are the largest of the leucocytes and they can move out of the blood to form macrophages. They engulf pathogens by phagocytosis.

Lymphocytes are small leucocytes with very large nuclei that are vitally important in the specific immune response of the body.

Platelets are involved in the clotting mechanism of the blood.

Megakaryocytes are large cells that are found in the bone marrow and produce platelets.

Transporting oxygen and carbon dioxide

By the end of this section, you should be able to...

- explain the role of platelets and plasma proteins in the sequence of events leading to blood clotting
- explain the structure of haemoglobin in relation to its role in the transport of respiratory gases, including the oxygen dissociation curve of haemoglobin and the Bohr effect
- explain the similarities and differences between the structures and functions of haemoglobin and myoglobin
- explain the significance of the oxygen affinity of fetal haemoglobin as compared to adult haemoglobin
- explain the role of platelets and plasma proteins in the sequence of events leading to blood clotting including:
  - platelets form a plug and release clotting factors, including thromboplastin
  - prothrombin changes to its active form, thrombin
  - soluble fibrinogen forms insoluble fibrin to cover the wound

As you have seen, the erythrocytes in the blood are adapted for transporting oxygen. The blood also carries away the carbon dioxide produced during respiration by cells.

Transport of oxygen

The haemoglobin molecules that are packed in the red blood cells transport oxygen. Each haemoglobin molecule is a large globular protein made up of four peptide chains, each with an iron-containing prosthetic group, which can pick up four molecules of oxygen in a reversible reaction to form **oxyhaemoglobin**:

$$Hb + 4O_2 \rightleftharpoons Hb_4O_2$$
$$\text{haemoglobin} + \text{oxygen} \rightleftharpoons \text{oxyhaemoglobin}$$

The first oxygen molecule that binds to the haemoglobin alters the arrangement of the molecule making it easier for the following oxygen molecules to bind. The final oxygen molecule binds several hundred times faster than the first. The same process happens in reverse when oxygen dissociates from haemoglobin – it gets progressively harder to remove the oxygen.

The concentration of oxygen in the red blood cells when the blood enters the lungs is relatively low. Oxygen moves into the red blood cells from the air in the lungs by diffusion. Because the oxygen is picked up and bound to the haemoglobin, the free oxygen concentration in the cytoplasm of the red blood cells stays low. This maintains a steep concentration gradient from the air in the lungs to the red blood cells, so more and more oxygen diffuses in and is loaded onto the haemoglobin.

In the body tissues the oxygen levels are relatively low. The concentration of oxygen in the cytoplasm of the red blood cells is higher than in the surrounding tissue. As a result oxygen moves out into the body cells by diffusion down its concentration gradient. The haemoglobin molecules give up some of their oxygen. When you are at rest or exercising gently, only about 25% of the oxygen carried by the haemoglobin is released into your cells. There is another 75% reserve in the transport system for when you are very active.

As a result of the strong affinity of haemoglobin for oxygen, a small change in the proportion of oxygen in the surrounding air can have a big effect on the saturation of the blood with oxygen. So in the lungs the haemoglobin loads up rapidly with oxygen – and in the tissues, as the oxygen saturation of the environment falls, oxygen is released rapidly as well (see **fig A**).

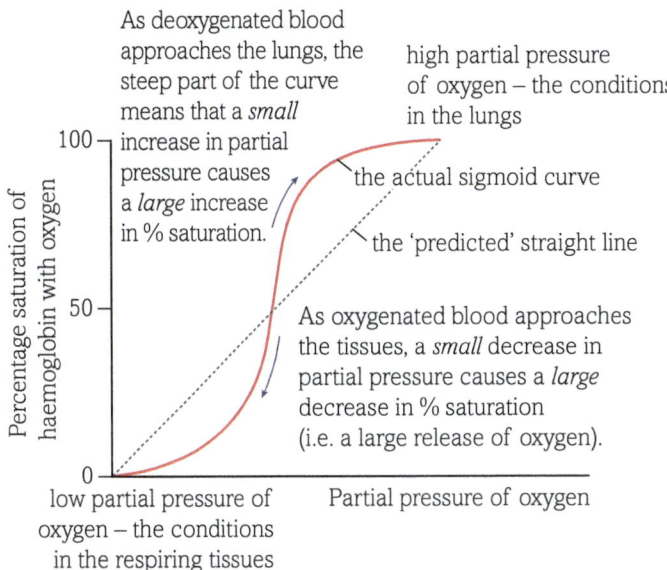

fig A Oxygen dissociation curve for human haemoglobin.

The Bohr effect

The way in which haemoglobin takes up and releases oxygen is also affected by the proportion of carbon dioxide in the tissues (see **fig B**). When the partial pressure of carbon dioxide is high, the affinity of haemoglobin for oxygen is reduced. In other words, haemoglobin needs higher levels of oxygen to become saturated and gives up oxygen much more easily. So in active tissues with high carbon dioxide levels, haemoglobin releases oxygen very readily. Carbon dioxide levels in the lung capillaries are relatively low, which makes it easier for oxygen to bind to the haemoglobin. The changes in the oxygen dissociation curve that result as the carbon dioxide level changes are known as the **Bohr effect**.

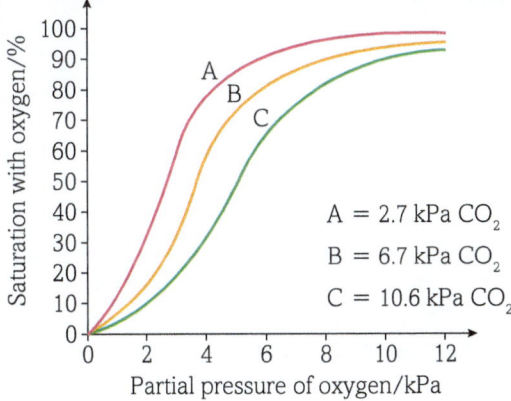

fig B As the proportion of carbon dioxide rises, the haemoglobin curve moves down and to the right – this is known as the Bohr effect.

Other respiratory pigments

Mammals have two more respiratory pigments as well as haemoglobin, each with a very specific function in the body:

- **Fetal haemoglobin** is found only in the developing fetus. When a fetus is in the uterus it is dependent on its mother to supply it with oxygen. Oxygenated blood from the mother runs through the placenta close to the deoxygenated fetal blood. If the blood of the fetus had the same affinity for oxygen as the blood of the mother very little oxygen would be transferred. But fetal

haemoglobin has a higher affinity for oxygen than that of the mother, and so can remove oxygen from the maternal blood (see **fig C**). The maternal and fetal blood also run in opposite directions so there is a counter current exchange system, maximising the oxygen transfer to the blood of the fetus.

- **Myoglobin** is a respiratory pigment found in the muscle tissue of vertebrates. It is a small, bright red protein, which gives red meat its strong colour. The structure of myoglobin is similar to a single haemoglobin chain, containing a haem group which binds oxygen. Myoglobin has a much higher affinity for oxygen than haemoglobin, so it easily becomes saturated with oxygen, and this affinity is not affected by the partial pressure of oxygen in the tissues. Once myoglobin is bound to an oxygen molecule it does not give up the oxygen easily, and so acts as an oxygen store. When the oxygen levels in very active muscle tissue get really low, and the carbon dioxide levels are correspondingly high, then myoglobin releases its store of oxygen when it is most needed (see **fig C**).

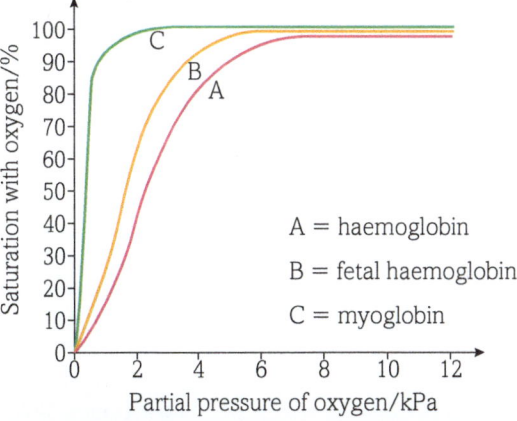

fig C Both fetal haemoglobin and myoglobin have higher affinities for oxygen than haemoglobin, so they can take up oxygen from haemoglobin in the body.

Transport of carbon dioxide

Waste carbon dioxide diffuses from the respiring cells of the body tissues into the blood along a concentration gradient. The reaction of the carbon dioxide with water is crucial. When carbon dioxide is dissolved in the blood it reacts slowly with the water to form carbonic acid, H_2CO_3. The carbonic acid separates to form the ions H^+ and HCO_3^-:

$$CO_2 + H_2O \rightleftharpoons H_2CO_3 \rightleftharpoons HCO_3^- + H^+$$

About 5% of the carbon dioxide is carried in solution in the plasma. A further 10–20% combines with haemoglobin molecules to form **carbaminohaemoglobin**. Most of the carbon dioxide is transported in the cytoplasm of the red blood cells as hydrogencarbonate ions. The enzyme **carbonic anhydrase** controls the rate of the reaction between carbon dioxide and water to form carbonic acid.

In the body tissues there is a high concentration of carbon dioxide in the blood, so carbonic anhydrase catalyses the formation of carbonic acid.

In the lungs the carbon dioxide concentration is low, so carbonic anhydrase catalyses the reverse reaction and free carbon dioxide diffuses out of the blood and into the lungs (see **fig D**).

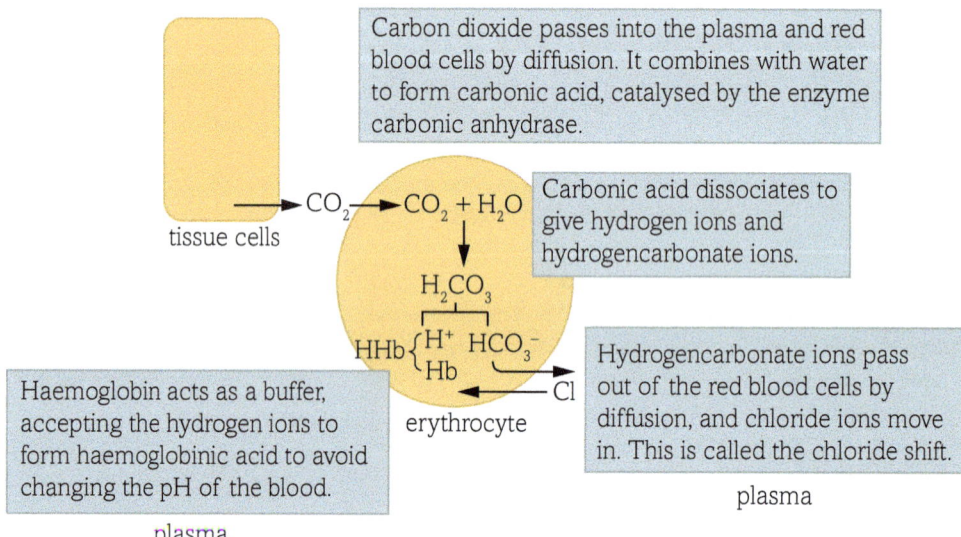

fig D The transport of carbon dioxide from the tissues to the lungs depends on the reaction of carbon dioxide with water, controlled by an enzyme in the red blood cells.

The clotting of the blood

You have a limited amount of blood. In theory, a minor cut could endanger life as the torn blood vessels allow blood to escape. First, and most immediately, your blood volume will fall. If you lose too much blood you will die. Second, pathogens can get into your body through an open wound. In normal circumstances your body has a damage limitation system in the clotting mechanism of the blood. This mechanism seals up damaged blood vessels to minimise blood loss and prevent pathogens getting in.

Forming a clot

Plasma, blood cells and platelets flow from a cut vessel. Contact between the platelets and components of the tissue (e.g. collagen fibres in the skin) causes the platelets to break open in large numbers. They release several substances, of which two are particularly important:

1 **Serotonin** causes the smooth muscle of the blood vessel to contract. This narrows the blood vessels, cutting off the blood flow to the damaged area.

2 **Thromboplastin** is an enzyme that sets in progress a cascade of events that leads to the formation of a clot (see **fig E**).

The blood clotting cascade

The blood clotting cascade is a very complex sequence of events in which there are many different clotting factors – you are looking at a simplified version here. Vitamin K is important in the production of many of the compounds in the blood clotting cascade, including prothrombin:

- Thromboplastin catalyses the conversion of a large soluble protein called **prothrombin** found in the plasma into another soluble protein, the enzyme called **thrombin**. This happens on a large scale at the site of a wound. Calcium ions need to be present in the blood at the right concentration for this reaction to happen.

- Thrombin acts on another soluble plasma protein called fibrinogen, converting it to an insoluble substance called **fibrin**. This forms a mesh of fibres to cover the wound.

- More platelets and blood cells pouring from the wound get trapped in the fibrin mesh. This forms a clot.

- Special proteins in the structure of the platelets contract, making the clot tighter and tougher to form a scab that protects the skin and vessels underneath as they heal.

In a cascade system such as clot formation, a relatively small event is amplified through a series of steps. However, sometimes the body's clotting mechanism is triggered in the wrong place, and this can lead to serious problems in the blood vessels. A clot in the vessels that supply your heart with blood can cause a heart attack and a clot in the brain can cause a stroke (see **Section 4.3.8**).

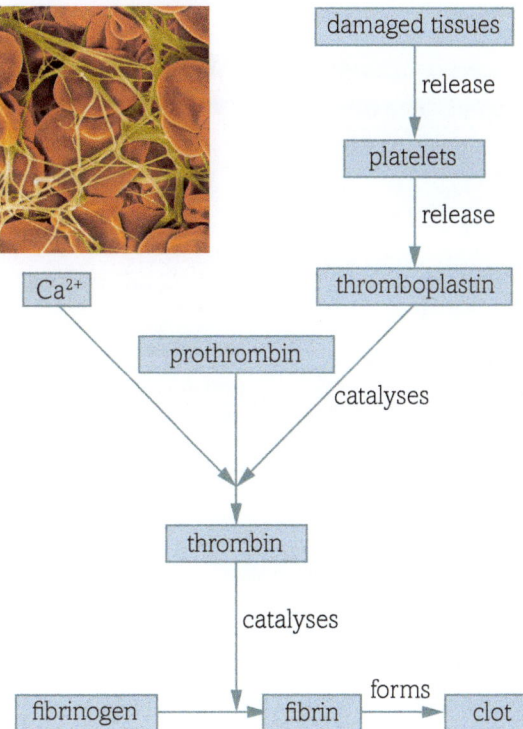

fig E The cascade of events that results in a life-saving clot. This seals the blood vessels and protects the delicate new tissues that form underneath.

Questions

1 Describe how oxygen is transported in the blood.

2 What is the importance of the oxygen affinity of fetal haemoglobin compared to adult haemoglobin?

3 Explain the role of diffusion in the transport of both oxygen and carbon dioxide around the body.

3 Prothrombin and fibrinogen are both precursors. Discuss the similarities and differences between these two proteins.

4 There is a rare condition in babies that causes excessive internal bleeding, which can cause brain damage and even death. Newborn babies in the UK are routinely given vitamin K either by injection or orally. Suggest how these two facts might be linked.

Key definitions

Oxyhaemoglobin is the molecule formed when oxygen binds to haemoglobin.

The **Bohr effect** is the name given to changes in the oxygen dissociation curve of haemoglobin that occur due to a rise in carbon dioxide levels and a reduction of the affinity of haemoglobin for oxygen.

Fetal haemoglobin is a form of haemoglobin found only in the developing fetus with a higher affinity for oxygen than adult haemoglobin.

Myoglobin is a respiratory pigment found in the muscle tissue of vertebrates with a higher affinity for oxygen than haemoglobin.

Carbaminohaemoglobin is the compound formed when carbon dioxide combines with haemoglobin.

Carbonic anhydrase is the enzyme that controls the rate of the reaction between carbon dioxide and water to form carbonic acid.

Serotonin is a chemical that causes the smooth muscle of the blood vessels to contract, narrowing them and cutting off the blood flow to the damaged area.

Thromboplastin is an enzyme that sets in progress a cascade of events that leads to the formation of a blood clot.

Prothrombin is a large, soluble protein found in the plasma that is the precursor to an enzyme called thrombin.

Thrombin is an enzyme that acts on fibrinogen, converting it to fibrin during clot formation.

Fibrin is an insoluble protein formed from fibrinogen by the action of thrombin that forms a mesh of fibres that trap erythrocytes and platelets to form a blood clot.

By the end of this section, you should be able to...

● describe the structure of the arteries, veins and capillaries

The blood vessels

The blood vessels that make up the circulatory system can be thought of as the biological equivalent of a road transport system. The **arteries** and **veins** are like the wide motorways carrying heavy traffic while the narrow town streets resemble the vast branching and spreading **capillary** network. In this capillary network substances carried by the blood are exchanged with cells in the same way as goods are uploaded from factories or offloaded to shops and homes. The structures of the different types of blood vessel closely reflect their functions in your body.

Arteries

Learning tip

You will look at the structure and the function of the types of blood vessel separately. However, you should remember that the vessels do not exist separately – they are all interlinked within the whole circulatory system.

Arteries carry blood away from the heart towards the cells of the body. The structure of an artery is shown in **fig A**. Almost all arteries carry oxygenated blood. The only exceptions are:

• the pulmonary artery – carrying deoxygenated blood from the heart to the lungs

• the umbilical artery – during pregnancy, this carries deoxygenated blood from the fetus to the placenta.

The arteries leaving the heart branch off in every direction, and the diameter of the **lumen**, the central space inside the blood vessel, gets smaller the further it is from the heart. The very smallest branches of the arterial system, furthest from the heart, are the **arterioles**.

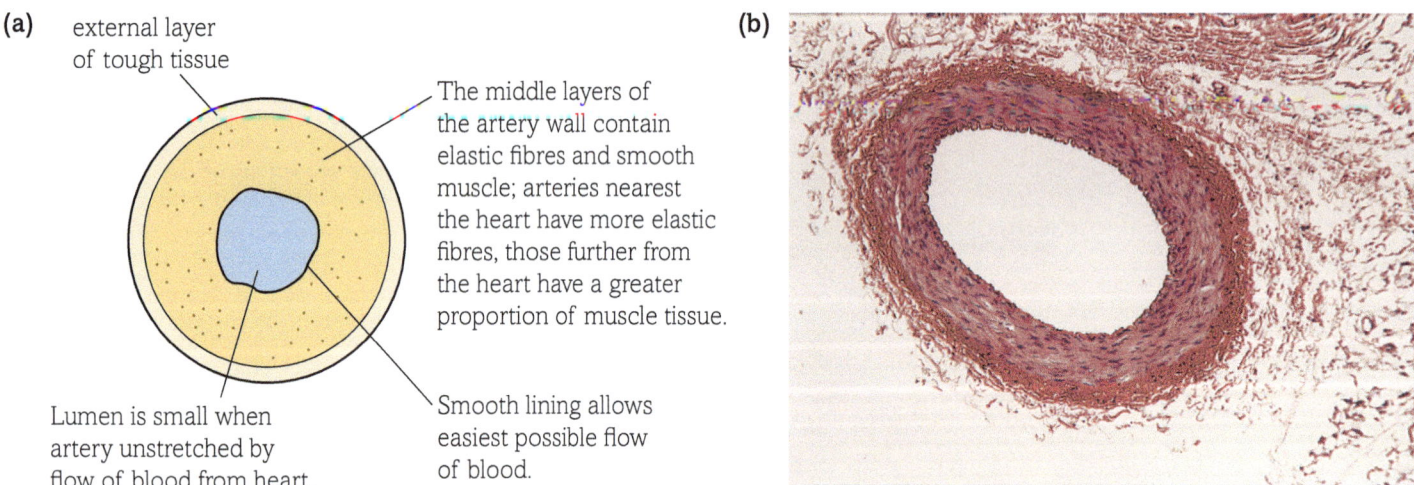

(a) external layer of tough tissue

The middle layers of the artery wall contain elastic fibres and smooth muscle; arteries nearest the heart have more elastic fibres, those further from the heart have a greater proportion of muscle tissue.

Lumen is small when artery unstretched by flow of blood from heart.

Smooth lining allows easiest possible flow of blood.

(b)

fig A The structure of an artery means it is adapted to cope with the surging of the blood as the heart pumps.

Blood is pumped out from the heart in a regular rhythm, about 70 times a minute. Each heartbeat sends a high-pressure surge of blood into the arteries. The major arteries close to the heart must withstand these pressure surges. Their walls contain a lot of elastic fibres so they can stretch to accommodate the greater volume of blood without being damaged (see **fig B**). Between surges the elastic fibres return to their original length, squeezing the blood and so moving it along in a continuous flow. The pulse you can feel in an artery is the effect of the surge each time the heart beats. The blood pressure in all arteries is relatively high, but it falls in arteries further away from the heart, known as the **peripheral arteries**.

In the peripheral arteries the muscle fibres in the vessel walls contract or relax to change the size of the lumen, controlling the blood flow. The smaller the lumen, the harder it is for blood to flow through the vessel. This controls the amount of blood that flows into an organ, so regulating its activity.

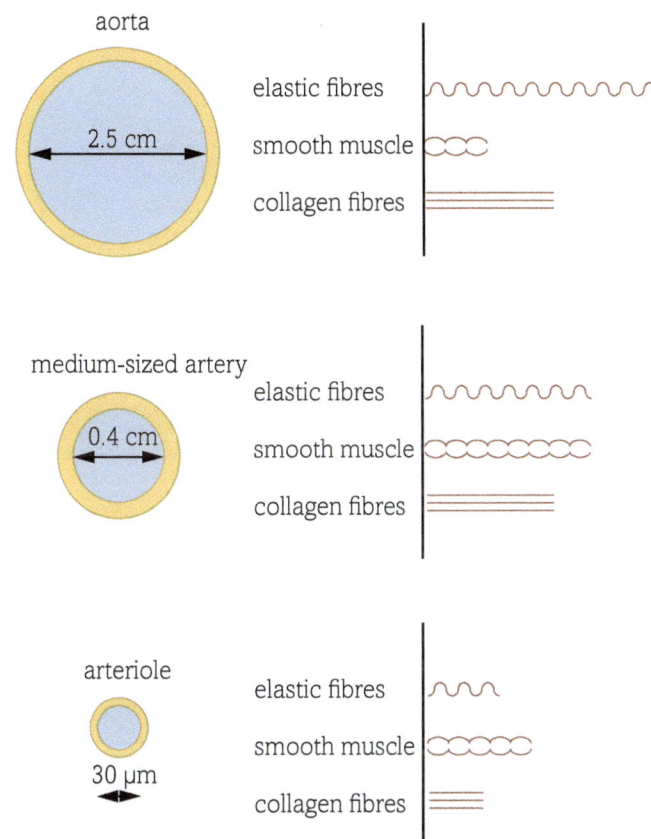

fig B The relative proportions of different tissues in different arteries. Collagen gives general strength and flexibility to both arteries and veins.

Capillaries

Arterioles feed into networks of capillaries. These are minute vessels that spread throughout the tissues of the body. The capillary network links the arterioles and the **venules**. Capillaries branch between cells – no cell is far from a capillary, so substances can diffuse between cells and the blood quickly. Also, because the diameter of each individual capillary is small, the blood travels relatively slowly through them, again giving more opportunity for diffusion to occur (see **fig C**). The smallest capillary is no wider than a single red blood cell.

Capillaries have a very simple structure well adapted to their function. Their walls are very thin, containing no elastic fibres, smooth muscle or collagen. This helps them fit between individual cells and also allows rapid diffusion of substances between the blood and the cells. The walls consist of just one very thin cell. Oxygen and other molecules quickly diffuse out of the blood in the capillaries into the nearby body cells, and carbon dioxide and other waste molecules diffuse in. Blood entering the capillary network from the arteries is oxygenated. By the time it leaves, it carries less oxygen and more carbon dioxide.

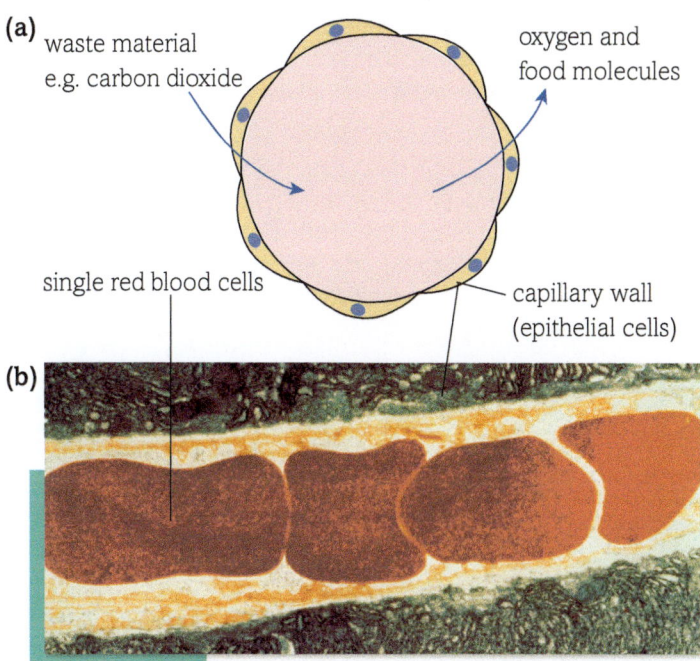

fig C The very thin walls of capillaries allow rapid diffusion of oxygen, carbon dioxide and digested food molecules. The lumen is just wide enough for red blood cells to pass through.

Veins

Veins carry blood back towards the heart. Most veins carry deoxygenated blood. The exceptions are:

- the pulmonary vein – carrying oxygen-rich blood from the lungs back to the heart for circulation around the body
- the umbilical vein – during pregnancy, it carries oxygenated blood from the placenta into the fetus.

Tiny venules lead from the capillary network, merging into larger and larger vessels leading back to the heart (see **fig D**).

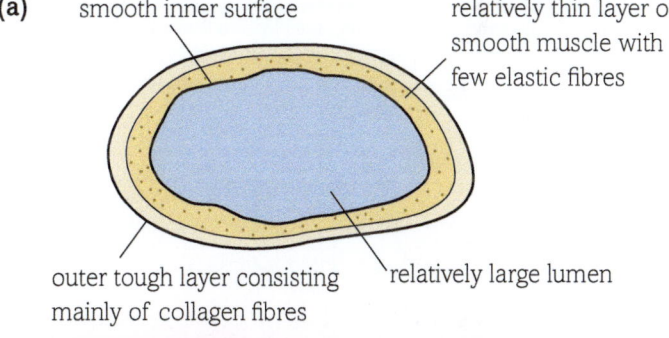

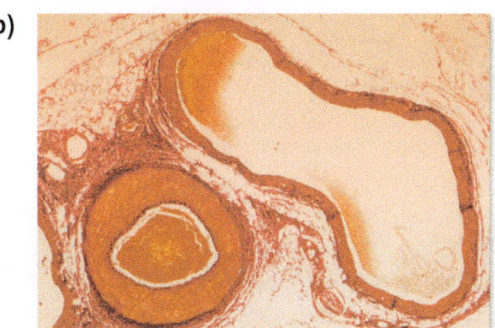

fig D The arrangement of tissues in a vein reflects the pressure of blood in the vessel.

Eventually only two veins carry the returning blood to the heart – the **inferior vena cava** from the lower parts of the body and the **superior vena cava** from the upper parts of the body.

Veins can hold a large volume of blood – in fact more than half of the body's blood volume is in the veins at any one time. They act as a blood reservoir. The blood pressure in the veins is relatively low – the pressure surges from the heart are eliminated as the blood passes through the capillary beds. This blood at low pressure must be returned to the heart to be oxygenated again and recirculated.

There are two main ways in which this is achieved:

- At frequent intervals throughout the venous system there are one-way valves. These are called **semilunar valves** because of their half-moon shape. They are formed from infoldings of the inner wall of the vein. Blood can pass through towards the heart, but if it starts to flow backwards the valves close, preventing any backflow (see **fig E**).
- Many of the larger veins are situated between the large muscle blocks of the body, particularly in the arms and legs. When the muscles contract during physical activity they squeeze these veins. With the valves keeping blood travelling in one direction, this squeezing helps to return the blood to the heart.

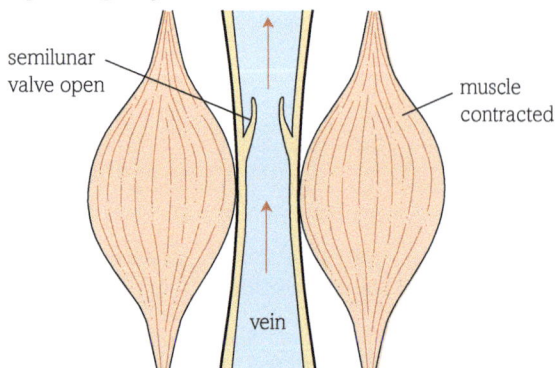

Blood moving in the direction of the heart forces the valve open, allowing the blood to flow through.

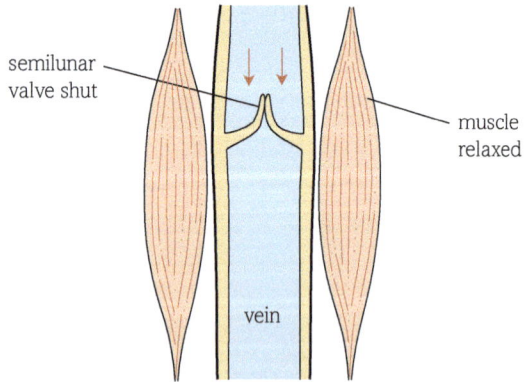

A backflow of blood will close the valve, ensuring that blood cannot flow away from the heart.

fig E Valves in the veins make sure blood only flows in one direction – towards the heart. The contraction of large muscles encourages blood flow through the veins.

The main types of blood vessels – the arteries, veins and capillaries – have very different characteristics. These affect the way the blood flows through the body, and the roles they play in the body. Some of these differences are summarised in **fig F**.

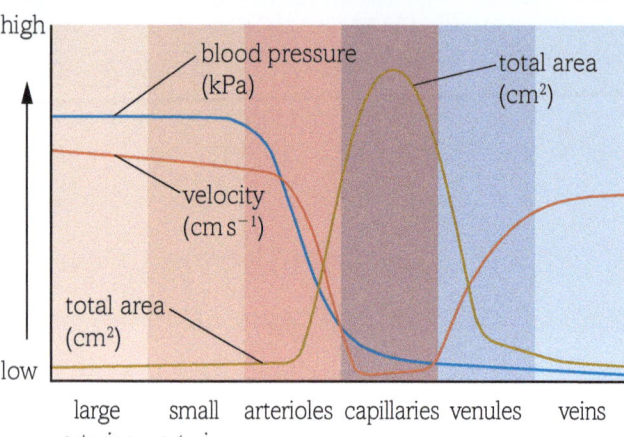

fig F Graph to show the surface area of each major type of blood vessel in your body, along with the velocity and pressure of the blood travelling in them.

Questions

1 Why are valves important in veins but unnecessary in arteries?

2 Compare the main structures and functions of arteries, veins and capillaries.

3 Look at the graph in fig **fig F**. Explain carefully what the different lines on the graph show you. How is this information linked to the functions of the different regions of the circulatory system?

Key definitions

Arteries are vessels that carry blood away from the heart towards the cells of the body.

Veins are vessels that carry blood towards the heart from the cells of the body.

Capillaries are minute vessels that spread throughout the tissues of the body.

The **lumen** is the central space inside the blood vessel.

Arterioles are the very smallest branches of the arterial system, furthest from the heart.

Peripheral arteries are arteries further away from the heart but before the arterioles.

Venules are the very smallest branches of the venous system, furthest from the heart.

The **inferior vena cava** is the large vein that carries the returning blood from the lower parts of the body to the heart .

The **superior vena cava** is the large vein that carries the returning blood from the upper parts of the body to the heart.

Semilunar valves are half moon-shaped one-way valves found at frequent intervals throughout the venous system to prevent the backflow of blood.

The human heart

By the end of this section, you should be able to...

- describe the structure of the heart
- describe the sequence of events of the cardiac cycle

In most animal transport systems the heart is the organ that moves the blood around the body. In mammals it is a complex, four-chambered muscular bag, that sits in the chest protected by the ribs and sternum. In an average lifetime your heart will beat about 3 000 000 000 (3×10^9) times and will pump over 200 million litres of blood – quite a work load.

The structure of the heart

The human heart, like other mammalian hearts (see **fig A**), is not a single muscular pump but two, joined together and working in perfect synchrony. The right side of the heart receives blood from the body and pumps it to the lungs. The left side of the heart receives blood from the lungs and pumps it to the body. The blood in each side of the heart does not mix with the blood from the other side. The two sides are separated by a thick, muscular **septum**. The heart is made of a unique type of muscle, known as **cardiac muscle**, which has special properties – it can carry on contracting regularly without resting or getting fatigued. The cardiac muscle has a good blood supply – the coronary arteries bring oxygenated blood to the tissue (see **fig B**). It also contains lots of myoglobin, which stores oxygen for the respiration needed to keep the heart contracting regularly (see **Section 4.3.3**).

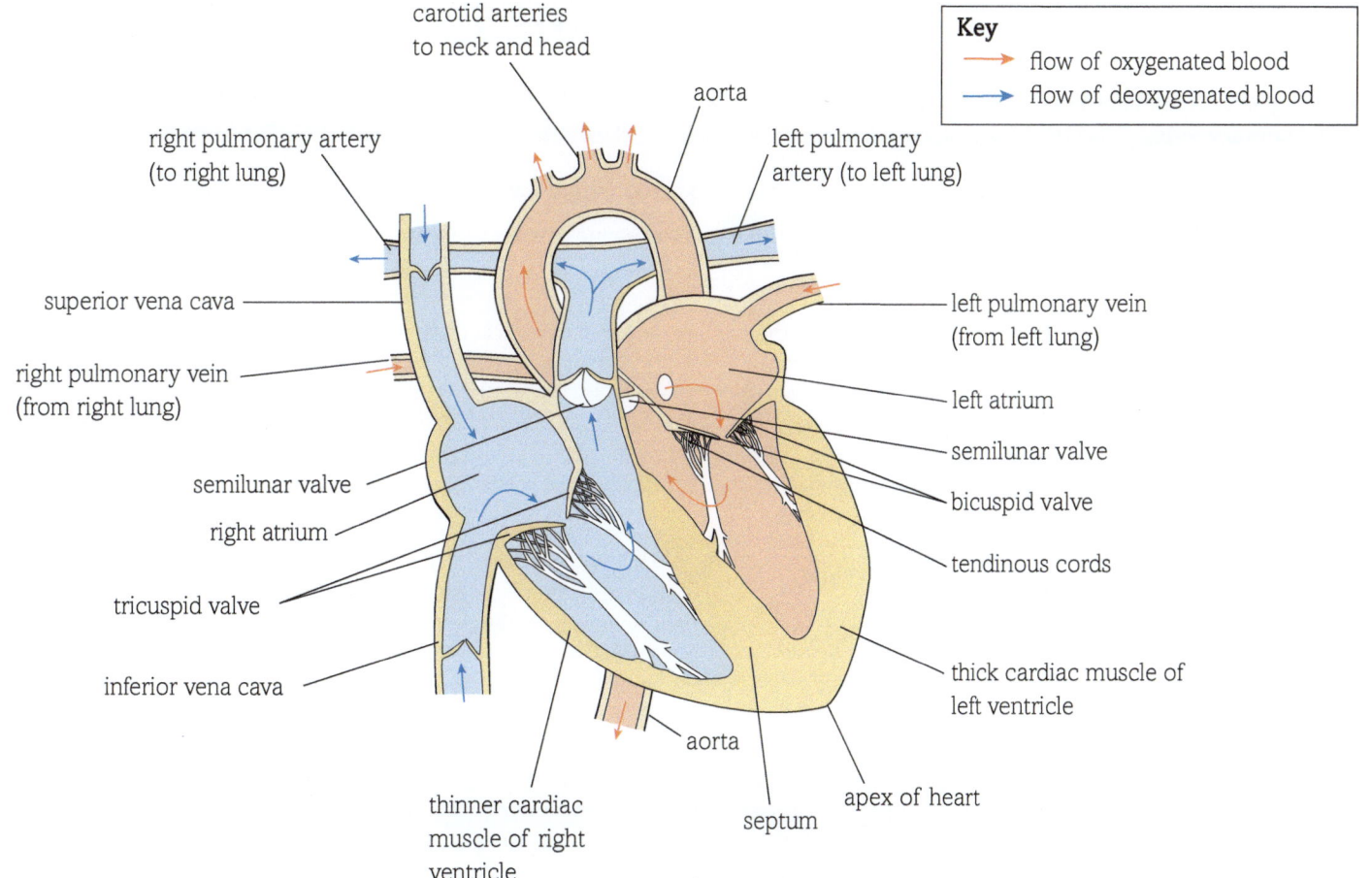

Key
→ flow of oxygenated blood
→ flow of deoxygenated blood

fig A The structure of the human heart.

- The inferior vena cava collects deoxygenated blood, from the lower parts of the body, while the superior vena cava receives deoxygenated blood from the head, neck, arms and chest. This blood is delivered to the **right atrium**.

- The right atrium receives the blood from the great veins. As it fills with blood, the pressure builds up and opens the tricuspid valve so the **right ventricle** starts to fill with blood too. When the atrium is full it contracts, forcing the blood into the ventricle. The atrium has thin muscular walls because it receives blood at low pressure from the two venae cavae and needs to exert relatively little pressure to move the blood into the ventricle. One-way semilunar valves, like the valves in veins described in **Section 4.3.5**, at the entrance to the atrium stop a backflow of blood into the veins.

- The **tricuspid valve** is made up of three flaps and is also known as an **atrioventricular valve** as it separates an atrium from a ventricle. The valve allows blood to pass from the atrium to the ventricle, but not in the other direction. The tough **tendinous cords**, also known as valve tendons or heartstrings, make sure the valves are not turned inside out by the pressure exerted when the ventricles contract.

- The right ventricle is filled with blood under some pressure when the right atrium contracts, then the ventricle contracts. Its muscular walls produce the pressure needed to force blood out of the heart into the **pulmonary arteries**. This carries the deoxygenated blood to the capillary beds of the lungs. As the ventricle starts to contract, the tricuspid valve closes to prevent blood flowing into the atrium. Semilunar valves, like those in veins, prevent the blood flowing back from the artery into the ventricle.

- The blood returns from the lungs to the left side of the heart in the **pulmonary veins**. The blood is at relatively low pressure after passing through the extensive capillaries of the lungs. The blood returns to the **left atrium**, another thin-walled chamber that performs the same function as the right atrium. It contracts to force blood into the **left ventricle**. Backflow is prevented by the **bicuspid valve**.

- As the left atrium contracts, the bicuspid valve opens and the left ventricle is filled with blood under pressure. As the left ventricle starts to contract the bicuspid valve closes to prevent backflow of blood to the left atrium. The left ventricle pumps the blood out of the heart and into the **aorta**, the major artery of the body. This carries blood away from the heart at even higher pressure than other major arteries that branch off from it. The muscular wall of the left side of the heart is much thicker than that of the right. The right side pumps blood to the lungs, which are relatively close to the heart. The delicate capillaries of the lungs need blood delivered at relatively low pressure. The left side has to produce sufficient force to move the blood under pressure to all the extremities of the body and overcome the elastic recoil of the arteries. Semilunar valves prevent the blood flowing back from the aorta into the ventricle.

- The septum is a thick wall of muscle and connective tissue between the two sides of the heart. It prevents the oxygenated blood mixing with the deoxygenated blood. In the embryo

there is a gap in the septum called the foramen ovale and the blood from the two sides of the heart can mix. This does not matter because the lungs of the fetus do not function and little blood flows to them. However, at birth this hole closes over. If it does not, the baby has a condition called patent foramen ovale or 'hole in the heart'. This may be so small it does not really matter. If it is large, it can be closed by surgery.

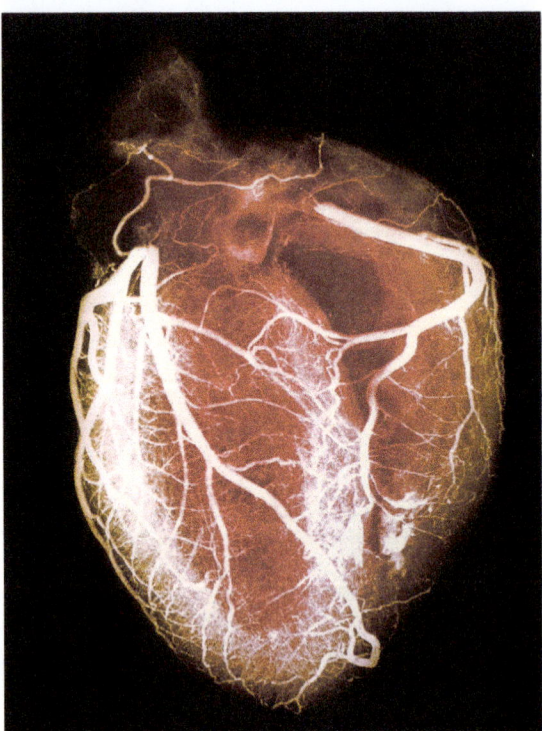

fig B The coronary arteries, which you can clearly see here, carry oxygenated blood from the aorta to the heart muscle, providing it with oxygen and digested food and removing carbon dioxide.

How your heart works

The beating of your heart produces the sounds that are your heartbeat. The sounds are made not by the contracting of the heart muscle but by blood hitting the heart valves. The two sounds of a heartbeat are often described as 'lub-dub'. The first sound comes as the blood is forced against the atrioventricular valves when the ventricles contract. The second sound comes as a backflow of blood hits the semilunar valves in the pulmonary artery and aorta as the ventricles relax. The rate of your heartbeat shows how fast your heart is contracting.

The cardiac cycle

Your heart is continuously contracting then relaxing. The contraction of the heart is called **systole**. Systole can be divided into **atrial systole**, when the atria contract together forcing blood into the ventricles, and **ventricular systole**, when the ventricles contract. Ventricular systole happens about 0.13 seconds after atrial systole, and forces blood out of the ventricles into the pulmonary artery and the aorta. Between contractions the heart relaxes and fills with blood. This relaxation stage is called **diastole**. One cycle of systole and diastole makes up a single heartbeat, which lasts about 0.8 seconds in humans. This is known as the **cardiac cycle** (see **fig C**).

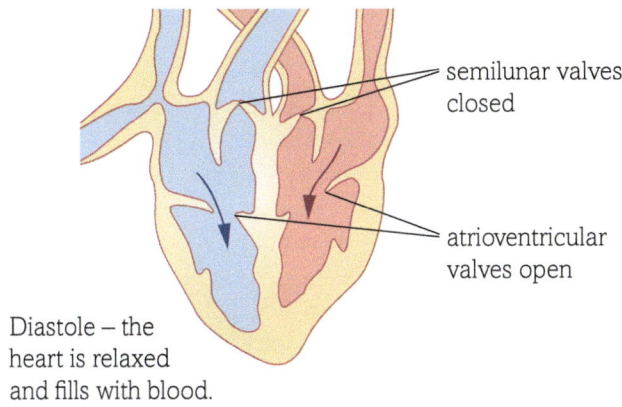

semilunar valves closed

atrioventricular valves open

Diastole – the heart is relaxed and fills with blood.

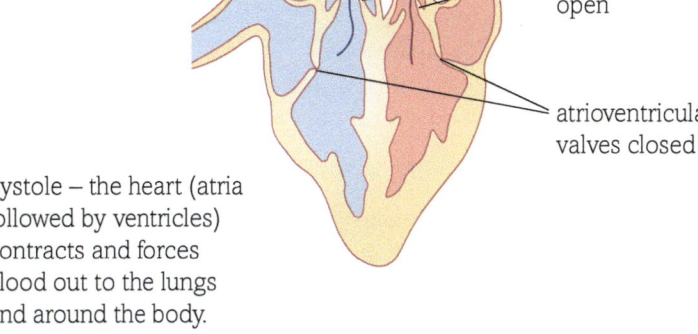

semilunar valves open

atrioventricular valves closed

Systole – the heart (atria followed by ventricles) contracts and forces blood out to the lungs and around the body.

fig C The cardiac cycle.

Questions

1   Describe the path of blood around the human body, identifying at which points the blood is oxygenated and where it is deoxygenated. Explain how this system efficiently supplies cells with the oxygen they need.

2   Discuss the relationship between structure and function for these parts of the heart:
   (a) semilunar valves
   (b) muscle wall (and its thickness) in the atria
   (c) ventricles
   (d) tendinous cords.

Key definitions

The **septum** is the thick muscular dividing wall through the centre of the heart that prevents oxygenated and deoxygenated blood from mixing.

Cardiac muscle is the special muscle tissue that makes up the bulk of the heart, which has an intrinsic rhythmicity and does not fatigue.

The **right atrium** is the upper right-hand chamber of the heart that receives deoxygenated blood from the body.

The **right ventricle** is the chamber that receives deoxygenated blood from the right atrium and pumps it to the lungs.

The **tricuspid valve (atrioventricular valve)** is the valve between the right atrium and the right ventricle that prevents backflow of blood from the ventricle to the atrium when the ventricle contracts.

Tendinous cords (also known as valve tendons or heartstrings) make sure the valves are not turned inside out by the great pressure exerted when the ventricles contract.

Pulmonary arteries are the blood vessels that carry deoxygenated blood from the heart to the lungs.

Pulmonary veins are the blood vessels that carry oxygenated blood back from the lungs to the heart.

The **left atrium** is the upper left-hand chamber of the heart that receives oxygenated blood from the lungs.

The **left ventricle** is the chamber that receives oxygenated blood from the left atrium and pumps it around the body.

The **bicuspid valve** is the valve between the left atrium and the left ventricle that prevents backflow of blood into the atrium when the ventricle contracts.

The **aorta** is the main artery of the body. It leaves the left ventricle of the heart carrying oxygenated blood under high pressure.

Systole is the contraction of the heart.

Atrial systole is when the atria of the heart contract.

Ventricular systole is when the ventricles of the heart contract.

Diastole is when the heart relaxes and fills with blood.

The **cardiac cycle** is the cycle of contraction (systole) and relaxation (diastole) in the heart.

By the end of this section, you should be able to...

● describe the structure of the heart

● explain the myogenic stimulation of the heart, including the roles of the sinoatrial node, atrioventricular node and bundle of His

● interpret data showing ECG traces and pressure changes during the cardiac cycle

Your heart beats continually throughout life, with an average of about 70 beats per minute, although in small children the heart rate is much higher. The heart can respond to need. During physical exercise, when the tissues need more oxygen, your heart beats faster to supply more oxygenated, glucose-carrying blood to the tissues and to remove the increased waste products. Stress can also raise the heart rate, while rest and relaxation can lower it.

The control of the heartbeat

Cardiac muscle contracts without any external stimulus from the nervous system – control of the basic heart beat is myogenic. A number of mechanisms have evolved that ensure the different parts of the heart contract in the right sequence and with an inbuilt frequency, so blood is always supplied to the body tissues. This myogenic stimulation gives rise to the **intrinsic rhythmicity** of the heart. An adult heart removed from the body will continue to contract as long as it is bathed in a suitable oxygen-rich fluid.

The intrinsic rhythm is around 60 beats per minute. This rhythm is maintained by a wave of electrical excitation similar to a nerve impulse that spreads through special tissue in the heart muscle (see **fig A**). This arises within the heart itself – no external nervous stimulus is needed. The control of the heart beat is a multi-step process:

• The area of the heart with the fastest intrinsic rhythm is a group of cells in the right atrium known as the **sinoatrial node (SAN)**, and this acts as the heart's own natural pacemaker.

• The sinoatrial node sets up a wave of electrical excitation (depolarisation) that causes the atria to start contracting. This initiates the heartbeat.

• The **annulus fibrosus** is a region of non-conducting tissue between the atrium and the ventricle. This prevents the excitation from the atrium spreading directly to the ventricles.

• The wave of excitation from the SAN spreads through the atria as they contract. It cannot pass through the annulus fibrosus but it stimulates another region of conducting tissue between the atrium and the ventricle called the **atrioventricular node (AVN)**.

• The AVN is stimulated but produces a slight delay before it passes the wave of depolarisation into the **bundle of His**, a group of conducting fibres in the septum of the heart. This ensures that the atria have stopped contracting before the ventricles start.

• The bundle of His splits into two branches and carries the wave of excitation on into the **Purkyne tissue**.

• The Purkyne tissue consists of conducting fibres that penetrate down through the septum spreading around the ventricles. As the depolarisation travels through the tissue it sets off the contraction of the ventricles, starting at the apex (bottom) and so squeezing blood out of the heart.

The speed at which the wave of excitation spreads through the heart, with the hesitation before the AVN stimulates the bundle of His, makes sure that the atria have stopped contracting before the ventricles start. It is these changes in the electrical excitation of the heart that cause the repeating cardiac cycle. The electrical changes are what is measured in an **electrocardiogram (ECG)**.

Because your heart has its own basic rhythm, body resources are not wasted on maintaining such a vital but continuous event. However, many people have a faster resting heart rate than this basic rhythm – around 70 beats per minute is the average. This is because lots of other factors including nerve impulses and hormones constantly influence the heart rate (see **Book 2 Topic 9**).

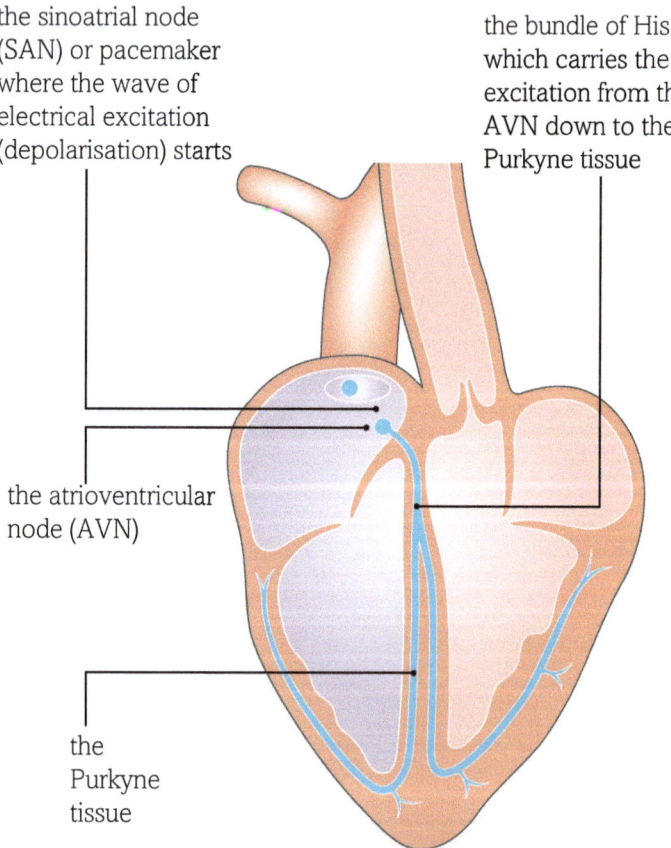

the sinoatrial node (SAN) or pacemaker where the wave of electrical excitation (depolarisation) starts

the bundle of His, which carries the excitation from the AVN down to the Purkyne tissue

the atrioventricular node (AVN)

the Purkyne tissue

fig A The area of the heart with the fastest intrinsic rhythm is a group of cells in the right atrium known as the sinoatrial node, and this acts as the heart's own natural pacemaker.

The use of the electrocardiogram (ECG)

An ECG is used to investigate the rhythms of the heart by producing a record of the electrical activity of the heart. As you know, the rhythm of the heart results from the spread of a wave of depolarisation (electrical activity) through specialised tissue within the heart muscle itself. This depolarisation in the heart causes tiny electrical changes on the surface of your skin. An ECG measures these changes at the surface of your skin.

To take an ECG, 12 electrodes and leads are attached to your body. This is a completely painless process. Your skin is wiped with alcohol to remove any grease or sweat so the electrodes make good contact. Sometimes a special gel is applied to the electrodes to make sure they conduct electricity as effectively as possible. As the recording is made, information is fed back from each of the electrodes, effectively giving 12 views of the heart.

An ECG can show you what is happening in a normal, healthy heart. However, it is often used to indicate different heart conditions, and to monitor patients with heart disease. It is usually done with the patient lying down and resting, but sometimes an ECG is carried out while a patient is exercising (known as a stress test) because some heart conditions show up only during exercise.

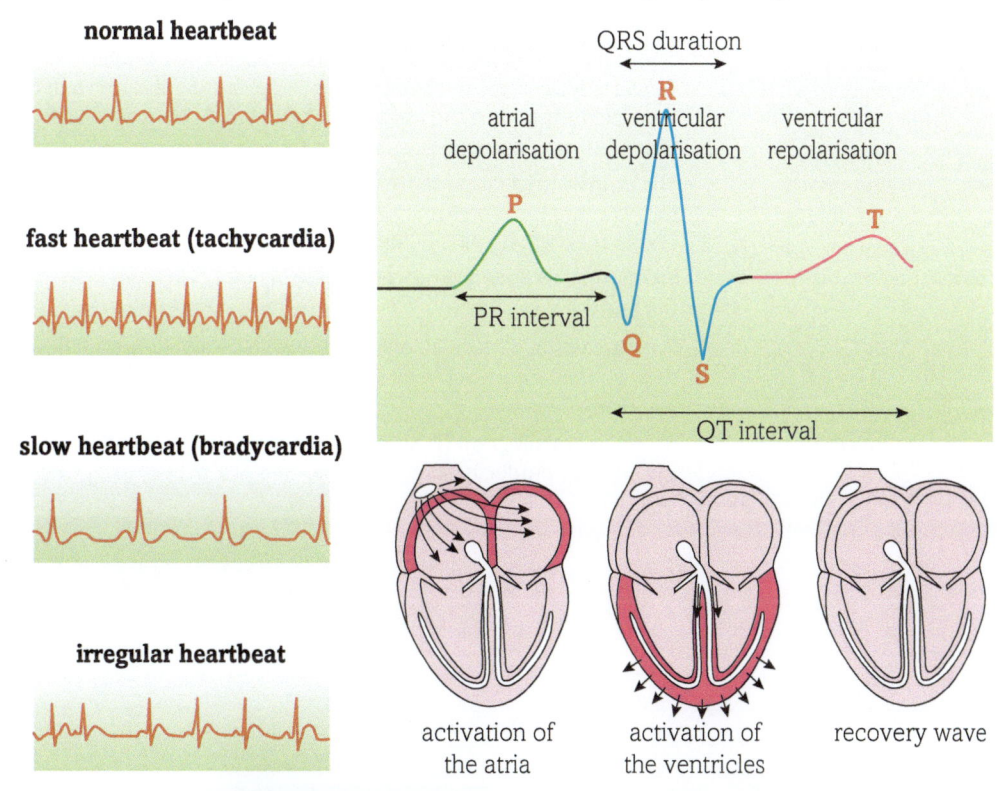

fig B The relationship between the electrical events in the heart and the heartbeat as shown by ECGs.

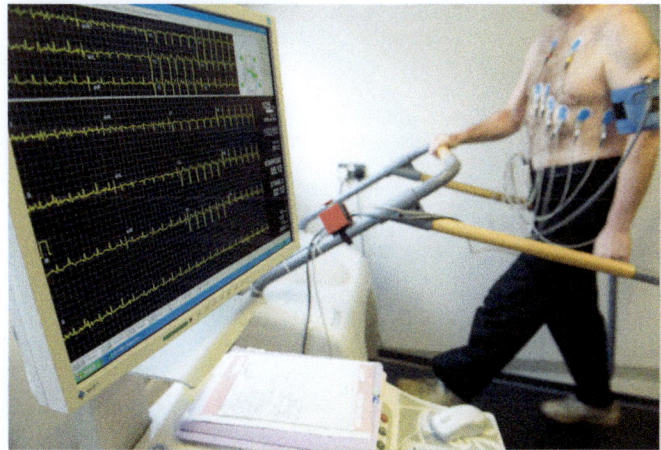

fig C Having an ECG does not take long and is not invasive, but it gives clinicians a lot of information about your heart.

It is possible to put together a wide range of readings to produce a diagram showing how the electrical activity recorded in an ECG and pressure changes measured in the different chambers of the heart work together during the cardiac cycle. In **fig D** you can see how differences in pressure in the regions of the left side of the heart cause the different valves to close. The same process is happening on the right-hand side of the heart at exactly the same time. You can also see how the stages of the cardiac cycle relate to the pressure changes and an ECG showing the electrical activity of the heart.

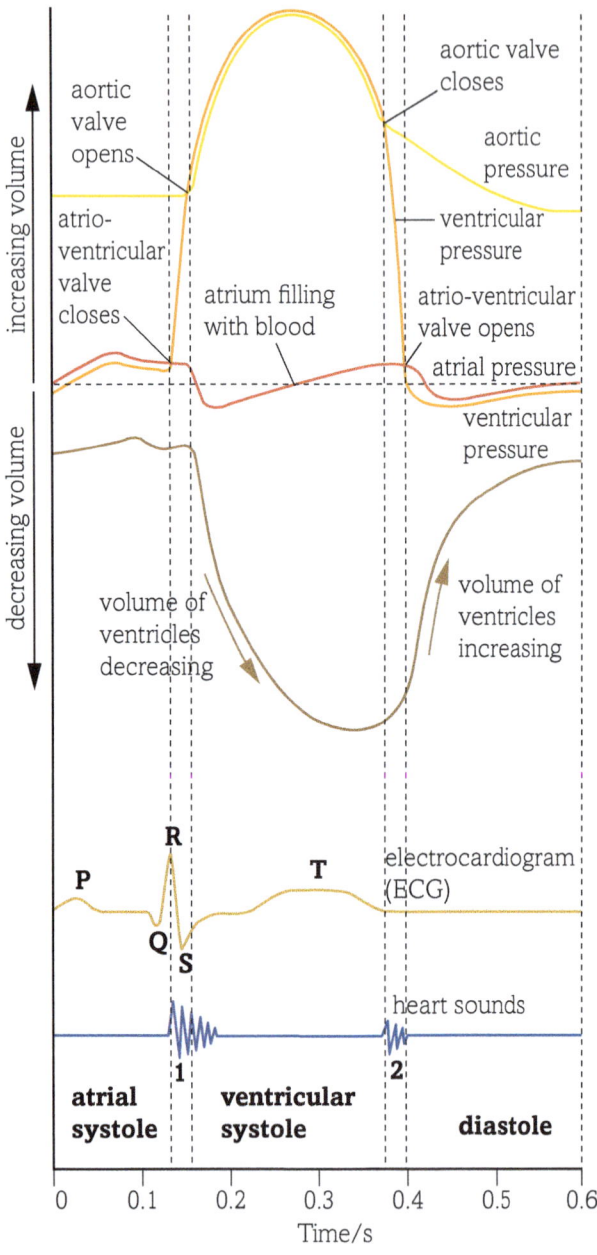

fig D Electrical and pressure changes in the heart during the cardiac cycle – and the causes of the heart sounds.

Questions

1 Explain how an artificial heart pacemaker, which delivers a regular electric shock to the right atrium, can help maintain a steady heart rate in people when the natural pacemaker is no longer working properly.

2 Explain how the changes in volume and pressure in the atria and ventricles, shown in **fig D**, relate to the events of the cardiac cycle.

Key definitions

Intrinsic rhythmicity refers to the intrinsic rhythm of contraction and relaxation in the cardiac muscle making up the heart.

The **sinoatrial node (SAN)** is a group of cells in the right atrium. It is the area of the heart with the fastest intrinsic rhythm and acts as the heart's own natural pacemaker.

The **annulus fibrosus** consists of non-conducting tissue that provides support for the heart and spreads between the atria and the ventricles, preventing the wave of excitation that spreads through the atria from passing directly on into the ventricles.

The **atrioventricular node (AVN)** is a group of cells stimulated by the wave of excitation from the SAN and atria. It imposes a delay before transmitting the impulse to the bundle of His.

The **bundle of His** is a group of conducting fibres in the septum of the heart.

Purkyne tissue is made up of conducting fibres that penetrate down through the septum of the heart, spreading between and around the ventricles.

An **electrocardiogram (ECG)** is used to investigate the rhythms of the heart by producing a record of the electrical activity of the heart.

By the end of this section, you should be able to...

● Explain the stages that lead to atherosclerosis, and its affect on health

Cardiovascular diseases in the UK

Problems with the cardiovascular system have serious consequences. Data from across the UK for 2010 shows that **cardiovascular diseases** were responsible for 33% of all deaths, which is more than any other cause. What is more, almost a third of these deaths were in people younger than 75.

Globally, more than 17 million people die from cardiovascular diseases each year. It is the biggest cause of death and disability. These deaths are known as premature deaths – people dying younger than expected. Many cardiovascular diseases are linked to a condition called **atherosclerosis**.

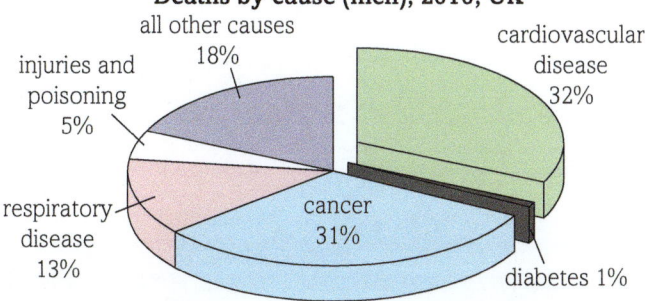

Deaths by cause (men), 2010, UK

all other causes 18%
injuries and poisoning 5%
respiratory disease 13%
cancer 31%
cardiovascular disease 32%
diabetes 1%

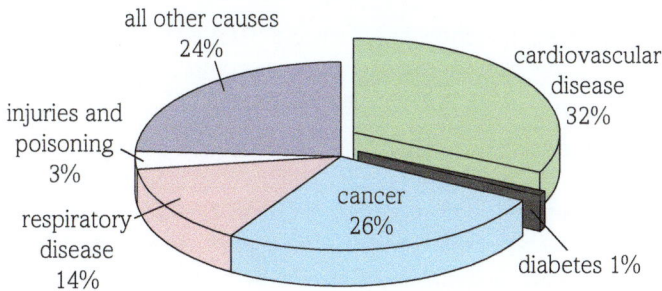

Deaths by cause (women), 2010, UK

all other causes 24%
injuries and poisoning 3%
respiratory disease 14%
cancer 26%
cardiovascular disease 32%
diabetes 1%

fig A The most common causes of death in England and Wales in 2010. Cardiovascular diseases, many of which are linked to atherosclerosis, kill more people than any other cause.

The formation of atherosclerosis

Atherosclerosis, which literally means hardening of the arteries, is a build-up of **plaques** (yellowish fatty deposits) on the inside of arteries. It can begin in late childhood and continues throughout life. A plaque can build up until it restricts the flow of blood through the artery or even blocks it completely. Plaques are particularly likely to form in the arteries of the heart (coronary arteries) and neck (carotid arteries). The development of a plaque follows a typical progression as shown in **fig B**.

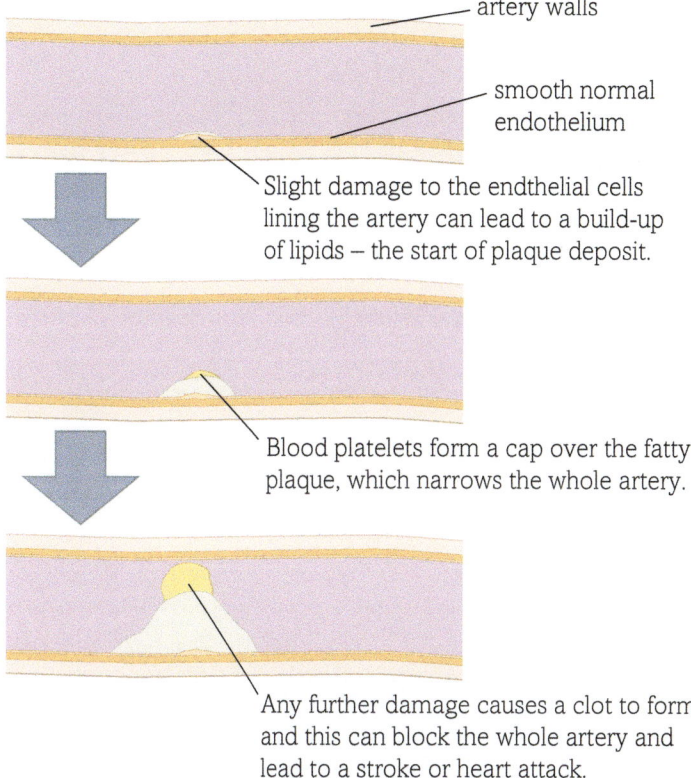

artery walls

smooth normal endothelium

Slight damage to the endthelial cells lining the artery can lead to a build-up of lipids – the start of plaque deposit.

Blood platelets form a cap over the fatty plaque, which narrows the whole artery.

Any further damage causes a clot to form and this can block the whole artery and lead to a stroke or heart attack.

fig B The development of atherosclerosis.

The damage to the endothelial lining of blood vessels that leads to plaque formation can be caused by several factors, such as high blood pressure and chemicals in tobacco smoke. Atherosclerosis usually occurs in arteries rather than in veins. This is because the blood in the arteries flows fast under relatively high pressure, which puts more strain on the endothelium lining the vessels and can cause small areas of damage. In the veins the pressure is lower so damage to the endothelium is much less likely.

Once the damage has occurred, the body's inflammatory response begins and white blood cells arrive at the site of the damage. These cells accumulate chemicals from the blood, in particular cholesterol. This leads to a plaque (also known as an **atherome**) forming on the endothelial lining of the artery. Fibrous tissue and calcium salts also build up around the atheroma, turning it into a hardened plaque. This hardened area means that part of the artery wall hardens, so it is less elastic than it should be. This is atherosclerosis.

The lumen of the artery becomes much smaller as a result of the plaque. This increases the blood pressure, making it harder for the heart to pump blood around the body. The raised blood pressure makes damage more likely in other areas of the endothelial lining and more plaques will form. This will make the blood pressure even higher, and so the problem gets worse. There are many factors that are linked to the development of atherosclerosis. You will look at these in more detail in **Section 4.3.9**.

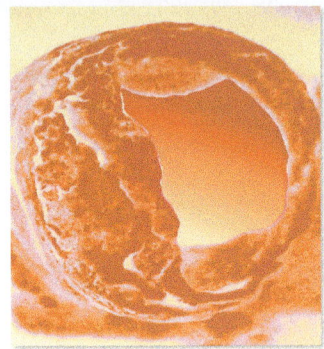

fig C Fatty deposits like these in an artery cause disease and death in thousands of people every year.

Effect of atherosclerosis on health

Atherosclerosis can have a number of serious affects on the health of an individual.

Aneurysms

If an area of artery is narrowed by plaque, blood tends to build up behind the blockage. The artery bulges and the wall is put under more pressure than usual, so it is weakened. This is known as an **aneurysm**. The weakened artery wall may split open, leading to massive internal bleeding. Aneurysms frequently happen in the blood vessels supplying the brain or in the aorta, especially when it passes through the abdomen. The massive blood loss and drop in blood pressure are often fatal, but if aneurysms are diagnosed they can treated by surgery before they burst.

Raised blood pressure

The arteries narrowed due to plaques on the walls cause raised blood pressure. This can cause serious damage in a number of other organs, including the kidneys, the eye and the brain. The high pressure damages the tiny blood vessels where your kidney filters out urea and other substances from the blood. If the vessels feeding the kidney tubules become narrowed, the pressure inside them gets even higher and proteins may be forced out through their walls. Doctors test for protein in your urine as a sign of kidney damage if you have high blood pressure. Similarly, the tiny blood vessels supplying the retina of your eye are easily damaged. If they become blocked or leak, the retinal cells are starved of oxygen and die and this can cause blindness. Bleeding from the capillaries into the brain also results in one type of stroke (see below).

Heart disease

There are many kinds of heart disease, but the two most common ones are **angina** and **myocardial infarction** (heart attack), both closely linked to atherosclerosis.

In angina, plaques build up slowly in the coronary arteries, reducing blood flow to the parts of the heart muscle beyond the plaques. Often symptoms are first noticed during exercise, when the cardiac muscle is working harder and needing more oxygen. The narrowed coronary arteries cannot supply enough oxygenated blood and the heart muscle resorts to **anaerobic respiration**. This causes a gripping pain in the chest that can extend into the arms, particularly the left one, and the jaw, and often also causes breathlessness. The symptoms of angina subside once exercise stops, but the experience is painful and frightening.

Fortunately, most angina is relatively mild. It can be helped by taking regular exercise, losing weight and not smoking. The symptoms can be treated by drugs that cause rapid dilation of the coronary blood vessels so that they supply the cardiac muscle with the oxygen it needs. However, if the blockage of the coronary arteries continues to get worse, so will the symptoms of the angina. Other drugs are then used to dilate the blood vessels and reduce the heart rate. Unfortunately, drugs cannot solve a severe problem permanently. A **stent** may be inserted into the coronary arteries to hold them open, or heart bypass surgery may be carried out.

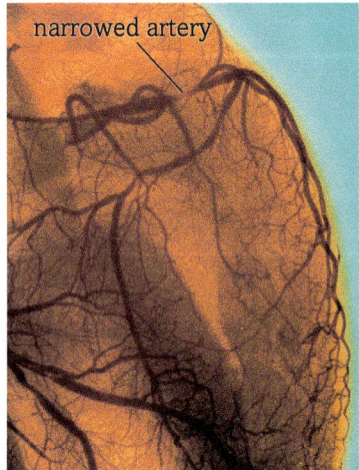

fig D Injecting the blood vessels with special dye allows doctors to see where the coronary arteries are narrowing due to atherosclerosis so they can treat the problem.

In a myocardial infarction, often called a heart attack, one of the branches of the coronary artery becomes completely blocked and part of the heart muscle is permanently starved of oxygen.

Many heart attacks are caused by a blood clot as a result of atherosclerosis. The wall of an artery affected by a plaque is stiffened, making it much more likely to suffer cracks or damage. Platelets come into contact with the damaged surface of the plaque and the clotting cascade is triggered (see **Section 4.3.4**). The plaque itself may rupture and break open, and the cholesterol that is released will also cause the platelets to trigger the blood clotting process. A clot may also form simply because the endothelial lining is damaged, for example by high blood pressure or smoking.

A clot that forms in a blood vessel is known as a **thrombosis**. The clot can rapidly block the whole blood vessel, particularly if it is already narrowed by a plaque. A clot that gets stuck in a coronary artery is known as a coronary thrombosis. The clot can block the artery, starving the heart muscle beyond that point of oxygen and nutrients, and this often leads to a heart attack (see **fig E**).

During a heart attack there is chest pain in the same areas as during an angina attack but it is much more severe. The pain may occur at any time, although exercise may trigger it, and it often lasts for several hours. Death may occur very rapidly with no previous symptoms, or it may happen after several days of feeling tired and suffering symptoms misinterpreted as indigestion.

It is very important to react quickly if you suspect someone is having a heart attack. Call an ambulance and give them two full-strength aspirin tablets to help stop the blood clotting.

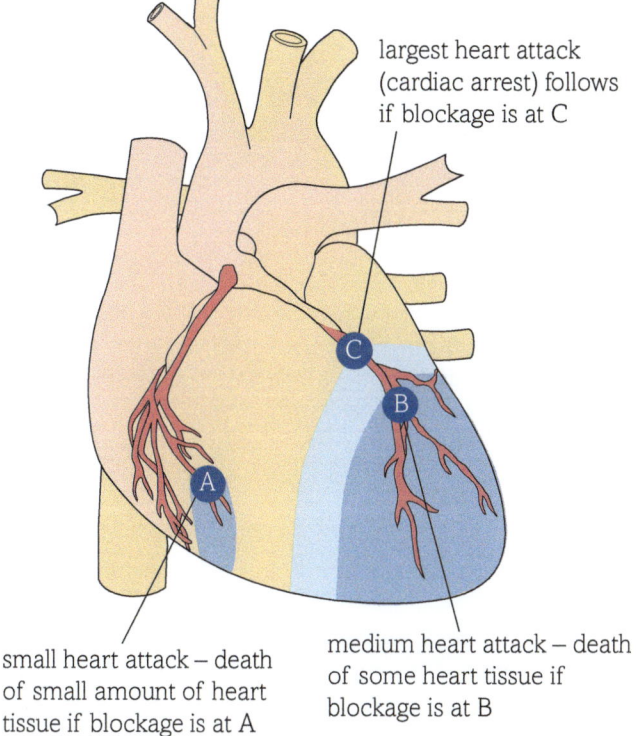

largest heart attack (cardiac arrest) follows if blockage is at C

small heart attack – death of small amount of heart tissue if blockage is at A

medium heart attack – death of some heart tissue if blockage is at B

fig E The size and severity of a heart attack is closely related to the position of the blockage in the coronary artery.

Strokes

A stroke is caused by an interruption to the normal blood supply to an area of the brain. This may be due to bleeding from damaged capillaries or a blockage cutting off the blood supply to the brain, usually caused by a blood clot, an atheroma or a combination of the two. Sometimes the blood clot forms somewhere else in the body and is carried in the bloodstream until it gets stuck in an artery in the brain. The damage happens very quickly. A blockage in one of the main arteries leading to the brain causes a very serious stroke that may lead to death. In one of the smaller arterioles leading into the brain the effects may be less disastrous.

The symptoms of strokes vary, depending on how much of the brain is affected. Very often the blood is cut off from one part or one side of the brain only. Symptoms include dizziness, confusion, slurred speech, blurred vision or loss of part of the vision (usually just in one eye) and numbness. In more severe strokes there can be paralysis, usually just down one side of the body.

The outcome of either a heart attack or a stroke usually depends on how soon the person is treated. The sooner patients are given treatment, including clot-busting drugs, the more likely they are to survive. For example, if treatment is given rapidly 75% of patients who survive the first week after a heart attack can expect to be alive five years later.

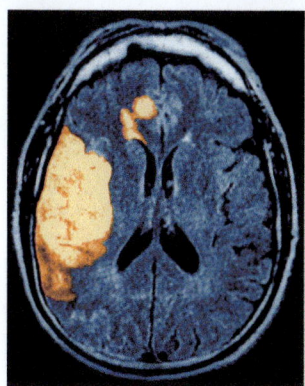

fig F The damage caused in the brain by a major stroke can be seen on the left of this MRI brain scan. The healthy part of the brain is shown in blue.

Questions

1 Plaque formation is one of the few examples in the human body of a positive feedback (where a change encourages even greater change, rather than returning to the normal state). Explain why this happens, and why it is so dangerous.

2 (a) Describe in detail the role of atherosclerosis in cardiovascular disease.

 (b) Summarise the similarities and differences between a heart attack and a stroke.

3 The build-up of a fatty plaque in the artery leads to changes in the blood flow and an increase in the blood pressure. Plan a way of modelling this that could be used on a television programme to explain high blood pressure to young people.

Key definitions

Cardiovascular diseases are diseases of the heart and circulatory system, many of which are linked to atherosclerosis.

Atherosclerosis is a condition in which yellow fatty deposits build up on the lining of the arteries, causing them to be narrowed and resulting in many different health problems.

Plaques are yellowish fatty deposits that form on the inside of arteries in atherosclerosis.

An **atheroma** is another term for a plaque formed on the arterial lining.

An **aneurysm** is a weakened, bulging area of artery wall that results from a build up behind a blockage caused by plaques.

Angina is a condition in which plaques build up and reduce the blood flow to the cardiac muscle through the coronary artery. It results in pain during exercise.

A **myocardial infarction** (heart attack) takes place when atherosclerosis leads to the formation of a clot that blocks the coronary artery entirely and deprives the heart muscle of oxygen so it dies. It can stop the heart functioning.

Anaerobic respiration is cellular respiration that takes place in the absence of oxygen.

A **stent** is a metal or plastic mesh tube that is inserted into an artery affected by atherosclerosis to hold it open and allow blood to pass through freely.

A **thrombosis** is a clot that forms in a blood vessel.

By the end of this section, you should be able to...

● Explain the factors that increase the risk of developing atherosclerosis

Risk describes the **probability** that a particular event will happen. Probability means the chance or likelihood of the event, calculated mathematically. For example the probability of throwing a 1, or of throwing any other number up to 6, can be expressed in one of three different ways, as 1 in 6, 0.166 66 recurring (0.17), or 17%. Just as you can calculate your 'risk' of throwing a 1 with a die, it is possible to work out your risk of developing certain diseases, or of dying from a particular cause.

The actual risk of doing something is not always the same as the perceived risk. Perception of risk is based on a variety of factors that include familiarity with an activity, how much you enjoy the activity and whether you approve of it. The mathematical risk may play very little part in building up your personal perception of risk.

Epidemiology

If you know the number of people in a population who are affected by a particular disease it is possible to calculate the average risk of a person within that population developing that disease. But the risk is higher for some people than others, depending on their lifestyle and the genes they have inherited.

By looking at people who have certain things in common, for example smoking, and comparing their risk of disease with the average risk for the whole population, it is possible to identify the factors that may be involved in causing that disease. This plays a very important role in identifying risk factors involved in causing a disease. Using these techniques, it appears that there are a number of factors that increase the likelihood that a person will develop atherosclerosis. It is a **multifactorial disease**, as many things influence your chances of being affected.

When there is a similarity in the pattern between the mortality from a disease such as atherosclerosis with a lifestyle factor such as smoking or lack of exercise, it suggests a link or relationship between the two. A link like this is called a **correlation**. This does not prove, however, that one is the cause of the other. They could both be caused by something else, which would explain why they change in the same way. Correlation is not the same as causation – further research is always needed to demonstrate a causal line.

The results from many epidemiological studies have identified a range of risk factors linked to atherosclerosis and the health problems of the cardiovascular system associated with it. These factors break down into two main groups – those you can not change and those you can do something about.

Non-modifiable risk factors for atherosclerosis

There are three main risk factors for atherosclerosis which you cannot (at present) control:

• Genes: studies show there is a genetic tendency in some families and in some ethnic groups to develop atherosclerosis or other cardiovascular diseases (CVDs). There are a variety of genetic tendencies – the arteries may be more easily damaged, there may be a tendency to develop hypertension or the cholesterol metabolism may be faulty.

• Age: as you get older, your blood vessels begin to lose their elasticity and to narrow slightly, making you more likely to suffer from atherosclerosis and CVDs, particularly heart disease.

- Sex: under the age of 50, men are statistically significantly more likely to suffer from atherosclerosis (and other CVDs) than women. The female hormone oestrogen appears to reduce the build-up of plaque, giving women some protection against atherosclerosis until the menopause when oestrogen levels fall.

Modifiable (lifestyle) risk factors for atherosclerosis

Your lifestyle can affect your risk of developing atherosclerosis in the future. Epidemiological studies have shown links with smoking, stress, diet, weight, lack of activity and high blood pressure. These are the factors we can do something about.

Smoking and atherosclerosis

Studies have shown that smokers are far more likely to develop atherosclerosis than non-smokers with a similar lifestyle, and 9 out of 10 people needing heart bypass surgery or stents as a result of atherosclerosis are smokers. In 2007, a Spanish study showed a clear correlation between smoking and the incidence of death from atherosclerotic heart disease. Causation was established by further research. For example, studies found that some of the chemicals in tobacco smoke can damage the artery linings, making the build-up of plaque more likely, and cause the arteries to narrow, raising the blood pressure and increasing the risk of atherosclerosis.

Exercise and atherosclerosis

Regular exercise helps lower blood pressure, prevent obesity and diabetes, lower blood cholesterol levels, balance lipoproteins and reduce stress. All of these also lower your risk of developing atherosclerosis. A study on 10 269 male Harvard University graduates aged between 45 and 84 showed that the men who changed from being inactive to taking regular exercise had a 23% lower mortality over the life of the study than their peers who did not exercise, and the main cause of the deaths was atherosclerosis and the linked CVDs. A study of 72 488 female nurses showed the same benefits for women – the more active women had a significantly lower risk of developing atherosclerosis and other CVDs. Various studies have shown that exercise both reduces the formation of plaques in the arteries and also keeps plaques that are present more stable and less likely to rupture.

Weight and atherosclerosis

An increasing number of studies suggest that being overweight does not directly affect your risk of developing atherosclerosis, but it is a very important *indicator* of risk. This is because other factors that are a direct result of being overweight *do* increase the risk of atherosclerosis. These include:

- High blood pressure – increases the risk of damage to blood vessel linings, and so of plaque formation

- Type 2 diabetes – this can result in damage to the lining of the blood vessels increasing the risk of plaque formation.

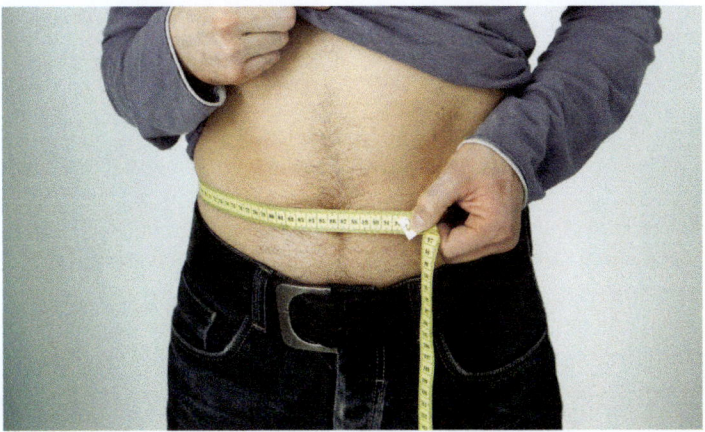

fig A Simply measuring your waist to hip ratio can give a good prediction of your risk of developing atherosclerosis. It should be below 0.8 for women and below 1.0 for men.

Stress and atherosclerosis

High stress levels increase the risk of atherosclerosis in a number of ways. Stress causes the release of cytokines that trigger an inflammatory response in the blood vessels, leading to plaque formation. It also tends to cause raised blood pressure that is both a cause and symptom of atherosclerosis.

Diet and atherosclerosis

There have been many studies on how diet is linked to atherosclerosis, some looking at general diet and some looking at specific foods. The evidence is very mixed and very difficult to interpret.

Over the last 50 years or so, a number of studies showed that a high intake of saturated fats seemed to be associated with high blood cholesterol levels. Cholesterol is involved in plaque formation in atherosclerosis, so this suggested a cause for the link between a high-fat diet and CVDs. But then, in 2014, a major study published by scientists from prestigious institutions including the universities of Cambridge, Oxford and Bristol, Imperial College and the Medical Research Council suggested that the links found between diets high in saturated fats and atherosclerosis and CVDs had been a correlation and nothing more. Looking at all of the data from 72 studies they found that diets high in saturated fats did not appear to be linked directly with increases in atherosclerosis and CVDs.

Our picture of the relationship between fat in the diet and cholesterol in the blood is further complicated by lipoproteins which transport lipids around the body:

- Low-density lipoproteins (LDLs) are formed from *saturated* fats, cholesterol and protein and bind to cell membranes before being taken into the cells. If levels of some forms of LDLs are high, your cell membranes become saturated and so more LDL cholesterol is left in your blood.

- High-density lipoproteins are formed from *unsaturated* fats, cholesterol and protein. They carry cholesterol from body tissues to the liver to be broken down, lowering blood cholesterol levels. HDLs can even help to remove cholesterol from fatty plaques on the arteries reducing the risk of atherosclerosis.

The balance of these lipoproteins in your blood is now recognised as an indication of your risk of developing atherosclerosis and the associated CVDs.

Blood cholesterol and LDL/HDL levels are not simply related to diet. The way your body deals with the fats you eat and with the levels of cholesterol and balance of lipoproteins in your blood are all linked to your genetic make-up. Some people can deal with almost any amount of fat and maintain a good balance of LDLs and HDLs. Other people cannot cope so well and almost any amount of fat in the diet is reflected in raised blood cholesterol levels.

Links between factors

Although it is difficult to separate out factors and identify those that are having an effect, many epidemiological studies are starting to find that an increased risk of developing a disease is due to a combination of factors. For example, evidence now suggests that smoking not only increases your risk of atherosclerosis because of its effect on your blood vessels and blood pressure, but it also increases the impact of raised LDL/HDL ratios on your risk of dying from atherosclerosis-related CVDs.

Preventing atherosclerosis

As epidemiological studies of links between risk factors and CVDs become more sophisticated, and as scientific research discovers more reasons why some factors could cause atherosclerosis, the advice about what is 'good' and what is 'bad' for us changes. Current evidence suggests that eating a balanced diet with a variety of fats and plenty of fruit and vegetables helps prevent atherosclerosis. Not smoking, maintaining a healthy weight to avoid high blood pressure and type 2 diabetes, reducing constant stress and getting plenty of exercise are also likely to help. It is important to take action as early as possible as there is clear evidence of the early signs of atherosclerosis in teenagers and even young children if known risk factors are already in place.

Questions

1 Explain the difference between risk, correlation and causation.

2 In the UK the mortality from heart disease is about three times greater for smokers than for non-smokers. Explain why this does not mean that an individual person who smokes will die from heart disease.

3 (a) What is the difference between modifiable and non-modifiable risk factors?

(b) Give two non-modifiable and two modifiable risk factors for atherosclerosis.

(c) For each factor you have chosen, explain how it increases the risk of developing atherosclerosis.

4 (a) Investigate the 2014 report that suggested that the link between fat in the diet and the risk of atherosclerosis and CVDs was a correlation but that it was not causative. Summarise the work and findings.

(b) This report was widely reported and many ordinary people felt confused and upset by it. Many others were very pleased by the findings. Suggest why people reacted in these ways.

(c) Discuss the importance of publishing scientific findings to both the scientific community and the general population. Explain current dietary advice to help avoid atherosclerosis.

Key definitions

Probability is a measure of the chance or likelihood that an event will take place.

A **multifactorial disease** results from the interplay of many different factors rather than having one simple cause.

A **correlation** is a strong tendency for two sets of data to vary together.

By the end of this section, you should be able to...

- explain how the interchange of substances occurs through the formation and reabsorption of tissue fluid, including the effects of hydrostatic pressure and oncotic pressure

- describe how tissue fluid that is not reabsorbed is returned to the blood via the lymph system

The blood is the main transport system of the body, but it does not deliver material directly to the cells. The movement of substances by diffusion into and out of the cells is from the **tissue fluid** that bathes every cell. Tissue fluid is formed from the blood.

The formation of tissue fluid

The junctions between the single cells that make up the capillary walls are not tight. As a result, the capillary walls are very permeable to everything in the blood except the erythrocytes and the large plasma proteins. Even white blood cells can squeeze out through the gaps between the epithelial cells of the capillary walls.

So in the capillary beds, a fluid that is basically plasma without the plasma proteins and the red blood cells moves out of the capillaries and bathes the individual cells of the body. This is the tissue fluid, from which substances diffuse into cells along concentration gradients and waste products diffuse out into the tissue fluid. But by the time the blood leaves the capillary beds, 90% of the tissue fluid is back inside the blood vessels. How does this work?

Two factors are involved:

- Water potential: the plasma proteins, and albumin in particular, exert an osmotic effect. They give the blood a relatively constant and fairly low water potential of −3.3 kPa. The tissue fluid surrounding the capillaries has a water potential of about −1.3 kPa. Water moves from an area of high water potential to an area of lower water potential, so the tendency is for water to move into the blood by osmosis. The pressure behind this movement is about −2 kPa. It is called **oncotic pressure**.)

water potential inside the capillary − water potential outside = tendency for water to move into or out of the capillary by osmosis (the oncotic pressure)

−3.3 kPa − (−1.3) kPa = −2.0 kPa

- **Hydrostatic pressure:** this is the residual pressure from the heartbeat that is still present as the blood enters the arterial end of the capillary beds, which tends to force fluid out through the leaky capillary walls.

The balance between oncotic pressure and hydrostatic pressure determines whether tissue fluid moves out of or into the capillaries.

The arterial end of the capillaries

As the blood flows from the arterioles into the capillary bed, the hydrostatic pressure tending to force fluid out of the capillary is relatively high at around 3.3 kPa.

The oncotic pressure of the plasma in the capillary is fairly constant at −2.0 kPa (see above):

−2.0 + 3.3 = 1.3 kPa

The hydrostatic pressure forcing water out is higher than the oncotic pressure moving water in, so fluid is squeezed out of the capillary and fills the spaces around all of the cells. This is the tissue fluid. It contains all of the dissolved substances present in the blood, apart from the large plasma proteins and the erythrocytes. This is where diffusion between the blood and the cells takes place.

The venous end of the capillaries

The blood moves steadily through the capillary system, and as it does so, the balance of forces between the hydrostatic and oncotic pressures changes. The hydrostatic pressure falls to around 1.0 kPa for two reasons:

- The pressure from the pulse is completely lost.

- Fluid moves out of the capillaries to form tissue fluid so the volume of blood in the capillaries is lowered.

However, the oncotic pressure of the blood is still around −2 kPa, due to the constant presence of the large plasma proteins:

−2.0 + 1.0 = −1.0 kPa

Now the pressure exerted by water moving into the capillaries by osmosis is greater than the hydrostatic pressure forcing fluid out, so water returns to the capillaries. By the time the blood enters the venules, most – but not all – of the tissue fluid has returned to the blood vessels.

To summarise:

When hydrostatic pressure is greater than the oncotic pressure (osmotic tendency for water to move into the capillary), fluid leaves the capillary and tissue fluid forms.

When hydrostatic pressure is lower than the oncotic pressure, water moves into the capillary and tissue fluid is lost.

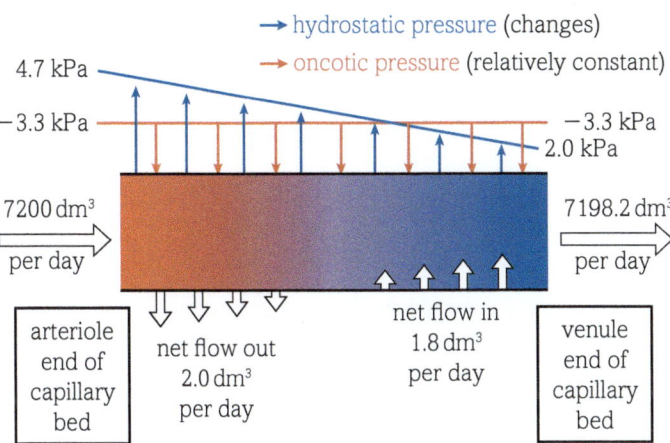

→ hydrostatic pressure (changes)
→ oncotic pressure (relatively constant)

fig A The formation of tissue fluid and the restoration of the blood volume through the capillary network.

The formation of lymph

The tissue fluid must be reabsorbed – it is being constantly formed and so it must be constantly removed or our tissues would swell up like balloons. Most of the tissue fluid returns to the capillaries. However, around 10% does not, as you will see from the calculations above. This remaining 10% of the tissue fluid drains into blind-ended tubes called **lymph capillaries**, and becomes **lymph**. Lymph capillaries join up to form larger and larger vessels. Lymph vessels are similar to veins in two ways – they have one-way valves that prevent the lymph flowing backwards and lymph is moved through the vessels by the contraction of the body muscles as we move about. The lymph is returned to the blood in the neck area, where it joins the left and right subclavian veins that are found under the collar bone.

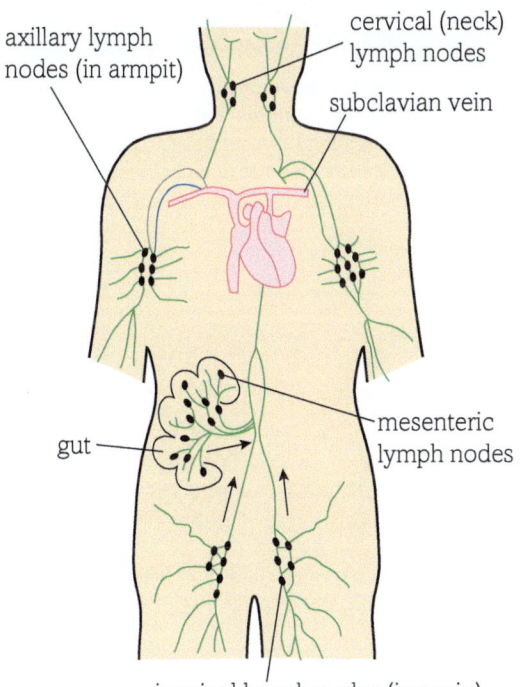

fig B The lymphatic system of vessels and glands is important for returning fluid to the blood and for destroying the pathogens that cause communicable diseases.

Lymph is very similar to tissue fluid, although it has fewer nutrients and less oxygen as these have been taken up by the cells from the tissue fluid. It may also have a high level of fatty acids that are absorbed directly into the lymph system in the villi of the small intestine.

An important feature of the lymph system is the **lymph glands**, which are found at intervals along the lymph vessels. This is where lymphocytes accumulate and produce antibodies that are used by the body to fight disease. The antibodies are emptied into the blood with the lymph at the subclavian veins. The lymph glands also remove bacteria and other pathogens to be taken in and destroyed by phagocytes. One sign that doctors look for in their patients is enlarged lymph glands in the neck, armpits, stomach or groin because enlarged lymph glands indicate that the body is fighting an infection.

Learning tip

Be very clear about the differences between plasma, tissue fluid and lymph.

Questions

1 What are the similarities and differences between plasma, tissue fluid and lymph?

2 If someone sits or stands for a long time they may develop swollen fingers and ankles. Suggest reasons for this based on your knowledge of the formation and removal of tissue fluid.

Key definitions

Tissue fluid is the fluid that surrounds all the cells in the body.

Oncotic pressure is the tendency for water to move into the capillaries by osmosis.

Hydrostatic pressure is the residual pressure from the heartbeat that is still present as the blood enters the arterial end of the capillary beds, that tends to force fluid out through the leaky capillary walls.

Lymph capillaries are the blind tubes that carry the lymph away from the tissues.

Lymph is the fluid that travels in the lymphatic system.

Lymph glands are the glands in the lymph system that contain lymphocytes that make antibodies.

AN ARTIFICIAL PACEMAKER

The sinoatrial node (SAN) of the heart produces a regular rhythm of contractions throughout life. However, if for any reason the SAN fails to maintain a regular heart rate, an individual may need an artificial pacemaker inserted. This takes over from the SAN and maintains the heart rate within set boundaries. The text below describes how an artificial pacemaker works.

HOW DOES A PACEMAKER WORK?

Introduction

Pacemaker implantation is a surgical procedure where a small electrical device called a pacemaker is implanted in your chest.

The pacemaker sends regular electrical pulses that help keep your heart beating regularly.

Having a pacemaker fitted can greatly improve your quality of life if you have problems with your heart rhythm, and the device can be lifesaving for some people.

Pacemaker implantation is one of the most common types of heart surgery carried out in the UK. During 2012–13 in England, more than 40 000 people had a pacemaker fitted.

How does a pacemaker work?

The pacemaker is a small metal box weighing 20–50g. It is attached to one or more wires, known as pacing leads, which run to your heart.

The pacemaker contains:

- a **battery**, which usually lasts six to 10 years depending on how advanced the device is (more advanced pacemakers tend to use more energy so have a shorter battery life)

- a **pulse generator**, a tiny computer circuit that converts energy from the battery into electrical impulses, which flow down the wires and stimulate your heart to contract

The rate at which these electrical impulses are sent out is called the discharge rate.

Almost all modern pacemakers work on demand. This means that they can be programmed to adjust the discharge rate in response to your body's needs. If the pacemaker senses that your heart has missed a beat or is beating too slowly, it sends signals at a steady rate. If it senses that your heart is beating normally by itself, it does not send out any signals.

Most pacemakers have a special sensor that recognises body movement or your breathing rate. This allows them to speed up the discharge rate when you are active. Doctors describe this as rate responsive.

Why do I need a pacemaker?

The heart is essentially a pump, made of muscle, which is controlled by electrical signals.

These signals can become disrupted for several reasons, which can lead to a number of potentially dangerous heart conditions, such as:

- an **abnormally slow heartbeat** (bradycardia) or an **abnormally fast heartbeat** (supraventricular tachycardia) – caused by damage to part of the heart called the sinoatrial node

- **heart block** – where your heart beats irregularly because the electrical signals that control your heartbeat are not transmitted properly

- **cardiac arrest** – when a problem with the electrical signals in the heart causes the heart to stop beating altogether

An implantable cardioverter defibrillator (ICD) is a device similar to a pacemaker. This sends a larger electrical shock to the heart that essentially reboots the heart to get it pumping again. Some devices contain both a pacemaker and an ICD.

ICDs are often used as a preventative treatment for people thought to be at risk of cardiac arrest at some point in the future. If the ICD senses that the heart is beating at a potentially dangerous abnormal rate, it will deliver an electrical shock to the heart. This can often help return the heart to a normal rhythm.

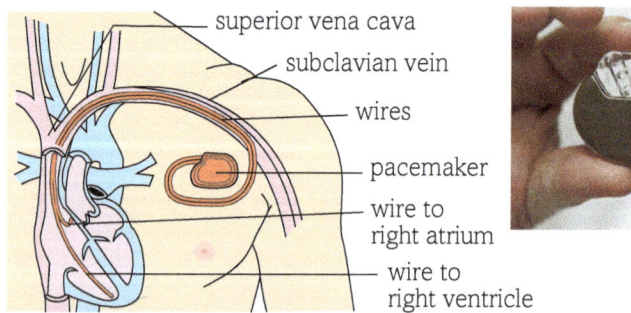

- superior vena cava
- subclavian vein
- wires
- pacemaker
- wire to right atrium
- wire to right ventricle

fig A A heart pacemaker can mean the difference between life and death for people with sick sinus syndrome – the SAN is diseased and only fires around 40 times a minute; with heart block – when the heart just stops for seconds at a time; and with severe and repeated tachycardias.

Where else will I encounter these themes?

1.1   1.2   1.3   1.4   2.1   2.2   2.3   2.4

Let us begin by looking at the material presented:

1. This information on how a heart pacemaker works comes from the NHS Choices website, which aims to provide clear and straightforward explanations on medical conditions and procedures for patients and their relatives and friends.

   a. Comment on whether the writers have achieved their aim of producing a clear explanation of how a heart pacemaker works.

   b. The way this source is worded makes it a suitable article for someone about to have a heart pacemaker fitted. Do you agree or disagree with this statement – and why?

   c. Think about the level of technical language in this piece. In the original, there are no pictures – the images were added for your benefit. Discuss the impact on the usefulness of the article without the picture.

Command word
When you are asked to comment on something, you need to think carefully about the data or information you are given and form your own judgement.

Now use your knowledge of the way the heart beats and the natural pacemaker system of the heart to answer the following questions:

2. How do doctors detect the type of heart arrhythmias that might need an artificial pacemaker?

3. Consider each of the three conditions describe in the caption to **fig A** and explain why a person with each of these problems would need an artificial pacemaker fitted.

4. What are the similarities and differences between the natural pacemaker system of the heart and an artificial heart pacemaker?

5. Discuss the potential limitations of an artificial pacemaker compared to the natural control of the heart.

6. Investigate why different types of heart pacemakers are needed and what they are used for. Present your findings in a table to make the comparisons easy to see.

Activity

Coronary heart disease

Certain arrhythmias of the heart can be treated effectively with an artificial pacemaker. However, some of the most serious heart conditions, affecting millions of people worldwide, cannot be treated with a pacemaker. In 2012, ischaemic heart disease killed about 7.4 million people around the world.

Using your knowledge of the structure and functions of the heart, along with your knowledge of the blood and the biochemistry of lipids and other molecules, research into this killer disease. Investigate the incidence, the risk factors, the symptoms of disease and how it can be treated and avoided. Produce a 5–10-minute presentation that informs your audience about how the healthy heart works and summarises all the key points about ischaemic heart disease.

● From the website of the NHS, www.nhs.uk/conditions/PacemakerImplantation/Pages/Introduction.aspx

Exam-style questions

1 (a) Draw a labelled diagram to show the structure of an artery. [3]

(b) Explain how the structure of an artery relates to its function. [2]

(c) Give **two** differences between the structure of a vein and the structure of a capillary. [2]

[Total: 7]

2 Many animals have hearts that pump blood through a network of blood vessels.

(a) The table below refers to blood flow in the four major blood vessels of the human heart. If the statement is correct, place a tick (✔) in the appropriate box and if the statement is incorrect, place a cross (✗) in the appropriate box.

Name of blood vessel	Carries blood away from the heart	Carries oxygenated blood
Aorta		
Vena cava		
Pulmonary artery		
Pulmonary vein		

[4]

(b) Humans and fish are both animals that have a heart and a network of blood vessels. However, there are some differences in their circulatory systems.

The diagrams below illustrate a human circulatory system and the circulatory system in a fish.

Human circulatory system

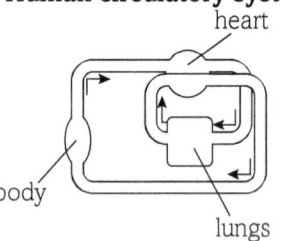

Fish circulatory system

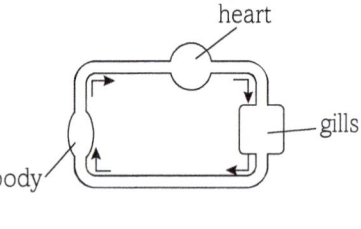

The arrows show the direction of blood flow.

(i) Using the information in the diagram, describe the circulation of blood in a fish. [3]

(ii) Using the information in both diagrams, suggest the advantages that the human circulatory system has compared with that of the fish. [2]

(c) The heart of an insect is a long tube. It pumps blood into the body cavity so that blood surrounds the cells. The blood then passes back into the heart from the body cavity.

The diagram below illustrates the circulatory system of an insect.

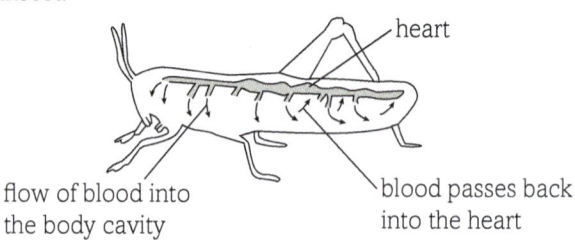

flow of blood into the body cavity

blood passes back into the heart

Suggest why the insect does not need blood vessels to transport its blood around the body. [2]

[Total: 11]

3 (a) Read through the following passage about the heart and its major blood vessels, then write on the dotted lines the most appropriate word or words to complete the passage.

The mammalian heart consists of four chambers, two upper chambers called .... and two lower chambers called ventricles.

The .... carries oxygenated blood away from the .... ventricle to the cells of the body and the pulmonary .... carries deoxygenated blood to the lungs.

The .... returns deoxygenated blood back to the heart from the body. [5]

(b) The diagram below shows the structure of the heart.

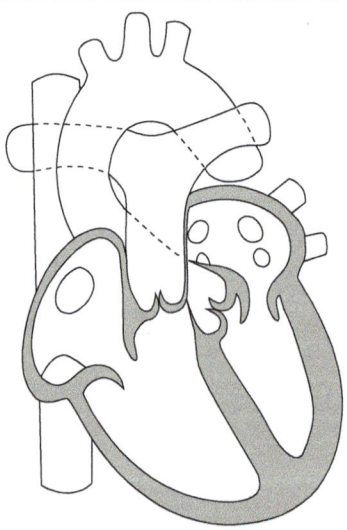

Suggest which stage of the cardiac cycle is shown in the diagram and give a reason for your answer. [2]

[Total: 7]

4 (a) The diagram below shows a section through the heart of a mammal.

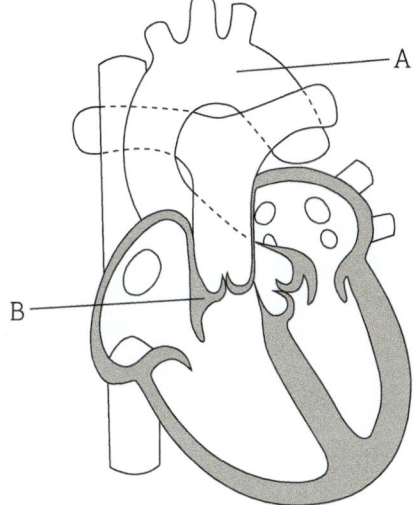

Name the parts labelled A and B.   [2]

(b) Heart muscle has a relatively high demand for oxygen. Explain how heart muscle is supplied with oxygen.   [3]

(c) During the cardiac cycle, the atria contract and then the ventricles contract.

Explain how this sequence of events is coordinated.   [3]

(d) The table below shows the effect of exercise on blood flow to the muscles of an adult man.

Blood flow at rest /dm³ min⁻¹	Blood flow during exercise/dm³ min⁻¹
1.0	16.0

Suggest an explanation for the change in blood flow as shown in the table.   [3]

[Total: 11]

5 (a) The graph below shows oxygen dissociation curves for
       (i) human myoglobin and
       (ii) human haemoglobin.

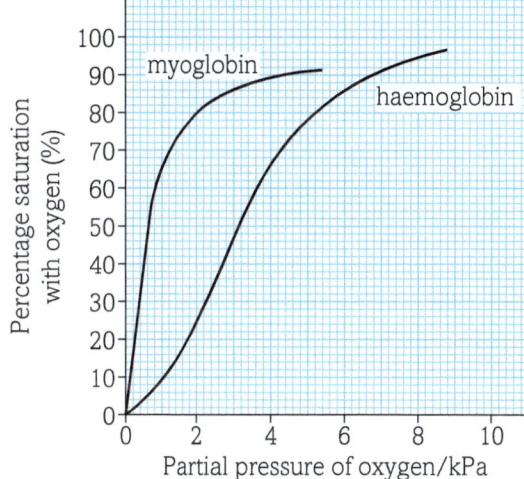

From the graph, find the partial pressures of oxygen at which myoglobin and haemoglobin are 50% saturated with oxygen.   [2]

(b) Describe the role of myoglobin.   [3]

(c) At increased partial pressures of carbon dioxide, the oxygen dissociation curve for haemoglobin moves to the right. This is known as the Bohr effect.

Explain the importance of the Bohr effect.   [4]

[Total: 9]

6 Tissue fluid is formed as a result of the blood pressure forcing water and dissolved solutes, such as glucose and mineral ions, out of capillaries.

(a) Explain how the structure of a capillary is related to the formation of tissue fluid.   [3]

(b) Suggest why proteins are present in the plasma but are not normally present in tissue fluid.   [2]

(c) Tissue fluid is reabsorbed back into capillaries as a result of the osmotic effect of the plasma proteins. Suggest why this reabsorption is reduced in a person on a protein-deficient diet.   [2]

[Total: 7]

TOPIC 4
Exchange and transport

CHAPTER

4.4 > Transport in plants

Introduction

Legumes such as peas, beans and clover have special nodules on their roots filled with *Rhizobium* spp. bacteria. These bacteria convert inert nitrogen gas from the air into biologically useful ammonia. A large proportion is absorbed by the plant and used to produce amino acids, the building blocks of proteins, from the carbohydrates made by photosynthesis. In return plants provide the bacteria with high levels of carbohydrates. Scientists have discovered that plants use their transport systems to select the most efficient bacteria and so get the maximum amount of ammonia. If the bacteria in a nodule produce a lot of ammonia, the plant transport system delivers lots of dissolved sugars for the bacteria to feed on. However, if the bacteria in a nodule are less efficient, the transport system to the nodule is cut off, depriving the bacteria of sugars and starving the nodule to death!

In this chapter you will be looking at how substances are transported around plants. You will be considering the two main transport tissues, the xylem and the phloem, and comparing their structure and functions.

You will also be looking at how water is moved into the plant from the soil, and then across the cells of the roots by diffusion and osmosis into the xylem. You will be learning about transpiration and the transpiration stream, and the forces that enable a plant to transport water from the roots to the shoots, sometimes 100 m or more in the air. Through practical investigations you will discover which factors affect the rate of water uptake – and so indirectly measure water movement through the plant.

You will also be looking at the translocation of sugars around a plant. The carbohydrates made by photosynthesis in the leaves and other green areas of the plant must be carried to every cell for cellular respiration to take place. Transport of sugars in the phloem, unlike the transport of water in the xylem, needs living cells for it to take place.

All the maths you need

- Recognise and make use of appropriate units in calculations (*e.g. rate of water uptake by a plant*)
- Find arithmetic means (*e.g. the mean number of stomata on different sides of a leaf, the mean rate of water uptake under different conditions*)
- Represent a range of data in a table with clear headings, units and consistent decimal places (*e.g. measurements of water uptake under different conditions*)
- Plotting variables from experimental data (*e.g. graph of temperature against rate of transpiration showing data from a practical investigation*)
- Understanding that $y = mx + c$ represents a linear relationship (*e.g. the relationship between rate of water uptake and the temperature of the surrounding air*)
- Determining the intercept of a graph (*e.g. graph showing data from a practical investigation into water loss from a plant*)
- Calculating rate of change from a graph showing a linear relationship (*e.g. rate of water uptake as volume taken up per unit time*)

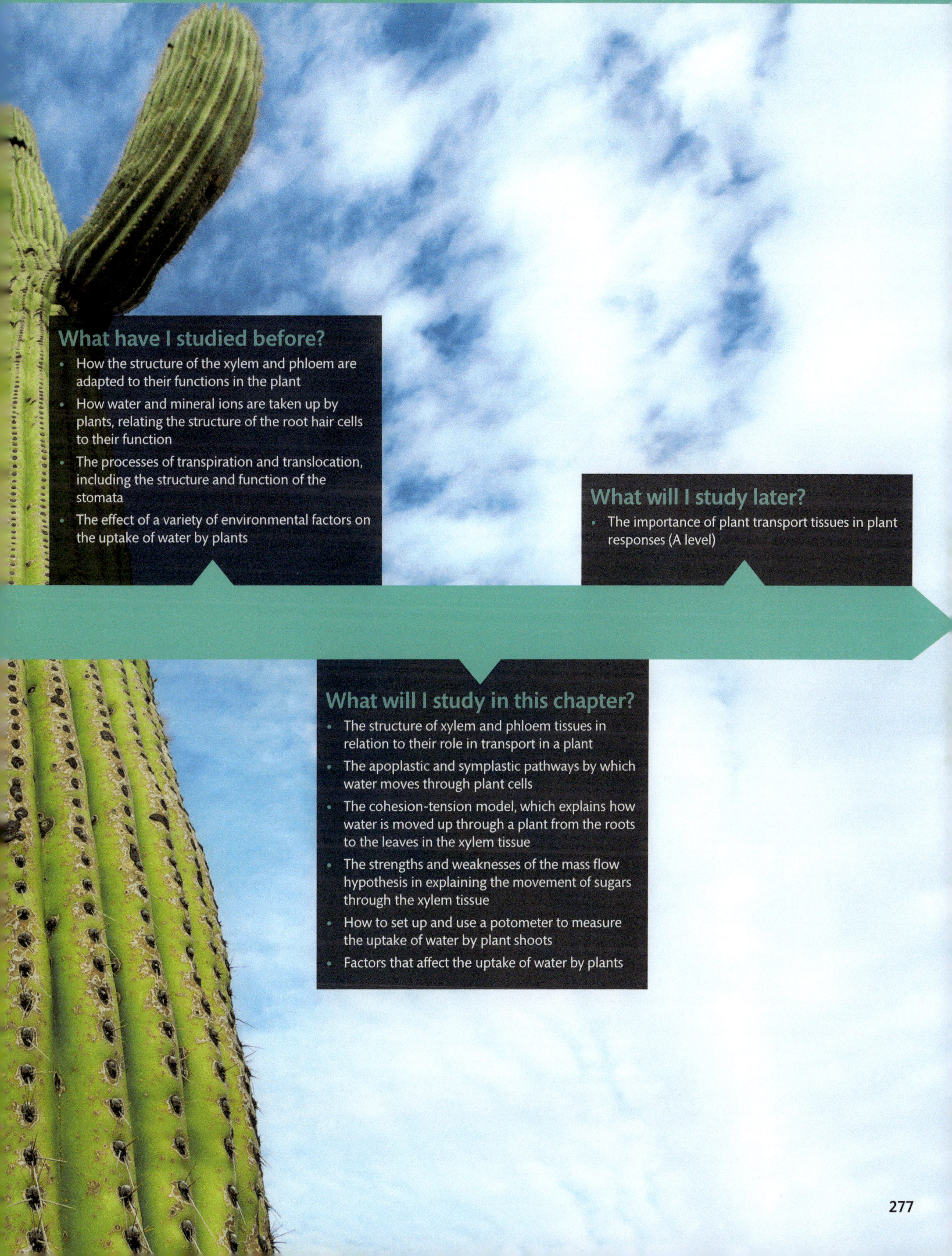

What have I studied before?

- How the structure of the xylem and phloem are adapted to their functions in the plant
- How water and mineral ions are taken up by plants, relating the structure of the root hair cells to their function
- The processes of transpiration and translocation, including the structure and function of the stomata
- The effect of a variety of environmental factors on the uptake of water by plants

What will I study later?

- The importance of plant transport tissues in plant responses (A level)

What will I study in this chapter?

- The structure of xylem and phloem tissues in relation to their role in transport in a plant
- The apoplastic and symplastic pathways by which water moves through plant cells
- The cohesion-tension model, which explains how water is moved up through a plant from the roots to the leaves in the xylem tissue
- The strengths and weaknesses of the mass flow hypothesis in explaining the movement of sugars through the xylem tissue
- How to set up and use a potometer to measure the uptake of water by plant shoots
- Factors that affect the uptake of water by plants

By the end of this section, you should be able to...

● explain how the structure of xylem and phloem tissues is related to their role in transport

Plants, like other large, multicellular organisms, need a system to transport substances to all the cells. The main transport tissues in plants are the **xylem** and **phloem** and they are found associated together in vascular bundles throughout the plant.

Xylem tissue carries water and dissolved minerals from the roots to the photosynthetic parts of the plant. The movement in the xylem is always upwards. The xylem is made up of several different types of cells, most of which are dead. Long tubular structures called xylem vessels are the main functional units of the xylem.

Plants make glucose by photosynthesis and convert this to sucrose to be transported (see **Section 1.2.1** for details of the different carbohydrates). Phloem is living tissue made of phloem cells that transport the dissolved product of photosynthesis from the leaves to where it is needed for growth or storage as starch. The flow through phloem can go both up and down within the plant.

Cambium is a layer of unspecialised cells that divide, giving rise to more specialised cells that in turn form both the xylem and the phloem.

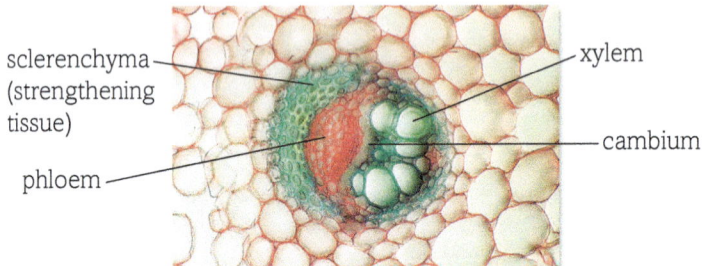

sclerenchyma (strengthening tissue)

phloem

xylem

cambium

fig A The xylem and phloem in a vascular bundle.

Xylem

The xylem starts off as living tissue. The first xylem to form is called the **protoxylem**. It is capable of stretching and growing because the walls are not fully lignified. The cellulose microfibrils in the walls of the xylem vessels are laid down more or less vertically in the stem, which increases the strength of the tube and allows it to withstand the compression forces from the weight of the plant pressing down on it. As the stem ages and the cells stop growing, increasing amounts of lignin are laid down in the cell walls. As a result the cells become impermeable to water and other substances. The tissue becomes stronger and more supportive, but the contents of the cells die. This lignified tissue is known as the **metaxylem**. The end walls between the cells largely break down so the xylem forms hollow tubes running from the roots to the tip of the stems and leaves.

Water and minerals are transported from the roots to the leaves and shoots in the transpiration stream (see **Section 4.4.2**). Water moves out of the xylem into the surrounding cells either through unlignified areas or through specialised **pits** (holes) in the walls of the xylem vessels. The lignified xylem vessels are very strong and play an important supportive role in the stems of plants, particularly larger plants. In smaller, non-woody plants, support comes mainly from the turgid **parenchyma** cells in the centre, and the **sclerenchyma** and **collenchyma**. This is why young plants wilt if too much water is lost. As woody plants grow older, more xylem tissue is lignified to increase support. In trees this is taken to the limit and lignified xylem makes up the bulk of the trunk of the tree (the wood). The living cells around the cambium are on the outside of the trunk of the tree, just under the bark. A new ring of vascular tissue is formed each year, so the growth rings of the tree are a record of the xylem produced in each growing season. You will look at how water moves through xylem tissue in more detail in **Sections 4.4.2** and **4.4.3**.

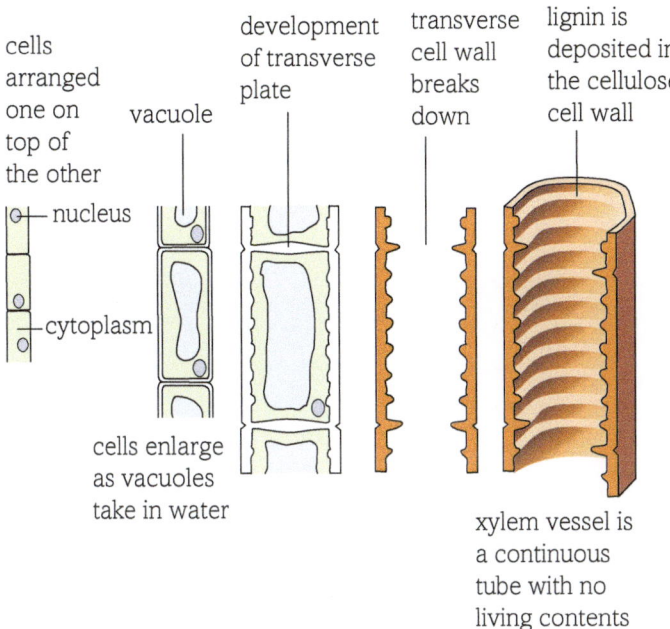

fig B As the xylem vessels develop, they change from living cells to non-living tubes of lignin.

Evidence for the movement of water through the xylem

- If the cut end of a shoot is placed in a solution of eosin dye, the dye can be seen being carried into the transport system and through to the vascular tissue of the leaves.

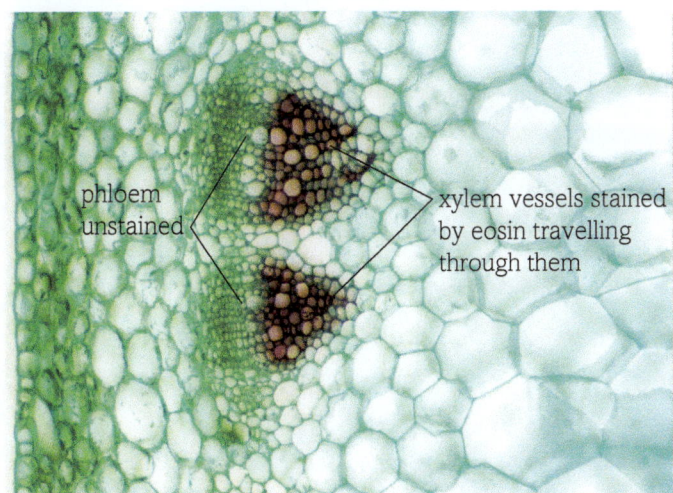

fig C Eosin dye is transported with water in the xylem. When sections of tissue are examined under the light microscope, the eosin is clearly seen in the xylem only.

- Ringing experiments involve removing a complete ring of bark, or killing a complete ring with a steam jet. This destroys the living phloem cells but not the xylem cells. Eosin dye placed in the water shows that the upward movement of water through the plant is unaffected.

- If the plant is provided with water containing radioactive isotopes, these can be traced by **autoradiography** as they move through the plant. Water is seen to travel up the xylem. The movement of minerals in the xylem can be followed in the same way. The technique of autoradiography is very useful for following the transport of substances around plants.

Autoradiography involves several steps:

1 The plant is given a radioactively labelled version of the substance being studied. For example, water containing deuterium (^2H, a radioactive isotope of hydrogen) instead of normal hydrogen can be used to investigate the movement of water through the xylem.

2 The radioactive substance is taken up in the same way by the plant as the normal isotope.

3 The substance can then be tracked by placing the plant against photographic film for a while to produce an autoradiograph. The labelled substance causes the photographic film to shadow, revealing the areas where it has accumulated. The radioactive label can also be traced by examining each area of the plant separately using a scintillation counter. This shows which parts of the plant, or even which organelles, have incorporated the radioactive substance.

The structure of the phloem

Mature phloem, unlike mature xylem, is a living tissue that transports food in the form of organic solutes around the plant from the leaves where they are made by photosynthesis to the tissues where they are needed. Materials in the phloem can be transported both up and down the stems.

The **phloem sieve tubes** (sieve tube elements) are made up of many cells joined together to make very long tubes that run from the highest shoots to the end of the roots. However, the phloem cells do not become lignified and so the contents remain living. The walls between the cells become perforated to form specialised **sieve plates** and the phloem contents flow through the holes in these plates. As the gaps in the sieve plates form, the nucleus, the tonoplast and some of the other organelles break down. The phloem sieve tube becomes a tube filled with phloem sap and the mature phloem cells have no nucleus. They survive because they have closely associated cells called **companion cells**. The companion cells are very active cells that have all the normal organelles, and they are linked to the sieve tube elements by many plasmodesmata. The cell membranes of companion cells have many infoldings that increase the surface area over which they can transport sucrose into the cell cytoplasm, and they have many mitochondria to supply the ATP needed for active transport. All the evidence suggests that companion cells support the sieve tube cells, which have lost most of their normal cell functions.

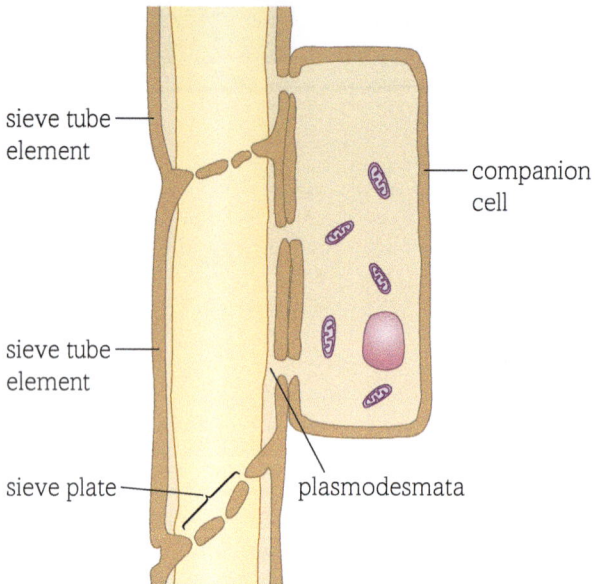

fig D Phloem sieve tubes and companion cells – the tissue that moves sugars around plants.

Questions

1　The bark of young trees contains a ring of vascular bundles. Forestry workers and gardeners protect young trees with plastic tubes around the lower part of their trunk. What do you think they are being protected from and why is this necessary?

2　What are the main similarities and differences between xylem and phloem?

3　Which of the methods of demonstrating the movement of water through the xylem would be best for the following situations? Explain your choice in each case:
(a)　in a sixth-form science investigation
(b)　in a university laboratory
(c)　in a year 7 science investigation.

Key definitions

Xylem is the main tissue transporting water around a plant.

Phloem is the main tissue transporting dissolved solutes around the plant.

Cambium is the layer of unspecialised plant cells that divide to form both the xylem and the phloem.

Protoxylem is the first xylem formed that can stretch and grow because the walls are not fully lignified

Metaxylem consists of mature xylem vessels made of lignified tissue.

Pits are specialised holes in the walls of the xylem vessels through which water moves out into the surrounding cells.

Parenchyma are relatively unspecialised plant cells that act as packing in stems and roots to give support.

Sclerenchyma are plant cells that have very thick lignified cell walls and an empty lumen with no living contents.

Collenchyma are plant cells with areas of cellulose thickening that give mechanical strength and support to the tissues.

Autoradiography is a useful technique for following the transport of substances around plants.

Phloem sieve tubes are the main transport vessels of the phloem, made up of many living cells joined together to make very long tubes that run from the highest shoots to the end of the roots, divided at intervals by sieve plates.

Sieve plates are the perforated walls between phloem cells that allow the phloem sap to flow.

Companion cells are very active cells closely associated with the sieve tube elements that supply the phloem vessels with everything they need and actively load sucrose into the phloem.

By the end of this section, you should be able to...

● explain how water is moved through plant cells by the apoplastic and symplastic pathways

● explain how the cohesion-tension model explains the transport of water from plant roots to shoots

● explain how temperatures, light, humidity and movement of air affect the rate of transpiration

Plants need water – it is vitally important both for photosynthesis and to maintain turgor in the cells which support the tissues of the plant. They absorb the water they need from the soil.

Water from the soil

An important component of soil is soil water. Even soil that is fairly dry has a thin film of water around the soil particles that plants can take in through their roots. Water is absorbed mainly by the younger parts of the roots where the majority of the **root hairs** are found. These microscopic hairs are extensions of the membranes of the outer cells of the root, and they greatly increase the surface area for absorption. The root hairs allow close contact with the soil particles.

At its simplest, uptake of water by the roots depends on the concentration gradient across the root from the soil water to the xylem. Water moves from the soil into a root hair cell down a concentration gradient by osmosis. This makes the root hair cell more dilute than its neighbour, so water moves from cell to cell by osmosis across the root to the xylem. However, this model is very simple – the detailed mechanism is more complicated. There is a concentration gradient across the root from the root hair cells to the cells closest to the xylem. This is the result of two effects. Water is continually moved up the xylem by transpiration, and the solute concentration increases in the cells across the root towards the xylem. But the water does not simply flow from one cell to another – there appear to be two alternative routes into the xylem vessels (see **fig A**).

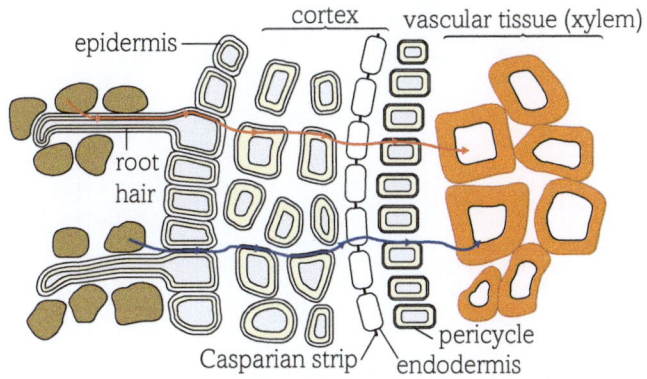

symplast pathway – the plasmodesmata form a continuous pathway between cells, and water passes through the cytoplasm by diffusion

apoplast pathway – water passes freely through the cellulose cell walls

fig A The cells of the roots are organised into a system that is very efficient at taking up water from the soil.

In the **symplast pathway** water moves by diffusion down the concentration gradient from the root hair cells to the xylem through the interconnected cytoplasm (symplast) of the cells of the root system. It moves through the plasmodesmata, gaps in the cellulose cell walls that allow strands of cytoplasm to pass through them, so the contents of the two cells are in contact.

In the **apoplast pathway** water is pulled by the attraction between water molecules across adjacent cell walls (the **apoplast**) from the root hair cell to the xylem. Because of the loose, open-network structure of cellulose, up to half of the volume of the cell wall can be filled with water. As water is drawn into the xylem, attraction between the molecules ensures that more water is pulled across from the adjacent cell wall and so on (see **Section 4.4.3**). Water entering the root hair from the soil has mineral ions dissolved in it, and they are drawn through the apoplast pathway too. The water moves across the cells of the root in the cell walls until it reaches the endodermis, that contains a waterproof layer called the **Casparian strip**.

Whichever route the water and minerals have taken across the root, once they reach the Casparian strip they enter the cytoplasm of the cell temporarily. Minerals may need to enter the cytoplasm up a concentration gradient, involving active transport. This seems to be a way by which the cells control the amount of water and minerals moving from the soil into the xylem. In spite of the barrier of the Casparian strip, the end result of all the pathways is a continuous stream of water across the root to the xylem.

Translocation of water

The movement of substances around plants is usually called **translocation**. Plants do not have mechanical systems like a heart to force materials along the narrow tubes of the xylem and phloem – they use a variety of physical processes instead. The translocation of water and mineral ions takes place in the xylem and it is a passive process. The translocation of organic solutes such as sucrose takes place in the phloem and it is an active process (see **Section 4.4.3**).

Plants have to move water up from the roots where it is absorbed to the aerial parts. The xylem vessels are dead tubes with an inner diameter of only 0.01–0.20 mm, so there is a great resistance to movement through them. Yet water has been shown to move up through the xylem vessels at speeds from 1 to $8\,m\,h^{-1}$, and to heights of up to 100 m above the ground in the tallest trees.

Transpiration and the transpiration stream

The movement of water in the xylem of plants depends on **transpiration**, which is the loss of water vapour from the surface of the plant, mainly from the leaves. Once water reaches the leaves, it moves from the xylem in the veins of the leaves into the spongy mesophyll cells by osmosis (see **fig B**). Water then evaporates from the cellulose walls of the spongy mesophyll cells into the air spaces. The water vapour moves through open stomata into the external air along a diffusion gradient. Even on a windy day, each leaf has a layer of still air around it. The thickness of this layer varies with the wind speed. The water vapour diffuses through this still layer before it is swept away by the mass of moving air. The amount of water lost by a plant due to transpiration can be surprisingly large. A sunflower may transpire 1–2 dm^3 in a day, whilst a large oak tree can lose up to 600 dm^3 in the same period.

A giant redwood tree regularly raises a column of water more than 30 m in its xylem. Water is moved up to 100 m in the tallest trees. What makes this possible?

When water is lost by transpiration from the leaves, it moves by osmosis across the leaf from cell to cell, all the way from the xylem. When molecules of water leave the xylem to enter a cell by osmosis, this creates tension in the column of water in the xylem, and this tension is transmitted all the way down to the roots. This is due to the **cohesion** of the water molecules. Because of their polar nature and the hydrogen bonds that form between them, water molecules 'stick together' giving the column of water a high **tensile strength** – it is not likely to break. So the loss of a molecule of water by evaporation from the surface of a spongy mesophyll cell causes a tension all the way through the plant because of the cohesiveness of the water molecules, and so more water is pulled up the xylem to replace what is lost. This is the **cohesion-tension theory** of transpiration.

The molecules also adhere strongly to the walls of the narrow xylem vessel and, probably more importantly, to the millions of tiny channels and pores within the cellulose cell walls of the leaf. **Adhesion** is the attraction between unlike molecules and it is sufficient to support the entire column of water in the xylem. The combination of adhesive and cohesive forces pulls the whole column of water in the xylem upwards. More water is continuously moved into the roots by osmosis from the soil to replace that lost from the leaves by transpiration (see **fig B**). This is the **transpiration stream**.

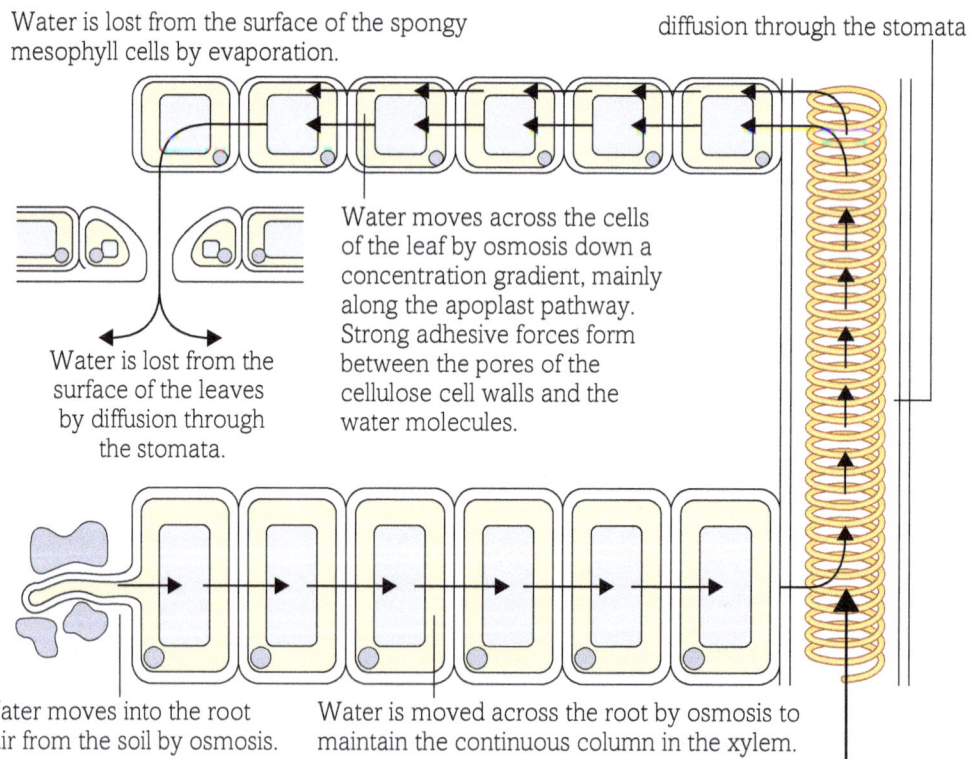

Water is lost from the surface of the spongy mesophyll cells by evaporation.

diffusion through the stomata

Water is lost from the surface of the leaves by diffusion through the stomata.

Water moves across the cells of the leaf by osmosis down a concentration gradient, mainly along the apoplast pathway. Strong adhesive forces form between the pores of the cellulose cell walls and the water molecules.

Water moves into the root hair from the soil by osmosis.

Water is moved across the root by osmosis to maintain the continuous column in the xylem.

As water molecules are lost by evaporation and moved out of the xylem, cohesion between the water molecules means that the whole column of water in the xylem is pulled upward.

fig B The transpiration stream is the result of physical processes and it can develop a pressure of around 4000 kPa. That is around 30 000 mmHg or 250 times higher than your arterial blood pressure, more than enough to supply water to the top of the tallest tree.

Modelling transpiration

In 1893 Josef Böhm used a model to demonstrate neatly the effect of evaporation on a column of water. Using a porous pot, he showed that adhesive forces between the water molecules and the pores of the pot are strong enough to support an enormous column of water, and cohesive forces between the water molecules stop the column breaking under the strain (see **fig C**). Drawn by the evaporation of water in the experiment, the column of mercury rises to over 1000 mm. It is calculated that if there was only water in the system (instead of the mercury) the column could be pulled to a height of more than 1 km – far greater than the height above ground of any plant. This passive, physical process alone is enough to explain what happens in the transpiration stream.

Measuring water movements

You can demonstrate the loss of water from the surfaces of a plant very easily. First seal the pot of a potted plant in a plastic bag to prevent evaporation of water from the soil surface interfering with the experiment. Then seal the plant in a bell jar. As water is lost a colourless liquid collects on the glass of the bell jar. You can show that this contains water by using cobalt chloride or copper sulfate paper.

It is not as easy to measure the amount of transpiration taking place. However, you can measure the uptake of water by a plant. As most of the water taken up by a plant is used for transpiration, this can be considered a close estimate. Uptake of water is demonstrated using a **potometer** (see **fig D**). Factors that affect the rate of transpiration can also be demonstrated using this apparatus.

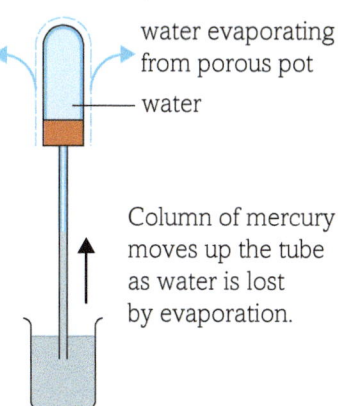

water evaporating from porous pot

water

Column of mercury moves up the tube as water is lost by evaporation.

fig C This simple experimental set-up gives us a good working model of transpiration.

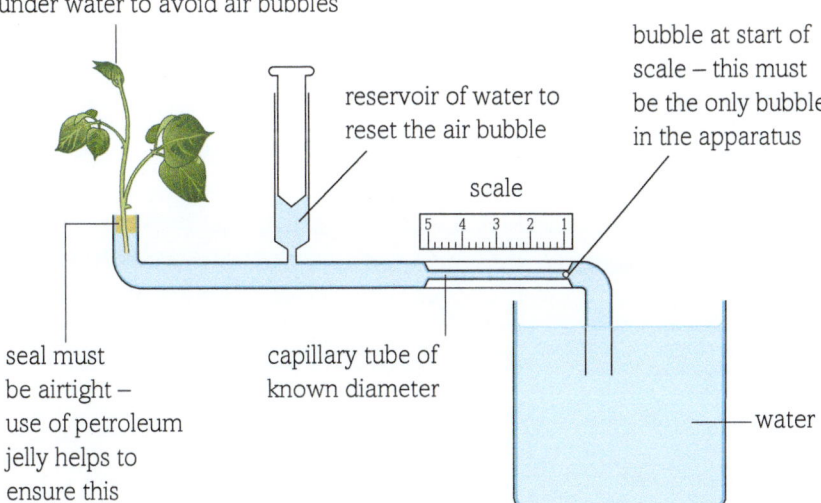

leafy shoot – must be fresh, with stem cut under water and then transferred to apparatus under water to avoid air bubbles

reservoir of water to reset the air bubble

bubble at start of scale – this must be the only bubble in the apparatus

scale

seal must be airtight – use of petroleum jelly helps to ensure this

capillary tube of known diameter

water

fig D A potometer demonstrates transpiration indirectly by looking at water uptake. It is important to prevent air entering the apparatus.

Learning tips

Translocation is a general term applied to the movement of substances around a plant.

Transpiration relates specifically to the evaporation of water from the surface of the spongy mesophyll cells and the loss of water by diffusion down a concentration gradient from a leaf.

Factors affecting transpiration

The rate of water loss from the leaves of a plant by transpiration can be affected by a variety of factors.

- Light: stomata open in the light for photosynthetic gas exchange, and most are closed in the dark. As a result, transpiration rates increase with light intensity until all of the stomata are open. At this point transpiration is at a maximum and will not be affected by any further increase in light intensity.

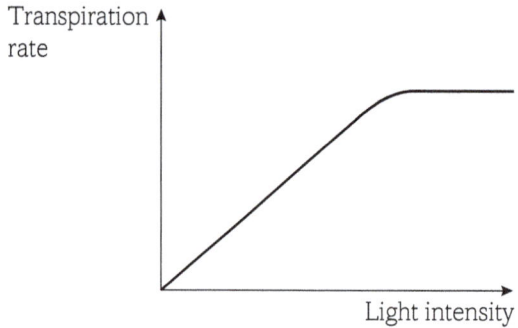

fig E The effect of light intensity on the rate of transpiration at a given temperature, humidity and air speed.

- Temperature: at a given light intensity, an increase in temperature will increase the amount of evaporation from the surfaces of the spongy mesophyll cells. It will also increase the amount of water vapour the air can take before it becomes saturated. Both of these factors increase the concentration gradient between the air inside and outside the leaf, increasing the rate of transpiration. An increase in temperature also increases the rate of movement of the molecules, which in turn increases the rate of diffusion of water vapour out of the leaf, increasing the rate of transpiration. Eventually another factor will become limiting – usually light intensity. If the light intensity is increased, the rate of transpiration at a given temperature will increase as well.

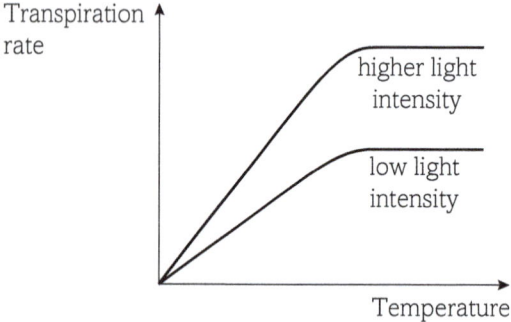

fig F The effect of temperature on the rate of transpiration at a given light intensity.

- Air movement or wind: this increases the rate of transpiration because it reduces the shell of still air around the stomata. This in turn increases the diffusion gradient between the inside and outside of the leaf, increasing the rate of transpiration.

fig G Transpiration rates will be much higher on a windy day than on a still one at a given temperature and light intensity.

- Air humidity: this is the concentration of water vapour in the air. A high air humidity lowers the rate of transpiration because of the reduced concentration gradient between the inside of the leaf and the air. Very dry air, with a low humidity, has the opposite effect and increases the transpiration rate.

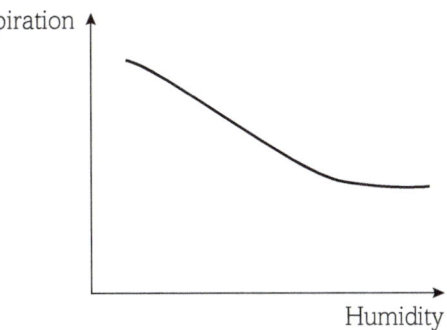

fig H The effect of humidity on the rate of transpiration at a given light intensity, temperature and air speed.

Using graphs such as these plotted from observed data, it is possible to calculate the rate of water loss from plants in different conditions and hence infer the effect of changing conditions on the rate of transpiration.

Many plants have adaptations that minimise water loss. These include curled, hairy and grooved leaves that trap still, moist air around the stomata and so reduce water loss by transpiration.

Root pressure

Transpiration is the major way by which water is moved through the xylem. Transpiration seems to be a passive process, but aspects of water transport are affected by metabolic inhibitors and lack of oxygen, both of which suggest a more active mechanism.

During the night, when transpiration rates can be extremely low, drops of water may be forced out of the leaves in a process known as **guttation**. In some plants, if the plant is cut off from the root, root sap will continue to ooze from the root xylem. This is a result of **root pressure**.

In tomatoes and some other plants, quite a strong root pressure can be measured when the top of the plant is cut off. This pressure disappears if the root cells are killed by steam or poisoned. This suggests that root pressure is based on active transport. The current model is that root pressure is produced by the active secretion of salts from the root cells into the xylem sap, increasing the concentration gradient across the root. This increases the movement of water into the cells by osmosis. The root pressure generated is about 100–200 kPa. This is not enough to explain all of the water movement in the xylem of many plants, but it certainly contributes, particularly in situations when the transpiration rate is low.

fig I Sometimes water appears at the edges of leaves when transpiration rates are low. This guttation is part of the evidence for the role of root pressure in water movements through the xylem of plants.

Learning tip

Make sure you can explain the difference between the transpiration stream and root pressure.

Questions

1 Summarise the main similarities and differences between the routes by which water appears to move from the soil into the xylem.

2 Explain how the structure of root hairs is adapted to their function.

3 Design an experiment that you could use to investigate the effect of one environmental condition on the rate of transpiration in a plant.

4 How might root pressure be measured? Why is the presence of root pressure alone not enough to explain the movement of water up from the roots to the leaves of a plant?

Key definitions

Root hairs are microscopic hairs that are extensions of the membranes of the outer cells of the root, greatly increasing the surface area for absorption of water and minerals.

The **symplast pathway** is the route by which substances, e.g. water and sucrose, can move by diffusion through the interconnected cytoplasm (symplast) in the plasmodesmata connecting cells in a plant.

The **apoplast pathway** is the route by which substances, e.g. water and sucrose, can move by the attraction between the molecules across adjacent cell walls (the apoplast) in a plant.

The **apoplast** is the free diffusional space outside of the cell surface membrane – mainly the cell walls and cell spaces.

The **Casparian strip** is the waterproof layer of waxy tissue in the walls of cells in the endodermis.

Translocation (verb) is the movement of substances around plants.

Transpiration is the loss of water vapour from the surface of the plant that has evaporated from the surface of the spongy mesophyll cells mainly within the leaves.

Cohesion is the attraction between like molecules.

Tensile strength is the strength of a material when it is stretched.

The **cohesion-tension theory** is the theory of transpiration based on the idea that the stream of water molecules stick together by cohesive forces, so that when a molecule is lost by evaporation it puts a tension on the column and another molecule of water is pulled up to replace it.

Adhesion is the attraction between unlike molecules.

The **transpiration stream** is the movement of water up from the soil through the root hair cells and across the roots to the xylem, then up the xylem, across the leaf until it is lost by evaporation from the spongy mesophyll cells and diffuses out of the stomata down the concentration gradient.

A **potometer** is a piece of apparatus used to measure the uptake of water by a shoot.

Guttation is a process in which drops of water may be forced out of the leaves as a result of root pressure.

Root pressure is the pressure that results when salts are actively secreted from the root cells into the xylem sap, increasing the concentration across the root and moving more water into the xylem by osmosis.

By the end of this section, you should be able to...

● explain the strengths and weaknesses of the mass flow hypothesis in explaining the movement of sugars through phloem tissue

The movement of water through the xylem of plants is usually known as the transpiration stream. The movement of solutes in the phloem is simply called translocation.

Translocation in the phloem

The glucose made by photosynthesis in the leaves is needed all over the plant for cellular respiration. The glucose is converted to sucrose for transporting around the plant in the phloem, because sucrose has less of an osmotic effect than glucose. It is then converted back to glucose or to starch for storage in the target cells, or used to make other molecules such as amino acids or lipids. The substances transported in the phloem are called **assimilates** and the main assimilate is sucrose.

The transport of assimilates in the phloem is from **sources** to **sinks**. Sources of sucrose in plants are the green parts (the leaves and stems), storage organs such as tubers, for example potatoes, or tap roots, which give biennial or perennial plants a boost at the start of the growing season, and the food stores of seed when they germinate. Sucrose from these sources is loaded into the sieve tube elements and transported to the sinks. Sinks are any plant tissues that need sucrose, for example the actively dividing cells in the meristems, fruits, seeds and storage organs as they build up food stores, and roots that are growing or actively absorbing mineral ions from the soil. Phloem loading and translocation is an active process, using ATP from respiration.

Scientists are still not entirely clear about the process of translocation – research continues – but the current model is:

Phloem loading

Sucrose is loaded into the phloem sieve elements from the surrounding cells. This is a very effective process. The sucrose content of phloem sap is between 20 and 30% m/v, compared with cell sap which is around 0.5% m/v sucrose. Scientists sometimes use percentage solutions instead of molarity. A 0.5% (m/v) solution means that 0.5 g (m is mass) of sucrose is dissolved in 100 cm^3 of water (v is volume).

There are two main loading routes into the phloem sap:

1 The symplast pathway (see **Section 4.4.2**) – sucrose moves by diffusion down a concentration gradient from the source cells through the cytoplasm of a number of cells into the companion cells and on into the phloem sieve tubes. This is largely a passive process. The high sucrose concentration in the phloem means water then moves into the sieve tubes by osmosis, causing a positive hydrostatic pressure that moves the phloem sap towards the sinks. Because the sucrose is moved in the phloem sap to the sinks, a constant diffusion gradient is maintained. In some species of plants this is the main way of phloem loading and is largely passive.

2 The apoplast pathway (see **Section 4.4.2**) – sucrose moves by diffusion down a concentration gradient through the cellulose cell walls and cell spaces to the companion cells. In the companion cells sucrose is moved from the wall spaces across the membrane into the cytoplasm by active transport using ATP, producing a high sucrose concentration in the cytoplasm of the companion cells and so in the sieve tubes – the sucrose passes into the sieve tubes through the many plasmodesmata. Water moves into the companion cells by osmosis as a result of the high sucrose concentration, producing a positive hydrostatic pressure that moves assimilates and water into and through the phloem sieve tubes. This hydrostatic pressure moves the sap towards the sucrose sinks, where the hydrostatic pressure is lower as there is less sucrose in the companion cell sap and so less movement of water into the sieve tubes by osmosis. The diffusion gradient for sucrose into the companion cells and sieve tubes is maintained because sucrose is removed by mass flow of the cell sap to the sucrose sinks.

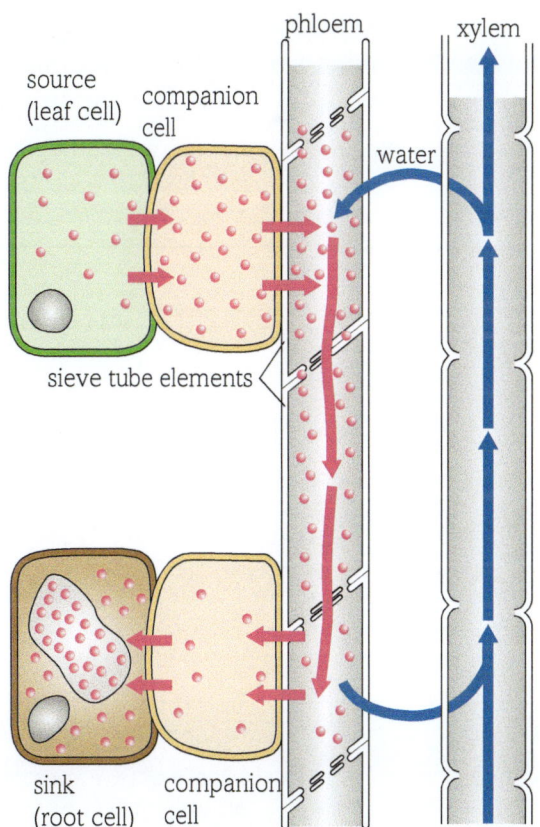

fig A Sucrose may be moved into the phloem by either the symplast or the apoplast routes. It travels from a source to a sink.

It is the accumulation of sucrose in the source phloem that leads to an increase in turgor pressure that in turn forces the sap to regions of lower pressure in the sucrose sinks. The pressures generated in the phloem of plants can be very high, which is how assimilates can be moved many metres up or down a plant, depending on the position of the source and the sink.

Phloem unloading

The phloem seems to be unloaded into the sink cells by a process of diffusion down a concentration gradient from the sieve tubes to the surrounding cells. The sucrose then moves rapidly on into other cells by diffusion or may be converted into other compounds. This maintains the sucrose diffusion gradient between the phloem and the sink cells. As the sucrose moves out of the sieve tubes, the water potential of the sap rises and so water also moves out by osmosis down a concentration gradient. Some of this water then moves into the xylem.

Evidence for translocation

- Radioactive isotopes of carbon (carbon-14) can be made available to the leaves of a plant so the glucose made is radioactively labelled. The sucrose made from this glucose is also labelled, and the path it takes in the plant can be traced using autoradiography. The labelled sucrose is found in the phloem.

- If a jet of steam is used to kill a complete outer ring of bark in a young shoot, just below a leaf, the movement of solutes from the leaf through the phloem to areas below the region of dead tissue stops. The movement of water in the xylem is not affected.

- Aphids penetrate the phloem with their mouthparts. Sometimes the pressure of the fluid in the phloem is so great that it moves right through the digestive system of the aphid and appears as a droplet at the end of the body. If the insect is removed from the plant, the contents of the phloem ooze out of the mouthparts and can be analysed, both for their content and the rate of flow.

fig B Aphids are a useful tool for investigating translocation.

Mass flow and pressure flow hypotheses

Mass transport systems involve the transport of materials from one point to another in a transport system with a transport medium and a pressure or force to bring about movement. Translocation fulfils all these requirements – the phloem sap is the transport medium, the phloem sieve tubes are the system, sucrose and other assimilates are the materials to be moved and turgor pressure (hydrostatic pressure) in the system provides the force for movement.

In 1930, a German plant physiologist called Ernst Münch (1876–1946) developed his simple mass flow hypothesis to explain the movement of solutes in the phloem of plants.

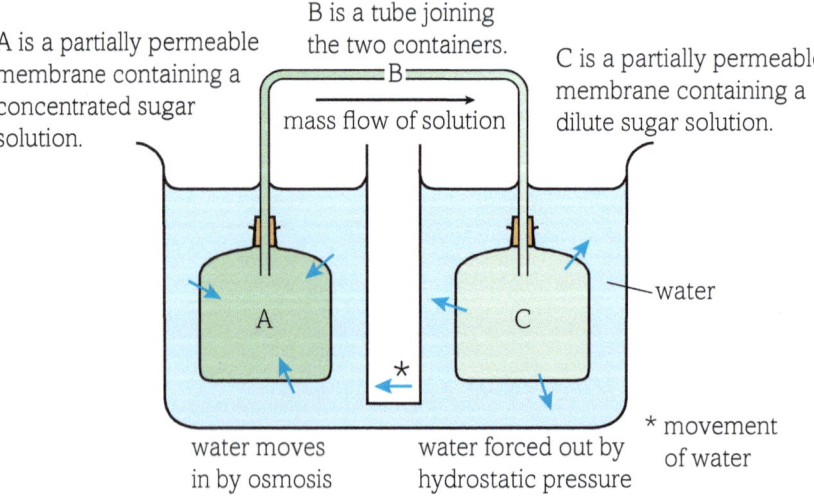

A is a partially permeable membrane containing a concentrated sugar solution.

B is a tube joining the two containers.

mass flow of solution

C is a partially permeable membrane containing a dilute sugar solution.

water

water moves in by osmosis

water forced out by hydrostatic pressure

* movement of water

fig C Münch's model for mass flow in the sieve tubes.

Initially water moves into both containers by osmosis. As A contains a much higher concentration of water than C, water will move into A more rapidly and there will be a flow of solution through the tube from A to C. The hydrostatic pressure this creates forces water out of C. This flow continues until the concentration of the solutions in A and C is the same.

The pressure flow hypothesis

When Münch developed his mass flow model suggesting passive transport in the phloem, there were no electron microscopes to show the structure of the companion cells, or instruments to measure pressure changes in individual cells. We now know that active processes are also involved in translocation. The main limitations of a passive mass flow model include:

- It does not take into account active loading of sucrose into the phloem sieve tube elements by the companion cells at the source. This changes the concentration gradient and so the rate of osmosis and can even change the direction of flow.
- Translocation is continuous – it does not end with equal concentrations – but the model does not take into account the continuous loading of sucrose at the sources and removal at the sinks that make this possible.
- Water can move into the tubes at any point by osmosis.
- The return route for water to the cells is through the xylem.

However, a modified version of the mass flow hypothesis, known as the pressure flow hypothesis, gives us a very good model of how assimilates are transported in the phloem.

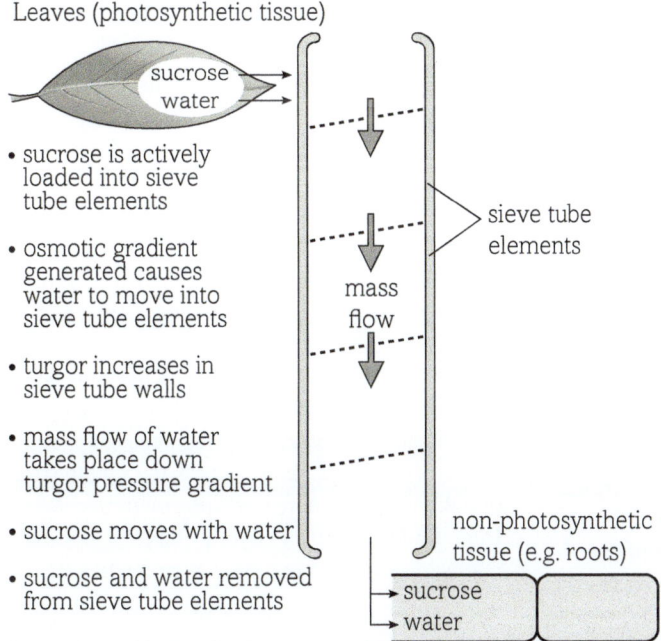

fig D The pressure flow hypothesis for translocation is still a good model for translocation in the phloem.

Questions

1   What are the roles of diffusion, osmosis and active transport in translocation?

2   How is the structure of a companion cell adapted for its functions in translocation?

3   What are the strengths and weaknesses of the mass flow hypothesis in explaining the movement of sugars through the phloem tissue?

Key definitions

Assimilates are the substances transported in the phloem, mainly sucrose.

A **source** is a region of a plant that is very high in sucrose and loads sucrose into the phloem.

A **sink** is a region of a plant that is low in sucrose and removes sucrose from the phloem.

Phloem loading is the process by which sucrose is moved into the phloem sieve elements from the surrounding cells.

Phloem unloading is the process by which sucrose and other assimilates are loaded into the phloem by the companion cells.

THINKING BIGGER

TURGOR – PRESSURE WHICH POWERS PLANTS

Turgor is vitally important in plants. It supports them, allows them to grow and makes it possible to open and close the stomata – and that is just for starters. In this activity you are going to find out more about what turgor can do.

TURGOR-DEPENDENT PROCESSES IN PLANT CELLS

fig A Plants in all their amazing shapes and forms rely on turgor in a variety of ways.

To study turgor pressure we have to be able to measure it. Often it is calculated from psychrometer measurements of ψ_w or π_i using eqn [1]. It can now be measured directly in single cells using the pressure probe. In using this device a glass microcapillary is inserted into a cell while observing with a microscope. Cell turgor drives the contents into the microcapillary; the pressure in the capillary is increased using a motor-driven plunger that forces the cell sap back into the cell. The balancing pressure required to do this is measured by an attached measuring device and is the cell turgor pressure. Techniques like the pressure probe have revealed that turgor in plant cells can be substantial.

A growing root cell has a hydrostatic pressure of around 6 bars (0.6 MPa), over three times that in a car tyre. Epidermal cells in a leaf can be 15–20 bars (1.5–2.0 MPa) depending on environmental conditions. Compare these with pressure in animal systems: high blood pressure might only be 0.03 bars (0.003 MPa). Thus plants operate at high pressures that are essential to the way they interact with their environment. Many examples can be found ranging from the subcellular to the whole plant.

The activity of the transport proteins pumping solutes into the cell can be regulated by cell turgor. If turgor pressure drops, pumping can be increased, increasing osmotic pressure and maintaining turgor. This can be important under drought conditions where, without this response, turgor would decrease. Turgor regulation of transport is also important for cells accumulating solutes such as in developing fruits.

Cell extension drives nearly all plant growth and occurs because, like a piston, turgor pressure extends the cell wall.

The forces that can be generated are large, as suggested by the observation that roots can force their way through compacted soils and even concrete. In causing the expansion of fruits and other produce turgor-driven growth is of major agricultural importance. Turgor pressure also provides a hydrostatic skeleton to support leaves and stems, dramatically demonstrated when plants wilt when they have insufficient water. Although plants are sedentary they can move: leaves and flowers move to track the sun, maximizing photosynthesis or pollination. In more extreme environments leaves turn away from the sun to prevent overheating. Such movements are caused by reversible variation in turgor on different sides of the stem or petiole.

An important turgor-driven process is the opening and closing of stomatal pores regulating water loss from, and carbon dioxide entry into, leaves.

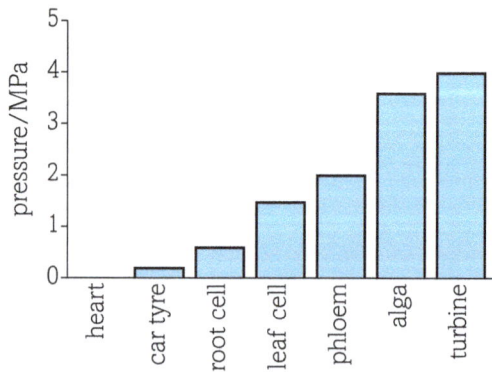

fig B Approximate hydrostatic pressures (turgor) found in a range of situations for comparison with those in the cells of vascular plants.

Where else will I encounter these themes?

1.1 1.2 1.3 1.4 2.1 2.2 2.3 2.4

Let us begin by thinking about the nature of the writing in this extract. The extract comes from an article published in the Encyclopedia of Life Sciences (eLS) by Dr Jeremy Pritchard:

1. What are the clues that tell you this is a piece of writing for a scientific publication?

2. You already know something about turgor and transport in plants from your studies earlier in the course. What more have you learned from reading this extract?

3. The graph conveys information about hydrostatic pressures in living and non-living things. Discuss whether it is an effective way of sharing this information.

> Look for clues such as technical language used, specific data, units of measurement and references.

Use your knowledge of turgor and plant transport systems to answer the following questions.

4. Produce a flow diagram to explain how turgor pressure can be measured directly in a plant and suggest why it is such an advantage to be able to make these direct measurements.

5. Look at the data in **fig B**:

   a. Which plant tissue develops the highest hydrostatic pressure?

   b. Suggest reasons why this tissue can develop such high pressures.

   c. Look at the data and display it differently – then assess whether your idea is more or less informative than the original format.

6. The pressure in both a root cell and a leaf cell is considerably higher than the pressure in a car tyre:

   a. What causes the high pressure in the plant cells?

   b. How is this adaptation important to plants?

> **Thinking scientifically**
> As you read these articles think of everything you have learnt so far in the course and how it fits together to inform your understanding as a scientist. By now you should be able to use correct terminology, analyse sources, write scientifically, and think like a scientist.

Activity

Moving parts

Plants are not known for being very mobile, but some plants can move very fast indeed, and many plants move on a 24-hour cycle. Much of this plant movement is the result of changes in turgor. These movements are different to the directional movement of plant tropisms, which involve growth. Write up to 300 words for an AS/A level textbook on turgor-dependent movement in plants, with plenty of examples, which could include:
- Venus fly trap
- Mimosa pudica (the sensitive plant)
- Sundews
- Daily opening and closing of flowers

● From the following journal article: Pritchard, Jeremy. 'Turgor Pressure.' eLS (2001)

1 Water travels through the stem of a plant in the xylem. Adhesion and cohesion are important in this process.

(a) Explain the difference between adhesion and cohesion. [2]

(b) Name the group of soluble substances transported from the roots to the leaves in the xylem. [1]

(c) When the flow of water through the xylem increases, the diameter of the stem is slightly reduced. When the flow is reduced, the stem returned to its original size.

Suggest why these changes take place. [1]

[Total: 4]

2 (a) Describe the pathways taken by water as it travels from the epidermis of a root to a xylem vessel. [4]

(b) In an investigation, the rate of transpiration of a sunflower plant was measured at 2-hourly intervals between 08.00 hours and 22.00 hours.

The results of this investigation are shown in the table below.

Time of day/ 24 hour clock	Rate of transpiration/ arbitrary units
08.00	5
10.00	32
12.00	42
14.00	50
16.00	41
18.00	30
20.00	5
22.00	3

(c) (i) Describe the changes in the rate of transpiration from 08.00 hours until 22.00 hours. [3]

(ii) Suggest an explanation for the changes in the rate of transpiration during each of the following time intervals.
- 08.00 hours until 14.00 hours
- 14.00 hours until 20.00 hours [2]

[Total: 9]

3 (a) The photograph below shows a transverse section through part of a root, as seen using a light microscope.

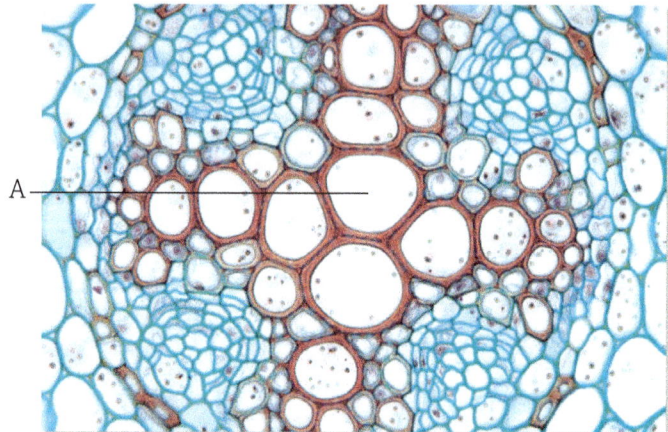

Name the part labelled A. [1]

(b) Describe the role of the **endodermis**. [3]

(c) An investigation was carried out into the effect of temperature on the uptake of potassium ions in barley roots. Pieces of barley roots were placed in solutions containing potassium ions, and kept at a range of temperatures. The initial concentration of potassium ions in the solutions was 8 mmol per dm^3.

After 10 hours, the concentrations of potassium ions in the root cells were determined.

The results are shown in the table below.

Temperature/°C	Potassium concentration in root cells/mmol dm^{-3}
5	32
10	28
15	57
20	80
25	100

(i) Describe the relationship between temperature and the concentration of potassium ions in the root cells. [2]

(ii) What do these results suggest about the mechanism for the uptake of potassium ions by barley roots? Give an explanation for your answer. [3]

[Total: 9]

4   A student investigated the water uptake by plants using the potometer illustrated in an enclosed school laboratory.

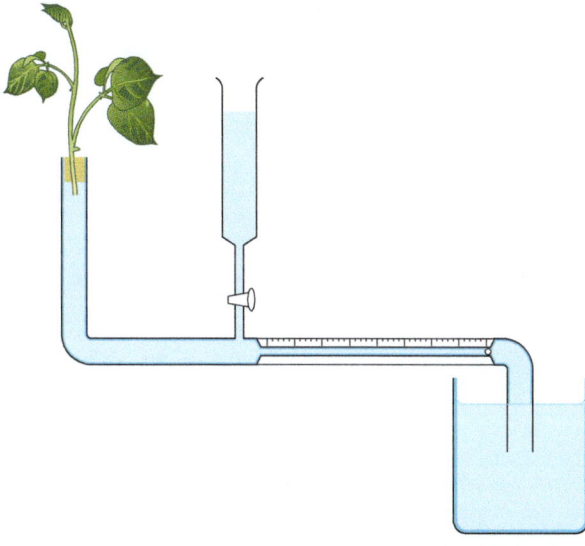

The following results were obtained:

Experiment	Environmental condition	Time taken for bubble to move 10 cm/minutes	Rate of bubble movement/ cm min^{-1}
A	High light intensity	6	
B	High light intensity and high humidity	18	
C	High light intensity and windy	2	
D	Dark and windy	19	
E	Dark and low humidity	20	

Data from pp. 168–169 *Complete Biology WR Pickering*, Oxford 2000

(a) For each environmental condition calculate the rate of bubble movement and fill in the third column. [2]

(b) Explain the difference in the rate of bubble movement for experiments A and B. [2]

(c) Explain the difference in the rate of bubble movement for experiments B and E. [2]

(d) Explain the difference in the rate of bubble movement for experiments C and D. [2]

[Total 8]

Maths skills

In order to be able to develop your skills, knowledge and understanding in Biology, you will need to have developed your mathematical skills in a number of key areas. This section gives more explanation and examples of some key mathematical concepts you need to understand. Further examples relevant to your AS/A level Biology studies are given throughout the book.

Arithmetic and numerical computation

Using standard form

Dealing with very large or small numbers can be difficult. To make them easier to handle, you can write them in the format $a \times 10^b$. This is called standard form.

To change a number from decimal form to standard form:
- Count the number of positions you need to move the decimal point by until it is directly to the right of the first number which is not zero.
- This number is the index number that tells you how many multiples of 10 you need. If the original number was a decimal, your index number must be negative.

Here are some examples:

Decimal notation	Standard form notation
0.000 000 012	1.2×10^{-8}
15	1.5×10^1
1000	1×10^3
3 700 000	3.7×10^6

Using ratios, fractions and percentages

Ratios, fractions and percentages help you to express one quantity in relation to another with precision. Ratios compare like quantities using the same units. Fractions and percentages are important mathematical tools for calculating proportions.

Ratios

A ratio is used to compare quantities. You can simplify ratios by dividing each side by a common factor. For example $12:4$ can be simplified to $3:1$ by dividing each side by 4.

WORKED EXAMPLE

Divide 180 into the ratio $3:2$

Our strategy is to work out the total number of parts then divide 180 by the number of parts to find the value of one part.

Total number of parts = 3 + 2 = 5

Value of one part = 180 ÷ 5 = 36

Answer = 3 × 36 : 2 × 36 = 72 : 108

Check your answer by making sure the parts add up to 180 : 72 + 108 = 180

Fractions

When using fractions, make sure you know the key strategies for the four operators:

To add or subtract fractions, find the lowest common multiple (LCM) and then use the golden rule of fractions. The golden rule states that a fraction remains unchanged if the numerator and denominator are multiplied or divided by the same number.

WORKED EXAMPLE

$$\frac{1}{2} + \frac{1}{5} = \frac{5}{10} + \frac{2}{10} = \frac{7}{10}$$

To multiply fractions together, simply multiply the numerators together and multiply the denominators together.

WORKED EXAMPLE

$$\frac{2}{7} \times \frac{4}{9} = \frac{8}{63}$$

To divide fractions, simply invert or flip the second fraction and multiply.

WORKED EXAMPLE

$$\frac{2}{3} \div \frac{7}{9} = \frac{2}{3} \times \frac{9}{7} = \frac{18}{21} = \frac{6}{7}$$

Percentages

When using percentages, it is useful to recall the different types of percentage questions.

To increase a value by a given percentage, use a percentage multiplier.

WORKED EXAMPLE

Increase 30 mg by 23%

If we increase by 23%, our new value will be 123% of the original value. We therefore multiply by 1.23

Answer = 30 × 1.23 = 36.9 mg

To decrease a value by a given percentage, you need to focus on the part that is left over after the decrease.

WORKED EXAMPLE

Decrease 30 mg by 23%

If we decrease by 23%, our new value will be 100 – 23 = 77% of the original value. We therefore multiply by 0.77

Answer = 30 × 0.77 = 23.1 mg

To calculate a percentage increase, use the following equation:

$$\text{Percentage change} = \frac{\text{difference between values}}{\text{original value}} \times 100$$

To calculate percentage decrease, use the same equation but remember that your answer should be negative.

WORKED EXAMPLE

The volume of a solution increased from 40 ml to 50 ml. Calculate the percentage increase.

Change in volume = 10 ml

Percentage increase = $\frac{10}{40} \times 100 = 25\%$

Algebra

Using algebraic equations

Using algebraic equations is a very important skill for finding the value of an unknown quantity. In the real world, letters are used to symbolise important variables such as the blood sugar level of a diabetic or the irregular heartbeat of a patient.

The key rule to remember when using equations is that any operation that you apply to one side of the equation must also be applied to the other side.

WORKED EXAMPLE

Find the value of x in the following equation: $7x - 6 = 36$

Adding 6 to each side gives $7x = 42$

Dividing each side by 7 gives $x = 6$

Changing the subject of an equation

It can be very helpful to rearrange an equation to express the variable that you are interested in in terms of other variables. Always remember that any operation that you apply to one side of the equation must also be applied to the other side.

WORKED EXAMPLE

The diameter of a cell measured under the light microscope at magnification ×100 is 2 mm. Calculate the actual size.

You may remember the equation image size = actual size × magnification but note the question is asking us to find the actual size given the image size and magnification. We can rearrange the equation to suit our needs:

image size = actual size × magnification

$\frac{\text{image size}}{\text{magnification}}$ = actual size

So actual size = $\frac{2}{100}$ = 0.02 mm

Handling data

Using significant figures

Often when you do a calculation, your answer will have many more figures than you need. Using an appropriate number of significant figures will help you to interpret results in a meaningful way.

Remember the 'rules' for significant figures:

The first significant figure is the first figure which is not zero.

- Digits 1–9 are always significant.
- Zeros which come after the first significant figure are significant unless the number has already been rounded.
- Here are some examples:

Exact number	To one s.f.	To two s.f.s	To three s.f.s
45 678	50 000	46 000	45 700
45 000	50 000	45 000	45 000
0.002 755	0.003	0.002 8	0.002 76

Understanding the terms mean, median and mode

There are three different measures of average that you should know how to calculate:

- The **mean** is calculated by adding up all of the values in the data set and dividing them by the number of values. It is sometimes called the arithmetical average. The mean takes into account each number of the data set equally and can be used for further statistical analysis such as calculating a standard deviation. However, a disadvantage of the mean is that it may be affected by extreme values.
- The **median** is the middle value when the values are arranged in order. The median of a data set is found by putting the values in order from lowest to highest and then finding the middle value. If there is an even number of values, the median is found by calculating the mean of the two middle values.
- The **mode** is the value that occurs most often. The mode of a data set is found by identifying the most frequent value. It may not be possible to calculate the mode if there are two or more values with the same highest frequency.

WORKED EXAMPLE

Find the mean, median and mode of the following data set: 7, 12, 18, 6, 2, 12

To find the mean, we add up all of the values in the data set and divide them by the number of values:

$$\text{Mean} = \Sigma \frac{x}{n}$$
$$= \frac{(7 + 12 + 18 + 6 + 2 + 12)}{6}$$
$$= \frac{57}{6}$$
$$= 9.5$$

To find the median, we need to arrange the values in increasing order: 2, 6, 7, 12, 12, 18

Since there is an even number of values, we need to look at the two middle values and find the mean. The third value is 7 and the fourth value is 12.

$$\text{Median} = \frac{(7 + 12)}{2} = 10.5$$

To find the mode, we need to identify the value that occurs most frequently. The only number that occurs more than once is 12.

$$\text{Mode} = 12$$

Calculating the mean from frequency data

The mean can be calculated from frequency data by finding the sum of the individual values multiplied by their respective frequencies and then dividing by the total frequency.

WORKED EXAMPLE

WORKED EXAMPLE

The table below shows the results of a survey looking into the number of units of alcohol consumed in a week by a sample of patients. Find the mean number of units of alcohol consumed per week.

Units of alcohol consumed in a week	Number of patients
0	4
2	7
4	12
6	9
8	15
10	23

$$\text{Mean} = \frac{(0 \times 4) + (2 \times 7) + (4 \times 12) + (6 \times 9) + (8 \times 15) + (10 \times 23)}{60}$$

$$= \frac{0 + 14 + 48 + 54 + 120 + 230}{60}$$

$$= \frac{466}{60}$$

$$= 7.7666...$$

$$= 7.8 \text{ to 1 d.p.}$$

Understanding measures of dispersion including standard deviation and range

Two different sets of data may have similar averages but statisticians are interested in looking deeper into the data for meaningful differences in dispersion. For example, if one data set refers to patients who are given a new cancer drug and a second data set refers to patients who are given a placebo drug, it is very important to look for key differences in the dispersion of data, such as standard deviation and range, and not just at measures of average.

Range

The range of a set of data is the difference between the highest and lowest values in the set. To find the range, subtract the smallest value in the set from the largest value in the set.

Standard deviation

Standard deviation is a measure of the dispersion or 'spread' of data around the mean.

- A low standard deviation indicates that the data have a narrow range and the points are closely grouped to the mean. This could indicate greater reliability.

- A high standard deviation indicates that the data points have a larger range and are less well grouped. This might indicate lower reliability.

To calculate the standard deviation, use the formula:

$$s = \sqrt{\frac{\Sigma(x - \bar{x})^2}{n}}$$

where s = standard deviation, x is an individual value, $\bar{x}$ = the mean value, n = the number of values.

Technique

1. Calculate the mean of the data set by finding the sum of the values and then dividing by the number of values. This is $\bar{x}$.
2. For each data value, calculate the difference between the data value and the mean. Record these figures in a table.
3. Find the square of each of these differences. Record these figures in a new column in your table.
4. Find the sum of these squares. This is $\Sigma(x - \bar{x})^2$.
5. Divide this figure by the number of items in the data set. This is $\dfrac{\Sigma(x - \bar{x})^2}{n}$
6. Find the square root of your answer. This is the standard deviation.

WORKED EXAMPLE

A pupil investigates the effect that two newly developed fertilisers (A and B) have on the growth of potato crops. Multiple 10 m² areas of a field were sectioned off and treated with either fertiliser A or B. The table below shows the yields of potatoes from the test areas following harvest.

(a) Calculate the mean and standard deviation for the test plot yields for fertilisers A and B.

(b) Interpret the results of your answers to (a).

Fertiliser	Test plot yield/kg						
	Plot 1	Plot 2	Plot 3	Plot 4	Plot 5	Plot 6	Plot 7
A	25	27	34	18	21	26	28
B	17	35	42	19	35	22	44

(a) To calculate the mean yield for A:
25 + 27 + 34 + 18 + 21 + 26 + 28 = 179
179/7 = 25.6 kg to 1 d.p.

To calculate the standard deviation for A:
$(25 - 25.6)^2 = 0.36$ $(27 - 25.6)^2 = 1.96$
$(34 - 25.6)^2 = 70.56$ $(18 - 25.6)^2 = 57.76$
$(21 - 25.6)^2 = 21.16$ $(26 - 25.6)^2 = 0.16$
$(28 - 25.6)^2 = 5.76$

Sum of squares = 157.72
$\sqrt{(157.72/7)}$ = 4.8 kg to 1 d.p.

To calculate the mean yield for B:
17 + 35 + 42 + 19 + 35 + 22 + 44 = 214
214/7 = 30.6 kg to 1 d.p.

To calculate the standard deviation for B:
$(17 - 30.6)^2 = 184.96$ $(35 - 30.6)^2 = 19.36$
$(42 - 30.6)^2 = 129.96$ $(19 - 30.6)^2 = 134.56$
$(35 - 30.6)^2 = 19.36$ $(22 - 30.6)^2 = 73.96$
$(44 - 30.6)^2 = 179.56$

Sum of squares = 741.72
$\sqrt{(741.72/7)}$ = 10.3 kg to 1 d.p.

(b) Fertiliser B produces a greater yield of potato crop (19% increase from fertiliser A), however the variation in crop yield (as shown by the standard deviation) of plots treated with fertiliser B is much greater and so fertiliser A produces a more consistent crop yield.

Understanding simple probability

The term probability is used to talk about the likelihood of an event happening on a scale of 0 to 1. A probability of 0 means that it is impossible that an event will occur. A probability of 1 means that it is certain that an event will occur. You should be comfortable interpreting probabilities in a scientific context, such as the probability of developing a disease or inheriting a particular gene.

Interpreting a scattergram

A scattergram is a useful way of representing the relationship between two variables. To draw a scattergram, first choose appropriate scales and label both axes. Then, use a pencil to draw a small point (a cross or sharp dot) for each pair of variables.

A scattergram can be used to interpret whether there is correlation between two variables. We say that there is correlation between two variables if when one variable changes, there is also a change in the other variable.

- If the points are distributed tightly around a line, the variables are strongly correlated.
- If the points are loosely distributed around a line, the variables are weakly correlated.
- If there is no pattern in the distribution of points, there is no correlation.

Correlation can be positive (as one variable increases, the other also increases) or negative (as one variable increases, the other decreases).

A scattergram may include one or more points that lie outside of the main spread of values. Such a point is called an outlier and it can be ignored.

To draw a line of best fit, use a ruler to draw a straight line that passes as close as possible to all of the points. You can use a line of best fit to make estimations. This is called interpolation. The more closely correlated the variables, the more accurate your estimate is likely to be.

WORKED EXAMPLE

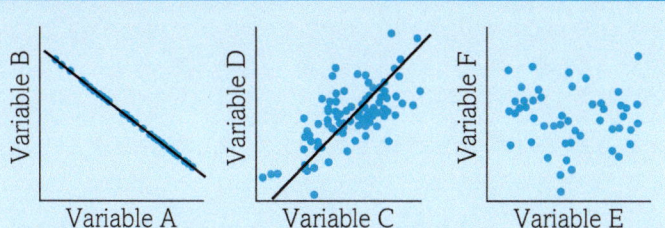

There is a strong negative correlation between variables A and B.

There is a weak positive correlation between variables C and D.

There is no correlation between variables E and F.

Constructing histograms

Constructing frequency tables and histograms is often the first step to looking carefully at a set of raw continuous data and helps us to begin to look for patterns and behaviours in a data set. Histograms are very similar to bar charts but there are a few key differences:

- In a bar chart, each column represents a discrete category. The columns are of equal widths and usually separated.
- In a histogram, the columns represent continuous data. The width of the columns is usually the same for each category. However, for more advanced work the widths may vary. The columns are usually adjacent.

Technique

1 Find the range of your values.

2 Choose the categories that you will use. Make sure that they are continuous i.e. there are no gaps and there is no overlap between categories.

3 Create a frequency table.

4 Plot your data, ensuring that frequency is represented on the y-axis and that the categories are represented on the x-axis.

WORKED EXAMPLE

The weights of field mice found in a specified area of farmland to the nearest gram are shown below. Draw a histogram to represent the weights of field mice.

Weights of field mice (g): 42, 66, 75, 44, 52, 56, 60, 81, 64, 54, 37, 59, 47, 79, 66, 76, 53, 35, 40, 63, 56, 28, 43, 78, 83, 50, 38, 67, 68, 47, 52, 49, 32, 46, 72, 58, 58

To choose our categories, we first identify the range of the data. The highest value is 83 g and the lowest value is 28 g. The categories we choose need to at least cover this range. One sensible way of splitting this range is to use four categories each covering an interval of 20 g:

Weight/g	Frequency
20 ≤ W < 40	5
40 ≤ W < 60	18
60 ≤ W < 80	12
80 ≤ W < 100	2

We can now use this frequency data to draw a histogram.

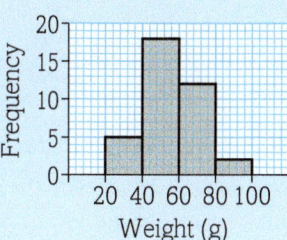

Principles of sampling

When a scientist studies a population, it is not possible to study each organism in detail. Scientists therefore use sampling to estimate characteristics of the whole population by looking at a subset of individuals in the population. It is important that the sample chosen is representative of the habitat.

Once a suitable sample has been selected, it can be analysed. A measure of biodiversity that takes into account both the species richness and the species abundance of an area can be calculated using the following formula.

$$D = \frac{N(N-1)}{\sum n(n-1)}$$

where n is the number of individuals of a particular species (or the percentage cover for plants), and N is the total number of all individuals of all species (or the total percentage cover for plants).

Graphs

Understand that $y = mx + c$ represents a linear relationship

Two variables are in a linear relationship if they increase at a constant rate in relation to one another. If you plotted a graph with one variable on the x-axis and the other variable on the y-axis, you would get a straight line. Any linear relationship can be represented by the equation $y = mx + c$ where the gradient of the line is m and the value at which the line crosses the y-axis is c. An example of a linear relationship is the relationship between degrees Celsius and degrees Fahrenheit, which can be represented by the equation $F = \frac{9}{5}C + 32$ where C is temperature in degrees Celsius and F is temperature in degrees Fahrenheit.

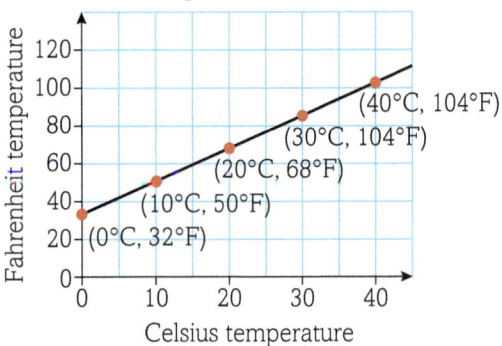

Conversion of Celsius temperatures to Fahrenheit

Calculate a rate of change from a graph showing a linear relationship

The rate of change from a graph showing a linear relationship is the gradient, or steepness, of the line. It is a measure of the rate of change of one variable, represented on the x-axis, in relation to the other variable, represented on the y-axis.

Technique

1 Draw a right-angled triangle anywhere on the line.

2 Use the following equation to calculate the rate of change:

$$\text{gradient} = \frac{\text{difference on } y\text{-axis}}{\text{difference on } x\text{-axis}}$$

3 State the unit for your answer.

Draw and use the slope of a tangent to a curve as a measure of a rate of change

Sir Isaac Newton was fascinated by rates of change. He drew tangents to curves at various points to find the rates of change of graphs as part of his journey towards discovering the calculus – an amazing branch of mathematics. He argued that the gradient of a curve at a given point is exactly equal to the gradient of the tangent of a curve at that point.

Technique

1 Use a ruler to draw a tangent to the curve.

2 Calculate the gradient of the tangent using the technique given for a linear relationship. This is equal to the gradient of the curve at the point of the tangent.

3 State the unit for your answer.

Applying your skills

You will often find that you need to use more than one maths technique to answer a question. In this section, we will look at two example questions and consider which maths skills are required and how to apply them.

WORKED EXAMPLE

Hydrogen peroxide is a toxic by-product of respiration and is made in all living cells. Cells make the enzyme catalase in order to convert the toxin into water and oxygen. In order to study the effect of temperature on catalase activity, an experiment was set up using the equipment shown in fig B. The volume of oxygen released in 30 seconds was measured using the gas syringe. The results of the experiment are shown in the graph below.

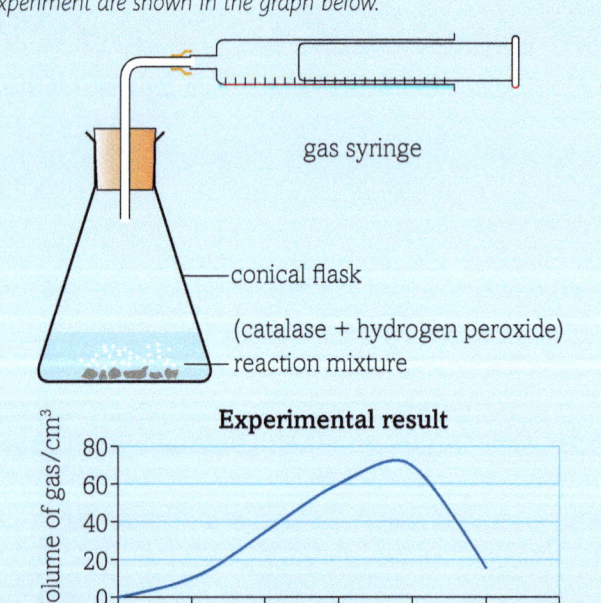

(a) Calculate the percentage increase in volume of oxygen produced by 30 seconds from 10 °C to 40 °C.

(b) Calculate the rates of each temperature test at 20, 40 and 50 °C and interpret the results.

(c) A further experiment is carried out where the volume of oxygen is recorded over the entire time of the reaction at 10 °C and 40 °C. The results are shown below:

Temperature	Total volume of oxygen released/cm³									
	0 s	10 s	20 s	30 s	40 s	50 s	60 s	70 s	80 s	90 s
10 °C	0	2	5	9	16	22	28	33	35	35
40 °C	0	7	15	27	33	35	35	35	35	35

Display both sets of results on an appropriately scaled graph.

(d) Calculate the difference in rate between both reactions at 15 s.

(a) The volume of gas at 10 °C = 10 cm³

The volume of gas at 40 °C = 70 cm³

The percentage increase = $70 - \frac{10}{10} \times 100 = 600\%$ increase

(b) $20 °C = \frac{35}{30} = 1.17 \text{ m}^3 \text{ s}^{-1}$

$40 °C = \frac{70}{30} = 2.33 \text{ m}^3 \text{ s}^{-1}$

$50 °C = \frac{15}{30} = 0.5 \text{ m}^3 \text{ s}^{-1}$

The rate has doubled between 20 °C and 40 °C, but at 50 °C, the rate has decreased.

(c)

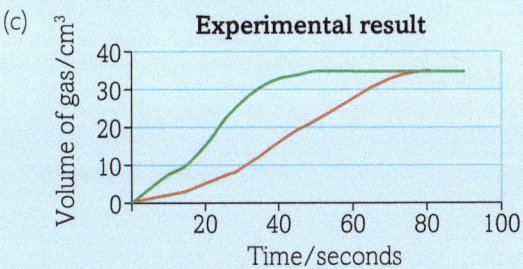

Experimental result

(d) We draw a tangent to each curve at 15 seconds so that we can use the gradient of the curve to calculate the rate.

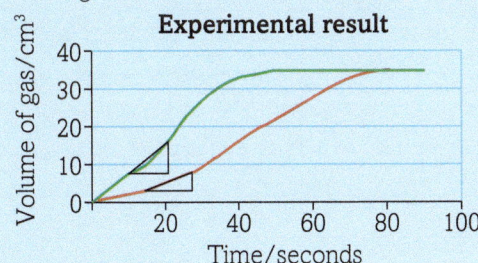

Experimental result

We can then use the equation the following equation to calculate gradient:

$$\text{gradient} = \frac{\text{difference on } y\text{-axis}}{\text{difference on } x\text{-axis}}$$

Rate of reaction at 10 °C at 15 seconds = $\frac{3}{10} = 0.3 \text{ m}^3 \text{ s}^{-1}$

Rate of reaction at 40 °C at 15 seconds = $\frac{8}{10} = 0.8 \text{ m}^3 \text{ s}^{-1}$

Difference in rate between reactions at 15 seconds
= 0.8 – 0.3 = 0.5 m³ s⁻¹

WORKED EXAMPLE

A photomicrograph of a T helper cell was taken using an electron microscope set at a magnification of ×50 000. In the image, several organelles were clearly identified and measured.

(a) Calculate the actual object length of each organelle.

Organelle	Image length/mm	Object length/μm
nucleus	240	
endoplasmic reticulum	360	
lysosome	10	
mitochondrion	120	

(b) A lysosome is a spherical organelle. Calculate the surface area and volume of a lysosome.

(c) Calculate the surface area to volume ratio of a lysosome.

(a) The question tells us that the magnification is ×50 000.

We know that image size = actual size × magnification

To make it easier to use, we can rearrange this equation as

$$\text{actual size} = \frac{\text{image size}}{\text{magnification}}$$

Actual length of nucleus = $\frac{240}{50\,000}$ = 0.004 8 mm

Actual length of ER = $\frac{360}{50\,000}$ = 0.007 2 mm

Actual length of lysosome = $\frac{10}{50\,000}$ = 0.000 2 mm

Actual length of mitochondrion = $\frac{240}{50\,000}$ = 0.002 4 mm

Before we can put these figures in the table, we need to convert to μm. 1 mm = 1000 μm so we need to multiply each figure by 1000.

Organelle	Image length/mm	Object length/μm
nucleus	240	4.8
endoplasmic reticulum	360	7.2
lysosome	10	0.2
mitochondrion	120	2.4

(b) Recall the following formulae, where r is radius:

Surface area of sphere = $4\pi r^2$

Volume of sphere = $\frac{4}{3}\pi r^3$

From (a) you know that the diameter of the lysosome is 0.2 μm. This means that the radius must be 0.1 μm.

Surface area of lysosome = $4\pi(0.1)^2 = 4\pi \times 0.01 = 0.125\,7 \text{ μm}^2$ to 4 d.p.

Volume of lysosome = $\frac{4}{3}\pi(0.1)^3 = \frac{4}{3}\pi \times 0.001 = 0.004\,2 \text{ μm}^3$ to 4 d.p.

(c) It is simplest and most accurate to use the exact expressions from (b) involving π, rather than the final answers which have been rounded.

Surface area to volume ratio = $4\pi \times 0.01 : \frac{4}{3}\pi \times 0.001$

We can simplify by multiplying each side by 1000 and dividing each side by π:

Surface area to volume ratio = $40 : \frac{4}{3}$

Now we can divide each side by 4 and multiply by 3 to get:

Surface area to volume ratio = 120 : 4 = 30 : 1

Preparing for your exams

Introduction

The way that you are assessed will depend on whether you are studying for the AS or the A level qualification. Here are some key differences:

- AS students will sit two exam papers, each covering 50% of the content of the AS specification.

- A level students will sit three exam papers, each covering content from both years of A level learning. The third paper will include synoptic questions that may draw on two or more different topics.

- A level students will also have their competency in key practical skills assessed by their teacher in order to gain the Science Practical Endorsement. The endorsement will not contribute to the overall grade but the result (pass or fail) will be recorded on the certificate.

The tables below give details of the exam papers for each qualification.

AS exam papers

Paper	Paper 1: Core Cellular Biology and Microbiology	Paper 2: Core Physiology and Ecology
Topics covered	Topics 1–2	Topics 3–4
% of the AS qualification	50%	50%
Length of exam	1 hour 30 minutes	1 hour 30 minutes
Marks available	80 marks	80 marks
Question types	multiple-choice short open open-response calculation extended writing	multiple-choice short open open-response calculation extended writing
Experimental methods?	Yes	Yes
Mathematics	A minimum of 10% of the marks across both papers will be awarded for mathematics at Level 2 or above	

A level exam papers

Paper	Paper 1: Advanced Biochemistry, Microbiology and Genetics	Paper 2: Advanced Physiology, Evolution and Ecology	Paper 3: General and Practical Principles in Biology
Topics covered	Topics 1–7	Topics 1–4 Topics 8–10	Topics 1–10
% of the A level qualification	30%	30%	40%
Length of exam	1 hour 45 minutes	1 hour 45 minutes	2 hours 30 minutes
Marks available	90 marks	90 marks	120 marks
Question types	multiple-choice short open open-response calculation extended writing	multiple-choice short open open-response calculation extended writing	short open open-response calculation extended writing synoptic
Experimental methods?	No	No	Yes
Mathematics	A minimum of 10% of the marks across all three papers will be awarded for mathematics at Level 2 or above		
Science Practical Endorsement	Assessed by teacher throughout course. Does not count towards A level grade but result (pass or fail) will be reported on A level certificate.		

Exam strategy

Arrive equipped

Make sure you have all of the correct equipment needed for your exam. As a minimum you should take:

- pen (black ink or ball-point pen)
- pencil (HB)
- ruler (ideally 30 cm)
- rubber (make sure it's clean and doesn't smudge the pencil marks or rip the paper)
- calculator (scientific).

Ensure your answers can be read

Your handwriting does not have to be perfect but the examiner must be able to read it! When you're in a hurry it's easy to write key words that are difficult to decipher.

Plan your time

Note how many marks are available on the paper and how many minutes you have to complete it. This will give you an idea of how long to spend on each question. Be sure to leave some time at the end of the exam for checking answers. A rough guide of a minute a mark is a good start, but short answers and multiple choice questions may be quicker. Longer answers might require more time.

Understand the question

Always read the question carefully and spend a few moments working out what you are being asked to do. The command word used will give you an indication of what is required in your answer.

Be scientific and accurate, even when writing longer answers. Use the technical terms you've been taught.

Always show your working for any calculations. Marks may be available for individual steps, not just for the final answer. Also, even if you make a calculation error, you may be awarded marks for applying the correct technique.

Plan your answer

In questions marked with an *, marks will be awarded for your ability to structure your answer logically showing how the points that you make are related or follow on from each other where appropriate. Read the question fully and carefully (at least twice!) before beginning your answer.

Make the most of graphs and diagrams

Diagrams and sketch graphs can earn marks – often more easily and quickly than written explanations – but they will only earn marks if they are carefully drawn.

- If you are asked to read a graph, pay attention to the labels and numbers on the x and y axes. Remember that each axis is a number line.
- If asked to draw or sketch a graph, always ensure you use a sensible scale and label both axes with quantities and units. If plotting a graph, use a pencil and draw small crosses or dots for the points.
- Diagrams must always be neat, clear and fully labelled.

Check your answers

For open-response and extended writing questions, check the number of marks that are available. If three marks are available, have you made three distinct points?

For calculations, read through each stage of your working. Substituting your final answer into the original question can be a simple way of checking that the final answer is correct. Another simple strategy is to consider whether the answer seems sensible. Pay particular attention to using the correct units.

Sample answers with comments

Question type: multiple choice

The genetic pedigree diagram below shows the inheritance of Tay-Sachs disease in one family.

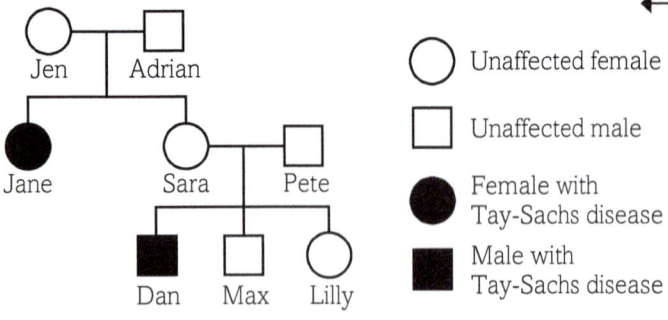

Jen — Adrian

○ Unaffected female

□ Unaffected male

● Female with Tay-Sachs disease

■ Male with Tay-Sachs disease

Jane Sara Pete

Dan Max Lilly

Put a cross (☒) in the box that correctly completes the statement.

The female whose genotype cannot be identified from the diagram is…

A Jane □
B Jen   □
C Lilly □
D Sara □   [1]

Question analysis

- Multiple choice questions look easy until you try to answer them. Very often they require some working out and thinking.
- In multiple choice questions you are given the correct answer along with three incorrect answers (called distractors). You need to select the correct answer and put a cross in the box of the letter next to it.
- If you change your mind, put a line through the box (☒) and then mark your new answer with a cross (☒).

Average student answer

B Jen   ☒

Verdict

This is an incorrect answer because:

- The student did not do the necessary working to find the correct answer. For a question like this, you should write the genotypes of each person on the diagram.

Question type: short open

Cystic fibrosis and albinism are examples of recessive genetic disorders. Tay-Sachs disease is another example of a recessive genetic disorder.

Explain the meaning of the term recessive genetic disorder.   [2]

Question analysis

- Generally one piece of information is required for each mark given in the question. There are two marks available for this question so make sure you make two distinct points.
- Clarity and brevity are the keys to success on short open questions. For one mark, it is not always necessary to write complete sentences.

Average student answer

Is that the only way to you are able to get the disease if both your parents had the disease or both your parents are carriers. You have to be homozygous, two alleles the same. The recessive allele codes for the disease.

> Misreading the question can lose your marks, as can answering in insufficient detail. One recessive allele does not code for the disease, but simply codes for a faulty protein.

Verdict

This is an average answer because:

- The student will get one mark for remembering that people only suffer from a recessive genetic disorder if they inherit two copies of the recessive allele from their parents.
- The student has not explained what made the allele potentially cause a disorder: the version of the gene is faulty and does not code for a protein properly.

Question type: open response

Molecules are transported across the cell membrane in a number of different ways.

Describe the structure of a cell membrane. [3]

> The command word in this question is describe. This means that you need too give an account of something. You do not need to include a justification or reason. Three marks are available so three distinct points need to be made. Remember that you can use bullet points or diagrams in your answer.

Question analysis

- With any question worth three or more marks, think about your answer and the points that you need to make before you write anything down. Keep your answer concise, and the information you write down relevant to the question. You will not gain marks for writing down biology that is not relevant to the question (even if correct) but it will cost you time.

Average student answer

A cell membrane is made up of a phospholipid bilayer. Within this bilayer there are some proteins that span the membrane and others that are free to move within the membrane. Other features of the membrane include cholesterol, which sit within the bilayer, glycoproteins and glycolipids which are on the outer layers of the membrane and attached to either a protein or a lipid.

> At this level, your answers need technical terms and clarity in expression otherwise you will find yourself losing marks.

Verdict

This is an average answer because:

- The student has made five points, three of which met the criteria needed to get full marks.
- The last sentence is poorly phrased: glycoproteins are short carbohydrates which are already attached to proteins.

Question type: extended writing

An investigation was carried out to study the effect of caffeine on the heart rate of a chicken embryo. The heart from a chicken embryo was removed and placed in a glucose solution. The heart rate was determined and recorded as a base heart rate. The experiment was repeated using glucose solutions containing five different concentrations of caffeine. The heart rate was determined and recorded as a percentage of the base heart rate for each solution.

Four marks are available so four points need to be made. If you have carried out the practical and written it up carefully (or corrected your write up using your teacher's feedback) then you should be well-prepared for this question.

Describe how this investigation could be carried out using Daphnia instead of chicken embryos. [4]

Question analysis

- There will be questions in your exams which assess your understanding of practical skills and draw on your experience of the core practicals. For these questions, think about:
  - how apparatus are set up
  - the method of how the apparatus are to be used
  - how readings are to be taken
  - how to make the readings reliable
  - how to control any variables.

- It helps with extended writing questions to think about the number of marks available and how they might be distributed. For example, if the question asked you to give the arguments for and against a particular case, then assume that there would be equal numbers of marks available for each side of the argument and balance the viewpoints you give accordingly. However, you should also remember that marks will also be available for giving an overall conclusion so you should be careful not to omit that.

- It is vital to plan out your answer before you write it down. There is always space given on an exam paper to do this so just jot down the points that you want to make before you answer the question in the space provided. This will help to ensure that your answer is coherent and logical and that you don't end up contradicting yourself. However, once you have written your answer go back and cross these notes out so that it is clear they do not form part of the answer.

Average student answer

By placing a daphnia under a microscope, you will be able to determine the bpm by counting the heart beats in one minute. This will give you a control to compare against Then by adding caffeine to the slide that the daphnia is placed on, in regular increasing concentrations of caffeine, you should be able to calculate the heart rate of the daphnia at different caffeine concentrations. By comparing against the control, this will allow you to note the differences in heart rate in relation to the concentration of caffeine.

Notice that the question says 'describe how this experiment could be carried out using Daphnia'. You need to adapt what you know already and apply it to this new situation.

Verdict

This is an average answer because:

- Some important details have been missed. The student does not mention repeating the experiment to check for anomalies or controlling variables to ensure the results are valid.
- The student has not detailed how the heart rate is to be counted. The heart will beat between 100–200 times a minute so a sensible method would be by using a felt pen to place a dot on a piece of paper every time the heart beats.

Question type: calculation

Age and gender are two other factors that may influence the development of heart disease in an individual. The graph below shows the results of a survey in America on the incidence of heart disease in adults aged 18 and older.

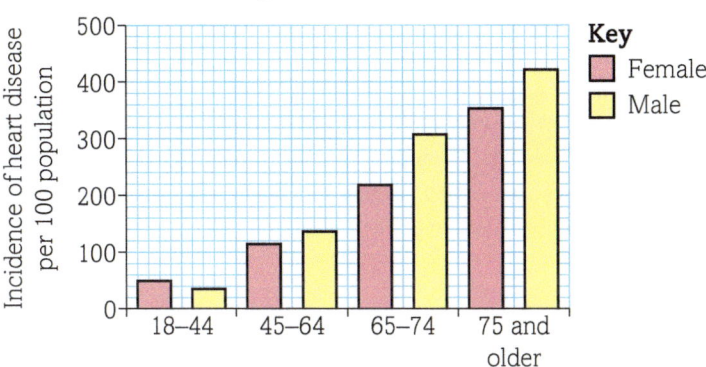

Key
- ▨ Female
- ☐ Male

Calculate the increased risk that a man who is 75 or older has of developing heart disease, compared to a man aged between 18 and 44 years old.   [2]

Question analysis

- The important thing with calculations is that you must show your working clearly and fully. The correct answer on the line will gain all the available marks, however, an incorrect answer can gain all but one of the available marks if your working is shown and is correct.

- Show the calculation that you are performing at each stage and not just the result. When you have finished, look at your result and see if it is sensible.

- At some point during your answer you will need to do some kind of sum, and the skills are to decide
  - ○ which numbers you need
  - ○ which operation you need.

Average student answer

$410 - 15 = 395. \frac{395}{15} = 26.3$, so 26.3 times.

Verdict

This is an average answer because:

- The student has misread the graph so has used incorrect figures in the calculation.

- However the correct technique has been used to work out the increased risk, so a mark would be awarded for using the correct method.

> The command word here is calculate. This means that you need to obtain a numerical answer to the question, showing relevant working. If the answer has a unit, this must be included.
>
> Finding the numbers requires you to read the graph really carefully, paying close attention to the increments on the y-axis, as well as choosing the correct set of bars.

> The student has not read the graph correctly. Five of the small sections on the y-axis equal 100, so each small section of the y-axis corresponds to 20. Therefore the calculation of difference is actually 420 − 35, which is 385.
>
> To work out the increased risk you divide the difference by the risk for 18–44 year old men.
>
> It's a good idea when you finish the question to check whether you need to put in the units. In this question there are none as it's simply a case of 'multiples of risk'.

Glossary

70S ribosomes are the ribosome found in the mitochondria and chloroplasts of eukaryotic cells and in prokaryotic organisms.

80S ribosomes are the main type of ribosome found in eukaryotic cells, consisting of ribosomal RNA and protein, made up of a 60S and 40S subunit. They are the site of protein synthesis.

The **acrosome** is the region at the head of the sperm that contains enzymes to break down the protective layers around the ovum.

Activation energy is the energy needed for a reaction to get started.

An **active site** is the area of an enzyme that has a specific shape into which the substrate(s) of a reaction fit.

Active transport is the movement of substances into or out of the cell using ATP produced during cellular respiration.

Adaptive radiation is a process by which one species evolves rapidly to form a number of different species that all fill different ecological niches.

Adenine is a purine base found in DNA and RNA.

Adenosine diphosphate (ADP) is a nucleotide formed when ATP loses a phosphate group and provides energy to drive reactions in the cell.

Adenosine triphosphate (ATP) is a nucleotide that acts as the universal energy supply molecule in cells. It is made up of the base adenine, the pentose sugar ribose and three phosphate groups.

Adhesion is the attraction between unlike molecules.

Agranulocytes are leucocytes that do not have granules to take up stain in their cytoplasm. They have unlobed nuclei and include monocytes and lymphocytes.

An **allele** is a version of a gene, a variant.

Allele frequency is the frequency with which a particular allele appears within a population.

Allopatric speciation is speciation that takes place when populations are physically or geographically separated and there can be no interbreeding or gene flow between the populations.

Allosteric enzymes are enzymes that have a site separate to the active site where another molecule can bind to have either an activating or inhibitory effect.

Amino acids are the building blocks of proteins consisting of an amino group ($-NH_2$) and a carboxyl group ($-COOH$) attached to a carbon atom and an R group that varies between amino acids.

Amylopectin is a complex carbohydrate made up of glucose monomers joined by both 1,4-glycosidic bonds and 1,6-glycosidic bonds so the molecules branch repeatedly.

Amyloplasts are plant organelles that store amylopectin, a polysaccharide used to form starch.

Amylose is a complex carbohydrate containing only glucose monomers joined together by 1,4-glycosidic bonds so the molecules form long unbranched chains.

An **anabolic reaction** is the reaction that builds up (synthesises) new molecules in a cell.

Anaerobic respiration is cellular respiration that takes place in the absence of oxygen.

Analogous features are features that look similar or have a similar function, but are not from the same biological origin.

Anaphase is the third stage of active cell division where the centromeres split so chromatids become new chromosomes. They are moved to the opposite poles of the cell, centromere first, by contractions of the microtubules of the spindle.

An **anatomical adaptation** is an adaptation involving the form and structure of an organism.

Aneuploidy is when a cell contains too few or too many chromosomes.

An **aneurysm** is a weakened, bulging area of artery wall that results from a build up behind a blockage caused by plaques.

Angina is a condition in which plaques build up and reduce the blood flow to the cardiac muscle through the coronary artery. It results in pain during exercise.

An **anion** is a negative ion.

The **annulus fibrosus** consists of non-conducting tissue that provides support for the heart and spreads between the atria and the ventricles, preventing the wave of excitation that spreads through the atria from passing directly on into the ventricles.

Anthers are male sex organs in plants that produce the male gametes, pollen.

The **anticodon** is a sequence of three bases on tRNA that correspond to the bases in the mRNA codon.

The **antisense strand** is the DNA strand that codes for proteins.

The **aorta** is the main artery of the body. It leaves the left ventricle of the heart carrying oxygenated blood under high pressure.

The **apoplast** is the free diffusional space outside of the cell surface membrane – mainly the cell walls and cell spaces.

The **apoplast pathway** is the route by which substances, e.g. water and sucrose, can move by the attraction between the molecules across adjacent cell walls (the apoplast) in a plant.

Apoptosis is cell suicide – the breakdown of worn out, damaged or diseased cells by the lysosomes.

The **Archaea** domain is made up of bacteria-like prokaryotic organisms found in many places including extreme conditions and the soil. They are thought to be early relatives of the eukaryotes.

Archaebacteria are ancient bacteria thought to be the oldest form of living organism.

Arteries are vessels that carry blood away from the heart towards cells of the body.

Arterioles are the very smallest branches of the arterial system, furthest from the heart.

Asexual reproduction is the production of genetically identical offspring from a single parent or organism.

Assimilates are the substances transported in the phloem, mainly sucrose.

An **atheroma** is another term for a plaque formed on the arterial lining.

Atherosclerosis is a condition in which yellow fatty deposits build up on the lining of the arteries, causing them to be narrowed and resulting in many different health problems.

ATPase is an enzyme that catalyses the formation and the breakdown of ATP, depending on conditions.

Atrial systole is when the atria of the heart contract.

The **atrioventricular node (AVN)** is a group of cells stimulated by the wave of excitation from the SAN and atria. It imposes a delay before transmitting the impulse to the bundle of His.

Autoradiography is a useful technique for following the transport of substances around plants.

Autotrophs are organisms that can make their own food, either by photosynthesis or chemosynthesis.

Bacilli are rod-shaped bacteria.

Bacteriophages are viruses that attack bacteria.

Basophils are part of the non-specific immune system. They have a two-lobed nucleus and produce histamines in inflammation and allergic reactions.

A **behavioural adaptation** is an adaptation involving programmed or instinctive behaviour making organisms better adapted for survival.

Behavioural isolation happens when changes occur in the courtship ritual, display or mating pattern so that some animals do not recognise others as being potential mates. This might be due to a mutation that changes the colour or pattern of markings.

The **bicuspid valve** is the valve between the left atrium and the left ventricle that prevents backflow of blood into the atrium when the ventricle contracts.

A **bilayer** is a double layer of closely packed atoms or molecules.

Binary fission is the splitting of one individual to form two new individuals as a result of mitosis.

Biodiversity is a measure of the variety of living organisms and their genetic differences.

A **biodiversity hotspot** is an area with a particularly high level of biodiversity.

Bioinformatics is the development of the software and computing tools needed to organise and analyse raw biological data, including the development of algorithms, mathematical models and statistical tests that help us to make sense of the enormous quantities of data being generated.

A **blastocyst** is a hollow ball of cells formed around five days after fertilisation.

The **Bohr effect** is the name given to changes in the oxygen dissociation curve of haemoglobin that occur due to a rise in carbon dioxide levels and a reduction of the affinity of haemoglobin for oxygen.

Breathing or **ventilation** is the process in which physical movements of the chest change the pressure so that air is moved in or out.

Budding is the production by mitosis of an outgrowth from the parent organism that develops into a small independent organism.

The **bundle of His** is a group of conducting fibres in the septum of the heart.

Cambium is the layer of unspecialised plant cells that divide to form both the xylem and the phloem.

Capillaries are minute vessels that spread throughout the tissues of the body.

The **capsid** is the protein coat of a virus.

Capsomeres are the repeating protein units that make up the capsid of a virus.

Carbaminohaemoglobin is the compound formed when carbon dioxide combines with haemoglobin.

Carbonic anhydrase is the enzyme that controls the rate of the reaction between carbon dioxide and water to form carbonic acid.

The **cardiac cycle** is the cycle of contraction (systole) and relaxation (diastole) in the heart.

Cardiac muscle is the special muscle tissue that makes up the bulk of the heart, which has an intrinsic rhythmicity and does not fatigue.

Cardiovascular diseases are diseases of the heart and circulatory system, many of which are linked to atherosclerosis.

The **cardiovascular system** is the mass transport system of the body made up of a series of vessels with a pump (the heart) to move blood through the vessels.

Carrier proteins are proteins that move a substance through the membrane in active transport – usually linked to an ATPase to break down ATP.

The **Casparian strip** is the waterproof layer of waxy tissue in the walls of cells in the endodermis.

A **catabolic reaction** is a reaction which breaks down substances within a cell.

A **catalyst** is a substance that speeds up a reaction without changing the substances produced or being changed itself.

A **cation** is a positive ion.

The **cell cycle** is a regulated process of three stages (interphase, mitosis and cytokinesis) in which cells divide into two genetically identical daughter cells.

Cell determination is the predestination of cells to become particular types of tissue from early in development of the embryo.

Cell sap is the aqueous solution that fills the permanent vacuole.

The **cell surface membrane** is the membrane that forms the outer boundary of the cytoplasm of a cell and controls the movement of substances into and out of the cell.

A **cell wall** is a freely permeable wall around plant cells, made mainly of cellulose.

Cellulose is a complex carbohydrate with β-glucose monomers held together by 1,4-glycosidic bonds. It is very important in plant cell walls.

Centrioles are bundles of tubules found near the nucleus and involved in cell division by the production of a spindle of microtubules that move the chromosomes to the ends of the cell.

The **centromere** is the region where a pair of chromatids are joined and which attaches to a single strand of the spindle structure at metaphase.

Chiasmata are the points where the chromatids break during recombination.

Chlorophyll is the green pigment that is largely responsible for trapping the energy from light, making it available for the plant to use in photosynthesis.

A **chloroplast** is an organelle adapted to carry out photosynthesis, containing the green pigment chlorophyll.

A **chromatid** is one strand of the replicated chromosome pair that is joined to the other chromatid at the centromere.

Chromatin is the granular combination of DNA bonded to protein found in the nucleus when the cell is not actively dividing.

Chromosomal mutations are changes in the position of entire genes within a chromosome.

Circulation is the passage of blood through the blood vessels.

A **class** is a group of orders that all share common characteristics.

Cleavage is a process involving a special type of mitosis with no interphase that results in a mass of small, undifferentiated cells.

Clones are genetically identical offspring produced as a result of natural or artificial asexual reproduction.

Cocci are spherical bacteria.

A **codon** is a sequence of three bases in DNA or mRNA.

Cohesion is the attraction between like molecules.

The **cohesion-tension theory** is the theory of transpiration based on the idea that the stream of water molecules stick together by cohesive forces, so that when a molecule is lost by evaporation it puts a tension on the column and another molecule of water is pulled up to replace it.

Collagen is a strong fibrous protein with a triple helix structure.

Collenchyma are plant cells with areas of cellulose thickening that give mechanical strength and support to the tissues.

Companion cells are very active cells closely associated with the sieve tube elements that supply the phloem vessels with everything they need and actively load sucrose into the phloem.

Competitive inhibition is inhibition in which the inhibitor molecule is similar in shape to the substrate molecule and competes with it for the active site of the enzyme (affected by both inhibitor and substrate concentrations).

A **complementary strand** is the strand of RNA formed that complements the DNA acting as the coding strand.

Conception is the term used for fertilisation of the ovum in humans.

A **condensation reaction** is a reaction in which a molecule of water is removed from the reacting molecules as a bond is formed between them.

Conservation refers to maintaining and protecting a living and changing environment.

Contractile vacuoles are vacuoles that can fill and empty to help control the concentration of the cytoplasm of simple freshwater animals.

A **correlation** is a strong tendency for two sets of data to vary together.

A **countercurrent exchange system** is a system in which two fluid components flow in opposite directions and some properties are exchanged between the two fluids.

Covalent bonds are formed when atoms share electrons. Covalent molecules may be polar if the electrons are not shared equally.

Cristae are the infoldings of the inner membrane of the mitochondria which provide a large surface area for the reactions of aerobic respiration.

Crossing over (recombination) is the process by which large multi-enzyme complexes cut and rejoin parts of the maternal and paternal chromatids at the end of prophase I.

Cyanide is a metabolic poison that stops mitochondria working.

Cyclin-dependent kinases (CDKs) are enzymes involved in the control of the cell cycle by phosphorylating other proteins, activated by attachment to cyclins.

Cyclins are small proteins that build up during interphase and are involved in the control of the cell cycle by their attachment to cyclin–dependent kinases.

Cytokinesis is the final stage of the cell cycle before it enters interphase again – division of the cytoplasm at the end of mitosis to form two independent, genetically identical cells.

Cytoplasm is a jelly-like liquid that makes up the bulk of the cell and contains the organelles.

Cytosine is a pyrimidine base found in DNA and RNA.

The **cytoskeleton** is a dynamic, 3D web-like structure made up of microfilaments and microtubules that fills the cytoplasm and gives it structure, keeping the organelles in place and enabling cell movements and transport within the cell.

A **deletion** is a type of point mutation in which a base is completely lost.

Denaturation is the loss of the 3D shape of a protein, e.g. as a result of changes in temperature or pH.

Deoxygenated blood is blood that has given up its oxygen to the body's cells.

Deoxyribonucleic acid (DNA) is a nucleic acid that acts as the genetic material in many organisms.

Deoxyribose is a pentose sugar that makes up part of the structure of DNA.

Diastole is when the heart relaxes and fills with blood.

To **differentiate** means to develop into specific types of tissues.

Diffusion is the movement of the particles in a liquid or gas down a concentration gradient from an area where they are at a relatively high concentration to an area where they are at a relatively low concentration.

A **dipeptide** is two amino acids joined by a peptide bond.

Diploid (2n) signifies a cell with a nucleus containing two full sets of chromosomes.

A **dipole** is the separation of charge in a molecule when the electrons in covalent bonds are not evenly shared.

Directional selection is natural selection showing a change from one dominant phenotype to another in response to a change in the environment – one phenotype is selected for over all the others.

A **disaccharide** is a sugar made up of two monosaccharide units joined by a glycosidic bond, formed in a condensation reaction.

A **disulfide bond** is a strong covalent bond formed as a result of an oxidation reaction between sulfur groups in cysteine or methionine molecules, which are close together in the structure of a polypeptide.

DNA helicase is an enzyme involved in DNA replication that unzips the two strands of the DND molecules.

DNA ligase is an enzyme involved in DNA replication that catalyses the formation of phosphodiester bonds between nucleotides.

DNA polymerase is an enzyme involved in DNA replication that lines up the new nucleotides along to DNA template strands.

DNA profiling is the process by which the non-coding areas of DNA are analysed to identify patterns.

DNA sequencing is the process by which the base sequences of all or part of the genome of an organism is worked out.

DNA viruses are composed of DNA as the genetic material.

DNA-directed RNA polymerase (RNA polymerase) is the enzyme that polymerises nucleotide units to form RNA in a sequence determined by the antisense strand of DNA.

Domains are the three largest classification categories, including the Eukaryota, the Bacteria and the Archaea.

Double circulation system is a circulation that involves two circulatory systems, one of deoxygenated blood flowing from the heart to the gas exchange organs and back oxygenated to the heart, and one of oxygenated blood leaving the heart and flowing around the body, returning deoxygenated to the heart.

Double fertilisation is the process that occurs in plants in which one male nucleus fuses with the nuclei of the two polar nuclei to form the endosperm nucleus and the other fuses with the egg cell to form the diploid zygote.

Dry mass is the mass of the body of an organism with all the water removed from it.

Ebola is a highly infectious viral disease that causes fever and internal bleeding and death in about 50% of cases.

Ecological isolation occurs when two populations inhabit the same region, but develop preferences for different parts of the habitat.

The **ecological species model** is a species definition based on the ecological niche occupied by an organism.

Ecology is the study of the interactions of organisms with each other and with the environment in which they live.

Ecosystems services are services provided by the natural environment that are of benefit to people.

An **electrocardiogram (ECG)** is used to investigate the rhythms of the heart by producing a record of the electrical activity of the heart

An **electron microscope** is a tool that uses a beam of electrons and magnetic lenses to magnify specimens up to 50 000 times life size.

Embryonic stem cells are cells in the early embryo that have the potential to form many other types of cells.

End product inhibition is a control system in many metabolic pathways in which an enzyme at the beginning of the pathway is inhibited by one of the end products of the reaction.

An **endemic** species is a species that evolves in geographical isolation and is found in only one place.

Endocytosis is the movement of large molecules into cells through vesicle formation.

The **endoplasmic reticulum** is a 3D network of membrane-bound cavities in the cytoplasm that links to the nuclear membrane and makes up a large part of the cellular transport system as well as playing an important role in the synthesis of many different chemicals.

An **endosymbiont** is an organism that lives inside the cells or the body of another organism.

The **endosymbiotic theory** is a theory that suggests that mitochondria and chloroplasts originated as independent prokaryotic organisms that began living symbiotically inside other cells as endosymbionts.

An **envelope** is a coat around the outside of a virus derived from lipids in the host cell.

Enzyme inhibitors are substances that slow down enzymes or stop them from working.

Enzymes are proteins that have a very specific shape as a result of their primary, secondary, tertiary and quaternary structures. They act as biological catalysts and each enzyme will only catalyse a specific reaction or group of reactions.

Eosinophils are part of the non-specific immune system. They are stained red by eosin stain and are important in the response of the body against parasites, in allergic reactions and inflammation, and in developing immunity to disease.

Epithelial tissues are tissues that form the lining of surfaces inside and outside the body.

An **ester bond** is a bond formed in a condensation reaction between the carboxyl group (–COOH) of a fatty acid and one of the hydroxyl groups (–OH) of glycerol.

Esterification is the formation of ester bonds.

Eubacteria are true bacteria (prokaryotic organisms).

Eukaryotes are a group of organisms with cells that have the genetic material contained in a membrane-bound nucleus and also contain a number of membrane-bound organelles such as mitochondria and chloroplasts.

Evolution is the process by which natural selection acts on variation to bring about adaptations and eventually speciation.

The **evolutionary species model** is a species model based on shared evolution between groups of organisms.

Exhalation is breathing out.

Exocytosis is the energy-requiring process by which a vesicle fuses with the cell surface membrane so the contents are released to the outside of the cell.

Ex-situ conservation is the conservation of components of biological diversity (living organisms) outside their natural habitats.

External fertilisation is the process of fertilisation in which the female and male gametes are released outside of the parental bodies to meet and fuse in the environment.

Extracellular enzymes are enzymes that catalyse reactions outside of the cell in which they were made.

Extremophiles are bacteria that can survive extreme conditions of heat, cold, pH, salinity and pressure.

Facilitated diffusion is diffusion that takes place through carrier proteins or protein channels.

Facultative anaerobes are organisms that use oxygen if it is available, but can respire and survive without it.

A **family** is a group of genera that all share common characteristics.

A **fatty acid** is an organic acid with a long hydrocarbon chain.

Fertilisation is the fusing of the haploid nuclei from two gametes to form a diploid zygote in sexual reproduction.

The **fertilisation membrane** is the tough layer that forms around the fertilised ovum to prevent the entry of other sperm.

Fetal haemoglobin is a form of haemoglobin found only in the developing fetus with a higher affinity for oxygen than adult haemoglobin.

Fibrin is an insoluble protein formed from fibrinogen by the action of thrombin that forms a mesh of fibres that trap erythrocytes and platelets to form a blood clot.

Fibrous proteins are proteins that have long, parallel polypeptide chains with occasional cross-linkages that form into fibres but with little tertiary structure.

Flagella are many-stranded helices of the contractile protein flagellin found on some bacteria. They move the bacteria by rapid rotations.

The **fluid mosaic model** is the current model of the structure of the cell membrane including floating proteins forming pores, channels and carrier systems in a lipid bilayer.

Fragmentation is the use of mitosis to regenerate a whole organism from a fragment of the original.

Gametes are haploid sex cells produced by meiosis that fuse to form a new diploid cell (zygote) in sexual reproduction.

Gametogenesis is the formation of the gametes by meiosis in the sex organs.

The **gametophyte generation** is the haploid generation in plants that gives rise to the gametes by mitosis.

Gated channels are protein channels through the lipid bilayer of a membrane that are opened or closed, depending on conditions in the cell.

Gel electrophoresis is a method of separating fragments of proteins or nucleic acids based on their electrical charge and size.

A **gene** is a sequence of bases on a DNA molecule. It contains coding for a sequence of amino acids in a polypeptide chain that affect a characteristic in the phenotype of the organism.

A **gene pool** is all of the variants of all of the genes in a population.

A **generative nucleus** is the male nucleus that will fuse with the female nucleus.

Genetic diversity is a measure of the level of difference in the genetic make-up of a population.

The **genetic species model** is a species model based on DNA evidence.

A **genome** is the entire genetic material of an organism.

The **genotype** is the genetic make-up of an organism with respect to a particular feature.

A **genus** is a group of species that all share common characteristics.

Geographical isolation occurs when a physical barrier such as a river or a mountain range separates individuals from an original population.

Germination (pollen) is the process by which a pollen tube starts to grow out of the pollen grain to transfer the male nuclei to the ovule.

Gills are the organs of gas exchange in a fish.

Globular proteins are large proteins with complex tertiary and sometimes quaternary structures, folded into spherical (globular) shapes.

Glucose is a hexose sugar.

Glycerol is propane-1,2,3-triol, an important component of triglycerides.

Glycogen is made up of many α-glucose units joined by 1,4-glycosidic bonds but also has 1,6-glycosidic bonds, giving it many side branches.

A **glycoprotein** is a protein with a carbohydrate prosthetic group.

A **glycosidic bond** is a covalent bond formed between two monosaccharides in a condensation reaction, which can be broken down by a hydrolysis reaction to release the monosaccharide units.

Golgi apparatus consists of stacks of membranes that modify proteins made elsewhere in the cell and package them into vesicles for transport, and also produce materials for plant cell walls and insect cuticles.

Gonads are the sex organs in animals.

Gram staining is a staining technique used to distinguish types of bacteria by their cell wall.

Gram-negative bacteria are bacteria that have no teichoic acid in their cell walls. They stain red with Gram staining.

Gram-positive bacteria are bacteria that contain teichoic acid in their cell walls and stain purple/blue with Gram staining.

Granulocytes are leucocytes that have granules in the cytoplasm of the cells that take up stain and are obvious under the microscope. They have lobed nuclei and include neutrophils, eosinophils and basophils.

Guanine is a purine base found in DNA and RNA.

Guard cells are the cells that open and close the stomatal pores, controlling the rate of gas exchange.

Guttation is a process in which drops of water may be forced out of the leaves as a result of root pressure.

Haemoglobin is a large conjugated protein involved in transporting oxygen in the blood and gives the erythrocytes their red colour.

Haploid (*n*) signifies a cell with a nucleus containing one complete set of chromosomes.

Heterotrophs are organisms that cannot make their own food and have to eat other organisms.

A **hexose sugar** is sugar with six carbon atoms.

Histones are positively charged proteins involved in the coiling of DNA to form dense chromosomes in cell division.

Homologous chromosomes describe a set of one maternal chromosome and one paternal chromosome that pair up during meiotic cell division.

Homologous structures are structures that genuinely show common ancestry.

Hybridisation is the production of offspring as a result of sexual reproduction between individuals from two different species.

Hydrogen bonds are weak electrostatic intermolecular bonds formed between polar molecules containing at least one hydrogen atom.

Hydrolysis is a reaction in which bonds are broken by the addition of a molecule of water.

Hydrophilic molecules dissolve readily in water.

Hydrophobic molecules will not dissolve in water.

Hydrostatic pressure is the residual pressure from the heartbeat that is still present as the blood enters the arterial end of the capillary beds, that tends to force fluid out through the leaky capillary walls.

A **hypertonic** solution is a solution in which the osmotic concentration of solutes is higher than that in the cell contents.

A **hypertonic solution** is a solution with a higher concentration of solutes and lower concentration of water (solvent) than the surrounding solution.

Incipient plasmolysis is the point at which so much water has moved out of the cell by osmosis that turgor is lost and the cell membrane begins to pull away from the cell wall as the protoplasm shrinks.

Independent assortment (random assortment) is the process by which the chromosomes derived from the male and female parent are distributed into the gametes at random.

The **induced-fit hypothesis** is a modified version of the lock-and-key hypothesis for enzyme action where the active site is considered to have a more flexible shape. Once the substrate enters the active site, the shape of that site is modified around it to form the active complex. Once the products have left the complex, the enzyme reverts to its inactive, relaxed form.

Industrial melanism is the evolution of dark-coloured individuals in a habitat that has been made darker by industrial pollution, e.g. soot.

The **inferior vena cava** is the large vein that carries the returning blood from the lower parts of the body to the heart .

Inhalation is breathing in.

The **initial rate of reaction** is the measure taken to compare the rates of enzyme controlled reactions under different conditions.

An **insertion** is a type of point mutation in which an extra base is added into a gene, which may be a repeat or a different base.

In-situ conservation is the conservation of ecosystems and natural habitats, and the maintenance and recovery of viable populations of species in their natural surroundings.

Internal fertilisation is the fertilisation of the female gamete by the male gamete, which takes place inside the body of the mother.

Interphase is the period between active cell divisions when cells increase their size and mass, replicate their DNA and carry out normal metabolic activities.

Intracellular means inside the cell.

Intracellular enzymes are enzymes that catalyse reactions within the cell.

Intrinsic rhythmicity refers to the intrinsic rhythm of contraction and relaxation in the cardiac muscle making up the heart.

Ionic bonds are formed when atoms give or receive electrons. They result in charged particles called ions.

Irreversible inhibition is inhibition of the action of an enzyme that is permanent and cannot be undone. It is never used within cells to control the rate of reactions.

Isomers are molecules that have the same chemical formula, but different molecular structures.

An **isotonic** solution is a solution in which the osmotic concentration of the solutes is the same as that in the cells.

A **karyotype** is a way of displaying an image of the chromosomes of a cell to show the pairs of autosomes and sex chromosomes.

A **kingdom** is the classification category smaller than domains. There are six kingdoms: Archaebacteria, Eubacteria, Protista, Fungi, Plantae and Animalia.

Latent is the state of the non-virulent virus within the host cell.

The **left atrium** is the upper left-hand chamber of the heart that receives oxygenated blood from the lungs.

The **left ventricle** is the chamber that receives oxygenated blood from the left atrium and pumps it around the body.

Lenticels are spongy areas with loosely packed cells that are the site of gas exchange in woody stems and roots.

A **light microscope (optical microscope)** is a tool that uses a beam of light and optical lenses to magnify specimens up to 1500 times life size.

Lignin is a chemical that impregnates cellulose cell walls in wood and makes it impermeable.

Lipids are a large family of organic molecules that are important in cell membranes and as an energy store in many organisms. They include triglycerides, phospholipids and steroids.

A **lipoprotein** is a protein with a lipid prosthetic group.

The **lock-and-key hypothesis** is the model that explains enzyme action by an active site in the protein structure that has a very specific shape. The enzyme and substrate slot together to form a complex as a key fits in a lock.

The **lumen** is the central space inside the blood vessel.

Lung surfactant is a special phospholipid that coats the alveoli and makes breathing easier.

Lymph is the fluid that travels in the lymphatic system.

Lymph capillaries are the blind tubes that carry the lymph away from the tissues.

Lymph glands are the glands in the lymph system that contain lymphocytes that make antibodies.

Lymphocytes are small leucocytes with very large nuclei that are vitally important in the specific immune response of the body.

Lysogeny is the period when a virus is part of the reproducing host cell, but does not affect it adversely.

A **lysosome** is an organelle full of digestive enzymes used to break down worn out cells or organelles, or digest food in simple organisms.

Magnification is a measure of how much bigger the image you see is than the real object.

Marsupials are mammals that give birth to very immature young and then protect them in pouches.

A **mass transport system** is an arrangement of structures by which substances are transported in the flow of a fluid with a mechanism for moving it around the body.

The **mate-recognition species model** is a species definition based on unique fertilisation systems, including mating behaviour.

Mating is the process by which a male animal transfers sperm from his body directly into the body of the female.

Mechanical isolation happens when a mutation occurs that changes the genitalia of animals, making it physically possible for them to mate successfully with only some members of the group, or it changes the relationship between the stigma and stamens in flowers, making pollination between some individuals unsuccessful.

A **megagamete** is the female gamete, the egg cell, in plants.

Megakaryocytes are large cells that are found in the bone marrow and produce platelets.

Megaspores are the result of meiosis in plants that develop into the female gametes, ovules.

Meiosis is a form of cell division in which the chromosome number of the original cell is halved, leading to the formation of the gametes.

The **meristem** is the region of mitosis and growth in a plant shoot or root.

Mesosomes are infoldings of the cell membrane of bacteria.

Messenger RNA (mRNA) is the RNA formed in the nucleus that carries the genetic code out into the cytoplasm.

A **metabolic chain (metabolic pathway)** is a series of linked reactions in the metabolism of a cell.

Metabolism is the sum of the anabolic and catabolic processes in a cell.

Metaphase is the second stage of active cell division where a spindle of overlapping protein microtubules forms and the chromatids line up on the metaphase plate.

The **metaphase plate (equator)** is the region of the spindle in the middle of the cell along which the chromatids line up.

Metaxylem consists of mature xylem vessels made of lignified tissue.

A **micelle** is a spherical aggregate of molecules in water with hydrophobic areas in the middle and hydrophilic areas outside.

Microfilaments are protein fibres that make up part of the structure of the cytoskeleton.

A **microgamete** is the male gamete produced in plants, the pollen grain.

Microspores are the result of meiosis in plants that develop into the male gametes, pollen.

Microtubules are tiny protein tubes about 20 nm in diameter that make up part of the structure of the cytoskeleton.

The **middle lamella** is the first layer of the plant cell wall to be formed when a plant cell divides, made mainly of calcium pectate (pectin) that binds the layers of cellulose together.

Mitochondria are rod-like structures with inner and outer membranes that are the site of aerobic respiration.

Mitosis is the process by which a cell divides to produce two genetically identical daughter cells.

Molecular activity (turnover number) is the number of substrate molecules transformed per minute by a single enzyme molecule.

Molecular phylogeny is the analysis of the genetic material of organisms to establish their evolutionary relationships.

The **Monera** is a kingdom in the five-kingdom classification system that contains the Archaea and Eubacteria.

Monocytes are part of the specific immune system. They are the largest of the leucocytes and they can move out of the blood to form macrophages. They engulf pathogens by phagocytosis.

A **monolayer** is a single closely packed layer of atoms or molecules.

A **monomer** is a small molecule that is a single unit of a larger molecule called a polymer.

A **monosaccharide** is a single sugar monomer.

Monosomy is when only one member of a pair of chromosomes is present in a cell.

Monotremes are primitive mammals that lay eggs and feed their offspring with milk from mammary glands.

A **monounsaturated fatty acid** is a fatty acid with only one double covalent bond between carbon atoms in the hydrocarbon chain.

The **morphological species model** is a species definition based solely on the appearance of the organisms observed.

Morphology is the study of the form and structure of organisms.

The **mortality rate** is a measurement of the number of deaths in a given population or due to a specific cause.

A **multifactorial disease** results from the interplay of many different factors rather than having one simple cause.

A **mutagen** is anything that increases the rate of mutation.

A **mutation** is a permanent change in the DNA of an organism.

The **mutation rate** is the rate at which mutations naturally occur.

A **myocardial infarction** (heart attack) takes place when atherosclerosis leads to the formation of a clot that blocks the coronary artery entirely and deprives the heart muscle of oxygen so it dies. It can stop the heart functioning.

Myoglobin is a respiratory pigment found in the muscle tissue of vertebrates with a higher affinity for oxygen than haemoglobin.

Natural selection is the process by which the organisms that are best adapted in a particular environment are most likely to survive and reproduce, passing on their advantageous alleles to their offspring.

Neutrophils are part of the non-specific immune system. They have multi-lobed nuclei and engulf and digest pathogens by phagocytosis. Up to 70% of all leucocytes are neutrophils.

A **niche** is the role of an organism within the habitat in which it lives.

Non-competitive inhibition is inhibition in which the inhibitor does not compete for the active site but forms a complex with the enzyme or enzyme/substrate complex and changes the shape of the active site so it can no longer catalyse the reaction (affected only by concentration of inhibitor).

Non-disjunction is the process that occurs when members of a pair of chromosomes fail to separate during the reduction division of meiosis, resulting in one gamete with two copies of a chromosome and one gamete with no copies of that chromosome.

Non-reducing sugars are sugars that do not react with Benedict's solution.

Non-virulent is a term used to describe a microorganism that is not disease-causing.

Nucleic acids are polymers made up of many nucleotide monomer units that carry all the information needed to form new cells.

A **nucleoid** is the area in a bacterium where we find the single length of coiled DNA.

A **nucleolus** is an extra dense area of almost pure DNA and protein found in the nucleus involved in the production of ribosomes and control of growth and division.

Nucleosomes are dense clusters of DNA wound around histones.

Nucleotides are molecules with three parts – a 5-carbon pentose sugar, a nitrogen-containing base and a phosphate group – joined by condensation reactions.

The **nucleus** is an organelle containing the nucleic acids DNA (the genetic material) and RNA, as well as protein, surrounded by a double nuclear membrane with pores.

Obligate aerobes are organisms that need oxygen for respiration.

Obligate anaerobes are organisms that can only respire in the absence of oxygen and are killed by oxygen.

Oligosaccharides are molecules with 3–10 monosaccharide units.

Oncotic pressure is the tendency for water to move into the capillaries by osmosis.

Oogenesis is the formation of the ova in the ovaries.

The **operculum** is the bony protective flap that covers the gills of bony fish.

An **order** is a group of families that all share common characteristics.

An **organ** is a structure made up of several different types of tissues grouped together to carry out a particular function in the body.

An **organ system** is a group of organs working together to carry out particular functions in the body.

Organelles are sub-cellular bodies found in the cytoplasm of cells.

Osmosis is a specialised form of diffusion that involves the movement of solvent molecules down a concentration gradient through a partially permeable membrane.

Osmotic concentration is a measure of the concentration of the solutes in a solution that have an osmotic effect.

Osmotic potential (π) is a measure of the potential of a solution to cause water to move into the cell across a partially permeable membrane as a result of dissolved solutes.

Ova are the haploid female gametes in animals.

Ovaries are the female sex organs in both animals and plants. They produce the female gametes called ovules in plants and ova in animals.

Ovules are the haploid female gametes in plants.

Oxygenated blood is blood that is carrying oxygen.

Oxyhaemoglobin is the molecule formed when oxygen binds to haemoglobin.

A **pandemic** is an epidemic that takes place in several countries at once.

Parenchyma are relatively unspecialised plant cells that act as packing in stems and roots to give support.

Parthenogenesis is the process by which an unfertilised egg cell develops into a new individual.

Passive transport is transport that takes place as a result of concentration, pressure or electrochemical gradients and involves no energy from a cell.

A **pathogen** is a microorganism that causes disease.

Pectin is a polysaccharide that holds cell walls of neighbouring plant cells together and is part of the structure of the primary cell wall.

A **pentose sugar** is a sugar with five carbon atoms.

A **peptide bond** is the bond formed by condensation reactions between amino acids.

Peptidoglycan is a large, net-like molecule found in all bacterial cell walls made up of many parallel polysaccharide chains with short peptide cross-linkages.

Peripheral arteries are arteries further away from the heart but before the arterioles.

Phagocytosis is the active process when a cell engulfs something relatively large such as a bacterium and encloses it in a vesicle.

Phenotypes are the physical traits (including biochemical characteristics) expressed as a result of the interactions of the genotype with the environment.

Phloem is the main tissue transporting dissolved solutes around the plant.

Phloem sieve tubes are the main transport vessels of the phloem, made up of many living cells joined together to make very long tubes that run from the highest shoots to the end of the roots, divided at intervals by sieve plates.

Phloem loading is the process by which sucrose and other assimilates are loaded into the phloem by the companion cells.

Phloem unloading is the process by which sucrose is moved into the phloem sieve elements from the the surrounding cells.

A **phosphodiester bond** is the bond formed between the phosphate group of one nucleotide and the sugar of the next nucleotide in a condensation reaction.

A **phospholipid** is a chemical in which glycerol bonds with two fatty acids and an inorganic phosphate group.

A **phylum (division** for plants) is a group of classes that all share common characteristics.

A **physiological adaptation** is an adaptation involving the way the body of the organism works, including differences in biochemical pathways or enzymes.

Pili (fimbriae) are thread-like protein projections found on the surface of some bacteria.

Pinocytosis is the active process by which cells take in tiny amounts of extracellular fluid by tiny vesicles.

Pits are specialised holes in the walls of the xylem vessels through which water moves out into the surrounding cells.

The **placenta (plant)** is the pad of special tissue that attaches the plant ovule to the ovary wall.

Placental mammals are mammals that provide for the developing fetus during gestation through a placenta.

Plant fibres are long cells with cellulose cell walls that have been heavily lignified so they are rigid and very strong.

Plaques are yellowish fatty deposits that form on the inside of arteries in atherosclerosis.

Plasmids are small, circular pieces of DNA that code for specific aspects of the bacterial phenotype.

Plasmodesmata are cytoplasmic bridges between plant cells that allow communication between the cells.

Plasmolysis is the situation when a plant cell is placed in hypertonic solution when so much water leaves the cell by osmosis that the vacuole is reduced and the protoplasm is concentrated and shrinks away from the cell walls.

Platelets are involved in the clotting mechanism of the blood.

Pluripotent means a cell is able to develop into most different cell types.

Pluripotent embryonic stem cells are embryonic stem cells that can form most, but not all, adult cell types.

A **point mutation (gene mutation)** is a change in one or a small number of nucleotides affecting a single gene.

Polar lipids are lipids with one end attached to a polar group, e.g. a phosphate group that makes one end of the molecule hydrophilic and one end hydrophobic.

A **polar molecule** is a molecule containing a dipole.

Pollen is the haploid male gametes in plants.

Pollen sacs are the parts of the anthers where the pollen grains develop.

A **pollen tube** is a tube that grows out of a pollen grain down the style, into the ovary and through the micropyle of the ovule to carry the two male nuclei to the ovule.

Pollination is the transfer of pollen from the anther to the stigma, often from one flower to another.

A **polymer** is a long chain molecule made up of many smaller, repeating monomer units joined together by chemical bonds.

A **polypeptide** is a long chain of amino acids joined by peptide bonds.

Polypoidy is when a cell or an organism has more than two sets of chromosomes.

A **polysaccharide** is a polymer made up of long chains of monosaccharide units joined by glycosidic bonds.

Polysomes are groups of ribosomes, joined by a thread of mRNA, that can produce large quantities of a particular protein.

Polysomy is when a cell contains three or more rather than two chromosomes of a particular type.

Polyspermy is the fertilisation of an egg by more than one sperm.

A **polyunsaturated fatty acid** is a fatty acid with two or more double covalent bonds between carbon atoms in the hydrocarbon chain.

A **potometer** is a piece of apparatus used to measure the uptake of water by a shoot.

Pressure potential is a measure of the inward pressure exerted by the plant cell wall on the protoplasm of a cell, opposing the entry of water by osmosis. It usually has a positive value.

The **primary cell wall** is the first very flexible plant cell wall to form, with all the cellulose microfibrils orientated in a similar direction.

Primordial germ cells are the cells that divide by meiosis to ultimately form the sperm and ova.

Probability is a measure of the chance or likelihood that an event will take place.

Prokaryotes are a group of organisms including bacteria and blue-green algae (cyanobacteria) that have few organelles and do not have the genetic material contained in a membrane-bound nucleus.

Prophase is the first stage of active cell division where the chromosomes are coiled up and consist of two daughter chromatids joined by the centromere. The nucleolus breaks down.

A **prosthetic group** is the molecule that is incorporated in a conjugated protein.

A **protease** is a protein-digesting enzyme.

Prothrombin is a large, soluble protein found in the plasma that is the precursor to an enzyme called thrombin.

The **Protista** is a kingdom in the five-kingdom classification system that contains all single-celled organisms, green and brown algae and slime moulds.

Protoplasm is the cytoplasm and nucleus combined.

Protoxylem is the first xylem formed that can stretch and grow because the walls are not fully lignified

A **provirus** is the DNA that is inserted into the host cell during the lysogenic pathway of reproduction in viruses.

Pulmonary arteries are the blood vessels that carry deoxygenated blood from the heart to the lungs.

The **pulmonary circulation** carries deoxygenated blood to the lungs and oxygenated blood back to the heart.

Pulmonary veins are the blood vessels that carry oxygenated blood back from the lungs to the heart.

A **purine base** is a base found in nucleotides that has two nitrogen-containing rings.

Purkyne tissue is made up of conducting fibres that penetrate down through the septum of the heart, spreading between and around the ventricles.

A **pyrimidine base** is a base found in nucleotides that has one nitrogen-containing ring.

Reducing sugars are sugars that react with blue Benedict's solution and reduce the copper(II) ions to copper(I) ions giving an orangey-red precipitate.

Reduction/oxidation (redox) reactions are reactions in which one reactant loses electrons (is oxidised) and another gains electrons (is reduced).

Regeneration is the use of mitosis to regrow a body part that has been lost.

Relative species abundance refers to the relative numbers of species in an area.

When a DNA molecule **replicates**, it copies itself exactly.

A **resistant** bacterium is not affected by an antibiotic.

Resolution (resolving power) is a measure of how close together two objects must be before they are seen as one.

Retroviruses are a special type of RNA virus that control the production of DNA corresponding to the viral RNA and insert it into the host cell DNA.

Reverse transcriptase is an enzyme synthesised in the life cycle of a retrovirus that makes DNA molecules corresponding to the viral RNA genome.

Reversible inhibition is inhibition of the action of an enzyme by an inhibitor that does not permanently affect the functioning of the enzyme and can be removed from the enzyme. It is often used to control reaction rates within a cell.

Ribonucleic acid (RNA) is a nucleic acid which can act as the genetic material in some organisms and is involved in protein synthesis.

Ribose is a pentose sugar that makes up part of the structure of RNA.

Ribosomal RNA (rRNA) is RNA that makes up about 50% of the structure of the ribosome.

Ribosomes are the site of protein synthesis in the cell.

The **right atrium** is the upper right-hand chamber of the heart that receives deoxygenated blood from the body.

The **right ventricle** is the chamber that receives deoxygenated blood from the right atrium and pumps it to the lungs.

RNA viruses are composed of RNA as the genetic material.

Root hairs are microscopic hairs that are extensions of the membranes of the outer cells of the root, greatly increasing the surface area for absorption of water and minerals.

Root pressure is the pressure that results when salts are actively secreted from the root cells into the xylem sap, increasing the concentration across the root and moving more water into the xylem by osmosis.

Rough endoplasmic reticulum (RER) is endoplasmic reticulum that is covered in 80S ribosomes and which is involved in the production and transport of proteins.

A **saturated fatty acid** is a fatty acid in which each carbon atom is joined to the one next to it in the hydrocarbon chain by a single covalent bond.

Scanning electron micrographs (SEMs) are micrographs produced by the electron microscope that have a lower magnification than TEMs, but produce a 3D image.

Sclerenchyma are plant cells that have very thick lignified cell walls and an empty lumen with no living contents.

Seasonal isolation occurs when the timing of flowering or sexual receptiveness in some parts of a population drifts away from the norm for the group. This can eventually lead to the two groups reproducing several months apart.

The **secondary cell wall** is the older plant cell wall in which the cellulose microfibrils have built up at different angles to each other making the cell wall more rigid.

Selection pressure is the pressure exerted by a changed environment or niche on individuals in a population, causing changes in the population as a result of natural selection.

Semilunar valves are half moon-shaped one-way valves found at frequent intervals throughout the venous system to prevent the backflow of blood.

The **septum** is the thick muscular dividing wall through the centre of the heart that prevents oxygenated and deoxygenated blood from mixing.

Serotonin is a chemical that causes the smooth muscle of the blood vessels to contract, narrowing them and cutting off the blood flow to the damaged area.

Sex chromosomes are the chromosomes that carry the information that determines the sex of the individual. Human females have two X chromosomes (XX) and males have an X and a Y chromosome (XY).

In **sexual dimorphism** there is a great deal of difference between the appearance of the male and female of a species.

Sexual reproduction is the production of offspring that are genetically different from the parent organism or organisms by the fusing of two sex cells (gametes).

Sickle cell disease (sickle cell anaemia) is a human genetic disease affecting the protein chains making up the haemoglobin in the red blood cells.

Sieve plates are the perforated walls between phloem cells that allow the phloem sap to flow.

Single circulation system is a circulation in which the heart pumps the blood to the organs of gas exchange and the blood then travels on around the body before returning to the heart.

A **sink** is a region of a plant that is low in sucrose and removes sucrose from the phloem.

The **sinoatrial node (SAN)** is a group of cells in the right atrium. It is the area of the heart with the fastest intrinsic rhythm and acts as the heart's own natural pacemaker.

Smooth endoplasmic reticulum (SER) is a smooth tubular structure similar to RER, but without the ribosomes, which is involved in the synthesis and transport of steroids and lipids in the cell.

A **source** is a region of a plant that is very high in sucrose and loads sucrose into the phloem.

Speciation is the formation of a new species.

A **species** is a group of closely related organisms that are all potentially capable of interbreeding to produce fertile offspring.

Species richness refers to the number of different species in an area.

Specificity is the characteristic of enzymes that means that, as a result of the very specific shapes resulting from their tertiary and quaternary structures, each enzyme will only catalyse a specific reaction or group of reactions.

Spermatogenesis is the formation of the sperm in the testes.

Spermatozoa (sperm) are the haploid male gametes in animals.

A **spindle** is a set of overlapping protein microtubules running the length of the cell, formed as the centrioles pull apart in mitosis and meiosis.

Spiracles are openings along the side of the thorax and abdomen of an insect that are the site of the entry and exit of the respiratory gases. They may be opened or closed by sphincters.

Spirilla are bacteria with a twisted or spiral shape.

Spongy mesophyll cells are the cells inside the leaf of a plant where gas exchange takes place.

The **sporophyte** is the diploid main body of the plant.

The **sporophyte generation** is the diploid generation in plants that produces spores by meiosis.

Sporulation is the process involving mitosis in the production of asexual spores that can grow into new individuals.

Starch is a long chain polymer formed of glucose monomers.

A **stent** is a metal or plastic mesh tube that is inserted into an artery affected by atherosclerosis to hold it open and allow blood to pass through freely.

Stomata (singular **stoma**) are specialised pores found mainly in the epidermis on the underside of the leaf through which gases diffuse into and out of the cell.

Suberin is a chemical that impregnates cellulose cell walls in cork tissues and makes them impermeable.

A **substitution** is a type of point mutation in which one base in a gene is substituted for another.

A **substrate** is the molecule or molecules on which an enzyme acts.

Sucrose is a sweet tasting disaccharide formed by the joining of glucose and fructose by a glycosidic bond.

The **superior vena cava** is the large vein that carries the returning blood from the upper parts of the body to the heart.

Surface area to volume ratio (sa : vol) is the relationship between the surface area of an organism and its volume.

Sympatric speciation is speciation that takes place between populations of a species living in the same place. They become reproductively isolated by mechanical, behavioural or seasonal mechanisms and gene flow continues between the populations to some extent as speciation takes place.

The **symplast** is all of the material (cytoplasm, vacuole, etc.) contained within the surface membrane of a plant cell.

The **symplast pathway** is the route by which substances, e.g. water and sucrose, can move by diffusion through the interconnected cytoplasm (symplast) in the plasmodesmata connecting cells in a plant.

The **systemic circulation** carries oxygenated blood from the heart to the cells of the body where the oxygen is used, and carries the deoxygenated blood back to the heart.

Systole is the contraction of the heart.

Taxonomy is the science of describing, classifying and naming living organisms.

Teichoic acid is a chemical found in the cell walls of Gram-positive bacteria.

Telophase is the fourth stage of active cell division where a nuclear membrane forms around the two sets of chromosomes, the chromosomes unravel and the spindle breaks down.

The **temperature coefficient (Q_{10})** is the measure of the effect of temperature on the rate of a reaction.

Tendinous cords (also known as valve tendons or heartstrings) make sure the valves are not turned inside out by the great pressure exerted when the ventricles contract.

Tensile strength is the strength of a material when it is stretched.

Testes are the male sex organs that produce the male gametes – sperm.

Thrombin is an enzyme that acts on fibrinogen, converting it to fibrin during clot formation.

Thromboplastin is an enzyme that sets in progress a cascade of events that leads to the formation of a blood clot.

A **thrombosis** is a clot that forms in a blood vessel.

Thymine is a pyrimidine base found in DNA.

A **tissue** is a group of specialised cells carrying out a particular function in the body.

Tissue fluid is the fluid that surrounds all the cells in the body.

The **tonoplast** is the specialised membrane that surrounds the permanent vacuole in plant cells and controls movements of substances into and out of the cell sap.

Totipotent means a cell is able able to develop into all different cell types.

Tracheae (singular: **trachea**) are the largest tubes of the insect respiratory system, carrying air directly into the body for gas exchange with the cells. They run both into and along the body of the insect.

Tracheoles are minute tubes of diameter 0.6–0.8 µm that are the site of gaseous exchange in insects.

Transfer RNA (tRNA) molecules are small units of RNA that pick up particular amino acids from the cytoplasm and transport them to the surface of the ribosome to align with the mRNA.

Translation is the process by which proteins are produced, via RNA, using the genetic code found in the DNA. It takes place on the ribosomes.

A **translocation** (noun) is a mutation in which part of one chromosome breaks off and rejoins to another completely different chromosome. It may be balanced, if part of two chromosomes effectively swap, or it may be unbalanced if a piece simply breaks off one chromosome and joins another.

Translocation (verb) is the movement of substances around plants.

Transmission electron micrographs (TEMs) are micrographs produced by the electron microscope that give 2D images like those from a light microscope, but magnified up to 500 000 times.

Transpiration is the loss of water vapour from the surface of the plant that has evaporated from the surface of the spongy mesophyll cells mainly within the leaves.

The **transpiration stream** is the movement of water up from the soil through the root hair cells and across the roots to the xylem, then up the xylem, across the leaf until it is lost by evaporation from the spongy mesophyll cells and diffuses out of the stomata down the concentration gradient.

The **tricuspid valve (atrioventricular valve)** is the valve between the right atrium and the right ventricle that prevents backflow of blood from the ventricle to the atrium when the ventricle contracts.

A **triose sugar** is a sugar with three carbon atoms.

A **triplet code** is the code of three bases, and is the basis of the genetic information in the DNA.

A **tube nucleus** is the male nucleus that will control the production of the pollen tube in fertilisation.

Turgor is the state of a plant cell when the solute potential causing water to be moved into the cell by osmosis is balanced by the force of the cell wall pressing on the protoplasm.

Turgor pressure (P) is a measure of the inward pressure exerted by the plant cell wall on the protoplasm of the cell as the cell contents expand and press outwards, a force which opposes the entry of water by osmosis.

The **ultrastructure** is the detailed organisation of the cell, only visible using the electron microscope.

A **unit membrane** is a bilayer structure formed by phospholipids in an aqueous environment, with the hydrophobic tails in the middle and the hydrophilic heads on the outside.

An **unsaturated fatty acid** is a fatty acid in which the carbon atoms in the hydrocarbon chain have one or more double covalent bonds in them.

Uracil is a pyrimidine base found in RNA.

A **vacuole** is a fluid-filled cavity within the cytoplasm of a cell surrounded by a membrane enclosing food, water or air.

Vegetative propagation is the process by which a plant forms a structure by mitosis that develops into a fully differentiated, genetically identical new plant.

Veins are vessels that carry blood towards the heart from the cells of the body.

Ventricular systole is when the ventricles of the heart contract.

Venules are the very smallest branches of the venous system, furthest from the heart.

Vesicles are membrane 'bags' that hold secretions made in cells.

Vibrios are comma-shaped bacteria.

Virulent is a term used to describe a microorganism that is disease-causing.

Virus attachment particles (VAPs) are specific proteins (antigens) that target proteins in the host cell surface membrane.

Water potential (Ψ) is a measure of the potential for water to move out of a solution by osmosis.

A **whole-chromosome mutation** is the loss or duplication of a whole chromosome.

Xylem is the main tissue transporting water around a plant.

The **zona pellucida** is a layer of protective jelly around the unfertilised ovum.

A **zygote** is the cell formed when two haploid gametes fuse at fertilisation.

Index